DATE DUE			
Apr 1 '83			

Coelenterate Ecology and Behavior

Coelenterate Ecology and Behavior

Edited by
G. O. Mackie
University of Victoria
Victoria, B.C., Canada

Plenum Press · New York and London

Library of Congress Cataloging in Publication Data

International Symposium on Coelenterate Biology, 3d University of Victoria, Victoria,
    B. C., 1976.
    Coelenterate ecology and behavior.

    Includes index.
    1. Coelenterata – Ecology – Congresses. 2. Coelenterata – Behavior – Congresses.
I. Mackie, George Owen, 1929-   II. Title.
QL375.C63   593'.5   76-43273
ISBN 0-306-30991-2

Selected papers from the Third International Symposium on Coelenterate Biology,
held at the University of Victoria, Victoria, British Columbia, May 10-13, 1976

© 1976 Plenum Press, New York
A Division of Plenum Publishing Corporation
227 West 17th Street, New York, N.Y. 10011

Printed in the United States of America

Preface

The study of coelenterates is now one of the most active
fields of invertebrate zoology. There are many reasons for this,
and not everyone would agree on them, but certain facts stand
out fairly clearly. One of them is that many of the people who
study coelenterates do so simply because they are interested in
the animals for their own sake. This, however, would be true
for other invertebrate groups and cannot by itself explain the
current boom in coelenterate work. The main reasons for all this
activity seem to lie in the considerable concentration of research
effort and funding into three broad, general areas of biology:
marine ecology, cellular-developmental biology and neurobiology,
in all of which coelenterates have a key role to play. They are
the dominant organisms, or are involved in an important way, in a
variety of marine habitats, of which coral reefs are only one,
and this automatically ensures their claims on the attention of
ecologists and marine scientists. Secondly, the convenience of
hydra and some other hydroids as experimental animals has long
made them a natural choice for a variety of studies on growth,
nutrition, symbiosis, morphogenesis and sundry aspects of cell
biology. Finally, the phylogenetic position of the coelenterates
as the lowest metazoans having a nervous system makes them uniquely
interesting to those neurobiologists and behaviorists who hope to
gain insights into the functioning of higher nervous systems by
working up from the lowest level.

The Third International Symposium on Coelenterate Biology served
as a meeting ground for a good many of the most active coelenterate
workers from around the world, especially in the areas of ecology
and behavior, and the present volume, which consists of a selection
of papers presented at the symposium, reflects this emphasis.
The hydra-experimentalists are well represented, but not to the
dominating extent which would have been the case if sessions on
growth, differentiation, etc. had been included in the planning.
The decision to concentrate on ecology and behavior stemmed from
the belief of the organizing committee (Mary Needler Arai,
Donald M. Ross and myself) that these two topics went naturally

together, that they were both hotbeds of exciting new work and that both were overdue for a gathering of the clan. (For a history of previous gatherings, see Werner.*)

The symposium, which was held at the University of Victoria in Victoria, British Columbia during May, 1976, exceeded the expectations of the organizers both in the numbers of attenders and in the variety and quality of the work offered. After four days of concentrated coelenterology some of us are still quivering from the impact of new personalities and ideas. And yet the meeting was more than a forum for the new wave of workers. Many of the established leaders contributed, people who only ten years ago may have been solitary pioneers but now find that they have to work to keep up with what their followers are doing.

How much of the success of the meeting was due to the efforts and planning of the organizers and how much to the generous level of funding we were lucky to be able to obtain is an intriguing question which need not be debated here! Undeniably, the conference benefitted enormously from the participation of overseas visitors, many of whom could only attend because of aid from the conference sponsors. In all, eighteen countries were represented. The bodies who supported the conference were: the Leon and Thea Koerner Foundation, the National Research Council of Canada, the Government of the Province of British Columbia, the University of Alberta and the University of Victoria. On behalf of the international fraternity in the field, I would like here to record our deep appreciation for this support.

I am personally grateful to those colleagues who helped with the selection, reviewing and editing of the manuscripts included in this volume, and to all those who otherwise assisted with the preparation of the work for the press, especially Wendy McPetrie, Conference Secretary, for typing and general assistance, and Peter Anderson, for help with the index. I also wish to thank Stephen Dyer of the Plenum Publishing Corporation for arranging for the rapid and efficient production of this book.

September, 1976   G.O. Mackie

*Werner, B. 1973. Introductory lecture, pp 1-6 in Recent Trends in Research in Coelenterate Biology (Proceedings of the Second International Symposium on Cnidaria), edited by T. Tokioka in collaboration with S. Nishimura, published as Volume 20 of the Publications of the Seto Marine Biological Laboratory.

Contents

BIOLOGY OF CTENOPHORES, SIPHONOPHORES AND MEDUSAE

REPRODUCTIVE BIOLOGY

ASSOCIATIONS

FUNCTIONAL MORPHOLOGY

Coelenterologists at Victoria, May, 1976

Water Flow in the Ecology
of Benthic Cnidaria

WATER FLOW IN THE ECOLOGY OF BENTHIC CNIDARIA

The following four papers were given in a special session of the Third International Symposium on Coelenterate Biology. They present some principles of water flow and the possible reactions of bodies to flow, and then apply them to studies of feeding and survival of certain benthic cnidarians.

Flow – along with light, temperature, pressure and salinity – is a major physical factor in aquatic environments. However, flow has received much less attention from researchers than has any other factor: for a review see Riedl (1971). In recent years simple devices have been designed to measure flow speed and direction as well as the forces on and the strains in stationary objects in the flow. Discussions of the principles of mechanical design as applied to biology are now available (Alexander, 1968; Wainwright et al., 1976).

The first paper here describes aspects of fluid flow that are important to organisms and then describes posture-maintaining mechanisms found in benthic cnidarians. In addition to the papers that follow on anemones and hydrozoans, a paper on a gorgonian was read (Leversee, 1976).

Alexander, R. M., 1968. Animal Mechanics. Univ. Washington Press, Seattle. 346 pp.

Leversee, G. J., 1976. Flow and feeding in fan-shaped colonies of the gorgonian coral, Leptogorgia. Biol. Bull., (in press).

Riedl, R., 1971. Water movement. O. Kinne, Ed., Marine Ecology, volume I. Wiley-Interscience, New York, 1244 pp.

Wainwright, S. A., W. D. Biggs, J. D. Currey, and J. M. Gosline, 1976. Mechanical Design in Organisms. Halsted Press (Wiley), New York, 423 pp.

THE NATURE OF FLOW AND THE REACTION OF BENTHIC CNIDARIA TO IT

S. A. Wainwright and M. A. R. Koehl

Department of Zoology, Duke University

Durham, North Carolina 27706, U. S. A.

INTRODUCTION

Fluid flow is an environmental factor acting upon organisms in its path. All organisms live in a fluid, liquid or gas. A fluid moving relative to a body imposes mechanical forces on that body. These forces, which tend to carry the body downstream, are known as drag forces. Organisms depending on moving fluid to transport them from place to place have various morphological features that maximize drag, whereas organisms that locomote through a fluid tend to have structures that minimize drag. Sessile organisms such as benthic cnidarians risk being dislodged or broken by drag forces, yet they depend on the fluid moving over them to bring them food and essential substances, to carry away their wastes, and to disperse their gametes or young. Various compromises between maximizing and minimizing the effects of flow can be recognized among the benthic cnidarians.

The reactions of sessile organisms to flow may be immediate and behavioral or they may occur at a slower rate over a longer time as the organism develops. Considered here will be the basic mechanical and structural properties of cnidarian polyps and colonies that are important to their appropriate reaction to flow. A more complete presentation of these and other mechanical principles of importance to organisms is given by Wainwright, et al., (1976).

WITHSTANDING AND UTILIZING FLOW

The two basic strategies of sessile organisms to both withstand and utilize the fluid moving around them are: 1) to avoid
rapid flow and have morphological and behavioral features which
compensate for the lack of external transport, and 2) to face
rapid flow and have features which affect the drag forces on the
organisms and their ability to resist those forces. Before dis-
cussing the tactics of various benthic cnidarians that employ these
strategies, we will review a few useful principles of fluid dynamics. A more thorough discussion of these principles can be
found in Rouse (1961) and in Shapiro (1961).

Reynolds Number and Drag

The drag force on an object depends upon the pattern of flow
around the object. There are three different basic patterns of
flow possible around an object of a given shape (Fig. 1).

A flow pattern is described as laminar when the fluid moves
smoothly around the body (Fig. 1,A). When flow is laminar, fluid
particles can be considered as moving in layers between which there
is no significant mixing.

Fluid in contact with the surface of an object does not slip
relative to that object. Thus, when flow is laminar, the fluid
velocity gradually increases with distance from the surface of the
body until it reaches mainstream velocity (Fig. 2). Fluid in the
velocity gradient near an object is subjected to shearing. Viscosity, defined as the stress required to produce a given shear
deformation at a given rate, is the resistance of the molecules in
a fluid to sliding past each other. When the flow pattern around
the object is laminar, drag is due to the viscosity of the fluid;
this type of drag is referred to as skin friction.

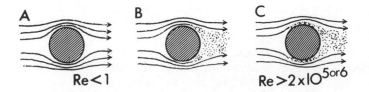

Fig. 1.   Fluid flow patterns around a solid object.

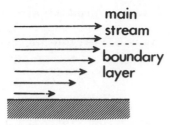

Fig. 2. Velocity gradient in a fluid near a solid surface.

If the velocity of fluid movement around the same object is increased, eventually a second pattern of flow is produced (Fig. 1,B). This second pattern of flow is characterized by a thin laminar layer of slow-moving fluid near the body surface (the boundary layer) and a wake of eddying fluid behind the body. An object in this flow pattern is subjected to form drag as well as skin friction.

According to Bernoulli's Law, as velocity increases along a streamline (the path travelled by a fluid particle), the pressure exerted by the fluid normal to that streamline decreases, and vice versa. When a fluid flows around the widest section of a solid body, the velocity increases and the pressure normal to the stream-lines decreases. Beyond the widest section of the body, the fluid particles slow down again and the pressure concomitantly rises. Viscosity causes fluid particles in the boundary layer to lose momentum as they travel around a solid body. When flow over the body is rapid enough, this viscous retardation may actually stop the downstream progression of the fluid particles in the boundary layer, which may then be pushed upstream by the increasing pressure of the decelerating fluid around them. Thus, behind the body a wake of eddying fluid separates from the main stream of flow. The pressure of the fluid on the rear of the body is then less than the pressure on the front of the body. This net downstream pressure on the body is known as form drag.

If the velocity of fluid around the object is increased be-yond a critical value, flow in the boundary layer becomes turbulent (Fig. 1,C). Turbulent flow is characterized by considerable inter-change of mass and momentum between different average streamlines. The drag on an object in turbulent flow is predominantly form drag.

Which of these three flow patterns occurs around an object of a given shape depends upon whether inertial or viscous forces pre-dominate. If an irregularity is produced in a stream of fluid, it

will persist if inertial forces predominate, but will be damped
out if viscous forces are more important. The ratio of inertial
to viscous forces for a flow situation is the Reynolds Number (Re),

$$Re = \frac{pVL}{\mu}$$

where V is the velocity, L is a linear dimension of the object,
and p is the density and μ is the viscosity of the fluid. Thus,
flow is laminar when Re is low and is turbulent when Re is high.
Flow situations characterized by the same Re are dynamically
similar. This means that the flow pattern will be the same regardless of the actual magnitudes of the various components of Re.

If a body is small or the flow over it is slow, the Re is
low. At low Re's (<1) the drag force (D) on an object due to
skin friction is

$$D = 1/2 \ (C_d \mu LV)$$

where C_d is the drag coefficient (dependent on the shape of the
object).

If a body is large or the flow over it is rapid, the Re is
high. At high Re's (>10^2 or 10^3) where pressure makes the greatest
contribution to the drag on a body,

$$D = 1/2 \ (C_d pV^2 S)$$

where S is the projected area of the body normal to the direction
of flow.

Note that the rules of the game are very different for high
Re situations than for low Re situations. For example, at high Re
drag is proportional to velocity squared, whereas at low Re drag is
proportional to velocity. At high Re any shape or orientation of
an object which reduces the size of the wake reduces form drag.
However, at low Re where drag is due to skin friction, orientation
makes no difference to drag. We should realize that the flow world
of a cnidarian colony may be very different from the flow world of
its polyps.

Flow Over Benthic Cnidarians

The flow over benthic cnidarians depends in part upon where the
animals live. Shallow-water coastal cnidarians are subjected to
wave action. A wave in a fluid is a surface shape that is transmitted some distance by local orbital movements of water particles
which themselves move only slightly in the direction of wave prop-

agation (Fig. 3,A). When a wave enters a shallow area (depth <
1/2 wave length), the circular orbits become compressed so that
near the substratum the water particles oscillate back and forth
parallel to the bottom (Fig. 3,B). Benthic cnidarians in such
areas are thus subjected to bidirectional flow with a period of
several seconds. Intertidal and high subtidal cnidarians are subjected to breaking waves and the surge and backwash of broken waves;
such cnidarians encounter turbulent or bidirectional flow of high
velocity. The predominant type of flow encountered by cnidarians
at depths greater than half the wave length and in coastal areas
protected from wave action is tidal current, a bidirectional flow
with a period of several hours. Benthic cnidarians living at still
greater depths, as well as those living in freshwater lakes and
streams, are subjected to more or less unidirectional flow or to
currents of unpredictable direction and duration. Remember that
velocities close to the substratum are lower than mainstream and are
modified by local topography.

Avoiding the Issue

Benthic cnidarians can avoid large drag forces by living in
calm bodies of water or in sheltered microhabitats within regions
exposed to rapid flow. Small or flat animals on the substratum can
"hide" in the slowly-moving fluid of the boundary layer. Thus,
newly-settled cnidarian larvae, or encrusting forms such as the
hydrocoral Allopora porphyra, are exposed to currents much slower
than mainstream. Larger animals such as sea anemones can remove
themselves from rapid flow by living in cracks and holes or in the
shelter of protrusions from the substratum.

Certain sea anemones clearly illustrate that the height of an
animal above the substratum and neighboring objects can determine
the current it encounters. Small (mean height above the substratum
= 1.1 cm, standard deviation (SD) = .4, number of measurements (n)
= 80) clonal Anthopleura elegantissima are sometimes found side-by-
side with the taller (mean height above the substratum = 7.1 cm,

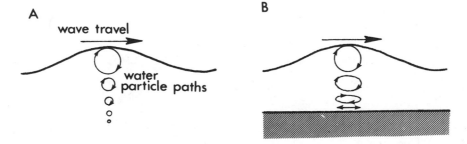

Fig. 3. Wave, A. In deep water, and B. In shallow water.

SD = 2.4, n = 25) solitary A. xanthogrammica, yet A. elegantissima encounter current velocities up to an order of magnitude lower (Fig. 4).

Sessile cnidarians found in slowly-moving water tend to have morphological, physiological, and behavioral adaptations which compensate for the lack of external transport. Examples of such adaptations of cnidarians are the active creation of currents by ciliary or peristaltic action, tolerating low oxygen levels (Sassaman and Mangum, 1972), minimizing burial by sediment settling out of slowly-moving water (Wood-Jones, 1909; Hubbard and Pocock, 1972), sitting in currents produced by other animals, "escaping" from the boundary layer by growing on other animals (Riedl, 1971) or by being tall, and feeding on motile prey or detritus.

Facing the Issue

Benthic cnidarians can benefit from the transport of food and other substances by flowing water if they can withstand the associated mechanical forces. Sessile cnidarians that are subjected to rapidly flowing water have morphological and behavioral features that affect the drag forces on the animals and their ability to resist those forces. Some cnidarians in flowing water are rigid enough to withstand the drag forces without being deformed significantly, whereas others are flexible and minimize breakage due to flow forces by bending.

Rigid Cnidarians. The stony corals are classic examples of rigid animals that withstand wave action and tidal currents. The flow over corals is generally at high Re, thus we expect corals in exposed localities to have most of their surface area parallel to the direction of flow, thus minimizing wake size and drag. Many corals in exposed localities tend to be more rounded and flattened

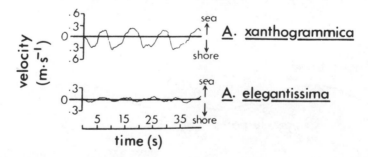

Fig. 4.   Current velocities measured across the oral disks of a solitary A. xanthogrammica and a clonal A. elegantissma sitting next to it during surge and backwash.

than members of the same species in protected areas (Wood-Jones, 1909; Stoddart, 1969). Not only do such shapes minimize wake size and hence drag, but they also are more likely than other coral shapes to land right-side-up if they are dislodged (Abbott, 1974). Calculations of drag on and bending stresses in branching corals indicate that drag-induced stresses are minimized when branches subjected to high velocities are oriented parallel to the flow direction (Chamberlain and Graus, 1975b). In a number of species of corals (e.g., Acropora palmata) the branches do grow parallel to the predominant directions of water movement (Stoddart, 1969; Wainwright, et al., 1976); the stronger the wave action, the more marked is this alignment of branches (Shinn, 1963).

The main problem with rigidly resisting flow forces as a stony coral does is that when a hurricane comes along and the stresses caused by large waves exceed the strength of the skeleton, the coral breaks catastrophically. Other cnidarians minimize breakage due to flow forces by bending with the forces rather than rigidly resisting them.

Flexible Cnidarians. Flexible sessile cnidarians, such as hydroids and sea anemones, pens, whips, and fans, are bent over by moving water. Such bending minimizes breakage in a number of ways. Bending moves an animal down closer to the substratum where velocities, and thus drag forces, are lower. Bending also reorients an animal so that more of its surface area is parallel to the direction of flow; this reduces drag at high Re by reducing wake size. Moving water must perform work on an animal to deform it; it takes more work (energy) to break an animal of a given strength (breaking force per cross-sectional area) if that animal is flexibly deformed by a force than if it rigidly resists the load.

When flow is predictably uni- or bi-directional over the substratum, flexible planar cnidarians are oriented with the plane of the colony normal to the direction of flow (Laborel, 1960; Théodor, 1963; Théodor and Denizot, 1965; Barham and Davies, 1968; Wainwright and Dillon, 1969; Svoboda, 1970; Riedl, 1971; Grigg, 1972; Rees, 1972; Kinzie, 1973; Velimirov, 1976; Leversee, 1976). Such orientation of planar animals is mechanically the most stable, thus planar animals tend to be pushed into this orientation by the water (Théodor and Denizot, 1965; Wainwright and Dillon, 1969). When planar organisms are oriented normal to the flow, the torque their axes must withstand is minimized, as is abrasion between branches (Grigg, 1972).

The branches or tentacles of bushy, flexible cnidarians are moved closer together as they are bent over by flowing water; this further reduces the projected area of the animal normal to the flow, hence reducing wake size and drag. Drag measurements on flexible

12   S.A. WAINWRIGHT AND M.A.R. KOEHL

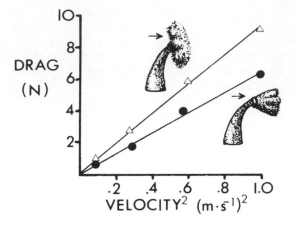

Fig. 5. Drag measurements made on a rigid (Δ) and on a flexible
(●) model of a M. senile in a flow tank.

and rigid models of the anemone Metridium senile in a flow tank
illustrate this effect (Fig. 5).

    Thus far we have been discussing the ways in which benthic cni-
darians react to the mechanical forces due to flow. Flow is also
important to these sessile animals for feeding.

                              Feeding

    Many benthic cnidarians are passive suspension feeders, i.e.
they depend upon ambient water currents to carry suspended food
across their food-capturing surfaces. The food-gathering ability
of such cnidarians is enhanced if they maximize the volume of food-
laden water they encounter per unit time, minimize the reprocessing
of water, and maintain optimum water velocities over the food-
capturing structures.

    If the feeding structures of a passive suspension feeder are
spread out in one plane normal to the direction of flow, then the
greatest volume of water per unit time can pass through the structure and the reprocessing of food-depleted water is minimized. It
has been suggested (Laborel, 1960; Barham and Davies, 1968) and
demonstrated (Leversee, 1976) that the planar form of certain cnidarians and their orientation normal to predictable currents enhances suspension feeding in this way. The same enhancement no
doubt occurs for hydrostatically-supported suspension-feeding
cnidarians such as the sea pen Scytaliopsis djiboutiensis (Magnus,

1966) and the anemone M. senile (Koehl, in prep.) which are
passively bent or twisted by ambient currents into positions with
their feeding structures normal to the flow direction.

When flow direction in an area is not predictable, a number of
octocoral and hydroid species are radial rather than planar in form
(Carlisle, et al., 1964; Riedl, 1971; Grigg, 1972; Rees, 1972;
Leversee, 1976; Velimirov, 1976) and thus can harvest food from
water flowing in any direction. A radial cnidarian may be able to
remove a greater percentage of the food carried in the water it
processes, as discussed by Leversee (1976).

Ambient currents can induce unidirectional flow through
organisms of certain configurations (Vogel and Bretz, 1972). Such
unidirectional flow through a passive suspension feeder minimizes
the reprocessing of food-depleted water. When a stream of water
encounters a branched coral colony, some of the water is acceler-
ated around the colony and some is decelerated as it passes into
the colony. The increase in velocity over the colony is concomitant with a decrease in pressure normal to the streamlines according to Bernoulli's Law. The food-depleted water within the colony
is sucked out of the top of the colony by this low pressure; this
results in a unidirectional pattern of flow into the upstream side
and out the top of the colony (Chamberlain and Graus, 1975a).

The likelihood of a food particle coming into contact with
and adhering to a feeding polyp or a tentacle depends upon the
flow pattern around that structure (Rubenstein and Koehl, 1976).
Therefore, the effectiveness of food capture by a given cnidarian
from the water flowing over it depends upon the velocity over the
individual tentacles or polyps. Such flow is likely to be of low
Re.

Colonial benthic cnidarians have a host of morphological
characteristics which modify ambient currents and in a sense regulate
the flow met by a polyp. Polyps may be within the boundary-layer of
slow-moving fluid that forms over the surface of a cnidarian colony.
The thickness of this boundary layer depends upon the surface tex-
ture and the shape of the colony. By retracting into or expanding
out of the boundary layer, a polyp can regulate the flow velocity
it encounters. Flexible polyps can be bent over by flowing water
into the boundary layer and into eddies behind colony branches;
such passive regulation of the velocity over polyps should allow
them to feed over a wide range of ambient velocities (Riedl and
Forstner, 1968).

The gross structure of a cnidarian colony also affects the
flow pattern over an individual polyp. For example, the diameter,
spacing, and arrangement of coral branches have been shown to change
the speed and direction of water movement within a colony; seemingly

very different colony configurations can produce the same flow pattern over the polyps within the colony (Chamberlain and Graus, 1975a). Flexible colonies sway back and forth in the oscillating flow associated with waves. Because the colonies are moving with the water, the relative velocity over their polyps is much lower than ambient (Wainwright, et al., 1976). Various patterns of arrangement of polyps around the branches of colonial cnidarians have been observed (Riedl, 1971; Muzik and Wainwright, 1976). Particular polyp arrangements can probably be correlated with the ways in which certain whole-colony morphologies modify ambient currents.

All the above examples of the ways in which benthic cnidarians affect the flow forces and feeding currents they encounter reveal the importance of the size and shape of these animals to their ability to utilize moving water. What happens to flow as a sessile animal grows?

Growth

As a benthic cnidarian grows, its Re increases, not only because the animal increases in size, but also because it encounters faster flow away from the substratum. Since the ability of a sessile cnidarian to capture a particular type of food depends upon the flow pattern over its food-catching structures, we expect either the diet or the morphology of a benthic cnidarian to change as it grows. If a colonial (or solitary) cnidarian is to feed on the same sort of material throughout its sessile life, colony (or whole-animal) morphology must be modified as the colony's (or animal's) Re increases in ways that would keep the Re of the polyps (or tentacles) the same. Shape becomes increasingly important to drag reduction at higher Re's where pressure drag predominates, thus we expect to find more pronounced wake-minimizing shapes and orientations in larger animals.

Thus far we have discussed what benthic cnidarians do in response to flow. We will now discuss how they do it.

RESISTANCE TO FLOW

Polyps and the branches of many colonial cnidarians are columnar, beam-like structures. The mechanical property that best describes a beam's ability to resist bending is flexural stiffness, EI, where E is Young's modulus of elasticity and I is the second moment of area of the cross section of the beam. E is a material property that arises from the strengths, angles of bonds and other interactions between molecules in the material. It is the stiffness

of the material: the amount of force per unit area that is required to produce a unit if deformation.

$$E = \frac{stress}{strain} = \frac{force/area}{\Delta\ length/original\ length}$$

E is independent of the size or shape of the beam's section.

I, on the other hand, is a purely morphological feature,
independent of the nature of the material. $I = \int y^2 dA$ where dA is
a unit of area of the section that lies distance y from the
centroid of the section. The further an increment of area lies
from the centroid, the greater is its contribution to the stiffness of the beam. Thus, one can make a beam stiffer in bending
either by using a material of higher modulus, E, or by arranging
the given material in a shape that puts the most material the
greatest distance from the centroid as possible. If the bending
force arrives from a single, predictable direction, an I-shaped
section is a most appropriate design whereas a round, hollow
section is best if bending in all directions is to be minimized.

Scleractinian corals are the most rigid large animals in the
world. Their skeletons are 94 to 99.9% by weight aragonite in
aggregates of submicroscopic crystals. There are also small
amounts of proteins and chitin and other organic materials as
well as submicroscopic voids. The material is linearly elastic to
failure and extremely rigid: $E > 10^{10}$ Nm^2 (Vosburgh, pers.
communication).

Alcyonids such as <u>Sinularia</u> form reefs with the bases of their
colonies. The material is apparently neither as rigid nor as
brittle as scleractinian coral skeleton. It is composed of
sclerites that have grown in the colonial mesoglea until they have
fused. Throughout the Cnidaria, sclerites are polycrystalline
aggregates of calcite.

Axial skeletons of antipatharians and holaxonian gorgonaceans
are made of rigid ($E \simeq 10^9$ Nm^{-2}) protein-polysaccharide complexes
that in some genera are associated with polycrystalline calcite.
Axes are layered and the polymeric material is laid down in lath-
shaped fibers such that all fibers in a layer are parallel to each
other but may not be parallel to fibers in adjacent layers.   In
<u>Leptogorgia</u> and others, the fibers lie in very steep helices having
low angles to the axial direction.   In <u>Ellisella</u>, the axis consists
predominantly of radially oriented calcite (Muzik and Wainwright,
1976). Although these are all comparatively rigid viscoelatic
materials, the axes they form are often quite flexible because of
their small diameter (low I). However, branches with axes greater
than 1 cm in diameter do not bend in the flow. An increase in
total diameter may be the simplest way to increase I, but it may

well not be the cheapest in terms of the energy of synthesis.

Figure 6 shows that the basal axis of the Caribbean sea fan, Gorgonia, is strongly elliptical, giving high resistance to sideways bending (high I) and thus good support for the fan-shaped colony and low resistance to bending (low I) with the flow normal to the fan. On the other hand, its finest, anastomozing branches are elliptical with their major axes parallel to the flow, thus streamlining the branches. This ellipticity is due less to axial shape that it is to the distribution of scleritic coenenchyme. Gorgonia thus accommodates to the direction of flow by optimizing I in appropriate places effecting appropriate mechanical functions. Axes of bushy Plexaurella colonies have rounder sections (Fig. 6) but the distribution of calcite in the axis is largely in one diametric plane permitting bending parallel to the flow and resisting bending normal to the flow. Gorgonia and Plexaurella share a habitat where the direction of flow is largely predictable. This brings to question the role, if any, of water flow and deformation in morphogenesis of such forms.

Hydrostatic Cylinders

In benthic cnidarians, hollow cylindrical polyps, branches and colonies are often supported by water taken in directly from the environment and kept at low pressures (Chapman, 1975). In a pressurized cylinder of homogeneous wall material, radially direct-

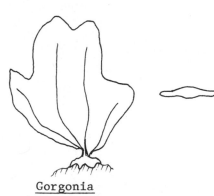

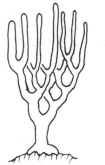

Gorgonia   Plexaurella

Fig. 6. Diagrams of the colony forms of Gorgonia and Plexaurella and of the cross sectional shapes of their basal axes. Stipple indicates distribution of microcrystalline calcite. Predominant direction of waves impinging on axial sections is parallel to the long axis of the page.

ed hoop stresses are twice longitudinal stresses. This could help explain predominant radial expansion in a hydrostat. Since cnidarian hydrostats are known instead for their great longitudinal extensibility, one may conclude that their walls are not homogeneous. In fact, they are bountifully reinforced with helically wound collagen fibers (Chapman, 1975).

Stresses are greater in inner layers of thick-walled cylinders than in outer layers. Koehl (this symposium) and Gosline (1971) have shown that the inner layers of mesoglea in _Anthopleura_ _xanthogrammica_ and _Metridium_ _senile_ contain a dense array of circumferentially oriented collagen fibers. According to this principle, shear forces in the interfibrillar matrix retard radial extension.

In cylinders that bend, helical reinforcement is superior to rectilinear reinforcement. Failure of a bent hollow cylinder occurs by kinking (also called local buckling). In a rectilinear array with N fibers/area of cylinder wall, only longitudinal fibers, N/2, will resist kinking, but each fiber contributes its modulus. In a helical array, all fibers, N, oppose kinking, but the contribution of each fiber lends a cosine function of its angle times the fiber's modulus to the resistance of kinking. Thus the fibers in a cylinder wall have their maximum antikinking effect when the fiber angle is low, and this is when the cylinder has maximum length but minimum diameter. This applies to all known cnidarian polyps except _Hydra_ whose fiber system is reported to be rectilinear by Hausman and Burnett (1969). Preliminary results of a study of tentacles of hydra indicate that these structures have a helical system of mesogleal fibers that maintains integrity of the tentacle over extensions that are 20X contracted length.

In a cylinder whose fibers are in a tightly linked crossed-helical array, a fully shortened cylinder can extend to a length equal to the circumference of the shortened cylinder. $l = d.\cos\alpha$, where d is the length of one turn of the helix around the cylinder. This means that hydra tentacles and stretchy anemones like _Metridium_ cannot have their collagen fibers in a tightly joined network: their mesogleal extensions must be accompanied by the slipping of overlapping fibers past one another. Here again shear stresses in the interfibrillar matrix will retard extension.

Joints

For all their simple structure and smoothly contoured shapes, some cnidarians possess the simplest possible kinds of joints in the animal kingdom. A joint is a relatively local decrease in the flexural stiffness of a beam, and can arise either from material

of low modulus E, or from a cross-section of relatively low I.
Pelmatohydra bends most readily at its peduncle and Metridium
bends just below the capitulum (see Koehl, this symposium, Fig. 2)
because their body diameters are minimal at these points, thus
reducing I. Minimization of body wall thickness will also reduce
I.

Isis and Melithaea are jointed scleraxonian gorgonaceans that
have interestingly different joint mechanisms. Branch axes of
Isis consist of rigid segments of fused calcite sclerites alternating with flexible joints of polymeric materials: flexibility is
by low E in the joints. Melithaea's axis is composed of fused
rod-shaped sclerites (Muzik and Wainwright, 1976). In the rigid
segments the sclerites are aligned parallel to the axis and are
fused along their lengths by additional $CaCO_3$. In the flexible
joints, sclerites form an open 3-dimensional framework (spaces
filled with mesoglea) in which sclerites are attached to each other
at their ends by an acidophil polymeric glue. Since the polyhedra
described by the sclerites in the framework are not tetra- or
octahedra, the framework is clearly designed to deform and the
flexible glue and mesoglea maintains viscoelastic integrity of the
framework throughout deformation. This type of flexural control
is not simply a matter of branch shape I or material E: it suggests a "modulus of structure" in which E and I of visible elements, their orientation and the nature of their bonding are
involved. Derivation of this modulus will constitute the statement of a previously unappreciated principle of mechanical morphology.

SUMMARY

Flowing water carries food to benthic cnidarians and subjects
them to drag forces. Body size and shape and behavior of these
animals affect the flow patterns and drag forces they encounter.
Predictive quantitative relationships exist between measurable
characteristics of a body and of fluid moving past it. However,
it is not possible to describe the flow pattern over an individual
polyp from measurements of mainstream velocity alone.

Polyps and branching cnidarians in situ are treated as beams
that tend to bend due to drag forces. Flexural stiffness of a
beam is determined by the elastic modulus of the material and by
the second moment of area of the cross section of the beam, a
property of the shape and size of the section. Some polyps and
colonies have flexible joints whose design is the simplest of
possible joints.

LITERATURE CITED

Abbott, B.M., 1974. Flume studies on the stability of model corals
    as an aid to quantitative palaeoecology. Palaeogeogr.
    Paleoclimatol. Palaeoecol., 15:1-27.

Barham, E.G., and I.E. Davies, 1968. Gorgonians and water motion
    studies in Gulf of California. Underwater Naturalist, Bull.
    Am. Littoral Soc., Winter: 24-28, 42.

Alexander, R.M., 1968. Animal Mechanics. Univ. Washington Press,
    Seattle. 346 pp.

Chamberlain, I.A., and R. R. Graus, 1975a. Water flow and hydro-
    mechanical adaptations of branched reef corals. Bull. Mar.
    Sci., 25:112-125.

Chamberlain, I.A., and R. R. Graus, 1975b. Adaptations in corals:
    How do corals withstand waves and currents? Abstracts with
    Programs, U.S. Geol. Soc. Ann. Meetings, Salt Lake City, Utah,
    1024.

Chapman, G., 1975. Versatility of hydraulic systems. J. Exp.
    Zool., 194: 249-270.

Gosline, J.M., 1971. Connective tissue mechanics of Metridium
    senile. J. Exp. Biol., 55: 763-774.

Grigg, R.W., 1972. Orientation and growth form of sea fans.
    Limnol. Oceanogr., 17: 185-192.

Hausman, R.E., and A. L. Burnett, 1969. The mesoglea of hydra.
    J. Exp. Zool., 171: 7-14.

Hubbard, J.A.E.B., and Y.P. Pocock, 1972. Sediment rejection by
    scleractinian corals: A key to palaeo-environmental re-
    construction. Geol. Rundschau., 61: 598-626.

Kinzie, R.A., III, 1973. The zonation of West Indian gorgonians.
    Bull. Mar. Sci., 23: 93-155.

Koehl, M.A.R., 1976. Effects of the structure of sea anemones on
    the flow forces they encounter. (In prep.).

laBorel, J., 1960. Contribution à l'étude directe des peuplements
    benthiques sciaphiles sur substrat rocheux en Meditérranée.

Rec. Trav. Stat. Mar. Endoume, 33: 117-173.

Leversee, G.J., 1976. Flow and feeding in fan-shaped colonies of
the gorgonian coral, Leptogorgia. Biol. Bull., (in press).

Magnus, D.B.E., 1966. Zur Ökologie einer nachtaktiven Flachwasser-
Seefeder (Octocorallia, Pennatularia) im Roten Meer. Veroff.
Inst. Meeresforsch. Bremerhaven, 2: 369-380.

Muzik, K.M., and S.A. Wainwright, 1976. Morphology and habitat of
five Fijian sea fans. (In press in Bull. Mar. Sci.)

Rees, J.T., 1972. The effect of current on the growth form in an
octocoral. J. Exp. Mar. Biol. Ecol., 10: 115-124.

Riedl, R.J., 1971. Water movement. O. Kinne, Ed., Marine Ecology,
volume I. Wiley-Interscience, London, 1244 pp.

Riedl, R., and H. Forstner, 1968. Wasserbewegung im Mikrobereich
des Benthos. Sarsia, 34: 163-188.

Rouse, H., 1961. Fluid Mechanics for Hydraulic Engineers. Dover
Publications, Inc., New York, 422 pp.

Rubenstein, D.I., and M.A.R. Koehl, 1976. The mechanisms of particle
capture by filter feeders: Some theoretical considerations.
(In press in Amer. Natur.).

Sassaman, C., and C.P. Mangum, 1972. Adaptations to environmental
oxygen levels in infaunal and epifaunal sea anemones. Biol.
Bull., 143: 657-678.

Shapiro, A.H., 1961. Shape and Flow: The Fluid Dynamics of Drag.
Doubleday and Co., Inc., Garden City, N.Y., 186 pp.

Shinn, E., 1963. Spur and groove formation on the Florida reef
tract. J. Sedimentary Petrology, 33: 291-303.

Stoddart, D.R., 1969. Ecology and morphology of recent coral
reefs. Biol. Rev., 44: 433-498.

Svoboda, A., 1970. Simulation of oscillating water movement in
the laboratory for cultivation of shallow water sedentary
organisms. Heloglander Wiss. Meeresunters, 20: 676-684.

Théodor, J., 1963. Contribution à l'étude des gorgones. III.
Trois formes adaptives d'Eunicella stricta en fonction de la
turbulence et du courant. Vie Milieu, 14: 815-818.

Théodor, J., and M. Denizot, 1965. Contribution à l'étude des
    gorgones I: A propos de l'orientation d'organismes marins
    fixés végétaux et animaux en fonction du courant. Vie
    Milieu, 16: 237-241.

Velimirov, B., 1976. Variation in forms of Eunicella cavolinii
    Koch (Octocorallia) related to intensity of water movement.
    J. Exp. Mar. Biol. Ecolo., 21: 109-117.

Vogel, S., and W.L. Bretz, 1972. Interfacial organisms: Passive
    ventilation in the velocity gradients near surfaces. Science,
    175: 210-211.

Wainwright, S.A., W.D. Biggs, J.D. Currey, and J.M. Gosline, 1976.
    Mechanical Design in Organisms. Halsted Press (Wiley),
    New York, 423 pp.

Wainwright, S.A., and J.R. Dillon, 1969. On the orientation of
    sea fans (genus Gorgonia). Biol. Bull., 136: 130-139.

Wood-Jones, F., 1909. On the growth forms and supposed species
    in corals. Proc. Zool. Soc. Lond., 2: 518-556.

MECHANICAL DESIGN IN SEA ANEMONES

M. A. R. Koehl

Department of Zoology, Duke University

Durham, North Carolina 27706, U. S. A.

INTRODUCTION

By applying principles of fluid and solid mechanics to biological structures, I have studied the morphological adaptations
of two species of sea anemones to the mechanical activities they
perform and the environmental forces they encounter. The two species
of anemones I used represent different extremes in mechanical behavior: Metridium senile, a calm-water species noted for the great
range of shapes and sizes it can assume (Batham and Pantin, 1950),
and Anthopleura xanthogrammica, a species occurring in areas exposed to extreme wave action (Hand, 1956; Dayton, 1971). I chose
anemones because they are simple in structure, thus differences in
morphology and function between species are easier to recognize and
characterize than they would be for more complex animals.

Which aspects of the structure of M. senile and A. xanthogrammica enable them to harvest food effectively from such different
flow regimes while they withstand the associated mechanical forces?
I expected these animals with the same basic body plan but with mechanically different life-styles to differ in structure on several
levels of organization: 1) the materials level (the structure and
composition of mesoglea), 2) the whole-organism level (the shapes
and dimensions of bodies), and 3) the environmental level (the ways
in which animals fit into their mechanical environments). The
environmental level of organization is the topic of this paper. Sea
anemones are sessile marine animals, thus water movement is the most
important aspect of their mechanical environment. I expected the
distribution, behavior, and structure of M. senile and A. xanthogrammica to determine the environmental mechanical forces they
encounter.

23

FLOW CONDITIONS IN THE FIELD

By SCUBA diving and working low tides, I surveyed a number of sites along the coast of Washington to determine the characteristic habitats of M. senile and A. xanthogrammica and to observe their posturing and feeding behavior. I also measured water velocities in order to characterize quantitatively the flow regimes of localities where the two species occur as well as of their microhabitats within those localities.

M. senile are subtidal animals occurring on rock in areas exposed to steady tidal currents typically on the order of 0.1 m·s^{-1}. These M. senile are tall (mean height above the substratum = 38.1 cm. standard deviation (SD) = 8.8, number of measurements (n) = 28) and thus are subjected to essentially mainstream current velocities (Figure 1). M. senile in flowing water are bent over (Figure 2); this bending orients their oral disks normal to the current. I have observed these anemones filtering zooplankton from the water passing through the meshwork of numerous small tentacles on their fluted oral disks. Gut content analyses confirm that M. senile are zooplanktivorous.

A. xanthogrammica carpet the bottoms of intertidal surge channels at rocky exposed coastal sites. At high tide they are subjected to the oscillating bottom flow as waves pass overhead; at low tide they are subjected to the shoreward surge and seaward backwash of waves breaking seaward of them. Mainstream velocities in surge channels are highest (often greater than 3 to 5 m·s^{-1})

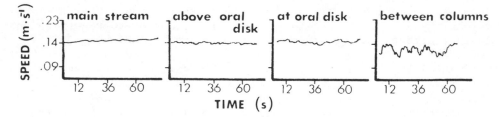

Figure 1. Water speeds measured using a thermistor flowmeter (LaBarbera and Vogel, 1976): "mainstream" = 60 cm above a bare vertical rock wall near a bed of M. senile; "above oral disk" = 10 cm above the oral disk of a M. senile 30 cm tall in the middle of the bed; "at oral disk" = 1 mm above the oral disk of the same M. senile; "between columns" = 20 cm above the substratum and 1 cm from the downstream side of the column of the same M. senile.

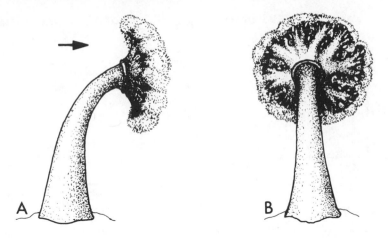

Figure 2. M. senile bent over in a tidal current. A.   Side view
(arrow indicates flow direction), and B.   Rear view.

during surge and backwash; the mainstream flow (Figure 3) is bidirectional with a period of about 10 s. A. xanthogrammica in surge
channels are short (mean height above the substratum = 2.5 cm,
SD = 0.9, n = 71). Flow over these anemones (Figure 3) is more
turbulent and of lower velocity (typically 1 m·s^{-1} maximum) than
mainstream. Hence, unlike M. senile, A. xanthogrammica in essence
hide from the maximum flow velocities which characterize the areas
where they occur by being short.

    A. xanthogrammica feed primarily on mussels which fall on the
anemone's oral disks after being ripped off the substratum by starfish, logs, and waves (Dayton, 1973). I found greater current velocities higher in the intertidal. Mussels occur higher in the
intertidal than do A. xanthogrammica, thus flow velocities are 3 to
5 times higher where the mussels are ripped off the rock than they
are where the anemones must catch and hold on to them.

    A few solitary A. xanthogrammica occur at sites exposed to
lower flow velocities (maximum mainstream velocities typically
1 m·s^{-1}) than those in surge channels. A. xanthogrammica in such
protected areas are taller (mean height above the substratum =
7.1 cm, SD = 2.4, n = 25) than those in exposed channels and thus
stick out into currents as rapid as those encountered by the short
A. xanthogrammica in exposed channels. Thus, anemones from apparently different flow habitats are actually in similar flow microhabitats by virtue of their heights above the substratum. This
illustrates that a knowledge of mainstream flow conditions does
not tell us the flow regime encountered by a particular organism.

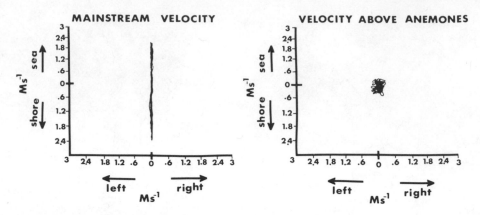

Figure 3. Flow velocities measured using an electromagnetic
flowmeter (EPCO Model 6130) in a surge channel during surge and
backwash: "mainstream" = 115 cm above the substratum; "above
anemones" = 3 cm above the oral disk of an A. xanthogrammica
surrounded by other A. xanthogrammica on the floor of the channel
(Flow into ("shore") and out of ("sea") the channel is indicated
on the vertical axis and flow from side to side ("right", "left")
in the channel is indicated on the horizontal axis.)

DRAG FORCES ON ANEMONES

When fluid moves relative to an object, it exerts a force
(termed drag) on that object tending to push the object downstream.
Drag increases as velocity increases. Although A. xanthogrammica
"hide" from mainstream currents and M. senile do not, A. Xanthogrammica encounter flow velocities generally an order of magnitude
higher than those met by M. senile. I therefore expected the shapes
of A. xanthogrammica to minimize drag more than those of M. senile.

I used my field measurements of flow velocities and of anemone
body dimensions to calculate the drag forces on M. senile in tidal
currents and A. xanthogrammica in waves. I calculated the forces
on M. senile-shaped objects using the standard drag equations for
cylinders in steady flow (Rouse, 1961). Calculation of the forces
on A. xanthogrammica subjected to wave action is complicated by the
fact that the water is continually accelerating and decelerating
in different directions around these anemones. Fortunately ocean
engineers have worked out a body of equations for predicting wave
forces on pilings (Keulegan and Carpenter, 1958; Wiegel, 1964;
Bretschneider, 1966). I used these equations to calculate the
forces on A. xanthogrammica-shaped pilings in surge and backwash.

My calculations indicate that although M. senile occur in calm
regions and A. xanthogrammica in exposed areas, the flow force on an
individual anemone of either species is nearly the same (on the
order of 1 N). Drag forces on A. xanthogrammica measured in the
field are in fact typically 1 N (Koehl, in prep., a).

How can the drag on a M. senile in a 0.1 m·s^{-1} current be the
same as the drag on a A. xanthogrammica in 1.0 m·s^{-1} flow? One way
to minimize the drag on an object is to minimize the size of the
wake that forms behind the object. One way to minimize wake size
is to present most of the surface area of an object parallel to the
direction of flow. (Think of the force on a flat plate oriented
parallel to the current versus on one oriented normal to the current.)
A. xanthogrammica in surge channels are short and wide (mean ratio
of pedal disk diameter to height = 2.5, SD = 0.7, n = 70) hence
most of their surface area is parallel to the direction of flow and
drag is minimized. M. senile on the other hand are tall (mean ratio
of pedal disk diameter to height = 0.3, SD = 0.1, n = 40), and are
bent over in currents so that their large oral disks are oriented
normal to the flow direction, hence drag is maximized. Flow tank
measurements of drag on models of anemones in various configurations
confirm that this effect of shape on drag is the case (Koehl, in
prep.,a). Thus, the drag on an individual anemone of either species
is essentially the same because of the respective shapes and orientations of the animals.

What types and magnitudes of stresses do such drag forces produce in the body walls of these anemones? (Stress is the force per
cross-sectional area over which that force is distributed.)

DRAG-INDUCED STRESSES IN ANEMONE BODY WALLS

The shape of an animal determines the distribution of stresses
within the animal for a given load distribution. The stiffness of
the materials composing an animal determines how much the animal
will deform in response to these stresses. Since the same load
that causes a M. senile to bend over does not noticeably deflect
an A. xanthogrammica, I predicted that the latter would have a more
stress-minimizing shape and a stiffer body wall than the former.

A sessile anemone can be considered as a cantilever supporting
a feeding apparatus in flowing water in the proper orientation for
food-capture. A cantilevered beam (sea anemone) subjected to a load
(drag) undergoes shearing and bending (Figure 4). The magnitude of
the shearing stress produced in a beam by a given load depends upon
the cross-sectional area of the beam, but is independent of the
shape of the beam. The magnitude of tensile stress in a bending
beam depends not only upon the cross-sectional area, but also upon
the length and the shape of the beam (see Wainwright and Koehl,

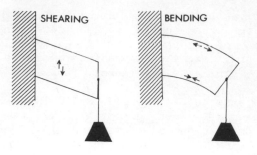

Figure 4.   Deformation of a cantilevered beam supporting a load.

this symposium).   The oral disk of an anemone deformed in shearing
remains parallel to the substratum and the flow direction whereas
the oral disk of an anemone deformed in bending will not.   I have
modelled sea anemones as continuously-loaded hollow cantilevers of
the proper shape and have used beam theory (Faupel, 1964) to calculate the shear stresses and the tensile stresses associated with
bending that environmental forces would produce in the body walls
of M. senile and A. xanthogrammica.

The calculated shear stresses in a M. senile are on the order
of 200 $N \cdot m^{-2}$ whereas the maximum tensile stresses associated with
bending are on the order of 4,000 $N \cdot m^{-2}$.   As expected, these maximum
stresses occur at the narrow region of the upper column of a M. senile.   It is not surprising that a M. senile in a current is bent at
this upper region of its column; the anemone's filtering oral disk
is thus oriented normal to the direction of flow and is held out in
the more rapidly flowing water away from the substratum.

The calculated shear stresses and maximum tensile stresses in
an A. xanthogrammica body wall are typically 120 $N \cdot m^{-2}$ and 100 $N \cdot m^{-2}$
respectively.   Thus, although an A. xanthogrammica is exposed to wave
action, because it is short and wide, the tensile stresses in its
body wall due to flow forces are an order of magnitude lower that
those in a tall, calm-water M. senile.   A. xanthogrammica do not
bend visibly in flowing water; their oral disks remain upright
where mussels are more likely to fall on them.

The elastic modulus, or stiffness (see Wainwright and Koehl,
this symposium), of A. xanthogrammica body wall is three times
greater than that of M. senile body wall when they are stretched
at biologically relevant rates (Koehl, in prep.,b).   Thus, these
two species of anemones differ in their mechanical response to flow
not only because of their respective shapes, but also because of the

materials from which they are built. Hence, the flexural stiffness, which depends upon both shape and material (see Wainwright and Koehl, this symposium), of A. xanthogrammica is two orders of magnitude greater than that of M. senile.

I used the same assumptions that I used in the above calculations to compute the deflections of the oral disks of individuals of each species when subjected to drag. Deflections predicted in this manner are consistent with deflections of the anemones observed in the field.

SUMMARY

Typical values for the current velocity, drag, stresses, flexural stiffness, and deflection of a M. senile and a A. xanthogrammica of the same pedal disk diameter (15 cm) are summarized in Table I.

The design of an organism can actually affect the flow forces it encounters as well as its response to those forces, as illustrated by the sea anemones M. senile and A. xanthogrammica. M. Senile occur in calm areas, but because they are tall, they are exposed to mainstream current velocities. Although A. xanthogrammica occur in areas exposed to wave action, they are short and effectively hide from mainstream velocities. Nonetheless, water velocities encountered by A. xanthogrammica are an order of magnitude greater

TABLE I

	Metridium senile	Anthopleura Xanthogrammica
TYPICAL VELOCITY	$.1 \text{ m} \cdot \text{s}^{-1}$	$1.0 \text{ m} \cdot \text{s}^{-1}$
FORCE ON AN INDIVIDUAL	.8 N	1.0 N
SHEAR STRESS	$180 \text{ N} \cdot \text{m}^{-2}$	$120 \text{ N} \cdot \text{m}^{-2}$
MAX BENDING TENSILE STRESS	$4520 \text{ N} \cdot \text{m}^{-2}$	$100 \text{ N} \cdot \text{m}^{-2}$
FLEXURAL STIFFNESS (EI)	$.08 \text{ Nm}^2$	10.56 Nm^2
DEFLECTION OF ORAL DISK	7 cm	.0004 cm

than those met by M. senile. However, the drag force on an individual of either species is about 1 N due to the respective shapes
of each species. Since the tensile stresses in a bending beam
increase with the length and decrease with the diameter of the
beam, these flow-induced stresses in the tall, slim M. senile are
an order of magnitude greater than in the short, wide A. xanthogrammica. Not only are the tensile stresses greater in M. senile,
but their body walls stretch more for a given stress than do those
of A. xanthogrammica.

The water currents encountered by these anemones and their
mechanical responses to the currents can be related to the manner
in which the anemones harvest food from flowing water. M. senile
bend over in currents and suspension-feed through their oral disks
whereas A. xanthogrammica remain upright in surge and catch mussels
which fall on their oral disks. I have used the case of these two
species of sea anemones to illustrate the importance of water flow
to sessile marine animals both in terms of their mechanical support
systems and their mechanisms of feeding.

ACKNOWLEDGMENTS

This work was supported by a Cocos Foundation Training Grant
in Morphology, a Duke University Travel Award, A Graduate Women in
Science Grant, a Sigma Xi Grant, and a Theodore Roosevelt Memorial
Fund of the American Meseum of Natural History Grant. I wish to
express my appreciation to S. A. Wainwright whose support and feedback made this work possible, and to S. Vogel who introduced me to
fluid mechanics. I thank the U. S. Coast Guard and R. T. Paine
for making it possible for me to work on Tatoosh Island, and M. Denny,
M. LaBarbera, K. Sebens, and T. Suchanek for their invaluable help
in the field.

LITERATURE CITED

Batham, E.J., and C. F. A. Pantin, 1950. Muscular and hydrostatic
     action in the sea-anemone Metridium senile (L.). J. Exp. Biol.,
     27: 264-289.

Bretschneider, C. L., 1966. The probability distribution of wave
     force and an introduction to the correlation drag coefficient
     and the correlation inertial coefficient. Coastal Engineering.
     Santa Barbara Specialty Conference, October, 1965. American
     Society of Civil Engineers, New York, pp. 183-217.

Dayton, P. K., 1971. Competition, disturbance, and community or-
     ganization: the provision and subsequent utilization of space
     in a rocky intertidal community. Ecol. Monogr. 41: 351-389.

Dayton, P. K., 1973. Two cases of resource partitioning in an
    intertidal community: making the right prediction for the
    wrong reason. Amer. Natur. 107: 662-670.

Faupel, J. H., 1964. Engineering Design: A Synthesis of Stress
    Analysis and Materials Engineering. John Wiley and Sons,
    Inc., New York, 979 pp.

Hand, C., 1955. The sea anemones of central California. Part II.
    The endomyarian and mesomyarian anemones. Wasmann J. Biol.,
    13: 37-99.

Keulegan, G. H., and L. H. Carpenter, 1958. Forces on cylinders and
    plates in an oscillating fluid. J. Res. National Bureau
    Standards, 60: 423-440.

Koehl, M. A. R., Effects of the structure of sea anemones on the
    flow forces they encounter. (in prep.,a).

Koehl, M. A. R., Mechanical adaptations of the connective tissues
    of hydrostatically supported organisms: Sea anemones.
    (in prep.,b.)

LaBarbera, M., and S. Vogel, 1976. An inexpensive thermistor flow-
    meter for aquatic biology. Limnol. Oceanogr. (in press).

Rouse, H., 1961. Fluid Mechanics for Hydraulic Engineers. Dover
    Publications, Inc., New York, 422 pp.

Wainwright, S. A., W. D. Bigg, J. D. Currey, and J. M. Gosline,
    1976. Mechanical Design in Organisms. Halsted Press (Wiley),
    New York, 423 pp.

Wainwright, S. A., and M. A. R. Koehl, 1976. The nature of flow
    and the reaction of benthic cnidarians to it. G. O. Mackie,
    Ed. Coelenterate Ecology and Behavior. Plenum Publishing
    Corporation, New York, (in press).

Wiegel, R. L., 1964. Oceanographical Engineering. Prentice-Hall,
    Inc., Englewood Cliffs, New Jersey, 532 pp.

HYDROID SKELETONS AND FLUID FLOW

Gordon R. Murdock

Department of Zoology, Duke University

Durham, North Carolina 27706, U. S. A.

INTRODUCTION

Most hydroid colonies are encased in a tubular chitinous exoskeleton, the perisarc. Hyman (1940, p. 402) observes of the
perisarc that "On stems it usually forms groups of rings or
annulations at definite points related to the branching . . . The
function of these is obscure, but it is generally supposed that
they lend flexibility." Perusal of the literature and conversa-
tions with biologists studying hydroids suggest that the annular
regions on hydroid perisarc are no better understood now than in
1940. Indeed, little more is known of the perisarc itself or of
the skeleton which it forms.

This paper will look at the structure and properties of the
annular perisarc and provide a simplified mechanical analysis of
it. It will also suggest that the annulations produce properties
which are desirable for a tubular exoskeleton in a dynamic fluid
environment.

It is apparent that we commonly mean two things when we say
that a tube is more or less flexible. The more flexible the tube,
the farther it will bend in response to a given load and the smaller the radius of curvature around which it can be bent without
suffering a collapse of its structure. Both of these follow from
the fact that as a tube (or a beam) is bent, mechanical stresses
build up within it which, together with their associated strains
(see below) store energy and tend to resist further deformation.
The more slowly these stresses build up, the more flexible we say
the tube (or beam) is.

I will argue that some annular regions in perisarc minimize
the build up of stresses in those areas when the perisarc is bent,
rendering it more flexible. Some of these regions become secondari-
ly modified however so that they no longer serve this function.

Two mechanical notions are particularly important to the present analysis and will be briefly explained. The first concerns
the properties of materials while the second concerns the geometry
of deformed structures. These are explained in readily accessible
terms by Wainwright et al. (1976) and are dealt with by most elementary engineering texts.

First, any material subjected to a force is deformed. The
magnitude of the deformation, expressed as a proportion of the
original size of the sample, is termed the strain. The force
associated with a strain will be distributed over some area in any
plane within the material which is perpendicular to the direction
of the force. The magnitude of the force divided by the area over
which it is distributed is termed the stress. As mentioned above,
these distributed forces tend to resist further deformation of the
material. If a material is elastic, the magnitudes of the stresses
within it and the strains undergone are not time dependent and are
related to each other by a constant, the elastic modulus, E. The
higher the elastic modulus of the material from which a beam is
made, the stiffer the beam.

Second, when a beam is bent, material toward the outside of
the bend is under tension while that on the inside of the bend is
under compression. It is obvious that there must be a surface somewhere between these inner and outer extremes which is neither
stretched nor compressed. This unstrained surface is termed the
neutral axis. The references here to inner and outer surfaces
have nothing to do with the inner and outer surfaces of a tube
as the beam described may as readily be solid as hollow. If the
beam is made of material which is homogeneous and has the same
properties in compression and extension, the unstrained surface
will pass through the center of mass of any cross section through
the beam. The tensile or compressive stress (or strain) experienced
by a particular part of a bent beam depends on the distance between
that part and the unstrained surface. The contribution of that
stress to resisting further bending of the beam will be proportional
to the square of that distance. The influence of the cross sectional geometry of a beam on its stiffness is given by a quantity
known as the second moment of area of the cross section, symbolised
as I. The flexural stiffness of a beam is given by the product of
the elastic modulus of the material in the beam, E, and the second
moment of area, I.

A flat sheet of any material is more flexible than is a straight
walled tube made from that sheet. This is because the unstrained
surface in each case lies within the structure, probably at its
center, so that all material in the sheet is very near this surface while much of the material in the tube is at a considerable
distance from it. In the bent sheet, then, the stresses and strains
are small while those in the walls of the tube are much larger.
A tube made with a wall which is folded like the bellows of a
partially opened concertina should be quite different, however. As
the tube is bent, the increased length needed along the outside of
the bend is provided by partial unfolding of the wall while
shortening of the wall along the inside of the bend is accommodated
by increased folding. Bending the tube therefore simply results in
bending the sheet of material from which the tube is made rather than
stretching or compressing that sheet as would happen in a straight
walled tube. A tube with folded walls therefore should develop
much smaller stresses as it is bent than would a straight walled
tube, and should be much more flexible.

Work done against stresses in deforming elastic material in
a structure may be stored as potential energy in the material; a
spring provides an example. If the store of potential energy becomes too great, however, it may be released inadvertently and do
work on the structure in which it was stored, destroying it. An
obvious example would be a dam burst by the pressure of the water
stored behind it. When stresses due to bending in hollow, thin
walled, circular cylinders reach a certain critical level, the
structure fails in a characteristic manner known as local buckling.
This is easily demonstrated by sharply bending any thin walled tube.

Two aspects of local buckling are especially important. First,
the lumen of the tube collapses as the walls of the tube pinch together. Second, two very sharp bends are produced in the walls of
the tube at the sides of the buckled region. Depending on the
material involved and the thickness of the tube wall, these bends
may cause the material in the tube wall to deform irreversably
or break. The maximum stress which a tube can withstand without
buckling locally is proportional to the elastic modulus of the
material in its wall and the thickness of the wall and is inversely
proportional to the diameter of the tube.

The annular regions in perisarc are regions of greater flexibility. They may facilitate passive orientation of parts of the
colony to water currents, a possible advantage to a suspension
feeder, or they may help prevent local buckling of the perisarc
tube which might interrupt the continuity of the enclosed tissue
and coelenteron.

MATERIALS AND METHODS

Three hydroid species were used. Living Tubularia crocea
were maintained in refrigerated aquaria filled with Instant Ocean
brand synthetic sea water. Pennaria tiarella specimens had been
preserved in ethanol while Obelia sp. (gracilia?) were preserved
in formalin.

Specimens to be sectioned were embedded in Tissue-Tek O.C.T.
compound. Sections were cut at 10μ on an American Optical cryostat.
The sections were examined with the polarizing microscope, with
phase contrast optics and with bright field illumination. All
measurements were made with an optical micrometer which could be
read directly to 0.06μ.

RESULTS

Figures in the literature suggest that annular regions in the
perisarc typically occur distal to points of branching and at the
bases of zooids when those zooids are enclosed in thecae (i.e. in
perisarc). There are apparently many variations on this pattern
however and some substantial departures from it. The annuli on
the Obelia which I have used fit this pattern precisely. In
Pennaria tiarella, there are annuli just distal to the branch points,
but they occur at regular intervals along the hydrocaulus and
branches whether there is a branch point or not. The polyps of
P. tiarella are not enclosed in thecae and show no specific association with annular regions, though no polyp is far beyond the
most distal annular region. Annuli are occasionally but rarely
seen just basal to branch points in both these species. In T. crocea
there is no apparent association between annular regions and polyps
or branch points.

Figure 1 shows a diagrammatic longitudinal section through an
annular region in the wall of each of the three species used here.
In each case the figure represents an area well away from the
growing tip of the colony. The lumen of the tube is shown shaded.

The perisarc consists of concentric laminae which in P. tiarella
and Obelia sp. are farther apart where the outer diameter is large
than where it is small. In T. crocea the wall thickness is always
uniform locally, the surface of the lumen approximately paralleling
that of the outer surface. Laminae in annular regions in T. crocea
therefore always parallel the outer surface, while those of the
other two species show progressively less relief toward the center
of the tube.

At the growing tip of a hydrocaulus, a thin walled tube of perisarc is produced, the outer diameter of which will not subsequently change. As the colony grows past this point, additional laminae are deposited inside the tube, thickening the wall and decreasing the relief of the inner surface in annular regions. In specimens of P. tiarella used here, the hydrocaulus had an outer diameter of about 0.45mm, each region between annuli being five to ten percent larger distally than basally. Within two to three millimeters of the tip of the hydrocaulus, the perisarc was 0.02mm thick, increasing to 0.06mm thick near the base of the hydrocaulus.

Superficially the perisarc of P. tiarella appears dark brown over most of the colony. It is paler near growing tips, however, being colorless at the tips. In cross section, an outer colorless region is seen with a dark region just inside it. Each of these bands is about 5μ thick in my specimen. Inside these bands, all the material is light brown. A similar pattern is seen along the hydrocaulus of Obelia or T. crocea though the perisarc is never as dark as in P. tiarella and the radial differences were not noticed.

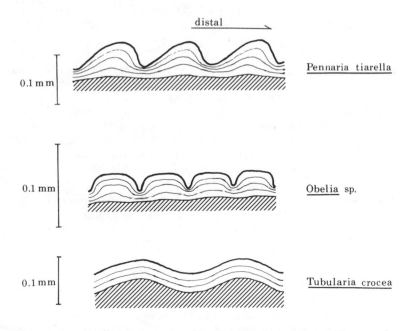

Figure 1. Diagrammatic longitudinal sections through annular regions in the perisarc well away from the growing tip of the hydrocaulus in three hydroids. The lumen of the tube is shaded. Light lines indicate the contours of selected laminae which are produced sequentially, from the outside inwards.

If the hydrocaulus of P. tiarella is bent manually, nearly all
of the deflection near the tip is due to bending at the annular regions. Near the base, the perisarc is only very slightly more
flexible in the annular regions. Annular regions near the tip of
the hydrocaulus could be bent 90° without buckling, while those
near the base buckled at such large bends. The annular regions
are much more flexible than non annular regions only where the
perisarc is thin.

Much the same behavior was seen in Obelia as in P. tiarella,
though here the annular pedicels beneath the polyps and gonangia
were always flexible. The perisarc of T. crocea is not obviously
much more flexible in annular regions than elsewhere.

DISCUSSION

It is clear that distally, where the walls are essentially
folded sheets, annular regions in perisarc certainly do provide
flexibility. This is doubtless due to the ability of the walls
to fold and unfold as suggested in the introduction. As the
perisarc is thickened, the inner surfaces of the annular regions
come more and more to resemble straight walled tubes both morphologically and mechanically.

The color of the perisarc is probably of mechanical significance though hard data are not available. Knight (1968, 1970,
1971) has shown that quinone tanning takes place in the perisarc
of at least one hydroid as had been suggested (Manton, 1941). The
morphologically similar material in podocyst cuticles is also
tanned by phenolic substances (Chapman, 1968; Blanquet, 1972). The
only mechanical data on these materials, however, seem to be my unpublished values for the elastic modulus (E) of the perisarc at
about midlength on the caulome (stem) of Tubularia crocea. The
values I obtained (10^8N/m^2 at a strain rate of 0.04%/sec and 5.5
x 10^8N/m^2 at a strain rate of 4%/sec.) are reasonable in light of
published values for insect cuticle (Neville, 1975) suggesting a
similar material. It is a reasonable hypothesis, therefore, that
darkening of cnidarian cuticle is associated with an increase in
the elastic modulus as it is in insect cuticle.

As the flexural stiffness of a beam is given by the product
of the elastic modulus and the second moment of area, placing high
modulus material at the periphery of a tubular beam will substantially stiffen it. More work will be necessary before it is certain that this is the correct model to apply to the dark layer seen
in the perisarc of P. tiarella, but if it is, that layer may dictate
much of the mechanical behavior of the perisarc tube.

The biological importance of the flexibility of annular regions requires further investigation. As suggested, it probably helps prevent local buckling. It remains to be shown what effect local buckling has on parts of hydroid colonies, though some suggestions were made above. Flexibility at the bases of polyps may permit them to passively orient with respect to water currents. Such orientation may be important in helping the polyps catch suspended food, but again the importance, while reasonable, is undemonstrated.

The thickening of the perisarc of the hydrocaulus stiffens it and tends to obliterate the greater flexibility of the annular regions. These have the effect of making the hydrocaulus better able to support the colony (or part of it) off the substratum in an environmental water current. The taller the colony, the greater the ambient currents which it will experience and the greater the drag forces acting upon it. Those drag forces, too, act farther from the base of the colony and so have greater leverage to cause bending. The stiffening of the hydrocaulus as it grows would tend to oppose both of those factors.

The functional significance of the flexibility afforded by the annular regions in the hydroid perisarc and of the gradual stiffening of the hydrocaulus is hypothetical. The hypotheses are reasonable and have good basis in the morphology of the animals. Before the skeletal mechanics of hydroid skeletons can be well understood, however, field and laboratory studies are needed to test suggestions such as those made here and to gain insight into the biological environment of these animals in situ. The studies by Koehl (1976) and Svoboda (1976) in this volume show how this may be done.

SUMMARY

The annular regions in hydroid perisarc lend flexibility where the perisarc is thin and has the general form of a folded sheet. The flexibility is apparently due solely to the geometry of the regions. Flexible areas may aid passive orientation to water currents and help prevent local buckling of the perisarc tube. The gradual thickening of the hydrocaulus stiffens it, particularly at the annular regions, probably rendering it better able to support the colony in ambient water currents. Further work is needed to test the biological importance of each of these factors.

ACKNOWLEDGEMENT

This work was supported by a grant from the Cocos Foundation. Thanks are due Dr. S. A. Wainwright for the use of his laboratory and equipment.

REFERENCES

Blanquet, R., 1972. Structural and chemical aspects of the podocyst cuticle of the scyphozoan medusa, Chrysaora quinquecirrha. Biol. Bull., 142: 1-10.

Chapman, D. M., 1968. Structure, histochemistry and formation of the podocyst and cuticle of Aurelia aurita. J. Mar. Biol. Ass. U. K., 48: 187-208.

Hyman, L. H., 1940. The Invertebrates Vol. 1. McGraw-Hill, New York, 726 pp.

Knight, D. P., 1968. Cellular basis for quinone tanning of the perisarc in the thecate hydroid Campanularia (= Obelia) flexuosa Hinks. Nature, 218: 584-586.

Knight, D. P., 1970. Sclerotization of the perisarc of the calyptoblastic hydroid, Laomedea flexuosa 1. The identification and localization of dopamine in the hydroid. Tissue & Cell, 2: 467-477.

Knight, D. P., 1971. Sclerotization of the perisarc of the calyptoblastic hydroid, Laomedea flexuosa 2. Histochemical demonstration of phenol oxidase and attempted demonstration of peroxidase. Tissue & Cell, 3: 57-64.

Koehl, M. A. R., 1976. Mechanical design in sea anemones. This volume.

Manton, S. M., 1941. On the hydrorhiza and claspers of the hydroid Myriothela cocksi (Vigurs). J. Mar. Biol. Ass. U. K. 25: 143-150.

Neville, A. C., 1975. Biology of the Arthropod Cuticle. Springer-Verlag, Berlin, 448 pp.

Svoboda, A. 1976. The orientation of Aglaophenia fans to current in laboratory conditions (Hydrozoa, Cnidaria). This volume

Wainwright, S. A., W. D. Biggs, J. D. Currey, and J. M. Gosline. 1976. Mechanical Design in Organisms. John Wiley & Sons, New York, 423 pp.

THE ORIENTATION OF AGLAOPHENIA FANS TO CURRENT IN LABORATORY CONDITIONS (HYDROZOA, COELENTERATA)

Armin Svoboda

Lehrstuhl für Spezielle Zoologie

Ruhr-Universität Bochum, F.R.G.

INTRODUCTION

The orientation of fan shaped sessile organisms perpendicular to the current is well known and has often been discussed, REES 1972, RIEDL 1966, THEODORE and DENIZOT 1965, VELIMIROV 1973. The possible trophic benefit of this orientation in fan shaped coelenterates was pointed out by RIEDL 1966 and RIEDL and FORSTNER 1968. WAINWRIGHT and DILLON 1969 described the increased mechanical stability of Gorgonia against twisting moments. The effect of current on bilaterally symmetric coelenterates until now has been studied only by MAGNUS 1966, working on the nocturnal sea pen Scytaliopsis and by SVOBODA 1970, 1973 in the marine hydroid Aglaophenia.

Like all Plumulariidae, the fans of the Aglaophenia species show a bilaterally symmetric construction. The branches insert alternately on the frontal side of the stem and the cross section of the fan is v-shaped in the profile. The hydranths face the frontal side. The fans are oriented parallel to each other in the field and perpendicular to the prevailing current (SVOBODA 1973). Aglapheniids therefore offer an interesting model for the orientation of sessile bilaterally symmetric cnidaria.

MATERIAL AND METHODS

The laboratory experiments were performed only on Aglaophenia picardi n. sp. (SVOBODA 1973, in press) which occurs

on the shady rocky sublittoral of the North Adriatic coast at
Rovinj, Jugoslavia. The transport and the maintenance of the
colonies and the experimental devices have been described by
SVOBODA 1970. In a flow chamber the colonies could be exposed to
reciprocating water movement. The specimens, growing on pieces of
rocky substrate, were mounted on a revolving device, which could
be turned by hand or by electrical devices for slow turning to
and fro within the range of 60° or for permanent rotation (1
rotation/min.), to simulate the various current conditions in
the field.

The developing fans were marked on photographic close-up
enlargements, covered by transparent plastic sheets and the
angle of the fanblade measured by the aid of an eye piece
reticule in a stereo microscope. The speed of the current was
determined by a method originally devised by AMBUHL 1959, and
modified by the author; the direction of current could be
demonstrated in electronic flash close-up photographs (fig. 1).

RESULTS

The colonies, which were not fed during the experiments,
always were inserted into the flow chamber perpendicular to
the orientation they had in the field before the experiment.
Within 1-2 weeks the hydranths began to degenerate and were
withdrawn into the coenosarc, which was absorbed by the stolon
afterwards. The remaining periderm skeleton was swept off by
the water movement 1-2 days later, leaving a short tubular
stump projecting from the stolon. The regeneration of new fans
always began from this stump after a few days, but sometimes
there was a delay of up to half a year. The stem of the
new colony was much thinner than the old one. Its tip
consisted of a bud of 300 μm in length without any tissue
differentiation, the transparent part below showing a
separation in ectoderm and entoderm layers.

The entoderm sits close to the frontal side of the stem,
where the branches will insert in the course of further
development. At this stage already showing the prospective
frontal orientation of the fan, in 40 experiments, the fan
plane was rotated by an angle of $30-60^o$. In 14 experiments,
12 fans turned even after the development of the first segment
of the stem, 2 fans grew up in the former direction. When 9
further specimens were rotated after the development of the
first polyp, only 2 fans attained the new direction. Three
fans of this experiment grew up twisted helically, the base
oriented to the former direction and the top to the new one.
Fans with 2 or more polyps could not be forced into a new growth

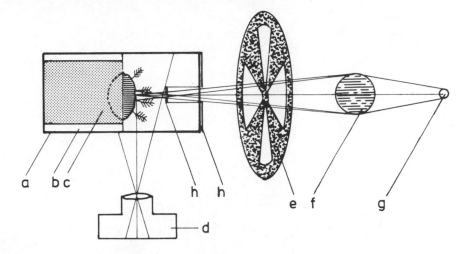

Fig. 1:   Photography of air bubble traces, cross section
through the flow chamber. a. Perspex wall, b. polystyrene
foam block, c. polystyrene foam cylinder, d. photocamera, e.
rotating segment shutter, f. cylinder lens, g. halogen quartz
lamp, h. diaphragm.

direction. Normally the fans grew up to a size of 7-13
branches with 66 polyps within 3 weeks. After reaching maximum
size the hydranths were again resorbed with the coenosarc into
the stolon.

93% of 198 fans which grew up without any disturbance by
rotation were oriented perpendicular to the current within a
deviation of less than 10^0. Only 3% of the fans showed a
deviation of more than 45^0 from the main direction (fig. 2a).
On substrate pieces which were turned slowly to and fro at a
range of 60^0 94 fans were regenerating. 65% of these fans
were oriented perpendicular to the resultant of this movement,
20% of them deviated more than 45% from this direction (fig.
2b). Seventy fans grew on a continuously turning mounting
device and therefore were exposed to a continuous change of
the flow direction. 60% of these fans regenerated in the same
orientation they showed in the field before the experiment,
25% of the specimens deviated more than 45^0 from this
direction (fig. 2c).

The fans that grew parallel to each other, on the average
did not show any preference in their frontal orientation within

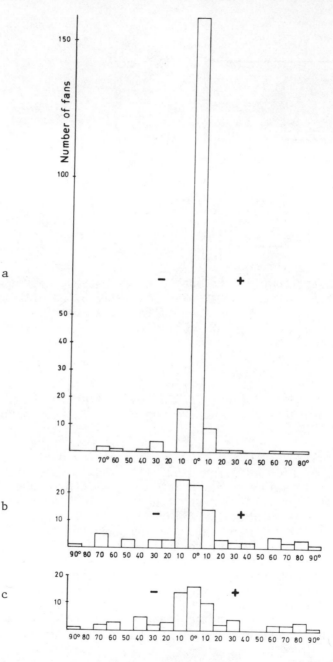

Fig. 2: Orientation angle of regenerating <u>A</u>. <u>picardi</u> fans. 0⁰ is normal to flow. +, clockwise deviation. -, counter-clockwise deviation. a, colonies were not turned during growth. b, colonies were turned to and fro 60⁰ during growth. c, colonies were rotated continuously during growth.

the flow chamber. A separate counting of the fans on the slope
regions and on the top region of the substrate pieces, where the
fans were growing showed that 161 fans were oriented frontally
away from the substrate margin of the slope, and only 26
specimens downward, however, on the top 153 were oriented to one
channel entrance and 155 to the other one. The fans
regenerating under the condition of a continuous change of flow
direction grew up in the same orientation as the fans they
replaced.

These results suggested a previously unsuspected flow
component, so the water flow in the current channel was
examined, using air bubbles as tracer particles (figs. 3a,b).
It was demonstrated that whirls rise on the lee side of the
substrate which reverse the current direction in the growth
zone on the substrate slopes. Therefore the fans were always
exposed to the flow coming from the margin of the substrate in
the slope regions--independent of the main current in the chamber
l-l but were exposed to the reciprocating change of the flow at
the top of the substrate. The current speed attained a ratio
of 1:10 between the margin and the top of the substrate (figs.
4a,b, 5).

CONCLUSIONS AND DISCUSSION

Observations in the field (SVOBODA 1973) have shown, that
there are often dead calms in the North Adriatic Sea where A.
picardi is common. In these periods, when no plankton can be
transported to the hydranths, the fans are resorbed into the
stolon and the resting stages survive for a long time at a low
energy level. Regeneration normally starts again with renewed
wave action. In unidirectional water flow the fans orient their
versal side to the current, a .very common position for bilaterally
symmetric filter feeders. This behavior was observed by the
author in several Plumulariidae, in the sea pens Pennatula
phosphorea (L.), Pteroides spinosum (ELLIS) and Virgularia
mirabilis (MÜLLER) and in the tentacles of the crinoid Antedon
mediterranea (LAMARCK). In many species the filter fans are
distorted passively when the flow direction is changed--as
was found in Aglaophenia harpago SCHENCK (SVOBODA 1973), the sea
pens, and as RIEDL 1966 described also for the filter baskets of
barnacles.

If the prevailing current direction changes during the
regeneration period of A. picardi fans, the new orientation is
perpendicular to the irritating forces. If the current on the
basis of the stem has another flow direction than that at the
top of the colony, the fan blade grows twisted helically. In

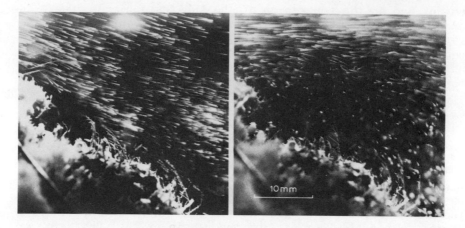

Fig. 3: Current flow direction close to substrate in reciprocal
water movement. a. Upstream slope: flow from right to left.
b. Lee slope: flow from left to right. Note counter current
close to the substrate, caused by eddies.

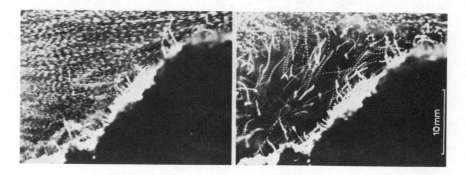

Fig. 4: Current flow and current speed in the upstream and the
lee of the substrate. a. Bubble traces in the upstream
slope region. b. Bubble traces in the lee slope region.

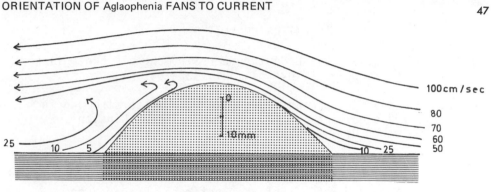

Fig. 5:   Isotachs in the flow chamber.

periods of absence of prevailing currents or dead calms, the
regenerating fans orient according to a "memory," which is
probably located in the periderm stump of the colony. Because
of the lack of any internal morphological structures, this
memory could well be a chemical marking of the periderm tube,
secreted by glands of the bilaterally symmetric coenosarc. This
hypothesis makes it easy to understand why regenerating fans,
growing up without prevailing flow direction, have the identical
orientation as the fans they replace. The hypothesis is an
explanation for the increasing deviations from the optimal
orientation of the fans which were moved slowly to and fro
during upgrowth, because they were inserted into the channel
perpendicular to their former positions in the field, which
they remember because of the lack of a strong prevailing
current. The bilaterally symmetrically constructed
Aglaopheniidae, especially the large branched species occurring
on deep bottoms, are optimally adapted to unidirectional water
flow. In contrast to many other thecate and athecate genera
this genus contains several species occurring in the extreme
surf zone too. On the one hand it was shown in the laboratory
that there are probably unidirectional currents in the microhabitat
of the surf zone too, on the other hand the Aglaophenia species
occurring here are very small, elastic to the bowing forces and
show an excellent mechanical protection of their hydranths by
their deep dentated hydrothecae (SVOBODA 1973). Therefore
Aglaophenia species seem to have a depth distribution range
which is not common in other hydrozoa.

LITERATURE CITED

AMBÜHL, H. (1959). Die Bedeutung der Strömung als ökologischer
    Faktor. Schweiz. Z. Hydrol., 21: 133-264.

AMBÜHL, H. (1962). Die Besonderheiten der Wasserströmung in
    physikalischer, chemischer und biologischer Hinsicht.
    Schweiz. Z. Hydrol., 24: 367-382.

MAGNUS, D. B. E. (1966). Zur Ökologie einer nachtaktiven
    Flachwasser-Seefeder (Octocorallia, Pennatularia) im
    Roten Meer. Veröff. Inst. Meeresforsch. Bremerhaven,
    Sonderband 2: 369-380.

REES, J. T. (1972). The effect of current on growth form in
    an octocoral. J. exp. mar. Biol. Ecol. 10: 115-123.

RIEDL, R. (1966). Biologie der Meereshöhlen. Verlag Paul
    Parey, Hamburg, 636 pp.

RIEDL, R. und H. FORSTNER. (1968). Wasserbewegung im
    Mikrobereich des Benthos. Sarsia, 34: 163-188.

SVOBODA, A. (1970). Simulation of oscillating water movement
    in the laboratory for cultivation of shallow water
    sedentary organisms. Helgoländer wiss. Meeresunters., 20:
    676-684.

SVOBODA, A. (1973). Beitrag zur Ökologie, Biometrie und
    Systematik der mediterranen Aglaophenia Arten. (Hydroidea)
    Ph.D. dissertation, Vienna.

THEODOR, J. et M. DENIZOT. (1965). Contribution à l'étude des
    Gorgones. I. A propos de l'orientation d'organismes marins
    fixés végétaux et animaux en fonction du courant. Vie
    Milieu, 16: 237-241.

VELIMIROV, B. (1973). Orientation in the sea fan Eunicella
    cavolinii related to water movement. Helgoländer wiss.
    Meeresunters., 24: 163-173.

WAINWRIGHT, S. A. and J. R. DILLON. (1969). On the
    orientation of sea fans (Genus Gorgonia). Biol. Bull.,
    136: 130-139.

General Ecology of
Bottom-Living Forms

COMPETITIVE INTERACTIONS AND THE SPECIES DIVERSITY OF CORALS

J.H. Connell

Department of Biological Sciences

University of California, Santa Barbara

The species diversity of a coral reef community is affected by many factors. If we consider only sessile organisms, for example plants or aquatic animals such as sponges, corals, etc., competition for space is obviously an important factor. On a small area of hard substrate with very little spatial variation in local physical conditions, one might expect that a single species would be superior to all others in competition and would eliminate them and eventually cover the surface. This unstable system could be symbolized by the diagram:

$$A \longrightarrow B$$
$$\searrow C \swarrow$$

in which species A, being competitively superior to B and C, eventually eliminates them.

Since it is obvious that many species that do compete for space also coexist on small areas of hard substrate, this simple model by itself is inadequate. Two general modifications that will stabilize the system and allow coexistence of competitors have been proposed. The first suggests that the hierarchy is not transitive, the arrows in the model being changed so that a species lowest in the hierarchy becomes superior specifically to the highest ranked one. Thus the model becomes:

$$A \longrightarrow B$$
$$\nwarrow C \swarrow$$

Jackson and Buss (1975) proposed this model, suggesting that it would operate if different mechanisms of competition applied. For example, if A overgrows B and B overgrows C, but C damages A by allelopathic chemicals, or some other mechanism, then the system could stabilize with three species.

51

In a second general modification, the competitive hierarchy remains transitive, but forces external to the system cause proportionately greater mortality to the species at the top of the hierarchy. Then competitive exclusion will be reduced and diversity maintained. This mechanism was first suggested by Charles Darwin (1859) who tested it with a field experiment. He showed that mowing turf (he suggested that grazing by quadrupeds would have similar effects) maintained high diversity of the plants. Evidence for the effects of such external disturbances, both physical and biological, in preventing competitive exclusion has been accumulated since then (see review of experimental evidence in Connell, 1975).

In this paper I will present evidence to test the original model and its two general modifications, using data from coral populations at Heron Island, Queensland. Since 1962 I have been observing permanently-marked quadrats by taking photographs at intervals. The quadrats were established in areas of low and high diversity; recruitment, growth and mortality were measured from the series of photographs (see Connell [1973] for details of techniques).

Model 1 (Non-transitive competitive networks). To see whether a low ranking species might be superior to a higher ranked one, I recorded all interactions that occurred between adjacent colonies of 12 species on a square meter on the north crest, between January 1963 and January 1972. In 32 of the 82 interactions (involving 55 colonies), the interaction was indirect, the branches of one colony growing above the edge of a neighboring colony, overshadowing though apparently not in contact with it. In 13 others, the interaction was direct, the edge of one colony dying, while the other grew toward that edge; the mechanism here may have involved extra-coelenteric digestion, as described by Lang (1973). In the remaining 37 cases, neither colony grew. These "stand-offs" lasted until one or both colonies died from other causes. All stand-offs lasted at least a year, in many cases much longer; one lasted for the entire nine year period of observation.

A matrix of the interactions observed is given in Table 1. By subtracting the number of losing interactions from the number of winning ones for each species, I constructed a ranking of competitive abilities as shown. Networks of the interactions, made separately for indirect and direct interactions, are given in Figure 1. Several patterns emerge. In general, the species rankings for both mechanisms are similar, with two striking exceptions. The massive Leptoria phrygia is the highest ranking in digestive ability, but never overgrows or is overgrown. It grows very slowly and cannot be regarded as an efficient competitor for space. It is relatively uncommon on this reef, as are most of the massive species which would be ranked high in ability to digest their neighbors. Lang (1973) found the same to be true on Caribbean reefs. Therefore I suggest that the

Table 1. Interactions between coral species. The arrows indicate that at least one interaction was initiated by the species from the left column against the species on the top row. Direct interaction ←; indirect interaction ←○.

Species Initiating	Code No.	Species Receiving 11	6	13	31	26	17	12	68	4	59	61	Number Initiated ←	Number Initiated ←○	Total	Total Initiated minus Received
Acropora hebes	11		←○						←○	←○	←○ ←	←○	2	3	5	5
Acropora digitifera	6	←		←○	←						← ←○	←	3	4	7	4
Acropora hyacinthus	13	←○					←○	←○		←○ ←○	←○	←○	0	6	6	3
Acropora valida	31	←		←							←○		1	1	2	1
Acropora squamosa	26												0	0	0	0
Acropora nasuta	17										←○		0	1	1	0
Acropora humilis	12									←		←○	1	1	2	-1
Stylophora pistillata	68												0	0	0	-1
Acropora palifera	4										←○	←○ ←	1	1	2	-2
Pocillopora damicornis	59											←○	0	1	1	-4
Porites annae	61												0	0	0	-6
Number ←		0	1	2	1	0	0	0	0	1	2	1				
Received ←○		0	2	1	0	0	0	3	1	3	3	5				
TOTAL		0	3	3	1	0	1	3	1	4	5	6				

Note: Species No. 48, Leptoria phrygia, has not been included; see text.

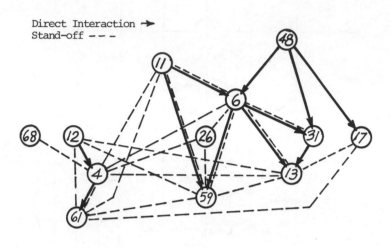

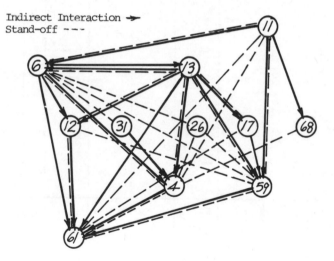

Figure 1.   Interactions between species; colonies which grew
close enough to each other to interact were ranked as shown
in Table 1.   The numbers refer to the different species given
in that table.   82 interactions were observed between 55
colonies of 12 species on a square meter at the North Crest
site over 9 years (1963 through 1971).

ability to digest neighbors should be regarded as primarily a defense rather than as a means of competitively securing the space at the expense of other species.

The other exception, Acropora hyacinthus (sp. 13) might fit this model in that it is able to overgrow most others yet loses to most in aggressiveness. However, it is not a clear dominant in ability to overgrow, being about equally matched to A. digitifera (sp. 6) and probably inferior to A. hebes (sp. 11). All of the other species are consistent in their rankings for both mechanisms of interaction.

All species ranked below the highest showed some ability to stand-off higher ranked species. In 12 of the 17 overgrowth inter- actions and in 6 of the 11 digestive interactions shown in Figure 1, there were also examples of stand-offs between the same species. There were five instances when an overgrowth and a stand-off occurred between the same pair of colonies. In one case there was a stand-off for a year, then one overgrew the other. In the other four cases one overgrew the other in part, while at the same time there was a stand-off along the rest of the borderline. In one of these, the first colony overgrew the second in one year, then the second grew back over the first during the next year; during both years there was a stand-off along another part of the border.

Another pattern is the scarcity of species-specific inter- actions. For the network model to work, the lowest ranked species must be superior to the highest ranked one. This clearly did not take place.

Lastly, when I looked for the circular network pattern that is the essential aspect of this model, I found only one possibility, as indicated by the following digram:

Species 6 directly dominated species 31 which dominated species 13 which overtopped species 6; so far this is a circular network. But species 6 also both overtopped and directly dominated sp. 13, can- celling out the network.

Lang (1973, Table 2) also found these patterns in laboratory tests among 11 species of highly aggressive corals. None were species-specific, some reciprocal interactions and stand-offs occurred and no circular networks were found. From all of the above findings I conclude that the network model of Jackson and Buss (1975) does not apply to interactions between these corals. In these circumstances, the principal mechanisms operating entirely within the system of competitors that reduce competitive exclusion

are reciprocal interactions of species of equal rank and the ability
of lower-ranked species to sometimes stand-off species of higher
rank.

   Model 2 (Transitive networks, external disturbances). In this
model, higher ranked species have the ability to out-compete all
lower ones (Jackson and Buss, 1975, term this a "linear ranked
hierarchy"). However, external forces reduce the higher-ranked
species proportionately more than they do lower-ranked ones, stabilizing the system.

   At Heron Island, the principal external forces causing mortality
were associated with the physical environment. These were either
exposure to air at low tide in shallow water on the reef flat and
crests, or damage in severe storms, from breakage, burial in sediment
and other catastrophic effects, in both shallow and deeper water.

   To see whether mortality from external disturbances fell proportionately more severely on the species that were competitively
superior, I first separated the species into two groups, on the
basis of the hierarchies in Table 1 and Figure 1. The branching
species of Acropora were all in the group of high rank; the low
ranked group included the encrusting Acropora palifera (=cuneata)
as well as Pocillopora damicornis, Stylophora pistillata and Porites
annae. Massive species such as Leptoria phrygia, although ranking
high in digestive dominance, were included in the latter group because this behavior was used as a defense rather than in aggressive
competition.

   The principal external disturbances whose effects could be
clearly estimated were the severe hurricanes that occur at intervals.
In two hurricanes, in January 1967 and in April 1972, the south crest
was only slightly damaged while the north crest was much more
severely damaged. The changes in living coral cover for two square
meters in each area for the period that included the hurricane of
1967, are given in Table 2. As can be seen, on the north crest
where damage was more severe, the highly-ranked species were greatly
reduced while the low-ranked species increased. In contrast, where
storm damage was slight on the south crest, the reverse was true.
On the south crest the decrease of the low-ranked species was due
mainly to their being overgrown by the high-ranked ones.

   The effect of the hurricane of 1972 was even more striking.
On the north crest, all colonies but one, the massive Leptoria,
were killed, whereas on the south crest there was little change.
It is clear that the effect of catastrophic storm damage conforms
to the model in that high-ranked species are reduced proportionately
more than low-ranked ones.

COMPETITIVE INTERACTIONS AND SPECIES DIVERSITY IN CORALS

Table 2. Changes in coverage of corals of different competitive abilities. In each area, two square meters were censused.

Reef areas whose degree of vulnerability to storm damage is:	HIGH (Branching _Acropora_ sp.)		LOW (Encrusting sp., massive sp., _Pocillopora_, _Stylophora_)	
	cover (cm^2) in 1965	% change in cover, 1965-69	cover (cm^2) in 1965	% change in cover, 1965-69
HIGH (north crest):	10,655	-39.1	3,071	+18.6
LOW (south crest):	7,027	+47.3	1,900	-12.5

SPECIES WHOSE COMPETITIVE ABILITIES ARE:

Another form of external disturbance is exposure to air at low tide. This happens when colonies grow above the water surface, and so occurs mainly on crests and reef flats. All species can be affected, but those that grow upwards and overgrow their neighbors will also tend to reach the surface first. Thus in very shallow water the species that usually win in competition are also the ones that tend to be reduced most in coverage as they project above the water surface.

Mortality from external forces such as storm damage and exposure to air not only reduces competition but provides open space into which new colonies are recruited. Thus diversity is maintained by external physical forces in two ways, by reducing competitive exclusion and by increasing the rate of recruitment of new colonies, some of which could be new species. These forces, plus the effect of stand-offs that greatly reduce the ability of one species to eliminate another, are probably sufficient to maintain the diversity of coral species on the reef crests at Heron Island.

REFERENCES CITED

CONNELL, J.H. 1973. Population ecology of reef-building corals. In: Biology and geology of coral reefs. Vol. II, Biology 1. Ed. by O. Jones and R. Endean. Academic Press, N.Y. pp. 205-245.

CONNELL, J.H. 1975. Some mechanisms producing structure in natural communities: a model and evidence from field experiments. In: Ecology and evolution of communities, ed. by M. Cody and J. Diamond. Belknap Press, Cambridge, Mass. pp. 460-490.

DARWIN, C. 1859. The origin of species. John Murray, London.

JACKSON, J.B.C. and L. BUSS. 1975. Allelopathy and spatial competition among coral reef invertebrates. Proc. Nat. Acad. Sci. U.S.A. 72: 5160-5163.

LANG, J. 1973. Interspecific aggression by scleractinian corals. 2. Why the race is not only to the swift. Bull. Mar. Sci. 23: 260-279.

HABITAT-RELATED PATTERNS OF PRODUCTIVITY OF THE FOLIACEOUS REEF CORAL, PAVONA PRAETORTA DANA

David S. Wethey and James W. Porter

Division of Biological Sciences and School of Natural

Resources, University of Michigan, Ann Arbor, MI, 48109

Reef corals contain mutualistic symbiotic zooxanthellae which photosynthesize and translocate photosynthate to the animal partner, although the animal is capable of tentacular feeding and other heterotrophic activities (Goreau, et al., 1971; Muscatine, 1973; Trench, 1974; Porter, 1974; Muscatine and Porter, 1976). Many workers have examined the photosynthesis of reef corals (McCloskey, et al., 1976) but habitat acclimation has generally been ignored.

The photosynthesis-radiant flux intensity responses of individuals from different depths on the same reef have been established for Pavona praetorta from Enewetak Atoll, Marshall Islands. Deep water individuals of this species saturate their photosynthetic apparatus at significantly lower ($p < 0.01$) radiant flux intensities than do those from shallow water (Wethey and Porter, 1976). In this paper we will provide evidence that the acclimation of deep water individuals compensates completely for the low available light. We will also argue that this foliaceous species which is morphologically specialized for autotrophy (Porter, 1976) is capable of a purely autotrophic existence down to 25 meters even under overcast conditions.

In situ 24 hour measurements of oxygen production and consumption as a function of ambient photosynthetically active radiation (400 nm - 700 nm) were carried out on individuals of Pavona praetorta on the Marine Pier Pinnacle Reef, 300 meters west of Enewetak Island (Site Fred), Enewetak Atoll, Marshall Islands. Biomass of the coral-algal complex was estimated as total chlorophyll a. The photosynthesis-radiant flux intensity relationship was established (Wethey and Porter, 1976) by a maximum-likelihood algorithm for fit-

TABLE 1

PARAMETER ESTIMATES FOR FITTED CURVES

Parameter	Units	Depth	Estimate	99 Percent Conf. Lim.	
P_{max}gross	$\dfrac{mg\ O_2}{mg\ chl\ \underline{a} \times hr}$	10 m	12.25	10.13	14.36
		25 m	8.74	7.42	10.07
K_m	$\dfrac{einstein}{m^2 \times hr}$	10 m	0.63	0.43	0.88
		25 m	0.26	0.18	0.37
R	$\dfrac{mg\ O_2}{mg\ chl\ \underline{a} \times hr}$	10 m	1.84	1.15	2.53
		25 m	1.41	0.77	2.05

ting data to the Michaelis-Menten equation (1), following an initial Lineweaver-Burk estimate (Bliss and James, 1966; Hanson, et
al., 1967)(Table 1).

$$P = \frac{P_{max} \times I}{K_m + I} \qquad \text{Michaelis-Menten Equation} \qquad (1)$$

A quantum sensor (Lambda Instruments) on shore recorded radiant
flux intensity over the daylight hours for three weeks. Radiant
flux intensity was recorded in situ during productivity experiments
on the reef. The ratio of surface radiation to that at 10 and 25
meters was calculated by comparison of measurements made simultaneously at the surface and at depth. Radiant flux at 10 and 25
meters was 28 percent and 15 percent of surface radiation respectively. The expected radiant flux at depth for cloudless and overcast days was calculated using these ratios and recordings from the
shore-based sensor (Figure 1). The theoretical daily radiant flux
curves were divided into 15 minute increments and the expected productivity was calculated for each increment from the regression
equations for shallow and deep photosynthesis-radiant flux intensity relationships. This analysis is similar to that of Franzisket
(1969). Oxygen production and consumption per unit chlorophyll a
was extrapolated to carbon assimilation and utilization assuming
photosynthetic and respiratory quotients of 1.0. The algae were
assumed to have a daytime gross photosynthesis to respiration ratio
of 4.0 (Humphrey, 1976). The animal respiration component was then
estimated by subtraction, and the proportion of gross photosynthetic
production which must be translocated by the algae to satisfy animal
maintenance requirements was calculated from the ratio of 24 hour
animal respiration to gross photosynthetic production (Muscatine
and Porter, 1976). This calculation is exact when the algal
respiration is one half of the total respiration; the greater the
deviation from this condition, the less reliable is the estimate.

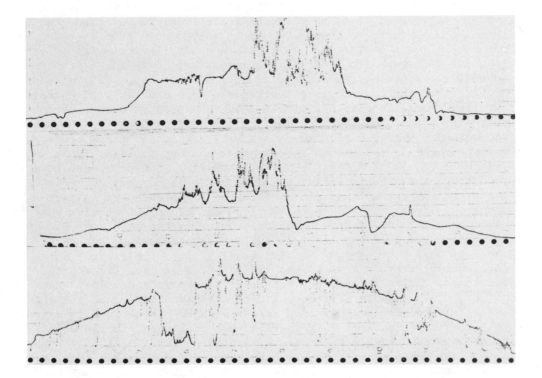

FIGURE 1. Daily radiant flux recordings from a shore-based, cosine-corrected photosynthetically active radiation (400nm - 700nm) detector. All three traces are on the same scale: the maximum steady radiant flux on the bottom trace is 8.64 $E/m^2/hr$. The unit of radiant flux is the einstein (E) = 6.02×10^{23} quanta. The upper trace corresponds to an overcast day. The middle trace is of an overcast day with intense storm activity during the afternoon. The bottom trace is of a cloudless day. A smoothed curve of the amplitude of the bottom trace was used to represent a sunny day in the regression analysis. A curve of approximately one half of the amplitude of the sunny day ideal curve was used to represent a cloudy day in the analysis (see footnotes to Table 2).

This estimate concerns only caloric parameters and not total nutrition, which includes vitamins and micronutrients (Johannes, et al., 1970; Porter, 1974; Trench, 1974).

The regression model of the photosynthetic response gives an estimate of gross photosynthesis and respiration over the 24 hour cycle. A ratio of gross photosynthesis to total respiration greater than 1.0 is indicative that more oxygen is produced than is consumed, but it is unclear how well this quantity indicates the degree to which the metabolic needs of the animal are satisfied by photosynthetic production (McCloskey, et al., 1976). These estimates suggest that shallow and deep individuals fare equally well under parallel weather conditions (Table 2a). The proportions of photosynthetic production that must be translocated in order to sustain the animal symbiont for 24 hours are well within the range reported for other corals and anemones (Muscatine and Cernichiari, 1969; Trench, 1971, 1974; Muscatine, et al., 1972). These authors have reported translocation efficiencies of 30 to 45 percent for most of the organisms tested, and the range 46 to 57 percent for Agaricia agaricites [the Caribbean morphological and ecological equivalent of Pavona (Wells, 1957)] from 5 meters. On cloudless days, the Pavona at both 10 and 25 meters need 30 percent translocation, and on overcast days they need 45 percent (Table 2b). On extremely dark days, during very severe storms, radiant flux is less (Figure 1), and translocation percentages required to support the animal are higher. Only on such days do the corals consume more than the algae are likely to provide, but such days are rare. These results suggest that the individuals of this species have acclimated to the worst conditions that they are frequently exposed to, rather than to the absolutely worst conditions ever encountered. The necessity of acclimation to the low radiant flux in deep water is apparent from the estimates of translocation percentage required by shallow water individuals if they were placed at 25 meters under overcast conditions (Table 2b). The shallow water, sun-acclimated photosynthetic machinery cannot cope with the low radiant flux in deep water and the required translocation percentage is far above any value reported in the literature, implying that shade acclimation is extremely important in deep water.

Pavona praetorta is morphologically at the extreme of development of surface area. The polyps are very poorly developed with less than 1.0 mm of surface relief. No tentacles are visible unless the coral is examined under a dissecting microscope. Suspension feeding with mucus nets has been documented for Agaricia agaricites (Lewis and Price, 1975), the Caribbean equivalent of Pavona, but the extent of the contribution of this form of heterotrophy to animal caloric maintenance needs is unknown. It is unlikely that tentacular-feeding-based heterotrophy could account for a large proportion of the energy needed to sustain the animal. This

TABLE 2a

ESTIMATED GROSS PHOTOSYNTHESIS TO RESPIRATION RATIO
[Estimate (99 Percent Confidence Limits)]

	Sunny Day [1]	Overcast Day [2]
Shallow	1.79 (1.38–2.26)	1.44 (1.06–1.91)
Deep	1.81 (1.40–2.13)	1.50 (1.15–1.92)
Shallow Individual at 25 Meters		1.08 (0.78–1.50)

TABLE 2b

ESTIMATED PERCENTAGE OF GROSS PHOTOSYNTHESIS NEEDED TO SUSTAIN
ANIMAL FOR 24 HOURS
[Estimate (99 Percent Confidence Limits)]

	Sunny Day [1]	Overcast Day [2]
Shallow	31 (19 – 48)	45 (23 – 70)
Deep	30 (22 – 47)	42 (27 – 62)
Shallow Individual at 25 Meters		68 (42 – 104)

1) Maximum radiant flux at surface = 8.64 E / m^2 / hr.

2) Maximum radiant flux at surface = 4.40 E / m^2 / hr.

potential specialization for autotrophy is consistent with the
analysis of Porter (1976). One might predict that this sort of
coral would have to depend almost exclusively on the endosymbiotic
algae, and that deep water individuals would have to be as autotrophic as those from shallow water, so the results of the regression
analysis are reasonable. What is surprising is that there is no
difference between the two depths in terms of the degree to which
the algae can support the animal, despite a two-fold radiant flux
change. Corroborative evidence for the acclimation phenomenon may
be found in Barnes and Taylor's (1973) observation that daily
carbon fixation in <u>Montastrea</u> <u>annularis</u> in Jamaica does not vary
between clones incubated <u>in situ</u> for 24 hours at 15 meters and 33
meters, despite a five-fold difference in illumination. McCloskey
(unpub.) also finds that the ratio of photosynthesis to respiration
over the 24 hour cycle does not change between clones incubated <u>in</u>
<u>situ</u> over the range 10 meters to 60 meters in another <u>Pavona</u> species.

If acclimation to ambient radiation is universal among coral
zooxanthellae, then a number of arguments as stated in the literature must be reconsidered. Goreau (1963), Roos (1967) and Barnes
(1973) have all postulated that the shape change of some corals
from rounded in shallow water to flattened in deep water is due to
a limitation of light-mediated calcification, with little reduction
in tissue growth as a result of heterotrophic maintenance of the
coral animal. If acclimation of the photosynthetic system (and
thereby presumably of the calcification system) to low light occurs,
the reason for the shift in morphology is no longer immediately
apparent, and we must look for other explanations of the phenomenon.
The shape change in <u>Montastrea</u> <u>annularis</u> begins at a depth of 18 to
20 meters on the fore reef in Jamaica (Barnes, 1973), where water
turbidity is lower than in the Enewetak lagoon. The angular distribution of radiation, expressed as percentage of the vertical component, is independent of depth down to 20 meters (Roos, 1967; Brakel,
1976) so depth-related skeletal morphology changes are unlikely to
be related to the directionality of incident radiation, except in
wall and overhang areas. In addition, extrapolations made from
photosynthesis and respiration measurements made in shallow water
to the compensation depth of the whole coral population on the reef
are totally unjustified. It is necessary to obtain measurements of
the depth distribution of the photosynthesis-radiant flux responses
of corals with depth-invariant and depth-dependent morphology, and
of potentially autotrophic and potentially heterotrophic morphology
in order to resolve this confusion. Such results will be of great
use in the interpretation of the distribution and abundance of corals
along depth gradients.

ACKNOWLEDGEMENTS

DSW is supported by a U. S. National Science Foundation Graduate
Fellowship and a grant from the Department of Zoology, University of
Michigan. A U. S. Energy Research and Development Administration
grant through the University of Hawaii - Mid Pacific Marine Labor-
atory Program to JWP supported the work at Enewetak. Robert K.
Trench assisted in the field work, and with Sharon L. Ohlhorst and
Susan S. Kilham, aided in discussion of the project. Roger Tyler,
Robert Reed, James Mite, Lawrence McCloskey, Lee Somers and Patrick
Blackburn aided in the design and testing of the instruments. The
regression program was obtained from the Yale University Computer
Center.

LITERATURE CITED

Barnes, D. J., 1973. Growth in colonial scleractinians. Bull.
    Mar. Sci., 23:280-298.
Barnes, D. J., and D. L. Taylor, 1973. In situ studies of calcifi-
    cation and photosynthetic carbon fixation in the coral
    Montastrea annularis. Helgoländer wiss. Meeresunters.,
    24:284-291.
Bliss, C. I., and A. T. James, 1966. Fitting the rectangular
    hyperbola. Biometrics, 22:573-602.
Brakel, W., 1976. The ecology of coral shapes: microhabitat
    variations in the growth form and skeletal structure of Porites
    on a Jamaican reef. Ph. D. Dissertation, Yale University.
Franzisket, L., 1969. The ratio of photosynthesis to respiration of
    reef building corals during a 24 hour period. Forma et Functio,
    1:153-158.
Goreau, T. F., 1963. Calcium carbonate deposition by coralline
    algae and corals in relation to their roles as reef builders.
    Ann. N. Y. Acad. Sci., 109:127-167.
Goreau, T. F., N. I. Goreau, and C. M. Yonge, 1971. Reef corals:
    autotrophs or heterotrophs? Biol. Bull., 141:247-260.
Hanson, K. R., R. Ling, and E. Havir, 1967. A computer program for
    fitting data to the Michaelis-Menten equation. Biochem.
    Biophys. Res. Commun., 29:194-197.
Humphrey, G. F., 1976. Project Report, C. S. I. R. O. Marine
    Biochemistry Unit, Sydney, Australia.
Johannes, R. E., S. L. Coles, and N. T. Kuenzel, 1970. The role of
    zooplankton in the nutrition of some scleractinian corals.
    Limnol. Oceanogr., 15:579-586.
Lewis, J. B., and W. S. Price, 1975. Feeding mechanisms and feeding
    strategies of Atlantic reef corals. J. Zool. Lond., 176:527-544.
McCloskey, L. R., D. S. Wethey, and J. W. Porter, 1976. The
    measurements and interpretation of photosynthesis and respir-
    ation in reef corals. Monogr. Oceanogr. Methods (SCOR-UNESCO),
    4 (in press).

Muscatine, L., 1973. Nutrition of corals. pp. 77-115. in: Jones,
    O. A., and R. Endean (eds.), The Biology and Geology of Coral
    Reefs. Vol. 2, Biology 1. Academic Press, New York.
Muscatine, L., and E. Cernichiari, 1969. Assimilation of photo-
    synthetic products of zooxanthellae by a reef coral. Biol.
    Bull., 137:506-523.
Muscatine, L., R. R. Pool, and E. Cernichiari, 1972. Some factors
    influencing selective release of soluble organic material by
    zooxanthellae from reef corals. Mar. Biol., 13:298-308.
Muscatine, L., and J. W. Porter, 1976. Reef corals: mutualistic
    symbioses adapted to nutrient-poor environments. Bioscience
    (in press).
Porter, J. W., 1974. Zooplankton feeding by the Caribbean reef -
    building coral Montastrea cavernosa. Proc. Second Int. Coral
    Reef Symp., 1:111-125.
Porter, J. W., 1976. Autotrophy, heterotrophy, and resource
    partitioning in Caribbean corals. Amer. Natur. (in press).
Roos, P. J., 1967. Growth and occurrence of the reef coral Porites
    astreoides Lamarck in relation to submarine radiance distri-
    bution. Dukkerij Elinkwijk, Utrecht. 72 pp.
Trench, R. K., 1971. The physiology and biochemistry of zooxanthel-
    lae symbiotic with marine coelenterates. I. Assimilation of
    photosynthetic products of zooxanthellae by two marine coel-
    enterates. Proc. R. Soc. Lond. B, 177:225-235.
Trench, R. K., 1974. Nutritional potentials in Zoanthus sociatus
    (Coelenterata, Anthozoa). Helgoländer wiss. Meeresunters.,
    26:174-216.
Wells, J. W., 1957. Coral reefs. Geol. Soc. Amer. Mem., 67:
    1087-1104.
Wethey, D. S., and J. W. Porter, 1976. 'Sun' and 'shade' differ-
    ences in the productivity of reef corals. Nature, 262:281-282.

THE ECOLOGY OF CARIBBEAN SEA ANEMONES IN PANAMA: UTILIZATION OF

SPACE ON A CORAL REEF

Kenneth P. Sebens

Department of Zoology, University of Washington

Seattle, Washington U.S.A. 98195

Despite the predominance of scleractinian corals as the constructors and the matrix of coral reefs, there exists a high diversity of other organisms on the substratum provided by these corals. The reef structure is dynamic and areas of cleared substratum are continuously supplied by the forces of physical collapse, bioerosion, and predation; yet sufficient evidence exists (Loya 1972, Porter 1972c, Connell 1973, Grigg and Maragos 1974) to illustrate the importance of space as a limiting resource on coral reefs.

With space a limiting factor, competition for space emerges as a salient feature of reef ecology. Spatial competition, and its mediation by predation, have been experimentally demonstrated in the temperate rocky intertidal community (Connell 1961, Dayton 1971, Paine 1966, 1974). On coral reefs, competitive interactions are highly developed among corals (Lang 1973) and may occur between benthic algae and corals (Connell 1973). Other benthic forms such as sponges and colonial ascidians can actively destroy and replace living coral by overgrowth (C. Birkeland, pers. comm.).

On some reefs, the activity of predators can preclude competitive exclusion by dominant coral species (Porter 1972a). The constant appearance of new primary substratum and the subsequent successional process give the reef a mosaic character and allow species which are usually outcompeted by corals to exist as fugitives (Connell 1973, Porter 1972a,b,c). Such processes contribute to the high diversity of sessile forms that is characteristic of coral reefs.

The sea anemones (including the corallimorpharians) belong to a guild of sessile carnivores. In comparison with other anthozoan forms, they are by far the largest individual polyps and, as in most coral reef anthozoans, they usually contain algal endosymbionts

67

(zooxanthellae) within their tissues. Thus their spatial requirements
are twofold; they must be situated so that they can make use of a
prey resource and they must occupy substratum in areas that are well
lighted. These general requisites are identical with those of the
scleractinian corals themselves and bring the anemones into direct
spatial competition with them.

The three dimensional quality of a coral reef offers several
substratum types suitable for sea anemone habitats and the anemones
display morphological adaptations appropriate to each. Although
generally sessile they can move when necessary, an ability that most
corals lack. It allows anemones access to newly created areas of
clear primary substratum as adults well in advance of other anthozoan
forms and allows them to change their location if local conditions
deteriorate. Anemones which form clones and carpet exposed surfaces
can most likely prevent coral settlement (Fishelson 1970) and if so
may be considered a true alternative climax state on certain parts
of a reef.

The research described here is an attempt to analyze the habitat
resource used by anemones and to investigate interactions among
anemone species as well as between anemones and corals. An understanding of competitive interactions of this nature will help to
explain the high diversity of coral reef communities and the
heterogenous character of substratum occupation.

METHODS

A quantitative analysis of habitat parameters was used to
investigate habitat partitioning between anemone species. Nine coral
reefs along the Caribbean coast of Panama were visited (October to
December 1974) (from 9°24'21"N; 79°52'18"W to 9°34'47"N; 78°43'11"W).
At each site, I surveyed the anemones using SCUBA and skindiving
techniques. When an anemone was encountered, the size, depth, type of
substratum and occurrence with coral species were noted. In addition,
I examined adjacent coral colonies to determine whether or not the
anemone tentacles were in contact with living or dead coral, and
whether the dead coral showed any pattern of damage which could
potentially be attributed to the anemone. An area of coral free of
live tissue following the outline of the anemone's expanded tentacular
crown was taken as an indication of damage caused by the anemone. The
clonal aggregations of corallimorpharians were treated as units and
parameters measured were recorded for the entire aggregation instead
of for individuals as was done for the large solitary species.

Only the nine most common species were investigated in this study
[Bartholomea annulata (Leseur), Condylactis gigantea (Weinland),
Heteractis lucida (Duchassaing and Michelotti), Lebrunia danae
(Duchassaing and Michelotti), Paradiscosoma neglecta (Duchassaing and
Michelotti), Phymanthus crucifer (Leseur), Ricordia florida

(Duchassaing and Michelotti), <u>Rhodactis</u> <u>sanctithomae</u> (Duchassaing
and Michelotti), <u>Stoichactis</u> <u>helianthus</u> (Ellis)].

Aggressive behavior has been reported for several temperate
anemone species (Bonnin 1964, Francis 1973). Because of field
observations indicating maintenance of space between individuals, I
conducted laboratory experiments designed to evoke aggressive behavior
if these anemones were capable of it. Specimens were collected and
kept in flowing sea water outdoor aquaria at the Smithsonian Tropical
Research Institute laboratory at Isla Galeta. They were allowed to
establish themselves on coral rock in the aquaria for several weeks.
Pairs of anemones were then brought together by moving the rocks to
which they remained attached. Any relocating movement or active
aggressive behavior was noted during a 24 hour period. Ten replicate
tests were conducted with each pair of species. However, not all of
the possible species combinations could be tested.

<u>Condylactis</u> <u>gigantea</u> showed indications of being able to damage
corals in the field. To verify that this anemone was the cause of the
observed damage, a field test was conducted at the Isla Galeta reef.
I transplanted whole coral colonies of several species and individuals
of the coral <u>Scolymia</u> <u>lacera</u> into positions where half of the colony
would contact an anemone's tentacles and half would be outside their
radius. Five anemones were used in this experiment, each with three
or four corals transplanted next to it. I observed the corals over
one week for signs of damage.

RESULTS

The nine subtidal sea anemones display three basic morphologies
which are adaptations for existence on exposed surfaces or in holes
and crevices in the reef structure (Fig. 1). Surface dwelling anemone

Figure 1. Body shape and posture of a surface form (A) <u>Rhodactis</u>
<u>sanctithomae</u>), sand pocket forms (B) <u>Phymanthus</u> <u>crucifer</u>
(C) <u>Bartholomea</u> annulata) and a hole dwelling form (D) <u>Heteractis</u>
<u>lucida</u>).

species (P. neglecta, R. florida, R. sanctithomae, S. helianthus) are
extremely flattened. The oral disc is wider than the pedal disc and
the tentacles are short or papillate and cover much of the oral disc.
All four surface dwelling species form clonal aggregations by asexual
longitudinal fission (Fig. 2b). They feed by capturing prey on the
tentacles, then folding the oral disc to bring the prey into the
mouth. Hole dwelling species (C. gigantea, H. lucida, L. danae)
exhibit an elongate column and long slender tentacles. They can
contract rapidly, pulling the entire polyp deep into the reef matrix.
The anemone species which occur in sand pockets (B. annulata, P.
crucifer) are similar to the hole dwelling forms in morphology. P.
crucifer, however, has a tentacular crown with reduced tentacles which
it spreads out on the sand surface, resembling somewhat the surface
dwelling species and showing a similar feeding behavior. B. annulata
and all of the hole dwelling forms feed by single or multiple tentacle
prey capture and tentacle wiping across the oral aperture to bring
food to the actinopharynx.

     The results of the habitat analysis are presented in Table 1.
2 X 2 tests of independence (G-test, Sokal and Rolfe 1969) were
performed for each parameter on each species pair. Hierarchies of
occurrence frequencies are given for each table with their significance levels. Table 1a demonstrates that very few anemones occurred
in contact with living coral tissue. B. annulata occurred most often
with living coral but is attached within sand pockets so that this
contact is limited to the distal parts of the tentacles. L. danae

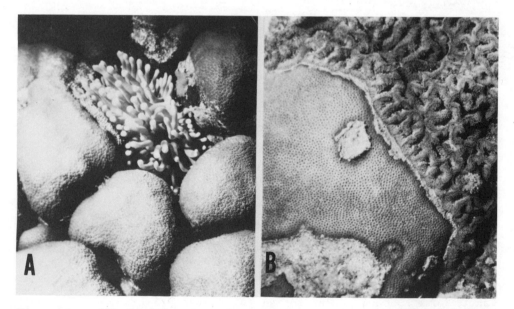

Figure 2. (A) Condylactis gigantea with coral, Montastrea annularis
surrounding it. (B) Ricordia florida growing on Siderastrea siderea.

Table 1. Habitat analysis of anemones (numbers and frequencies of occurrence) and hierarchies of occurrence with significance levels.

	a. OCCURRENCE WITH LIVING OR DEAD CORAL					b. OCCURRENCE WITH MASSIVE OR BRANCHING CORAL					c. OCCURRENCE WITH CORAL DAMAGE				
	no. live	no. dead	total	freq. live	freq. dead	no. mass.	no. bran.	total	freq. mass.	freq. bran.	no. dam.	no. undam.	total	freq. dam.	freq. undam.
C. gigantea	0	54	54	.00	1.00	38	16	54	.70	.30	31	19	50	.62	.38
L. danae	37	47	84	.56	.44	13	71	84	.16	.84	3	70	73	.04	.96
B. annulata	27	16	43	.63	.37	2	41	43	.05	.95	0	30	30	.00	1.00
H. lucida	30	143	173	.17	.83	5	168	173	.03	.97	17	151	168	.10	.90
P. crucifer	0	23	23	.00	1.00	4	19	23	.17	.83	not applicable				
S. helianthus	0	37	37	.00	1.00	28	9	37	.76	.24	21	11	32	.66	.34
R. florida	4	62	66	.06	.94	27	39	66	.41	.59	7	38	55	.31	.69
R. sanctithomae	0	21	21	.00	1.00	1	20	21	.05	.95	0	15	15	.00	1.00
P. neglecta	0	108	108	.00	1.00	0	108	108	.00	1.00	0	103	103	.00	1.00

Frequency of occurrence with living coral — $B.a. > L.d. > H.l.$ $p=0.01$ $p=0.005$ $p=0.005$ $>$ $\left[C.g. = P.c. = S.h. = R.f. = R.s. = P.n. = 0 \right]$

Frequency of occurrence with massive corals — $\left[S.h. = C.g. \right] > R.f. > L.d.$ $p=0.005$ $p=0.05$ $p=0.075$ $>$ $\left[B.a. = H.l. = R.s. = P.n. = 0 \right]$

Frequency of occurrence with coral damage — $\left[S.h. = C.g. \right] > R.f. > H.l.$ $p=0.005$ $p=0.005$ $p=0.005$ $>$ $\left[L.d. = B.a. = R.s. = P.n. = 0 \right]$

Figure 3. Coral colonies (A,B,C) transplanted around a Condylactis gigantea.

also occurred with living coral, attached to dead portions of the same coral colony. H. lucida occurred with live coral in only 17% of the cases and the other species almost never did so.

Table 1b illustrates the occurrence of the anemones with either massive and nodose corals (Diploria strigosa, D. clivosa, Montastrea annularis, M. cavernosa, Mycetophyllia lamarckana, Porites asteroides, Siderastrea siderea) or branching (ramose) and plating species (Acropora cervicornis, Agaricia agaricites, Porites furcata), and the hydrocorals (Millepora alcicornis and M. complanata). S. helianthus and C. gigantea occurred most often with the massive corals. R. florida occurred with them in 40% of the cases and L. danae in 16%. The other 5 species occurred almost exclusively with branching and plating forms.

Table 1c illustrates the occurrence of anemones with coral that shows damage potentially caused by the anemone. S. helianthus, C. gigantea and R. florida all occurred with coral that appeared to be damaged and their tentacles did not contact living tissue. H. lucida occurred with corals that appeared damaged in only 10% of the cases, while the other species never did so significantly. P. crucifer is not included in this analysis because it occurs in sand pockets and its tentacles do not normally contact coral which it could potentially damage, whereas those of B. annulata, also in sand pockets, do contact surrounding coral. C. gigantea (Fig. 2a) showed the most obvious occurrence with damaged coral (4 species), versus R. florida and H. lucida (3 species each), and S. helianthus (2 species).

The nine reefs studied can also be characterized by the extent of wave action. The reefs at Isla Galeta, Isla Verde and Porvenir (San Blas Isls.) receive heavy wave action while those at Buenaventura Is., Ironcastle Pt., and four reefs in the San Blas Isls. (Nalunega 1, Nalunega 2, Salar and Holandes Cays) are in very calm locations. Comparing reef areas of approximately the same size, those with heavy

Table 2.   Coral damage by Condylactis gigantea.

Coral Species	No. of contacts in the field	No. with damage in the field	No. colonies used in experiment	No. with damage in experiment
Siderastrea siderea	3	2	5	5
Agaricia agaricites	13	5	5	5
Montastrea annularis	25	22	0	0
Porites furcata	3	0	0	0
Diploria clivosa	4	2	3	3
Mycetophyllia lamarckana	0	0	2	2
Scolymia lacera	0	0	2	2

wave action had dense populations of C. gigantea and S. helianthus
(up to 5/m²) but lacked H. lucida and B. annulata. Calm reefs
contained all nine species but C. gigantea and S. helianthus were
rare (less than 1/100m²) or absent. C. gigantea and S. helianthus
consume macroscopic prey (gastropods, echinoids) and probably depend
on the heavy wave action to supply the prey. B. annulata and H.
lucida have long delicate tentacles and feed on zooplankton. Their
feeding may be impaired in heavy surge conditions. In aquaria, their
tentacles collapsed in rapidly flowing water and the anemones withdrew
into their holes.

Three of the anemone species displayed active interspecific
(never intraspecific) aggression (H. lucida, B. annulata, R. florida).
Attacked individuals suffered tissue damage and consistently moved
away from the aggressor (as in A. elegantissima (Francis, 1973)).
In many cases, even though aggression and tissue damage were not
evident, one species consistently moved away from another or, only in
the case of C. gigantea, from individuals of the same species. C.
gigantea never caused observable tissue damage to other species
although three of the species consistently moved away from it. This
could be a nonaggressive spacing behavior. Each of the groups of
anemones arranged by substratum type (surface, hole, sand) contained
only one aggressive species (R. florida, H. lucida and B. annulata
respectively). On the calm water reefs, the aggressive species in
each of the groups are by far the most common.

The field experiment designed to test whether or not C. gigantea
was the actual cause of the observed coral damage (Fig. 3) gave positive
results (Table 2). All 17 coral colonies used in the experiment were
damaged by C. gigantea only within the radius of the anemone's expanded tentacles. Outside the radius, the colonies were unharmed
(serving as a control). S. lacera are solitary polyps and thus were
placed completely within the radius. The remaining anemones which
may cause coral damage were not tested in this manner. H. lucida
does not occur at Isla Galeta where the experiments were most easily
conducted. S. helianthus and R. florida are surface dwelling forms
and transplants of coral into adjacent areas were not feasible. Wave
action would move the colonies and, because these anemones may damage
coral by slow overgrowth, the experiment will be a long-term project.
C. Birkeland (pers. comm.) has observed R. florida entering areas
where corals have been overgrown by the colonial ascidian Trididemnum
sp. which subsequently regresses, leaving an area of clear substratum.
Therefore, it is debatable at this time whether R. florida has actually caused the damage observed in the field or whether it is simply
filling areas such as those damaged by Trididemnum.

DISCUSSION

The nine species of anemones show a distinct partitioning of the
spatial resource on the coral reef. Although prey resources have not

been examined in detail, it is clear that all species except C. gigantea and S. helianthus are planktivores; the latter two feed on macroscopic prey washed off the reef platform by wave action. It appears that space is a limiting resource for coral reef anemones and habitat partitioning is finer than prey resource partitioning. The three main groups (inhabiting surfaces, holes, or sand) are distinct and non-overlapping due to the specialized morphology necessary to occupy each type of substratum.

Tables 1b and 1c show that the three species which occur most often with massive corals are also those which appear to directly or indirectly damage corals (C. gigantea, S. helianthus, R. florida). The massive coral skeletons are areas of high substratum stability (Connell 1973) but slow coral growth. They also contain few available holes in comparison with branching species. Anemones on massive coral are usually free of a branching superstructure which could obstruct the movement of dislodged macroscopic prey used by both species (C. gigantea, S. helianthus) occurring there. Anemones existing with massive corals must be able to cope with the coral's ability to grow over damaged areas (Fishelson 1973) and potential aggression by the coral (Lang 1973). This could be done either by shading of the coral, or by active aggression.

The three species which show active aggression toward other anemones occur in relatively stable areas of the reef, massive corals (R. florida) or areas of dead coral that have been cemented into the reef (H. lucida, R. florida) and sand pockets (B. annulata). P. crucifer appears nonaggressive and also occupies sand pockets. The remaining species (L. danae, P. neglecta, R. sanctithomae) occur in areas of rapid coral growth and collapse (e.g. Porites furcata and Agaricia agaricites beds). These are the least stable substratum types (Connell 1973) and the anemones may be forced to relocate themselves periodically. I consider these three to be fugitive species, avoiding direct competition with corals and other anemones by occupying areas of lower substratum persistence.

The possibility of aggression by corals against sea anemones was considered. No evidence for this was found in the field. However, due to the lack of skeleton in anemones, individuals killed by corals would not persist long to be observed. Field tests of this potential phenomenon should be conducted in the same manner as was done here with C. gigantea. The corals' ability for extracoelenteric digestion (Lang 1973) makes this a very possible occurrence.

In summary, there seem to be three options available to anemones on the coral reef. They can occupy the most persistent substratum, massive corals, and actively destroy or overgrow these corals. They can occupy areas of dead coral, a less persistent substratum in an earlier stage of succession. Anemones which occur in these areas are actively aggressive and must compete with other anemones for suitable holes or surfaces rather than with corals. Finally, they can exist as fugitives in areas which are the least stable, beds of live branching and plating corals. In these areas, individual movement or rapid clonal growth must keep pace with the changing nature of the habitat. Such areas, below a superstructure of live coral, offer the least exposure to light and probably also to currents bearing zooplankton.

The following diagram presents a gradient of substratum persistence (stability) and arranges the anemones in order of their occurrence along that gradient.

<---------------SUBSTRATUM PERSISTENCE--------------->

```
                     areas of
                   dead coral                      live branching
    massive coral  and coral rock   sand pockets      coral
 ┌───────────┐┌──────────────┐ ┌──────────────┐ ┌───────────────┐
┌ S.helianthus > R.florida > H.lucida > B.annulata = P.crucifer > L.danae   = ┐
└ = C.gigantea                                      R.sanctithomae =          ┘
                                                    P.neglecta
 └─────────────┘└─────────────────────────────┘ └──────────────────────────┘
   able to destroy       aggressive against         nonaggressive
 or overgrow corals          anemones
```

Fishelson (1970) has described the natural history of sea anemones at Eilat, Red Sea. The morphologies and habitat types are strikingly similar to those described here. Of 14 coral reef species at Eilat, 6 are congeneric with species in Panama and two others are members of very closely related genera, reflecting the eastern origin of the Caribbean fauna. Five species at Eilat occupy holes or crevices (6 in Panama), 5 on surfaces (5 in Panama) and 4 in sand pockets (2 in Panama). Aggressive behavior was not tested for the Red Sea species. Habitat partitioning in this guild of sessile carnivores may show a general pattern and research on other coral reefs is necessary to test this generality and determine whether or not similar interactions between corals and anemones are common and important occurrences.

I thank the Smithsonian Tropical Research Institute for their support of this research (Dougherty Foundation fellowship to the author), C. Birkeland, G. Hendler, E.G. Leigh, and D. Meyer for helpful discussions and suggestions on the research and L. Francis, A.J. Kohn and R.T. Paine for comments on the manuscript. J. den Hartog and C.E. Cutress aided in identification of anemone species.

SUMMARY

A quantitative comparison of habitat parameters was used to study the pattern of habitat partitioning among 9 species of sea anemones. Competitive interactions among anemone species and between anemones and corals were inferred from field observations. These interactions were demonstrated in the laboratory and in field experiments. The anemones, Heteractis lucida, Bartholomea annulata and Ricordia florida were actively aggressive toward, and caused visible damage to, other anemone species. Condylactis gigantea showed an ability to damage several species of coral near which it normally occurs. The anemone species can be arranged along a gradient of substratum persistence: anemones occurring with large massive corals maintain their position by damaging or overgrowing the corals. Species which occur in areas of dead coral actively attack other anemone species, while species in areas of rapid coral growth and collapse are nonaggressive.

REFERENCES

Bonnin, J.-P., 1964. Recherches sur la "reaction d'aggression" et sur le fonctionnement des acrorhages d'Actinia equina L. Bull. Biol. Fr. Belg., 1:225-250.

Connell, J.H., 1961. Effects of competition, predation by Thais lapillus and other factors on the natural population of the barnacle Balanus balanoides. Ecol. Monogr., 31:61-106.

_____, 1973. Population ecology of reef-building corals. In: Biology and Geology of Coral Reefs, D.A. Jones and P. Endean eds., Academic Press, New York, 360 pp.

Dayton, P.K., 1971. Competition, disturbance and community organization: The provision and subsequent utilization of space in a rocky intertidal community. Ecol. Monogr., 41:351-389.

Fishelson, L., 1973. Ecological and biological phenomena influencing coral-species composition on the reef tables at Eilat (Gulf of Aqaba, Red Sea). Mar. Biol., 19:183-196.

_____, 1970. Littoral fauna of the Red Sea: the population of non-scleractinian anthozoans of shallow waters of the Red Sea (Eilat). Mar. Biol., 6:106-116.

Francis, L., 1973. Intraspecific aggression and its effect on the distribution of Anthopleura elegantissima and some related sea anemones. Biol. Bull., 144:73-92.

Grigg, R.W., and J.E. Maragos, 1974. Recolonization of hermatypic corals on submerged lava flows in Hawaii. Ecology, 55:387-395.

Loya, Y., 1972. Community structure and species diversity of hermatypic corals at Eilat, Red Sea. Mar. Biol., 13:100-123.

Paine, R.T., 1966. Food web complexity and species diversity. Amer. Nat., 100:65-75.

_____, 1974. Intertidal community structure: Experimental studies on the relationship between a dominant competitor and its principle predator. Oecologia, 15:93-120.

Porter, J.W., 1972a. Predation by Acanthaster and its effect on coral species diversity. Amer. Nat., 106:487-492.

_____, 1972b. Ecology and species diversity of coral reefs on opposite sides of the Isthmus of Panama. Bull.Bio.Soc.Wash., 2:89-116.

_____, 1972c. Patterns of species diversity in Caribbean reef
    corals. Ecology, 53:745-748.
Sokal, R.R. and F.J. Rolfe, 1969. Biometry, the principles and
    practice of statistics in biological research, W.H. Freeman and
    Co., San Francisco, 776 pp.

GROWTH INTERACTIONS AMONG MORPHOLOGICAL VARIANTS OF THE

CORAL ACROPORA PALIFERA

D.C. Potts

Department of Environmental Biology, Australian

National University, Canberra, Australia

The material presented in this paper is part of a study of the responses of scleractinian corals to natural selection in different habitats and, more generally, of the ecological and genetical structure of coral populations. On the Heron Island reef, near the southern end of the Great Barrier Reef, there are at least five distinct forms of corals which I consider all belong to Acropora palifera Lamarck. Each form is most abundant in a particular physical and biological habitat, although all forms occur in low frequencies in most other habitats (Potts, 1976). In an earlier paper (Potts, 1976) I described an experiment in which different forms grew in contact with other corals for three months (March-June 1975). Each coral seemed able to respond in several ways depending on the identity of its neighbour. The present paper describes the subsequent history of that experiment (to March 1976) and examines relatively long-term interactions which may influence the partitioning of space within coral communities.

MATERIALS AND METHODS

Habitats and Forms

Inner Flat (I). This is a shallow area (approximately 25 cm at low tide) lying close to the island and covered with numerous, partly dead patches of coral. The fauna is dominated by an orange-brown form of A. palifera that branches continually and irregularly all over its surface. The bases of branches fuse rapidly, often giving an amorphous colony with a massive centre. Individual corallites are relatively small, tubular and closely packed.

79

Lagoon (L). An extensive lagoon about 5 km long and less than
5 m deep lies east of Heron Island. It contains isolated patch
reefs on which one of the more conspicuous corals closely resembles
Acropora palifera forma α as described by Wells (1954). This is a
yellow-brown form with regular branches (10-20 cm apart) which
remain separate. Corallites are well spaced, but are relatively
small, thick-walled and adpressed to the surface.

Outer Flat (O). This habitat is well developed 500-600 m
north of Heron Island and about 150 m behind the edge of the reef.
Extensive areas of coral are separated by sandy channels (0.5-1 m
deep at low tide). All forms of A. palifera grow here, but the
most abundant form has greenish-brown colonies very like those of
the Lagoon form, but with much more frequent branching and
corallites which tend to have thinner walls and be more closely
packed.

Crest (C). Around the outer edge of the reef there is a hard
limestone pavement with a dense cover of live corals. This zone is
subject to heavy wave action, and is completely exposed only on low
spring tides. The only form of A. palifera found on the Crest has
dark brown colonies which consist of large, encrusting basal discs
producing thick vertical plates. These plates are covered with
small, densely packed corallites adpressed to the surface.

Slope (S). From the Crest, the reef front slopes down steeply,
with many small vertical cliffs, to a sandy bottom in 15-20 m of
water. A. palifera usually grows here as dark purplish-brown cones
or colonies with a few heavy branches. It has relatively large,
thick-walled, tubular corallites which are well spaced over the
surface. This form closely resembles A. palifera forma β of Wells
(1954).

Growth Experiment

The growth experiment was set up from 15 to 25 March 1975. I
collected four typical colonies from each site, using corals at
least 10 m apart to minimize chances that they belonged to the
same clone. They were placed in shallow water near the laboratory
until needed, when they were transferred to large aerated tanks.
I broke up each colony and selected 8 similar, undamaged pieces
which were usually 4-6 cm long with at least one growing tip. They
were assigned randomly to one of three classes of treatments: 1)
one piece from each colony was paired with another piece from the
same colony, i.e. with a coral known to be phenotypically and
(presumably) genotypically identical; 2) two pieces were paired with
pieces from another, phenotypically similar, but genotypically
different ("alien") colony from the same site; 3) pieces were paired
with one of the morphologically different, alien colonies from each

of the other four sites.   There were four replicated pairs for each
treatment.   The two pieces to be grown together were mounted with
their tips in contact by embedding their broken ends in quick-setting
concrete in PVC cups,   As soon as the concrete was hard to the
touch, the corals were placed in the sea near the laboratory.   They
were never exposed to the air for more than 10 minutes at a time
and concreting was completed within 36 hours of collecting the
corals.   Potts (1976) contains further details of the design and
transplanting techniques.

All corals were photographed, and bolted 20 cm apart on steel
mesh grids which were attached to horizontal steel frames at the
Inner Flat experimental site about 200 m east of the laboratory,
and 25 m out from the base of the beach rock.   Eight pairs were
discarded because one or both pieces appeared adversely affected by
the concreting.   The corals were about 10 cm above the sand and
covered by 10-15 cm of water at low spring tides.   Frequent checks
over the next three months showed that all the corals recovered
rapidly from the effects of transplanting and grew vigorously.   In
June and November 1975, the corals were brought into the laboratory,
photographed, and returned to the experimental site within 24 hours.
They were collected for the last time in March 1976.   At each
collection, notes were made of the nature of any interactions
between the corals.   After the final collection, the nature of
tissue junctions across the boundaries between corals was examined
using electrical stimuli to test for nervous connections.   Vertical
sections were then cut through the corals with a diamond saw and
skeletal connections were examined macroscopically.

RESULTS

Photographs taken in June 1975, after 3 months growth, suggested that corals in contact with aliens were growing asymmetrically
with the fastest growth on the side facing the alien (Potts, 1976).
In contrast, where two pieces of the same colony were paired, each
piece seemed to grow symmetrically on all sides (see Fig. 1).   After
8 and 12 months growth I calculated the area (to nearest cm^2) of
the surface of contact between each pair of corals.   Table 1 shows
that, at both collections, contacts between alien corals were more
than twice the size of contacts between two pieces of the same
colony (t-tests; $p < .001$ in both collections) which suggests that
the presence of an alien stimulates growth.   The area of contact
between aliens was the same whether the corals belonged to the same
form, or to different forms.

Growth Responses

After three months growth there were three distinct categories

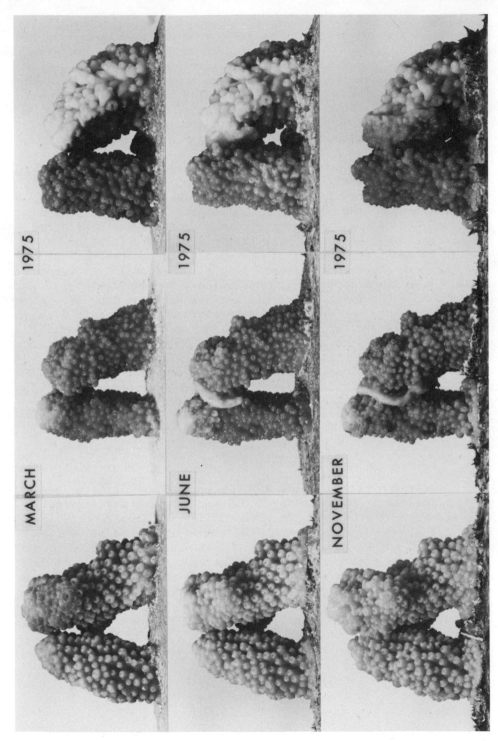

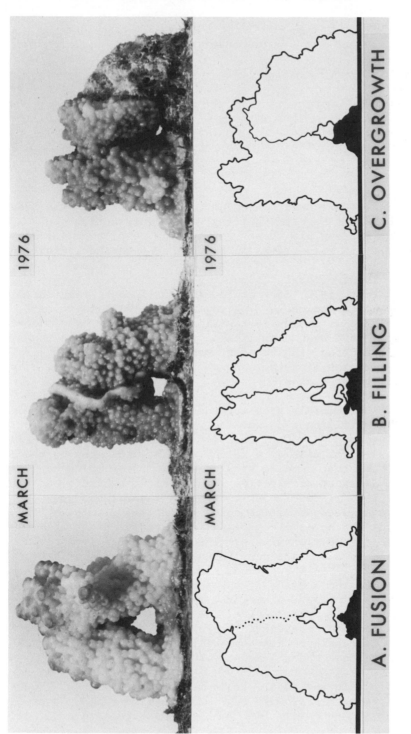

Fig. 1. Development of growth responses, and vertical sections after 12 months. A. Fusion of pieces of the same Lagoon coral. B. Filling between two different Inner Flat corals. C. Overgrowth of a Slope coral (right) by an Outer Flat coral (left). (The Slope coral died shortly before the final collection during prolonged bad weather.) Scale is 0.9X natural size.

Table 1

Mean area of contact (cm^2) between pairs of corals. Sample sizes
in parentheses

	November 1975	March 1976
Paired with self	2.2 (17)	4.8 (16)
Paired with similar aliens	5.3 (19)	10.3 (16)
Paired with different aliens	5.4 (27)	10.5 (27)

of boundaries formed between adjacent corals and during subsequent
growth (up to 12 months after pairing) every interaction could
still be classified within these groups. The development of each
type of junction is illustrated in Fig. 1 and described below.

Fusion. The two corals joined together with few visible signs
of a junction (Fig. 1A). In 8 of 17 cases of fusion, some parts of
the zone of contact showed as a narrow band of relatively small
corallites. These were usually restricted to the bottoms of clefts
where there may have been insufficient space for polyps to reach
full size. In section, the boundary was either not visible at all,
or showed as a region of rather amorphous skeletal material, often
with small irregular spaces.

Filling. Both corals grew towards each other and contributed
about equally to filling the space between them. At the contact,
each coral developed a growing edge similar to that at the expanding edge of the basal disc of a colony. Each piece grew up the
underside of the other coral's growing edge. The growing edge was
typically a thin sheet of skeleton 2-4 mm thick; behind a smooth
convex front, rows of developing corallites faced the edge.
Growth often continued after the space was filled forming a low
ridge (Fig. 1B) which, as the growing edges radiated out from the
point of contact, occasionally developed into a relatively thin
disc in a plane more or less perpendicular to the axis of contact
between the corals. In one natural contact the disc extended
nearly 2 cm above the surface of the surrounding coral. The crest
of the ridge was smooth (2-5 mm wide) with a central groove
separating the corals. In section, the contact showed as a solid
line which seemed to represent an unbroken wall formed by one or
both corals. The skeleton near the wall often contained abnormal,
irregularly sized spaces.

Overgrowth. In the most extreme response, unilateral growth
by one coral produced a growing edge which spread rapidly over the
surface of the second coral (Fig. 1C). The subordinate coral did
not produce a growing edge in the region of contact. In section,

the contact was very like that seen with filling: a discrete wall
with variable spaces nearby.

Filling and overgrowth may both result from a single process
in which the presence of an alien stimulates the production of a
growing edge. The only important difference seems to be that
filling involves similar and equal responses by both corals whereas
overgrowth is either a unilateral response, or one which also
involves suppression of growth in the subordinate coral by the
dominant one. Changes in responses during the first few months
provide more evidence that filling and overgrowth are related. 20%
of corals finally scored as filling began with overgrowth of one
coral by the other, and an additional 20% showed some signs of overgrowth in the early stages: this suggests that the time lag before
initiation of growth responses may vary between members of a pair.
Similarly, 45-65% of pairs finally scored as overgrowths had begun
by filling, which may indicate that the overgrowing coral takes
some time to exert its dominance.

Patterns of Growth Responses

Differences between the growth responses reported after three
months (Table 2 in Potts, 1976) and those assigned after 12 months
were most common in pairs derived from similar colonies of the
same form. In 55% of such pairs, the response altered, with
similar numbers changing from filling to overgrowth and vice versa.

Table 2

Summary of growth responses to the presence of neighbours by each
form of A. palifera. Entries are proportions of pieces of coral
and (N) gives the numbers of pieces

| Origin | Paired with self | | Paired with aliens | | | | | | | |
| | | | Similar form | | | | Different forms | | | |
	Fuse	(N)	Fill	Over +	Over -	(N)	Fill	Over +	Over -	Nil	(N)
Inner	1.00	(8)	.5	.25	.25	(8)	.79	.07	.14		(14)
Lagoon	1.00	(8)	.75	.13	.13	(8)	.62	.08	.23	.08	(13)
Outer	1.00	(6)	.5	.25	.25	(8)	.67	.17	.17		(12)
Crest	1.00	(8)	1.00			(8)	.20	.67	.07	.07	(15)
Slope	1.00	(4)	.5	.25	.25	(8)	.33	.08	.58		(12)
Total	1.00	(34)	.65	.18	.18	(40)	.52	.23	.23	.03	(66)

By contrast, the growth responses altered in only 24% of pairs between different forms. Thus, differential initiation of growth responses and delayed establishment of overgrowth dominance were both common between similar corals, whereas almost all changes involving different forms were associated with delayed establishment of dominance by overgrowing corals.

The final responses are summarized in Table 2 in which sample sizes refer to the number of pieces of coral used, not to the number of pairs. The data are expressed in this way so that the direction of overgrowth can be shown. The results for corals paired with themselves are unequivocal. In every one of 17 pairs there was complete fusion of skeletons and tissues. At no stage did any coral paired with itself give any indication of filling or overgrowth.

Responses to similar aliens involved a varied pattern of both overgrowth and filling in most treatments except among the Crest corals where every pair filled the space without even early indications of overgrowth. Similar corals were not only the least consistent through time (see above); they were also the pairs in which it was most difficult to separate filling from overgrowth, especially during the first few months. The distinguishing features were often poorly developed, and sometimes parts of the contact looked like filling, while other parts looked like overgrowth.

With different forms, results were always more consistent over time; in 50% of treatments, all pairs also had identical responses. Although the relative numbers of filling and overgrowth were not significantly different overall between similar and different forms of aliens (Chi^2, p > .05), certain responses were most commonly associated with particular forms. 79% of pieces of Inner Flat corals were involved in filling responses which contrasts with 20% and 33% respectively for Crest and Slope corals. The directions of overgrowths involving Slope and Crest corals differed: Crest corals were dominant in all but one pair; Slope corals were dominant in only one pair.

Tissue Connections

Nervous connections between pairs of corals were assumed if electrical stimulation of one coral caused contraction of polyps on both corals. Using techniques and procedures based on Pantin (1935) and Horridge (1957), each pair was transferred, at dusk, to a darkened laboratory where a probe was positioned on one coral about 5 mm from the junction of the two pieces. When the polyps were fully expanded the coral was stimulated repeatedly. In every set of corals paired with self, contractions continued across the boundary between the pieces with no detectable irregularity in the extent or rate of contraction along the contact. Conversely,

nervous connections were never detected across filled or overgrown
boundaries.   This fusion of skeletons also involved fusion of the
nervous system, and probably of other tissues as well, whereas
physiologically functional union of tissues was not present across
filled or overgrown boundaries after 12 months growth.   Hildemann
et al. (1974) described grafts between different colonies of other
Acropora spp. in which functional tissue connections persisted for
many weeks; they did not mention skeletal growth responses.   In the
present study, temporary tissue connections would not have been
detected.

DISCUSSION

A. palifera accurately distinguishes "self", to which it
responds with complete fusion of skeleton and soft tissues, from
"not-self", to which it responds by skeletal filling or overgrowth
without functional joining of soft tissues.   Filling involves mutual
growth stimuli, but does not give one coral an advantage over the
other.   Instead, the boundary seems to remain static (for over a
year, at least) after the corals divide the available space between
them.   Overgrowth seems caused either by stimulation only of the
dominant coral, or also by suppression of growth in the subordinate
coral.   Thus the dominant coral expands over the space occupied by
the other so that the boundary changes with time as the dominant
coral continues to grow.   In the field these growth responses are
not restricted to A. palifera.   Many other Acropora spp. produce
similar growth responses, especially in habitats such as the Crest
and Outer Flat where living coral cover is high.

Recent consideration of interactions among corals has concentrated on the hierarchy of extracoelenteric digestive aggression
described by Lang (1973).   Although Dustan (pers. comm.) has seen
intra-specific digestion in Montastrea annularis, digestive
aggression seems totally unimportant among the forms of Acropora
palifera, possibly because mesenteric filaments of all Acropora spp.
are very short (Lang, pers. comm.).   In the experiment described
here, I only twice saw possible signs of digestive aggression - a
narrow band of dead coral along the edge of its neighbour; this
condition did not persist and both were finally scored as filling.

The three interactions (fusion, filling, overgrowth) are not
a graded series of responses associated mainly with the genetic
distance between members of a pair.   Instead, the distribution of
responses is more closely linked with the original habitats of the
corals.   This may reflect different selective pressures influencing
the partitioning of space in each habitat.   Corals from the densely
covered, but essentially 2-dimensional pavements of the Crest were
usually dominant overgrowers.   On the Crest, free space may be in
short supply for long periods during which individual corals are

likely to be in direct contact with other corals: under these
circumstances, selection may favour individuals with highly developed
competitive mechanisms (e.g. overgrowth). Inner Flat corals usually
responded to aliens by filling, while Slope corals were usually
subordinate and were overgrown by all other forms. These observations are consistent with previous speculations about these habitats
(Potts, 1976). On the sparsely populated Inner Flat, an extremely
variable and unpredictable environment may favour vigorous growth
during short periods of favourable conditions, without strong
selection for competitive mechanisms. Thus filling may represent
ability to grow rapidly rather than competitive ability. In deeper
water, colonies may be overtopped by corals which grow above them
without physically contacting them. Therefore Slope corals may be
adapted primarily for persistence under continuously adverse physical
conditions (e.g. low light, high sedimentation) which provide little
opportunity for selection to favour such phenotypic characters as
vigorous growth or high competitive ability.

E. Lovell, P. Rowles, D. Bender, G. Calaresu, L. Tybrant and
A. Bothwell provided field assistance. I particularly want to
thank P. Morrow and L. Fox for their contributions to extensive
revisions of the manuscript.

LITERATURE CITED

HILDEMANN, W.H., DIX, T.G., and COLLINS, J.D., 1974. Tissue
    transplantation in diverse marine invertebrates. In
    Cooper, E.L., Ed., Contemporary Topics in Immunobiology, 4:
    141-150.
HORRIDGE, G.A., 1957. The co-ordination of the protective
    retraction of coral polyps. Phil. Trans. Roy. Soc. Lond.,
    Ser. B, 240: 495-529.
LANG, J., 1973. Interspecific aggression by scleractinian corals.
    2. Why the race is not only to the swift. Bull. mar. Sci.,
    23: 260-279.
PANTIN, C.F.A., 1935. The nerve set of the Actinozoa. I. Facili-
    tation. J. exp. Biol., 12: 119-138.
POTTS, D.C., 1976. Differentiation in coral populations. Atoll
    Res. Bull., in press.
WELLS, J.W., 1954. Recent corals of the Marshall Islands. U.S.
    geol. Survey, prof. Pap., 260-I: 385-486.

SETTLEMENT, MORTALITY AND RECRUITMENT OF A RED SEA SCLERACTINIAN CORAL POPULATION

Y. Loya

Department of Zoology, The George S. Wise Center for

Life Sciences, Tel Aviv University, Tel Aviv, Israel

INTRODUCTION

Connell (1974) reviewed different aspects concerning the distribution and abundance of hermatypic corals and pointed out the many gaps in our knowledge of the population ecology of corals. Thus, very little is known on the capacity of larval dispersal, survivorship, and conditions for settlement of hermatypic corals. These areas of ignorance are among the main obstacles to an understanding of life histories of these organisms (Grassle, 1974). Although the settlement behavior and population dynamics of coral planulae have not been studied in the field, the rate of settlement and early survival have been followed in two cases (Stephenson and Stephenson, 1933; Connell, 1974). Stephenson and Stephenson (1933) placed on the reef at Low Isles different artificial objects as settling surfaces for coral planulae. They used logs of wood, pieces of beach- sandstone, earthenware drain pipes, glass jars, and cleaned shells of clams. About 40 young colonies settled on these materials during a period of 11 months. Of 36 colonies recovered, 26 were Pocillopora bulbosa, which might indicate the opportunistic character and colonizing capacity of this species. Connell (1974) marked permanent quadrats on the reef flat at Heron Island, Great Barrier Reef and took photographs at intervals since 1962, for 12 years. He found that the average rate of recruitment for all species between 1962 and 1970 was about 5 new colonies per square meter annually. In all areas the commoner species had high rates of recruitment and mortality and maintained their abundance.

In the present study, I have followed the rates of settlement, mortality and recruitment of the scleractinian coral Stylophora

<u>pistillata</u> (Esper), in the northern Gulf of Eilat, Red Sea. In
spite of its small colony size (largest colonies reach 30-35 cm in
diameter), <u>S. pistillata</u> is among the most important frame builders
of the reef flat, because of its great abundance (Loya, 1972). The
sexual reproduction of this species was studied by Rinkevich (1975).
He found that <u>S. pistillata</u> is a hermaphrodite coral. Both ova and
testes are situated in one polyp. Female gonads appeared in July
and developed into ova in December. Male gonads appeared in October
and produced sperm in December. Release of planulae was observed
from December to August. While synchronization in breeding was
found within branches of the same colony, no such synchronization
was found within the entire population. Rinkevich (1975) further
postulated that the lack of synchronization in breeding enables this
species to extend its breeding season to 8-9 months.

     I have elsewhere termed <u>S. pistillata</u> as an r-strategist (Loya,
1976a), since it exhibits most of the r-characteristics summarized
by Pianka (1970). The purpose of the present paper is to quantify
some of these characteristics, namely, the opportunistic character
of <u>S. pistillata</u> in colonizing unpredictable and new habitats and
its high rate of mortality and great population turnover.

METHODS

     The area of work took place in shallow water (3-4 m depth),
across the Marine Biological Laboratory at Eilat (Fig. 1). This

     Fig. 1. The running sea-water system of the Marine Biological
Laboratory at Eilat has been intensively colonized by <u>S. pistillata</u>.
The lower pipe was installed in May 1967 ("old pipe"). The upper
pipe was installed in March 1974 ("new pipe"). Note the grazing
activities of the gastropod <u>Trochus dentatus</u> on the upper portions
of the new pipe.

area may be best described as a reef in a process of building, since
it has been only recently (in the last 8-10 years) heavily populated by corals. As a result of different research activities and
technical needs of the laboratory, many artificial objects such as
cement blocks, PVC pipes, iron frames, fish traps and others, were
placed underwater in different dates. All these objects were heavily colonized by S. pistillata (Figs. 1 and 2), while other scleractinian species were represented by 1-10 colonies, and there were
only four such species. The running sea-water system of the laboratory (two PVC pipes installed on cement blocks, Fig. 1), provided
a unique opportunity to study the rates of settlement, mortality and
recruitment of S. pistillata, as well as other aspects of population
dynamics of this species (Loya, in preparation). There was a time
interval of seven years between the installment of the two pipes:
The lower pipe (20 m long, 15 cm in diameter) was installed on May
1967 and the upper pipe (25 m long, 20 cm in diameter) on March
1974.

On May 17, 1973, I tagged and numbered 189 colonies of S. pistillata of various sizes on the old pipe and cement blocks, and

Fig. 2.   Intensive settlement of S. pistillata colonies on iron
frames across the Marine Biological Laboratory at Eilat. Approximately 150 colonies of different sizes were counted on this frame
(colonies smaller than 4.0 cm in diameter cannot be seen in this
picture).

studied their population dynamics for 33 months. The length, width and height of each colony were measured underwater every 2-3 months (Fig. 3). Length is defined as the distance across a coral between the tips of branches which are farthest apart; width is a measure perpendicular to the length axis; height is orthogonal to the width and length axes. Since the general shape of S. pistillata approximates a sphere (Fig. 3), the colonies were divided into size groups according to the geometric mean of their radius; $\bar{r} = (l \times w \times h)^{1/3}/2$ where l = length; w = width; h = height, in cm. The geometric mean radius ($\bar{r}$) served also as the criterion for growth rate and regeneration rate of S. pistillata (Loya, 1976a, 1976b).

On March 1974, the new pipe was placed on the old pipe (Fig. 1) and since that day, all the new colonies settling on the new pipe were periodically measured and the number of colonies which died within each time interval was recorded. Photographs were taken with a Nikonos II camera with a close-up lens. A metal frame fixed to the lens (24 x 17 cm) provided the opportunity to take a series of photographs of marked areas of the same colonies from a fixed distance and angle.

RESULTS

Figure 4 illustrates the distribution of S. pistillata colonies measured on May 17, 1973 on the old pipe and cement blocks (Fig. 1). The colonies were clustered in size groups according to the geometric mean of their radius ($\bar{r}$). The survivorship record of these colonies is presented in Fig. 5. Knowledge of the exact date when the pipe was installed, made it possible to estimate the corals population

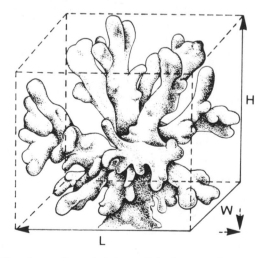

Fig. 3. The length, width and height of S. pistillata colonies were measured periodically. The geometric mean radius ($\bar{r}$) served as a criterion for dividing the population into size (age) groups.

age structure on the pipe (see text to Fig. 5), by dividing them into age groups. An extensive study of the growth rate of S. pistillata colonies in different size classes (Fig. 6), taking into account the variance within and between size groups (Loya, in preparation), indicated a significant linear correlation (r = 0.97,
P < 0.001) between the geometric mean radius of a colony and its
age, up to an age of 9-10 years. The estimates of the population
age structure on the old pipe, in the beginning of the study, fitted
quite closely to the much more accurate determination of age, based
on the growth rate study.

During the present study, the rate of mortality of S. pistillata
colonies varied between 70-95% (Fig. 5). The smallest size group
($\bar{r}$ = 0.01-1.00) exhibited the heaviest mortality. The five largest
colonies recorded at the beginning of measurements ($\bar{r}$ = 5.01-6.00;
age estimate of 5.5 years old) are not shown in Fig. 5, because
three of these colonies were completely broken during the installment of the new pipe. No significant conclusions could be drawn,
therefore, concerning the survivorship of this size group. The two
colonies left intact continued to grow during the entire study. Note
the steep drop in the percentage of surviving colonies between December 1973 and March 1974, and especially between March and May

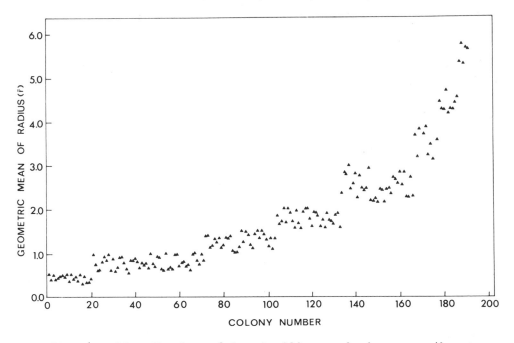

Fig. 4. Distribution of S. pistillata colonies according to
size groups (in cm) at the beginning of measurements (May 17, 1973).
The number of colonies within each size group was 20, 51, 32, 30,
33, 9, 9 and 5, in order of smallest to largest size group.

1974. The heaviest mortality occurred after two southern strong
storms in April and May 1974.

Our present knowledge of the life history of S. pistillata does
not provide any direct evidence on the life expectancy of its planulae. It may be expected, however, that the heaviest mortality occurs in the planula larvae of this species, from the time they are
released and during the early stages of their settlement and development. The installment of a new sea-water pipe, parallel to the
old pipe (Fig. 1), provided a unique opportunity to study the population dynamics of corals invading this "new environment" (in the
sense of Slobodkin and Sanders, 1969).

Table 1 summarizes the data on the rate of settlement, mortality and recruitment of new colonies of S. pistillata on the new
pipe. No other scleractinian corals settled on this pipe until
March 22, 1976. Successful settlement and development of S. pistillata colonies has mainly occurred on the lower portions of both
pipes (Figs. 6 and 7). On January 1, 1975, 45 colonies were recorded on the new pipe (Table 1) belonging to two size groups: 39

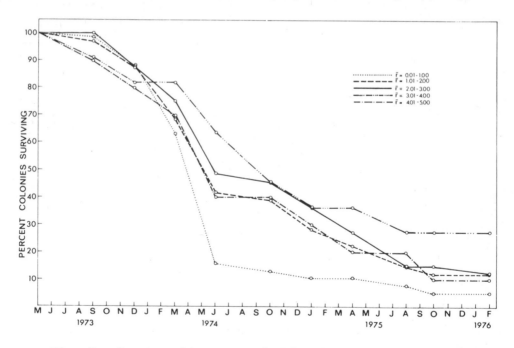

Fig. 5. Survivorship curves for S. pistillata colonies. The
number of colonies recorded within each size interval at the beginning of measurements was 71, 63, 33, 9 and 9, in order of smallest
to largest size group. Estimation of maximum age for each size
group in the beginning of measurements is 0.75, 1.5, 2.5, 3.5 and
4.5 years, in order of smallest to largest size group.

Fig. 6.   Differences in size (age) structure of S. pistillata
colonies on the old pipe.   Colony S-44 is approximately 6 years old
(15.6 x 12.8 x 9.3 cm).   The two small colonies to the right and
left of colony S-44 are approximately 3 months old.

Fig. 7.   Successful settlement and development of S. pistillata
colonies has mainly occurred on the lower portions of the pipes. The
picture was taken on April 1976, i.e., when the reproductive activity
of S. pistillata is most intensive.   Most of the numerous and very
small colonies observed on the pipes are 1-3 months old.

Table 1. Settlement, mortality and recruitment of S. pistil-lata colonies on the new pipe. The coral population was divided in-to size groups according to (r̄) in cm. Numbers in parentheses in-dicate the number of colonies dying within time interval and size interval.

Date of measurement	Geometric mean of radius (r̄)						Total number of colonies alive	% mortality within time interval
	0.10-0.25	0.26-0.50	0.51-1.00	1.01-1.50	1.51-2.00	2.01-2.50		
Jan. 1. 75	39 (1)	6					45	
								2.2
Apr. 8. 75	48 (2)	35	9 (2)				92	
								4.3
Aug. 10. 75	85 (40)	46 (15)	30 (8)	12 (7)			173	
								40.5
Oct. 13. 75	--	45 (30)	31 (12)	22 (10)	5 (2)		103	
								52.4
Dec. 11. 75	--	--	21 (2)	20	8		61	
								3.3
Jan. 22. 76	185 (54)	--	19 (1)	20 (2)	8		232	
								24.5
Mar. 22. 76	278	131	--	18	16	8	451	
% mortality within size interval	27.2	34.1	22.7	25.7	9.5			

colonies with r̄ = 0.10-0.25 and six colonies with r̄ = 0.26-0.50.
The age estimation for the smaller colonies is 1-2 months, and that
of the larger colonies is six months. That is, the first six colonies settled on the pipe, probably during July 1974, and the next
39 colonies during December 1974. Although the pipe was installed
in March 1974, which is a most intensive period in the reproduction
of S. pistillata (Rinkevich, 1975), no successful settlement has
occurred for at least four months. Also, there is a marked difference between the rate of colonization during the first and second
year since installation (Table 1). A total of 93 colonies settled
on the pipe during the first year, while 548 colonies settled during
the second year. It seems that a certain "conditioning" of a new
substrate is required before intensive colonization of scleractinian corals takes place. Harrigan (1972) thus found, that the young
larvae of Pocillopora damicornis postpone settlement on clean hard
substrates until a thin living algal diatomaceous-bacterial film is
created. High recruitment of new colonies occurred between December 1974 and August 1975 and especially between December 1975 and
January-March 1976 (Table 1). Conversely, no colonization was observed during the months August until November. These results confirm Rinkevich's (1975) study on the sexual reproduction cycle of
S. pistillata.

Similarly to the survivorship curves of the coral population on
the old pipe (Fig. 5), the smallest colonies on the new pipe (size

interval 0.10-1.50) showed a two-three fold higher mortality than
the larger size groups (1.51-2.50). The most extensive mortality of
colonies occurred during August-December, 1975 (Table 1) and was
mainly caused by strong southern storms. The number of new recruits
of S. pistillata on the new pipe, on March 1976 varied between 0-15
per 1/25 m² (the Nikonos areal frame). The average number of colonies calculated per square meter of the lower half-portion of the
new pipe, indicated an extremely dense population of 60 colonies/m².
No successful settlement of S. pistillata colonies has occurred on
the upper half-portions of both pipes. The newly settled colonies
tended to form aggregations, which were clearly indicated by the
high ratios obtained in calculations of variance/mean abundance per
1/25 m².

DISCUSSION

The present study has clearly shown that S. pistillata is characterized by a great population turnover, i.e., high rate of mortality (Fig. 5) and high rate of recruitment of new colonies (Table 1).
The shallow water area studied is periodically disturbed by the
strong southern storms, which seem to be the major causes of the
high mortality observed. Levin and Paine's (1974) model concerning
disturbance and patch formation predicts a reduced variety of patches
under conditions where the patch birth rate and extinction rate are
both high. They conclude that in such conditions fugitive or transition populations should predominate in the species list. The hermatypic coral community structure, in the area studied, closely fits
the predictions deduced from Levin and Paine's (1974) model.

Perhaps the most important feature of S. pistillata as an r-
strategist (Loya, 1976a), is that it is usually the pioneer coral
species colonizing unpredictable, short lived or unexploited habitats. The intensive and quick colonization of S. pistillata on the
newly introduced sea-water system, and the lack of colonization by
other scleractinian corals, elucidates the survivorship strategy of
this species. Wilson and Bossert (1971) postulated that, in a new
habitat, such a species succeeds by discovering the habitat quickly,
reproducing rapidly (to use up resources before other competing
species can exploit the habitat), and disperses in search of other
new habitats, as the existing one begins to grow unfavorable. S.
pistillata fulfils almost all these "requirements" in the northern
Gulf of Eilat.

The coral genus Pocillopora, which is taxonomically allied to
Stylophora (same family, Pocilloporidae), seems to exhibit a similar
strategy of survival in the Pacific to that of S. pistillata in the
Red Sea. Thus, Stephenson and Stephenson (1933) reported P. bulbosa
as a pioneer species colonizing unexploited habitats at Low Isles,
Great Barrier Reef. Grigg and Maragos (1974) described P. meandrina

as the pioneer species settling on the youngest underwater lava flows
at Hawaii. P. damicornis in the Pacific shows, however, the greatest
similarity to S. pistillata in its life history and strategy of survival. P. damicornis in Hawaii shows great population turnover,
small colony size, high growth rate (maximum size could be reached
in 5-6 years), high mortality rate (Maragos, 1972), high reproductive potential, pioneer in colonization of new substrates (Harrigan,
1972). The only feature that is contrary to that of S. pistillata
is the high digestive dominance (Lang, 1971) of P. damicornis (Porter, 1974), whereas S. pistillata is ranked among the lowest species
in the digestive (aggressive) hierarchy (Loya, 1976a).

Although physical factors seem to be the major causes of mortality in the S. pistillata population studied, several natural en-
emies were observed in this area, that might have contributed to
some of the high mortality observed. Thus, several species of par-
rot fishes, wrasses (Coris angulata and C. lunula) and trigger fishes
(Pseudobalistes fucus and Hemibalistes chrysoptera) have often been
observed to bite pieces of the coral branches. Other animals observed to slash the tissues or suck off bits of polyps include the
sea-urchin Diadema setosum and the gastropod Quoyula monodonta.
Boring organisms, such as the sponge Clione and the bivalve Litho-
phaga lessepsiana might weaken the skeleton of S. pistillata and
consequently decrease its ability to withstand destruction by physical factors. The starfish Acanthaster planci, which has been described as the major coral predator in many reefs, is quite rare in
the study area. The grazing activity of the gastropod Trochus den-
tatus (Fig. 1) and the sea-urchin D. setosum, has been observed to
be restricted to the upper portions of the sea-water pipes. It is
possible that a synergistic effect of intensive grazing and sedimentation prevents successful settlement of coral planulae, on the
upper portions of the pipes (Figs. 6 and 7).

The high rates of mortality and recruitment of new colonies of
S. pistillata, contribute significantly to the accumulation of coral skeletons to the reef framework of the area studied. After the
colonies die they are quickly encrusted by algae and turn into a
reef rock by lithification. New colonies of S. pistillata settle
on the lithified dead colonies and, thus, the reef, in this area,
has been growing at a surprisingly fast rate, mainly due to the
great population turnover of S. pistillata.

This research has been supported by the United States-Israel
Binational Science Foundation (BSF), Jerusalem, Israel. I thank
Ms. Horowitz and Mr. Y. Schlezinger for technical assistance and
Mr. S. Shefer for drawing Figure 3. Dr. A. Barash identified the
predatory gastropods and boring bivalve.

SUMMARY

The rates of settlement, mortality and recruitment of S. pis-tillata have been studied at the northern Gulf of Eilat, Red Sea. During 33 months of the study, the mortality within a population of 189 colonies of S. pistillata varied between 70-95%: The youngest colonies showed the heaviest mortality. The major causes of mortality seem to be periodical strong southern storms. S. pistillata is shown to be the pioneer coral in colonizing unexploited habitats in large numbers. The average number of new recruits settling on newly introduced artificial objects reached 60 colonies/m^2, during the peak of the reproductive period. S. pistillata is the major coral contributing to the reef framework in the area studied.

REFERENCES

Connell, J.H., 1974. Population ecology of reef building corals. Pages 205-245 in O.A. Jones, and R. Endean, Eds., Biology and Geology of Coral Reefs. Vol. II. Academic Press, New York, 480 pp.

Grassle, J.F., 1974. Variety in coral reef communities. Pages 247-270 in O.A. Jones, and R. Endean, Eds., Biology and Geology of Coral Reefs. Vol. II. Academic Press, New York, 490 pp.

Grigg, R.W., and J.E. Maragos, 1974. Recolonization of hermatypic corals on submerged lava flows in Hawaii. Ecology, 55:387-395.

Harrigan, J.F., 1972. The planula larva of Pocillopora damicornis: Lunar periodicity of swarming and substratum selection behavior. Ph.D Thesis, Univ. of Hawaii, 303 pp.

Lang, J.C., 1971. Interspecific aggression by scleractinian corals. 1. The rediscovery of Scolymia cubensis (Milne Edwards & Haime). Bull. Mar. Sci., 21:952-959.

Levin, S.A. and R.T. Paine, 1974. Disturbance, patch formation, and community structure. Proc. Nat. Acad. Sci., 71:2744-2747.

Loya, Y., 1972. Community structure and species diversity of hermatypic corals at Eilat, Red Sea. Mar. Biol., 13:100-123.

Loya, Y., 1976a. The Red Sea coral Stylophora pistillata is an r-strategist. Nature, 259:478-480.

Loya, Y., 1976b. Skeletal regeneration in a Red Sea scleractinian coral population. Nature, in press.

Maragos, J.E., 1972. A study of the ecology of Hawaiian reef corals. Ph.D Thesis, Univ. of Hawaii, 290 pp.

Porter, J.W., 1974. Community structure of coral reefs on opposite sides of the isthmus of Panama. Science, 186:543-545.

Rinkevich, B., 1975. On the reproduction of Stylophora pistillata (Esper) and harmful effects of oil pollution on its population. M.Sc Thesis, Univ. of Tel Aviv, Israel, 80 pp.

Slobodkin, L.B., and H.L. Sanders, 1969. On the contribution of en-
        vironmental predictability to species diversity. Brookhaven
        Symposia in Biology. Diversity and Stability in Ecological
        Systems, 22:82-95.
Stephenson, T.A., and A. Stephenson, 1933. Growth and asexual re-
        production in corals. Sci. Rep., Great Barrier Reef Exped-
        ition, 3:167-217.
Wilson, E.O., and W.H. Bossert, 1971. A Primer of Population Biology.
        Sinauer Associates, Stanford, Connecticut, 192 pp.

INTRASPECIFIC VARIABILITY OF ZOOPLANKTON FEEDING IN THE HERMATYPIC CORAL MONTASTREA CAVERNOSA

Howard R. Lasker

Department of the Geophysical Sciences
University of Chicago
Chicago, Illinois 60637

INTRODUCTION

Zooplankton feeding is a readily observed characteristic of many species of corals, but the importance of this mode of feeding has yet to be determined. The present study reports on the effects of intraspecific variability on zooplankton feeding in the reef coral Montastrea cavernosa and discusses the influence of this variability on the ecologic significance of planktivory.

The relative importance of zooplankton feeding is often discussed in terms of dietary input (Porter, 1974; Johannes and Tepley, 1974). Other approaches can be taken, however, which may be equally important in understanding the significance of this mode of nutrition. Independent of the question of dietary importance is the question of whether or not corals differ in their ability to utilize this food resource. The existence of a polymorphism in a single species of coral provides an opportunity to examine this question in populations which appear to be morphologically and ecologically distinct. If differences in patterns of zooplankton feeding can be demonstrated in a single species then such differences are probably widespread between species.

M. cavernosa forms large, massive colonies on reefs throughout the Caribbean. It captures prey through both active tentacular capture and mucus entrapment (Price and Lewis, 1975). Porter (1974) has demonstrated that feeding in this species is readily quantifiable.

Of particular interest to this study is the striking intraspecific variability found in M. cavernosa. In Panama the species may
be divided into at least two morphs (Lehman and Porter, 1972). I
am engaged in a detailed study of this polymorphism, and tentative
morph definitions similar to those used by Lehman and Porter are used
at this time. The diurnal morph is made up of colonies whose polyps
are continuously expanded, have calice diameters less than 6 mm and
are found in waters 10 m or less in depth. The nocturnal morph is
composed of colonies whose polyps are nocturnally expanded, have
diameters greater than 6 mm, and are usually found in waters deeper
than 10 m. The principal characteristics of the morphs are summarized in Table 1.

Other characters vary between the morphs and two of these are
pertinent to the study of zooplankton feeding. First, the nocturnal
morph has substantially longer tentacles than the diurnal morph, and
secondly it exhibits greater sensitivity to potential prey items
(Lehman and Porter, 1972).

The two morphs of M. cavernosa feed in much the same manner,
but differences in tentacle lengths and in reaction to prey items
suggest the existence of quantitative differences in feeding. This
possibility was tested in three series of experiments which determined feeding rates under conditions of varying time of day, varying
prey density and varying time spent feeding. These results are compared with data from field populations.

METHODS

Corals for which feeding measurements were to be made were
starved in filtered sea water for 24 hr prior to the experiment.
Approximately 2-4 hr before the start of the experiment the coral
was transferred to a 1.5 or 10 1 aquarium. The size of the aquarium
used was determined by both colony size and the desired prey density.
Day experiments were conducted between 1000 and 1500 hr and night
experiments between 2100 and 0100 hr. The exact time of the

Table 1. Defining characteristics of the diurnal and nocturnal
morphs of Montastrea cavernosa in Panama.

Character	Morph	
	diurnal	nocturnal
daytime polyp expansion	yes	no
polyp diameter	<6 mm	>6 mm
polyp density	high	low
tentacle length	short	long
night-time polyp expansion	yes	yes

experiment was determined by the expansion of the polyps. Experiments were not started until the majority of the polyps were fully
expanded. In some cases the same colony was used for several experiments. These colonies were allowed to recover for approximately one
week between experiments. This recovery period was sufficient for a
complete return to normal behavior. Comparisons of feeding rates of
the "fresh" and "used" colonies suggested no alteration in feeding
due to experimental use.

Once full expansion was reached a known volume of Artemia nauplii were added to the aquaria (volumetric determination of Artemia
after Slobodkin, 1964). Air bubbled continuously through the aquaria
established currents which uniformly distributed the nauplii. The
duration of the feeding period was set at 10, 30, or 60 minutes.
The colony was monitored at regular intervals, and the experiment
was halted and the results discarded if the colony exhibited unusual
behaviors not characteristically associated with feeding. At the
termination of the experiment the colony was rinsed with sea water.
This operation removed all Artemia from the colony surface.

Feeding rates were determined by counting the number of nauplii
in the guts of 24 polyps. Gut samples were collected and analyzed following a procedure slightly modified from that of Porter (1974). Each
sample was collected with a wide bore (13 g) needle and syringe inserted through the stomodaeum of the contracted polyp. Approximately
1 ml of fluid was removed from each polyp. The gut extract was stained
with a drop each of Neutral Red (1%) and Alcian Blue (1%). All samples were immediately preserved with 5-10% formalin. After staining
for 16 (+/- 4) hr the samples were filtered through 24 mm glass fiber
filters (Millipore type AP20 prefilters). These were mounted on
slides with polyvinyl lactophenol and scanned for Artemia at 90x.

Field samples were collected as in Porter (1974) and were
treated like laboratory samples. Glass fiber and 8 μm Millipore
filters were used for field samples.

                          RESULTS

In the first set of experiments (day/night experiments) colonies
were exposed to 0.04 ml Artemia/l sea water for a period of 10 minutes. The data from these experiments may be broken into four classes
or treatments according to the morph and the time of day (Table 2).
A nested analysis of variance (ANOVA) shows that the treatments result
in significantly different feeding rates ($p < 0.01$). The analysis indicates that the night-time feeding rate of the nocturnal morph is significantly greater than that of the diurnal morph during both the day
($p < 0.05$) and night ($p < 0.01$). There is no significant difference in the
day and night feeding rates of the diurnal morph, but significant
differences are found between the day and night feeding rates of the
nocturnal morph.

Table 2.   Results of the Day/Night comparisons. The first three
colonies are nocturnal and the last three diurnal. Tentacle rankings are in order of increasing tentacle length. PB-107 exhibited
nocturnal morph tentacle lengths during the night and shorter diurnal morph lengths during the day. Experiment numbers indicate the
order in which the experiments were conducted.

		DAY			NIGHT	
Coral	Exp. No.	Avg. No. Artemia ingested per polyp	Rank of tentacle length	Exp. No.	Avg. No. Artemia ingested per polyp	Rank of tentacle length
PB-110	31	0.04	1	32	1.50	12
PB-102	34	0.08	2.5	17	2.30	11
OC-103	37	0.04	2.5	12	1.10	9
PB-107	27	0.58	7	29	2.30	13
OC-106	35	0.50	10			
OC-108	16	0.75	7	33	0.38	7
OC-105	36	0.17	4	26	0.54	5

     The individuals used in this first set of experiments had ten-
tacle lengths which spanned the range found in M. cavernosa. It is
of interest, therefore, to examine the effects of tentacle length on
feeding. In Table 2 the corals used in the experiments are ranked
according to their tentacle lengths. A Spearman rank correlation of
those data with feeding success indicates that the number of ingested prey is significantly correlated with tentacle length (r_s=
0.89, $p<0.01$). As suspected, the colonies whose polyps had long
tentacles ingested the greatest number of prey.

     The second set of experiments (prey density experiments) com-
pared feeding in the two morphs under differing prey concentrations.
In each experiment the colony being tested was exposed to a measured
volume of nauplii for ten minutes. The results of the 21 experiments
are shown in Figure 1. A Spearman rank correlation of these data
indicates that prey density is significantly correlated with uptake
for both the nocturnal ($p<0.01$) and diurnal ($p<0.05$) morphs.

     In this series of experiments the data are grouped into six
treatments based on the two morphs and three ranges of prey density
(low, 0.03-0.1 ml Artemia/l sea water; intermediate, 0.25-0.3; high,
0.5-1.0). As in the day/night comparisons one finds significant differences which show that the nocturnal morph (when it feeds) is the
superior predator. It is the case for both morphs that there are no
significant differences between the feeding rates at low and

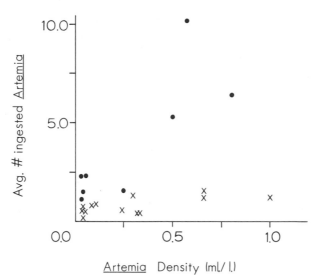

Figure 1. Effects of prey density on zooplankton feeding. Circles, nocturnal morph; crosses, diurnal morph.

intermediate densities. Feeding rates of colonies exposed to high prey densities are significantly greater than those of the low and intermediate densities (ANOVA, $p < 0.01$). When the two morphs are compared, one finds that the nocturnal morph ingests significantly greater numbers of prey at both the low and high prey densities ($p < 0.01$). At the intermediate density no significant difference was found, but this is most likely a result of the small sample size (only a single nocturnal morph experiment at this density).

Observations of corals indicate that individual polyps feed until they become satiated. The possibility remains, therefore, that the two morphs of M. cavernosa become satiated after ingesting equivalent volumes of food, and if presented with prey for a sufficiently long time period both morphs would consume equal numbers of prey. The final series of experiments (feeding time experiments) tests for these long term similarities in feeding rates.

In these experiments colonies were exposed to 0.04 ml Artemia/1 sea water for time periods of 10, 20, and 60 minutes. Observations of colonies made during experiments indicated that the corals rapidly started to feed at the introduction of the nauplii. After 10 minutes large numbers of polyps were observed in "capture posture", a puckered and skewed appearance with all or some of the tentacles withdrawn. After 30 minutes most polyps were fully re-expanded, but few instances of prey capture or of capture posture were noted.

The results of the feeding time experiments depicted in Figure 2 agree closely with these observations. The greatest number of prey in the gut is seen after 10 minutes followed by a decline at 30 and 60 minutes. At 10 minutes the nocturnal morph polyps have significantly greater numbers of prey in their guts than diurnal morph polyps (ANOVA, $p < 0.05$). For both the nocturnal and diurnal morphs average number of prey ingested and the number of polyps with prey have significant negative correlations with time (Spearman rank correlation, $p < 0.05$, and $p < 0.01$ respectively). The decline in gut contents is attributed to a reduction in the number of <u>recognizable</u> prey through digestion. At 30 and 60 minutes no significant differences can be found between the morphs, but the results at ten minutes again suggest that the nocturnal morph is the superior predator.

Laboratory experiments allow one to predict feeding rates in controlled stable environments. In nature, however, the many components kept constant in the lab vary, and this can have a great effect on the results.

Porter's (1974) measurements of zooplankton feeding in <u>M. cavernosa</u> were made in the Isla San Blas on nocturnal morph colonies (6 samples, 540 polyps). I have also collected feeding data on the nocturnal morph in the San Blas (6 samples, 105 polyps) and additional information on the diurnal morph at Galeta (13 samples, 237 polyps). These data, pooled with those of Porter, reveal that when the full range of field variability is dealt with, the feeding

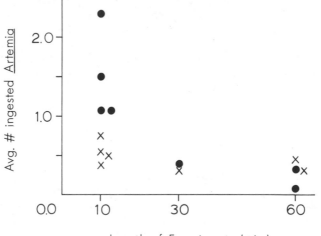

Figure 2.   Effects of feeding time on number of recognizable prey in the gut. Circles, nocturnal morph; crosses, diurnal morph.

differences seen in the lab are obscured. When partitioned into treatments similar to those of the day-night experiments the field data show no significant differences between treatments.

DISCUSSION

Laboratory results demonstrate the superior feeding abilities of the nocturnal morph. Why, therefore, aren't these differences observed in the field data? The lack of statistical significance between treatments of the field data is attributable to a large between-days component of variance. The factors leading to this variance are clearly a feature of major importance in interpreting the field results.

The between-days variance is caused by marked differences in the feeding rates of the same morph on different days. This in turn may be attributable to the variability of plankton abundance. Zooplankton densities are known to vary both temporally and spatially. Herman and Beers (1969) and Glynn (1973) quantified the temporal variation on two Caribbean reefs, but the magnitude of spatial variation in these environments is unknown. Data of Porter (1973) and those of Emery (1968) suggest that spatial variation is an important feature of the reef habitat. Porter (1974) has further speculated that M. cavernosa captures most of its prey during the peaks in plankton abundance which occur shortly after dusk and before dawn. A short sampling time (i.e., feeding period) coupled with spatial heterogeneity would lead to a high variance in the number of plankters supplied to any single colony. The laboratory experiments indicate that prey capture is proportional to prey density, yet Porter (1974) reports that similar relationships were not found when zooplankton were sampled from the general area of the coral being sampled. This further suggests that small scale spatial variability of zooplankton exists. If this variability is sufficiently great the potential differences in the feeding rates of the two morphs would go unrealized in nature. This may account for the anomalous field result.

The feeding time experiments also suggest an explanation for the field data. In those experiments rapid satiation and digestion obscured significant differences in only 30 minutes of feeding. Both Porter's samples and my own were taken during or shortly after the evening peak in zooplankton abundance. If satiation occurred early in the period of increasing abundance, samples taken at the peak prey density would be too thoroughly digested to detect any feeding differences.

The hypothesis of zooplankton variability and of satiation and digestion both account for the available data, but it is not yet possible to distinguish between these alternatives. The data do,

however, provide some guidelines for assessing the importance of zooplankton feeding. The field data indicate that no feeding differences exist between the two morphs. This represents a lower limit
of possible results and one which suggests that zooplankton feeding
plays no differential role in the ecologies of the two morphs.

The laboratory experiments, on the other hand, show that under
certain circumstances feeding differences do exist. Artemia nauplii
are slower than most naturally occurring prey and appear to be captured more readily than naturally occurring zooplankters. The ease
of capture should accentuate any differences which exist between the
morphs. The laboratory experiments, therefore, may indicate an upper
limit to the feeding differences. Given this upper limit it is difficult to suggest for a mechanism by which zooplankton feeding alone
can account for the maintenance of the polymorphism. It is likely
that any model capable of explaining the polymorphism in Montastrea
cavernosa must also include additional factors like autotrophy and
suspension feeding.

This research has been supported by a Grant-in-Aid of Research
from Sigma Xi, the Gurley Fund of the University of Chicago, and by
a summer fellowship from the Smithsonian Tropical Research Institute.
I thank G. Hendler for discussion and aid in the field, T. J. M.
Schopf and R. G. Johnson for comments and criticisms and J. W.
Porter for discussion of the gut sampling technique.

SUMMARY

On the Caribbean coast of Panama colonies of the reef coral
Montastrea cavernosa can be divided into diurnal and nocturnal
morphs. Characteristics of these morphs suggest that differences
exist in their relative abilities to capture zooplankton.

Laboratory experiments compared the abilities of the two morphs
to capture Artemia nauplii. Feeding was compared under conditions
of varying prey density, time of day, and length of exposure to prey.
Analysis of 32 experiments involving a total of 768 gut samples indicate that under all tested circumstances the nocturnal morph is
the superior predator. Different feeding rates of the morphs have
not been found in nature. Temporal and spatial variation of zooplankton is offered as one likely explanation of the field result.
A second explanation assumes that differences in the feeding rates
do exist but are masked by the rapid digestion of ingested prey.

Zooplankton feeding may contribute to the maintenance of the
M. cavernosa polymorphism, but it is incapable of independently
explaining the characteristics of the polymorphism.

REFERENCES

Emery, Alan R., 1968. Preliminary observations on coral reef
    plankton. Limnol. Oceanogr., 13:293-304.
Glynn, P.W., 1973. Ecology of a Caribbean coral reef. The Porites
    reef-flat biotope. Part II. Plankton community with evidence
    for depletion. Mar. Biol., 22: 1-21.
Herman, S.S., and J.R. Beers, 1969. The ecology of inshore plankton
    populations in Bermuda. Part II. Seasonal abundance and com-
    position of zooplankton. Bull. Mar. Sci., 19:483-505.
Johannes, R.E., and L. Tepley, 1974. Examination of feeding of the
    reef coral Porites lobata in situ using time lapse photography.
    Proc. Sec. Int. Coral Reef Symp., 1:127-131.
Porter, J.W., 1973. Biological, physical and historical forces
    structuring coral reef communities on opposite sides of the
    Isthmus of Panama. Ph.D. Thesis, Yale University. 160p.
Porter, J.W., 1974. Zooplankton feeding by the Caribbean reef
    building coral Montastrea cavernosa. Proc. Sec. Int. Coral
    Reef Symp., 1:111-125.
Price, W.S., and J.J. Lewis, 1975. Feeding mechanisms and feeding
    strategies of Atlantic reef corals. Jour. Zool., London,
    176:527-544.
Slobodkin, L.B., 1964. Experimental populations of Hydrida. J.
    Anim. Ecol., 33(Suppl.):131-148.

ACCUMULATION OF DISSOLVED CARBON BY THE SOLITARY CORAL

BALANOPHYLLIA ELEGANS - AN ALTERNATIVE NUTRITIONAL PATHWAY?

Peter V. Fankboner*

Bamfield Marine Station

Bamfield, British Columbia V0R 1B0 CANADA

INTRODUCTION

In classical treatments of their feeding biology, madreporarian corals are viewed as chiefly carnivorous (Hyman, 1940;
Yonge, 1930). However, some recent studies indicate that hard
corals may obtain their nutritional requirements by not one, but
several means including autotrophy, particulate feeding and accumulation of dissolved organic material. For example, coral species
which host endosymbiontic zooxanthellae accrue nutritional benefit
in the form of soluble carbohydrates released by their algal guests
(Lewis and Smith, 1971; Goreau, Goreau and Yonge, 1971). Sorokin
(1972 and 1973) has confirmed that some scleractinian corals feed
upon planktonic algae and bacteria. However, the nutritional role
of dissolved organic matter in corals is less certain than autotrophy and particulate feeding. Studies by Stephens (1962) upon
the hermatypic coral Fungia scutaria have demonstrated that under
laboratory conditions, F. scutaria is capable of accumulating
^{14}C-labelled glucose and amino acids from modest concentrations
made available in seawater. Nonetheless, the significance of
dissolved organic material in the nutrition of corals remains to
be established because, to date, there have been no experiments in
this regard carried out under natural field conditions. The
experiments reported here were undertaken to determine whether the
solitary coral Balanophyllia elegans Verrill will accumulate under
in situ conditions, dissolved carbon exuded by the "large kelp"
Macrocystis integrifolia.

*Permanent address: Department of Biological Sciences,
                     Simon Fraser University,
                     Burnaby, British Columbia V5A 1S6, CANADA

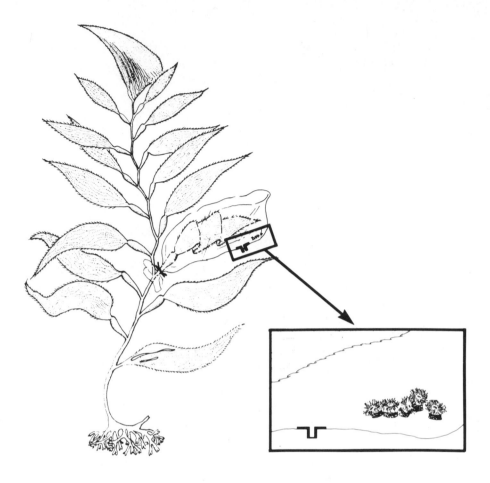

Figure 1. The "large kelp" <u>Macrocystis</u> <u>integrifolia</u> <u>in</u> <u>situ</u>
with the experimental serum-stoppered plastic bag containing
an algal blade plus specimens of the solitary coral <u>Balanophyllia</u>
<u>elegans</u>. The kelp stipe is approximately 3.0 meters in length.

MATERIALS AND METHODS

     The solitary cup coral <u>Balanophyllia</u> <u>elegans</u> was selected for
dissolved carbon accumulation experiments for two reasons. First,
<u>B</u>. <u>elegans</u> lives as a cohabitant with <u>Macrocystis</u> <u>integrifolia</u>
and would thus have access to the kelp's exudated dissolved carbon
under naturally occurring conditions. Second, <u>B</u>. <u>elegans</u> lives as
an ahermatypic predator and this condition allows a single additional pathway for the coral's nutrient acquisition, the taking
up of dissolved matter.

Experiments were conducted via SCUBA at inshore kelp beds adjacent to Bamfield, Vancouver Island, B.C. Five unencrusted, non-necrotic fronds of Macrocystis integrifolia were selected for each experiment and treated as follows:

Single blades from each algal frond were enclosed with 2.0 litres of ambient seawater in a clear, polyethylene bag fitted with a serum cap (Fig. 1). The bags were sealed at the base of the blade's pseumatocyst and 1.0 mCi of $Na_2^{14}CO_3$ (Atomic Energy of Canada) was injected through the serum cap into each bag. The algae were incubated in this labelled medium for 48 hours. Following incubation, the plastic bags were removed, and the blades were washed in fresh seawater to remove sorbed isotope.

For the accumulation experiments each blade was replaced in a new bag containing five adult specimens of Balanophyllia elegans to allow for the heterotrophic transfer of ^{14}C-labelled dissolved carbon, exudated by Macrocystis integrifolia, to the corals.

After an incubation period of 24 hours, blade tissue samples plus the five specimens of Balanophyllia elegans were taken from each bag. The corals were rinsed in fresh seawater, blotted, and weighed. Each coral's soft tissues were solubilized in 2.0 ml NCS Tissue Solubilizer (Amersham/Searle), neutralized with glacial acetic acid, and, after removal of the coral skeleton, were counted in 15.0 ml of Liquifluor toluene cocktail (New England Nuclear). Coral skeletons which had been removed from NCS Tissue Solubilizer, were weighed; skeleton weight was subtracted from earlier measurement of weight of living coral to yield weight of coral soft tissues. Control corals were killed with 0.01 m. solution of NaF in seawater and treated as were experimental corals. Algal blade tissue samples for scintillation counting were solubilized using the methods of Lobban (1974) and each was counted in 10.0 ml of Scintiverse cocktail (Amersham/Searle). All counting results were corrected for background quenching by the external standard method, and converted to disintegrations per minute (DPM).

RESULTS

In all cases (Table 1), dissolved carbon originating from the kelp Macrocystis integrifolia was accumulated in situ by the solitary cup coral Balanophyllia elegans. Moreover, the dissolved carbon was concentrated in the experimental corals by approximately one order of magnitude over the killed controls.

Table 1. ^{14}C activity in blades of Macrocystis integrifolia, soft tissues of experimental and control specimens of Balanophyllia elegans at the termination of 24 hours of incubation in situ. Concentration factors were calculated on the basis of activity per unit wet weight (DPM/mg) of the animal tissues divided by the ambient activity per unit wet weight (DPM/mg) remaining in the algal blades at the end of the experiments.

Sample Type	Sample Number	Activity Accumulation ($\bar{x}$ ± S.D.)	Concentration Factor
Experimental Corals	15	2841.51±1048.35 DPM/mg wet wt.	21.1
Control (dead) Corals	5	406.54± 127.14 DPM/mg wet wt.	3.0
Algal Blades	5	13487.31±3509.0 DPM/mg wet wt.	100.0

DISCUSSION

It is apparent from the results above that the heterotrophic transfer of algal exudated dissolved carbon from a kelp to a coral occurs under in situ conditions. These observations indicate that studies by Stephens (1962) on uptake of glucose and amino acids by Fungia scutaria, and his unpublished results on uptake of glycine by other corals including Acropora sp., Favia speciosa, and Dendrophyllia micranthus, are valid insofar as they demonstrate the ability of these corals to accumulate organic molecules. The conclusions arising from Stephens' observations on D. micranthus may be particularly relevant in this regard because of this coral's close taxonomic relationship to Balanophyllia elegans.

The nutritional significance of dissolved carbon to Balano-phyllia elegans cannot be established until the coral's total energy budget has been determined. However, a wide range of soft bodied marine invertebrates studied to date leak back to the medium dissolved organic carbon (Johannes, Coward and Webb, 1969). Thus, it is evident that any dissolved carbon that the organism is taking up positively affects the net flux of carbon from the organism and the medium. Further, in the specific case of B. elegans, the ability to accumulate dissolved carbon from seawater may be quite decisive to the coral's survival during the winter months when the zooplanktors upon which it normally feeds are either greatly reduced or altogether unavailable.

ACKNOWLEDGMENTS

This research was supported by Operating Grant A6966 from the National Research Council of Canada for which I am most grateful. Thanks are also due to my technician, Mr. Glenn Cota, Dalhousie University, who carried out the field experiments and to Dr. Louis Druehl, who conceived the idea of using the kelp Macrocystis inte-grifolia as an in situ source of ^{14}C-labelled dissolved carbon.

LITERATURE CITED

GOREAU, T. F., GOREAU, N. I., AND C. M. YONGE, 1971. Reef corals: autotrophs or heterotrophs? Biol. Bull., 141: 247-260.

HYMAN, L. H., 1940. The Invertebrates: Protozoa through Ctenophora. Volume I. McGraw-Hill Company, New York. 696 pp.

JOHANNES, R. E., COWARD, S. J., AND K. L. WEBB, 1969. Are dis-solved amino acids an energy source for marine invertebrates? Comp. Biochem. Physiol., 29: 283-288.

LEWIS, D. H., AND D. C. SMITH, 1971. The autotrophic nutrition
    of symbiotic marine coelenterates with special reference to
    hermatypic corals. Proc. Roy. Soc. London Series B, 178:
    111-129.
LOBBAN, C. W., 1974. A simple, rapid method of solubilizing
    algal tissue for scintillation counting. Limnol. Oceanogr.,
    19: 356-359.
SOROKIN, YU. I. 1972. Bacteria as food for coral reef fauna.
    Oceanology, USSR, 1: 169-177.
SOROKIN, YU. I., 1973. On the feeding of some scleractinian
    corals with bacteria and dissolved organic matter. Limnol.
    Oceanogr., 18: 380-385.
STEPHENS, G. C., 1962. Uptake of organic material by aquatic
    invertebrates. 1. Uptake of glucose by the solitary coral,
    Fungia scutaria. Biol. Bull., 123: 648-659.
YONGE, C. M., 1930. Physiology of reef corals: feeding mechanisms
    and food. Scient. Rep. Gt. Barrier Reef Exped., 1: 13-57.

REGULATION OF FREQUENCY OF PEDAL LACERATION IN A SEA ANEMONE

Nathan Smith III and Howard M. Lenhoff

Department of Developmental and Cell Biology

University of California, Irvine, California 92717

Many acontiate sea anemones (Subtribe Acontiaria) undergo pedal laceration as a primary method of asexual reproduction (Stephenson, 1920 and 1929). Pedal laceration consists of a radial spreading of the pedal disc followed by the separation of a more or less complete ring of tissue derived from the peripheral margin of the disc. This ring subsequently fragments into two to eight pieces of tissue, each of which develops in about two weeks into a small anemone capable of feeding.

Atoda (1954a, 1954b, 1955, and 1960) reported that pedal lacerates and experimentally excised pedal disc tissues gave rise to morphologically atypical animals. MacGinitie and MacGinitie (1949) stated that Sagartia undergo pedal laceration at a higher rate when fed frequently, but gave no data and details of their experiments. In this paper, using a clone of an unidentified sea anemone, we show that the frequency of pedal laceration is inversely related to the frequency of feeding, and that periods of fasting stimulate this form of asexual reproduction.

DESCRIPTION AND CULTURE OF THE ANEMONES

The anemones used in this work are members of a clone derived from one individual found attached to a floating Sargassum weed in Biscayne Bay, Florida, in 1966. They are approximately 1.0 by 0.2 cm when full size, pink in color, do not contain endosymbiotic zooxanthellae, and have about 6 short acontia in their cavities. They do not fit the description of any other known acontiate anemones. We have been maintaining a clone of animals for over nine years,

and will be pleased to send live specimens on request. The animals
are extremely hardy and are easy to rear. They were cultured in
19 x 30 cm Pyrex baking dishes containing either filtered sea water
or Instant Ocean (Aquarium Supplies, Inc., Wycliff, Ohio), and were
fed on freshly hatched Artemia salina nauplii.

EXPERIMENTAL PROCEDURES AND TERMINOLOGY

     Experimental anemones were removed from the culture trays and
divided into six groups of 20 anemones each. Subgroups of five
anemones each were placed into plastic Petri dishes 8.5 cm in diameter (volume 75 ml). All experimental anemones were fed every day
for seven days prior to instituting the feeding regimens described
below. Feeding was performed at the same time each day (0900) by
introducing an excess of Artemia nauplii into the individual dishes.
After 15 minutes, the water and free Artemia were removed and fresh
sea water was added. Mucus and adhering debris were removed with
a rubber policeman daily. Next, each anemone was examined to see
if it had eaten and if it had lacerated its pedal disc. The number
of fragments of each lacerate was recorded. Data was kept on each
individual anemone.

     In this paper, we use the following terminology: Pedal lacera-
tion refers to the process of asexual reproduction whereby (a) either
the complete outside rim of the flattened pedal disc, or parts of it,
detach from the rest of the anemone; (b) the detached rim fragments
into two or more smaller pieces; and (c) each fragment regenerates
into a fully developed small sea animal. Laceration event refers to
step (a) above. A lacerate is defined as the detached rim, or piece
or rim that exists before step (b) occurs. The term fragments refers
to the pieces formed in step (b).

     In order to help analyze data on laceration events that were
collected in periods between feedings, we found the following terms
useful: (a) 0-24 hour interval: the 24 hours immediately following
a meal. (b) 24-48 hour interval: a period of non-feeding that immediately followed a 0-24 hour interval. As a consequence, obviously
the animals in Group I, i.e. animals fed once each day, could not
have had a 24-48 hour interval. (c) 48-72 hour interval: the period
of non-feeding immediately following a combined 0-24 and 24-48 hour
interval. Accordingly, the animals in Groups I and II had no 24-48
hour interval. (d) 72-96 hour interval: the same kind of reasoning
applies to this period. (e) 96-120 interval: this interval occurred
only in one instance--the Group VIa animals referred to in Fig. 1
and Fig. 4.

RESULTS AND DISCUSSION

Effect of Graded Feeding Regimens on Fragment Production

In order to quantify our casual observation that fasted animals reproduce by pedal laceration more frequently than fed animals, we placed four groups of twenty animals on a series of graded feeding regimens as follows: All animals are fed daily for eight days at the beginning of the experiment. After day eight Groups I, II, III, and IV are fed on alternate 1st, 2nd, 3rd, and 4th days respectively.

In Fig. 1, we plot the average number of _fragments_ formed per animal in the various groups. The circles in the graphs indicate the days during which the animals were offered food. Casual inspection of the raw data for Group I, II, III, and IV (Fig. 1) indicates that the process of pedal laceration occurs more often in periods of absence of food. There might appear to be two inconsistencies to this generalization as seen in the increase in fragment formation in Group I at about day 27, and in Group II at about day 10. To the contrary, not shown in the graphs are the observations that in all our experiments, there were two instances in which most sea anemones had retracted tentacles and did not eat even when offered food. These instances were in Group I, when most anemones did not eat on day 26, and in Group II, when on day 7, all the anemones did not eat the nauplii offered to them. (Hence, Group II animals went without food on days 5, 6, and 7.) Furthermore, it was noted that in Group I on day 26, pedal laceration occurred only in those animals that did not eat.

In order to take into account these unexpected behavioral responses seen in Groups I and II, the data for Groups I, II, III, and IV were set at 0 at day 13, and normalized thereafter (Fig. 2). Also, because the formation of fragments occurred after the rim of the pedal disc separated, we record here the initial observable stage of pedal laceration, the _laceration events_. The results, shown in Fig. 2, show clearly that the onset of laceration events follows a period of non-feeding. Even in Group I, the laceration events observed followed the single instance when most anemones in that group did not eat food offered to them.

Effect of Random and/or Irregular Feeding Regimens

Two other feeding regimens (Groups V and VI) were set up to test further the conclusion that periods of fasting lead to pedal lacerations. As with the animals in Groups I-IV, these were also fed for eight days at the beginning of the experiment. Next, the animals in Group V were fed daily for seven days, and then fasted for three days; this cycle was repeated another one and a half

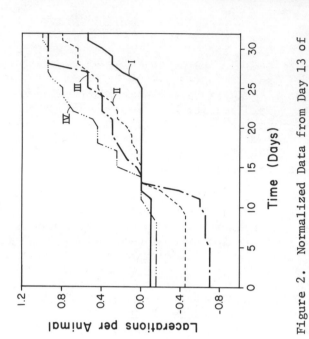

Figure 2. Normalized Data from Day 13 of Laceration Events in Animals Fed on Feeding Regimens I-IV as described in the text.

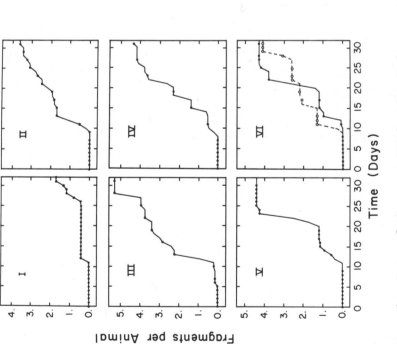
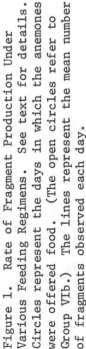

Figure 1. Rate of Fragment Production Under Various Feeding Regimens. See text for details. Circles represent the days in which the anemones were offered food. (The open circles refer to Group VIb.) The lines represent the mean number of fragments observed each day.

times. Note (Fig. 1) that after days 11 and 20, fragments were
formed. The animals in Group VI were divided into two subgroups
of ten animals each. After day eight, the animals in each subgroup
were fed on a schedule determined from tables of random numbers.
Even in these cases, it is apparent from the data (Fig. 1) that the
fragments began to form only after a period in which there was no
feeding.

           Sequence of Appearance of Laceration Events and Fragments
                        After Periods of Fasting

      Here we plot (Fig. 3) the average number of individuals that
lacerated and formed fragments during daily <u>intervals</u> following
ingestion of a meal (see preceding pages for precise terminology).
Examination of the data indicates that the largest number of laceration events and formation of fragments occurred during the 48-72
and 72-96 hour intervals after feeding.

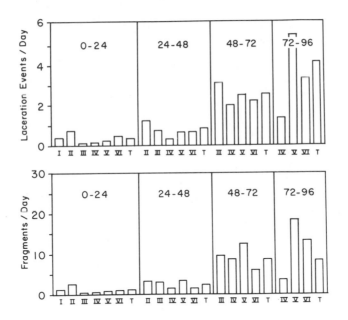

Figure 3. Appearance of Laceration Events and Fragments During Different 24 Hour Intervals of Fasting After a Meal. The Roman numerals
under the bars indicate the experimental group of anemones; the time
intervals (see text) are indicated by the Arabic numbers 0-24, 24-48,
etc.; T is described under Results. Not shown in this figure is one
96-120 period for Group VIa between days 17 and 22 because this one
instance cannot be compared with any other data.

The relatively low rate of laceration and fragmentation observed in the 72–96 interval for the Group IV animals (compared to the animals in Groups V and VI) might be explained by the fact that the animals in Groups V and VI were fed on a more regular basis in between those intervals they were without food. For example, animals in Group V were fed for seven consecutive days between four-day intervals, whereas those in Group IV were fed only every fourth day. However, Group IV animals may have had a smaller mass of tissue to lacerate. Such apparent discrepancies can be eliminated by averaging the data for each interval, and such an average is represented by the bars labeled T, for "total." This bar allows us to focus on the overall interval regardless of the feeding schedule.

Is Pedal Laceration a Random Event?

To answer this question, it is necessary to analyze for each experimental group of animals the percent of the laceration events occurring within each time interval of fasting, and then to compare these percentages with the calculated number of laceration events if pedal laceration occurred at random. Such an analysis is given in Fig. 4.

The dotted horizontal lines refer to percent laceration events if this phenomenon was a random event. For example, in Group III the events could occur in either of the three intervals; therefore, if pedal laceration was a random event, there is a 33-1/3% chance (dotted horizontal line) that the laceration events could take place in any of the three time intervals. The same reasoning is applied to the analysis of the data from each group.

From examination of the data, it can be seen that pedal laceration events usually occur in the time intervals farthest from the last feeding, and that these events do not occur at random. Nearly identical patterns were obtained by analyzing the number of fragments formed per interval.

Relationship Between Number of Fragments and Feeding Regimen

The number of fragments forming from the lacerates produced during the various feeding regimens were as follows for each Group: I = 2.6; II = 2.4; III = 3.1; IV = 3.8; V = 4.2; and VI = 3.3. (In all cases the standard error of the mean was about 0.3.) Two patterns can be noticed. First, it appears that lacerates from Groups III, IV, V form more fragments than do lacerates from Groups I and II; this somewhat greater appearance of fragments may be related to the fact that the animals in Groups III, IV and V had longer periods between feedings than did animals from Groups I and II. Second, the

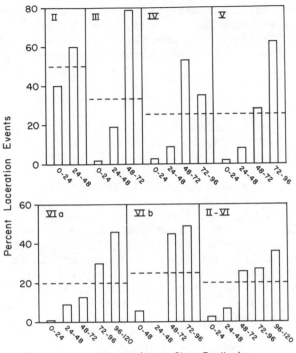

Figure 4. Percentage of Laceration Events Occurring in Each Interval as Compared to Random Phenomena. The percent laceration events refers to the number of laceration events occurring within a given interval times 100, divided by the total number of laceration events for a given group of experimental animals. Note that because in Group VIb no laceration events took place within the 24-48 hour interval, only three intervals were used in calculating the horizontal dotted line representing the random events.

most fragments are produced by lacerates from the Group V animals. Possibly the seven day feeding periods preceding the three day fasting periods resulted in larger animals and, hence, in larger lacerates which could give rise to more fragments.

Another influence on fragment production can be seen by comparing the feeding regimens of Groups II and VI. Because both groups were fed on the average of every other day, it might be assumed that the amount of food consumed by each group was nearly equal. The animals in Group VI, however, were fed on the random feeding schedule. This schedule led to longer intervals between many of these feedings (Fig. 1) and, hence, to more fragments per pedal lacerate than was formed by lacerates from Group II.

From these series of observations it might be concluded that
the same factors enhancing the production of pedal lacerates during
periods of prolonged fasting may also play a role in promoting a
further subdivision of the lacerates into fragments.

SUMMARY AND GENERAL DISCUSSION

The foregoing experiments show that in the acontiate sea anemone described in this paper: (a) the frequency of pedal laceration
is inversely related to the frequency of feeding; (b) periods of
fasting stimulate pedal laceration; and (c) the laceration events
usually occur in the time intervals farthest from the last feeding.

These results are of interest from a number of standpoints.
Developmentally, we have an intriguing situation whereby a developmental event is initiated (or "inhibited") by controlling the level
of food intake of the organism. At the moment it is not possible
to verify possible explanations of these events by such phenomena
as production of an inhibitor (or stimulator), mobilization of energy reserves, or regulation of the cell cycle of the participating
cells. But we do feel that our system, because of the experimental
ease with which it can be handled, offers promise for the study of
some developmental control mechanisms.

From the standpoint of the adaptive value of such control of
pedal laceration, we have a number of interesting possibilities.
Although on first thought it would appear paradoxical that asexual
reproduction of a species is stimulated by the absence of food, on
second thought it would seem very logical for survival of the species. That is, if food is limiting it would be advantageous to the
anemone to increase the number of "mouths" available to capture
nutriment from the sparse food supply. Furthermore, we have observed, for example, that following pedal laceration, both the
parent anemone and the regenerated fragments move from their original location; by dispersing in this manner, possibly these new
mouths can cover a larger territory.

A very attractive explanation for analogous events occurring
in another sea anemone is being put forth by Kenneth P. Sebens of
the University of Washington (Sebens, unpublished doctoral thesis).
While studying factors affecting longitudinal fission in Anthopleura xanthogrammica in both the laboratory and in the field,
Sebens observed that in a two month period more than 10 times the
number of fasted animals divided than did fed animals.

But even more convincing was his analysis of the relative
abilities of the "smaller mouths" to capture food compared to the
mouths of larger animals. He showed first that the relative surface area of the oral discs of smaller anemones to body weight was

much greater than that of the larger anemones. And, second, he
showed that the efficiency of prey capture was greater per surface
area of the oral disc in the smaller anemones. Sebens concludes
that "longitudinal fission increases the energy/time intake for the
'individual' (=clone) for [food] particles which can be captured by
any size of anemone." Likewise, Seben's conclusions should also
apply to the pedal lacerating anemone described herein. That is to
say, the more mouths a population of a single species has, the
greater are its chances for surviving when food is limiting.

LITERATURE CITED

Atoda, K., 1954a. The development of the sea anemone; _Diadumene
    luciae_, reproduced by the pedal laceration. Sci. Rep. Tohoku
    Univ., 4th Ser. (Biol.) 20: 123-129.
Atoda, K., 1954b. The development of the sea anemone, _D. luciae_.
    II. The individuals originated from the fragments without
    stripes by artificial laceration. Sci. Rep. Tohoku Univ.,
    4th Ser. (Biol.) 20: 362-369.
Atoda, K., 1955. The development of the sea anemone _D. luciae_.
    III. The individuals which originate from the fragments with
    one stripe by pedal laceration. Sci. Rep. Tohoku Univ., 4th
    Ser. (Biol.) 2: 89-95.
Atoda, K., 1960. The development of the sea anemone, _P. luciae_.
    IV. The correlation between the form of laceration between
    the forms of laceration pieces and those of their regene-
    rates. Bull. Mar. Biol. Sta. Asamushi, 10: 87-94.
MacGinitie, G.E. and MacGinitie, N., 1949. Natural History of
    Marine Animals. McGraw-Hill, New York: 139-140.
Stephenson, T.A., 1920. On the Classification of Actiniaria.
    Quart. J. Micros. Sci., 64: 421-574.
Stephenson, T.A., 1929. On methods of reproduction as specific
    characters in sea anemones. J. Mar. Biol. Assn. U.K., 16:
    131-172.

THE ACRORHAGIAL RESPONSE IN <u>ANTHOPLEURA</u> <u>KREBSI</u>:

INTRASPECIFIC AND INTERSPECIFIC RECOGNITION

Charles H. Bigger

Dept. of Biological Science, Florida State University

Tallahassee, Florida 32306

The acrorhagi (marginal sperules) of some sea anemones of the family Actiniidae are normally inconspicuous vesicles at the base of the tentacles, bearing atrich nematocysts (usually some basitrich nematocysts and spirocysts present), which have been considered a major characteristic of acrorhagi (Carlgren, 1949; Abel, 1954; Bonnin, 1964; and Francis, 1973b).

In studies with <u>Actinia equina</u>, Abel (1954) was the first to describe the acrorhagial response. All investigators (Abel, 1954; Bonnin, 1964; Francis, 1973b; unpublished personal observations) have presented the same general description of an acrorhagial response. Contact of an acrorhagi-bearing anemone with some con-specifics or certain other species causes the acrorhagi in the area of contact to swell and elongate markedly. The expanded acrorhagi are placed on the other animal, withdrawn, and then the application process repeated. Pieces of acrorhagial ectoderm (acrorhagial peel) often remain on the target animal (an animal which elicits acrorhagi application behavior or peeling) and cause a localized necrosis.

Bonnin (1954) examined the initiation of the acrorhagial response in <u>A. equina</u> and Francis (1973b) investigated it primarily in <u>Anthopleura elegantissima</u>. Both investigators showed the response had specificity in that it was only elicited by conspecifics and other zoantharians, not by hydrozoans, non-coelenterates, or various physical and chemical stimuli. Francis found that a 1 cm separation between anemones would prevent initiation of the response. Bonnin suggested the nematocysts of the target anemone served as the elicit-ing factor for the acrorhagial response.

This report will focus on the target recognition aspects of the acrorhagial response in Anthopleura krebsi. In particular, preliminary observations and experiments dealing with acrorhagial response components, their receptors, and eliciting factor(s) will be discussed.

MATERIALS AND METHODS

The Anthopleura krebsi (Duchassaing and Michelotti, 1860) used in this study were collected intertidally from rock groins at Anna Maria Island, Florida. The anemones were found in aggregations of contacting individuals, which will be referred to as groupmates. In the laboratory, A. krebsi commonly reproduce asexually and, although groupmates may in fact be members of the same clone, the term clonemate will be reserved for those anemones known by direct observation to have resulted from asexual reproduction.

The A. krebsi were collected as naturally occurring groups and transported to Florida State University where they were maintained by group, individually, or as clones on bivalve shells in 80 x 100 mm culture dishes. Each dish was individually aerated and the natural sea water was changed every other day. The dishes were illuminated by Gro-lux fluorescent lights. The anemones were fed Artemia nauplii and pieces of shrimp except for three days prior to and during an experiment.

With the exception of the Artemia (raised from San Francisco Bay Brand "eggs"), the other animals were collected from various areas of the Florida Gulf Coast and Keys. The Cassiopea scyphistomae (polyps) were from the same general region as the Cassiopea xamachana medusae (Key Largo) and fit the description of C. xamachana scyphistomae (Bigelow, 1900); however, I have not raised medusae from the scyphistomae to confirm the identification.

To arrange behavioral interactions, anemones attached to bivalve shells were placed together in the same dish. After re-expansion, the experimental animals were moved into contact. In previous observations of A. krebsi intraspecific interactions (unpublished), 98% of the acrorhagial responses were initiated within 15 minutes; therefore, behavioral interactions were observed for at least one hour.

All A. krebsi interactions were viewed through a Wild M5D stereomicroscope. In all trials, approximately the same lighting conditions were maintained. The water was replaced with aerated sea water after each trial.

A one second contact technique was used to provide more control over the eliciting stimuli. For this technique, an object (excised tentacle, portion of intact animal, glass coverslip, etc.) was lightly touched to an area of an anemone for about one second every thirty seconds for 15 minutes. The number of contacts required to elicit a response was considered the threshold.

Anthopleura krebsi and Bunodosoma cavernata acrorhagi and peels were examined with phase contrast light microscopy and scanning electron microscopy (SEM) using techniques described in Mariscal and Bigger (this volume).

RESULTS

The light and SEM study of the acrorhagi and peels of A. krebsi and B. cavernata revealed anvil-shaped spines arranged in three spiral rows (visible with the light microscope) on the threads of the nematocysts previously called atrichs by Carlgren and Hedgpeth (1952). These nematocysts, also seen by Schmidt (1969) in the acrorhagi of other Anthopleura, should more properly be called holotrichous isorhizas (holotrichs) following the terminology of Mariscal (1974).

The acrorhagial response of A. krebsi was not elicited by a groupmate or clonemate. During fifty one hour observations of groupmate interactions, no acrorhagial responses occurred. No acrorhagial responses have been seen between groupmates or clonemates living in contact for more than a year, during which time they did respond to non-groupmates.

To see if groupmates required constant interaction to maintain protection from an acrorhagial response, six sets of two groupmates were isolated in separate dishes and then put back in contact (Table I). The tentacles of the anemones interlaced, but the anemones showed no behavioral response.

Table I. The response of A. krebsi groupmates in contact after an extended isolation. -, no behavioral change from the behavior of a solitary anemone.

Trial	Time isolated (days)	Time in contact after isolation (min)	Response
1	20	96	-
2	47	390	-
3	47	390	-
4	41	60	-
5	41	60	-
6	206	70	-

Interactions between A. krebsi and twenty-one other species were observed to determine if the acrorhagial response of A. krebsi was used interspecifically. Representatives of all three coelenterate classes were placed in contact with three A. krebsi from different groups for one hour periods (Table II). Contact with the hydroid elicited a slight acrorhagial expansion and application behavior with no peeling from one A. krebsi and no response from the other two A. krebsi.

Not all anthozoans elicited an acrorhagial response. Although all the Actiniidae and some other anemones elicited an acrorhagial response, three anemones (Lebrunia danae, Bartholomea annulata, and Bunodeopsis globulifera) elicited little or no response. Excised tentacles of B. annulata (applied with 1 sec contact tech.) also failed to elicit an acrorhagial response from four A. krebsi of different groups. Although the ceriantharian, madreporarian, and pennatulacean elicited an acrorhagial response, the gorgonian did not.

The scyphozoan medusae and scyphistomae (Cassiopea) elicited different responses from A. krebsi. Cassiopea medusae elicited no behavioral response from A. krebsi. However, the scyphistomae provided the first known case of a non-anthozoan eliciting an acrorhagial response. Cassiopea medusae and scyphistomae were later examined in a light and SEM study (Mariscal and Bigger, this volume) where the same nematocysts were seen in both the medusae and the scyphistomae.

The interactions of A. krebsi (same three A. krebsi as in other interspecific interactions) with the stone crab Menippe mercenaria, the nudibranch Spurilla neapolitana, the snail Calliostoma jujubinum, and Artemia were examined. None of the four was known to interact with A. krebsi in the field but M. mercenaria was among the rocks where the A. krebsi was collected.

In the laboratory, S. neapolitana and M. mercenaria ate A. krebsi and could be considered to be predators. A. krebsi ate Artemia , nauplii and adults, so Artemia could be considered prey. None of the four species elicited an acrorhagial response nor were acrorhagial responses seen during the routine laboratory feeding.

To see if the factor eliciting the acrorhagial response was released in a water soluble form by the target animal, three sets of A. krebsi were separated by a permeable barrier (#1 Whatman filter paper). The tentacles and column of both animals touched opposing sides of the barrier and water soluble material could pass through or over the barrier. There was no acrorhagial response in this case (Table III). When the barrier was removed, the tentacles of the anemones immediately touched and an acrorhagial response followed.

Table II. The interspecific nature of the A. krebsi acrorhagial
response is shown by the interactions of three A. krebsi with other
coelenterates for 60 minute periods. +, the response in question
was elicited; -, the response in question was not elicited; 0, no
data; *, several contacts with an acrorhagus required to produce
a peel.

Coelenterate in contact with A. krebsi	A. krebsi response		
	acrorhagi expansion	application behavior	acrorhagial peel
Anthozoa			
Bunodeopsis globulifera	--0	--0	--0
Lebrunia danae	+--	---	---
Bunodosoma cavernata	+++	+++	+++
Anemonia sargassensis	+++	+++	+++
Condylactis gigantea	+++	+++	+++
Bunodactis texaensis	+++	+++	+++
Phymanthus rapiformis	++-	++-	++-
Calliactis tricolor	+--	+--	+--
Aiptasia pallida			
(pedal disc dia 12 mm)	+++	+--	+**
(pedal disc dia 2 mm)	++-	++-	---
Bartholomea annulata	+--	+--	---
Aiptasiomorpha texaensis	+--	---	+--
Ceriantheopsis americanus	+--	+--	+--
Astrangia danae	+--	+--	+--
Renilla mulleri	++-	++-	*--
Leptogorgia virgulata	---	---	---
Scyphozoa			
Cassiopea xamachana			
medusa	---	---	---
scyphistoma	++-	++-	++-
Hydrozoa			
Podocoryne carnea (selena)	+--	+--	---

Table III. The responses of A. krebsi both before and after removal
of a filter paper barrier between them. +, an acrorhagial response
occurred; -, an acrorhagial response did not occur; 0, no data.

	Barrier in place		Barrier removed	
Trial	Acrorhagial response	Time observed (hr)	Acrorhagial response	Time to response (sec)
1	-	12.5	0	0
2	-	6.7	+	110
3	-	6	+	100

The epidermal mucus of A. krebsi would not elicit an acrorhagial response. Strands of older mucus and swabs covered with fresh mucus from the tentacles and oral disc of an A. krebsi were applied with the 1 sec contact technique to a conspecific that had previously responded to the tentacles of that anemone. In six trials of each, neither old nor fresh mucus elicited an acrorhagial response. Following each trial, the tentacles of the mucus-supplying anemone (1 sec contact tech.) elicited an acrorhagial response.

The location of the factor(s) in A. krebsi and A. sargassensis responsible for eliciting the acrorhagial response was investigated by stimulating (1 sec contact tech.) A. krebsi with portions of each of these anemones. In four trials each, stimulating A. krebsi (tentacles for acrorhagial expansion and application and expanded acrorhagi for peeling) with the pedal disc, column, oral disc, and tentacle of intact A. sargassensis and pedal disc, column and tentacle of intact A. krebsi produced an acrorhagial response.

Excised Condylactis gigantea tentacles were turned inside out and applied to A. krebsi tentacles and expanded acrorhagi (1 sec contact tech.). The endoderm (1 cm from the severed end) elicited an acrorhagial response from four A. krebsi of different groups. Although requiring more than one contact, Aiptasia pallida acontia also elicited acrorhagial peeling from A. krebsi on two occasions.

Excised tentacles of A. sargassensis were used (1 sec contact tech.) to test the lower column, oral disc, acrorhagi, and tentacles of four A. krebsi for the location of the acrorhagial response receptors (Table IV). No response was elicited by stimulating the lower column or oral disc. As a control, an acrorhagial response was obtained by using the same excised tentacle to stimulate the

Table IV. The location of the acrorhagial response receptors as determined by stimulating portions of A. krebsi with an excised A. sargassensis tentacle (1 sec contact tech.). The results are listed as the average number of contacts required to elicit the response for four trials. -, the response in question was not elicited.

A. krebsi response	Oral disc	Lower $\frac{1}{4}$ column	Acrorhagus	Tentacle
Average threshold acrorhagi expansion	-	-	8	6
Average threshold application behavior	-	-	10	6
Average threshold acrorhagial peeling	-	-	9	-

Table V. The specificity of acrorhagial peeling was demonstrated by the response of expanded A. krebsi acrorhagi to contact with various objects. +, peel elicited; -, peel not elicited; 0, no data.

Stimulus (in sequential order)	Acrorhagial peel elicited
1. Tentacle of intact A. sargassensis	+ (11)
2. Tentacle of intact groupmate	- (11)
3. Clean coverslip	- (11)
4. Artemia extract coated coverslip	- (11)
5. 2nd presentation of A. sargassensis	+ (11)
6. Tentacle of intact C. gigantea	+ (2), 0 (9)

A. krebsi tentacles. Stimulating the tentacles and acrorhagi elicited acrorhagial responses.

To test the specificity of the acrorhagial response, various objects were applied (1 sec contact tech.) to A. krebsi tentacles. In ten trials with anemones from different groups, clean glass coverslips, Artemia extract coated coverslips (which elicited nematocyst discharge in tentacles), a metal probe, and excised groupmate tentacles all failed to elicit acrorhagi expansion or application. Following each failure, an excised A. sargassensis tentacle elicited an acrorhagial response from the A. krebsi.

To demonstrate the specificity of acrorhagial peeling, an A. sargassensis tentacle was used to elicit acrorhagial expansion from A. krebsi; then, following an acrorhagial peel, the A. sargassensis was removed. An A. krebsi groupmate, a glass coverslip, and an Artemia extract coated coverslip were each touched ten times to an expanded acrorhagus. The A. sargassensis tentacles were again touched to an expanded acrorhagus. In eleven trials (Table V) only the A. sargassensis tentacles elicited a peel (on first contact). In two trials, excised tentacles of C. gigantea were also touched to the expanded acrorhagi and peels resulted.

Only twice did a peel result from more than two hundred intra-specific contacts between A. krebsi in which its own or a groupmate's tentacle and column were touched with expanded acrorhagi.

DISCUSSION

This study suggests the acrorhagial response is not used among A. krebsi groupmates. The isolation experiment indicates continued contact is not required to protect groupmates from acrorhagial aggression and that there is no acclimation period. That A. krebsi do not recognize groupmates as targets might be expected if the groups were actually clones, as suggested for A. elegantissima by Francis (1973a). The acrorhagial response could not function if an anemone recognized itself as a target. Therefore, because all members of a clone presumably have the same genome, an A. krebsi would not recognize a difference between itself and a clonemate.

Three separate components seem to comprise the acrorhagial response in A. krebsi; acrorhagi expansion, application behavior, and acrorhagial peeling. Acrorhagi expansion and application behavior were observed when peeling could not be elicited (Table I). Acrorhagi expansion and peeling without application behavior were elicited by A. pallida and were not uncommon (unpublished personal observations) during intraspecific acrorhagial responses. Peeling only occurred with expanded acrorhagi but there is not enough evidence to indicate whether the peeling dependence on acrorhagi expansion is mechanical, physiological or both. More work is required to establish the physiological basis for coordination between the three components.

The interspecific acrorhagial response of A. krebsi conforms to that of A. equina and A. elegantissima (Bonnin, 1964; Francis, 1973b) in that it is not elicited by prey, predators, or hydroids but is elicited by other Actiniidae. However, A. krebsi's apparent lack of response to some anthozoans and a positive response to a scyphozoan are different and more data will be required to make general taxonomic statements about targets of acrorhagial responses.

Bonnin (1964) suggested that nematocysts were the factor that elicited the A. equina acrorhagial response. The cnidoms of A. pallida tentacles and acontia are different, yet both elicited an acrorhagial response from A. krebsi. Although the Cassiopea medusa and scyphistoma have the same nematocysts, the medusa did not elicit, and the scyphistoma did elicit an acrorhagial response. These data, along with the variety of animals with different cnidoms (and probably toxins) that elicited an acrorhagial response from A. krebsi suggest nematocysts do not serve as the eliciting factor for the acrorhagial response in A. krebsi.

The barrier experiment and tests of the mucus and portions of anemones indicate the A. krebsi acrorhagial response eliciting factor is surface bound and is located on most of the external surface of some target animals. In some cases it was also found on the endoderm. Recognition of a surface bound factor has also been found in hydroids (Ivker, 1972) and possibly in corals (Lang, 1973).

The scyphozoan scyphistoma, but not the medusa, elicited an acrorhagial response from A. krebsi. If the scyphistoma and medusa were in fact the same species it appears that the eliciting factor is not present in all stages of the life cycle. That there may be developmental differences (quantitative or qualitative) in the eliciting factor is also suggested by the difference in response to the two sizes of A. pallida (Table I).

The A. krebsi receptors for acrorhagi expansion and application behavior were found in the tentacles and acrorhagi, but were shown to be absent from the lower column and oral disc. The different thresholds for the three acrorhagial response components when the stimulus was applied to an acrorhagus suggest the possibility that each component has different receptors rather than the same receptors linked to three different effector systems. If such were the case, the three sets of receptors would be responding either to the same eliciting factor and have different thresholds or to different eliciting factors.

The receptors of all three systems show specificity, but more work is required to show whether the specificity is based on a qualitative or quantitative difference in eliciting factors and to elucidate the mechanism of receptor/eliciting factor interaction.

Anthopleura krebsi is anatomically a relatively simple animal. However, this investigation has shown it possesses a sensitive target recognition system. Because of this, A. krebsi will be useful for further investigations concerning recognition in the lower invertebrates.

This study was supported in part by a Sigma Xi Grant-In-Aid of Research to the author and by NSF Grant GB-40547 to Dr. Mariscal. Dr. Richard N. Mariscal's advice during this study and his critical reading of the manuscript were greatly appreciated.

LITERATURE CITED

Abel, E.F., 1954. Ein Beitrag zur Giftwirkung der Aktinien und Funktion der Randsackchen. Zool. Anz., 153: 259-268.

Bigelow, R.P., 1900. The anatomy and development of Cassiopea xamachana. Boston Soc. Natur. Hist. Mem., 5(6): 191-236.

Bonnin, J.P., 1964. Recherches sur la "reaction d'agression" et sur le fonctionnement des acrorrhages d'Actinia equina. Bull. Biol., 98: 225-250.

Carlgren, O., 1949. A survey of Ptychodactiaria, Corallimorpharia and Actiniaria. Kgl. Sv. Vetenskapsakad. Handl., 1: 1-121.

Carlgren, O. and J.W. Hedgpeth, 1952. Actiniaria, Zooantharia and Ceriantharia from shallow water in the northwestern Gulf of Mexico. Pub. Inst. Mar. Sci. Texas, 2(2): 143-172.

Francis, L., 1973a. Clone specific segregation in the sea anemone Anthopleura elegantissima. Biol. Bull., 144: 64-72.

Francis, L., 1973b. Intraspecific aggression and its effect on the distribution of Anthopleura elegantissima and some related sea anemones. Biol. Bull., 144: 73-92.

Ivker, F.S., 1972. A hierarchy of histo-compatibility in Hydractinia
     echinata. Biol. Bull., 143: 162-174.
Lang, J., 1973. Interspecific aggression by scleractinian corals.
     2. Why the race is not only to the swift. Bull. Mar. Sci.,
     23: 260-279.
Mariscal, R.N., 1974. Nematocysts. Pages 129-178 in L. Muscatine
     and H.M. Lenhoff, Eds., Coelenterate Biology: Reviews and
     New Perspectives. Academic Press, New York.
Schmidt, H., 1969. Die Nesselkapseln der Aktinien und ihre
     differentialdiagnostische Bedeutung. Helgoländer wiss
     Meeresunters., 19: 284-317.

ECOLOGICAL PHYSIOLOGY AND GENETICS OF THE COLONIZING ACTINIAN

HALIPLANELLA LUCIAE

J. Malcolm Shick

Department of Zoology, University of Maine

Orono, Maine 04473 USA

Since its presumed origin on the Pacific coast of Asia, Haliplanella luciae (Verrill) has become distributed throughout the northern hemisphere. The dispersal of the species has been effected through attachment to oysters shipped commercially (Verrill, 1898), transportation on ship bottoms (Stephenson, 1935), and attachment to floating seaweed (Williams, 1973a). Stephenson (1935) notes that the species may appear suddenly in a locality, flourish for a time, and then die out or disappear abruptly. It exhibits classic characteristics of a colonizing species.

H. luciae is highly variable in color and stripe pattern. Local populations may be exclusively of one morph, or a mixture of several (Uchida, 1932; Hand, 1955; Williams, 1973b). Further, as might be expected in a species having such an extensive geographic range, physiological races (sensu Stauber, 1950) have been described in isolated localities (Williams, 1973b).

The species apparently is dioecious, Davis (1919) and R. B. Williams (personal communication) having found sexually mature individuals of both sexes. Verrill (1898) reported seeing ciliated embryos swimming inside the tentacles of specimens from New Haven, Connecticut, USA, but embryonic and larval development have not been followed; indeed, despite an intensive search during different seasons, Davis (1919) was unable to find any newly-metamorphosed individuals. Asexual reproduction, both by longitudinal fission and by pedal laceration, has been more thoroughly studied (Torrey and Mery, 1904; Davis, 1919; Miyawaki, 1952; Atoda, 1973). Asexual fission has been suggested as the principal means of maintaining or increasing the population in a given region (Davis, 1919); such a mechanism is

137

obviously advantageous to a colonizing organism.

Another attribute of this successful colonizer is its resistance
to high summer temperatures, to encasement in ice in winter, and to
drastically reduced salinities, which has been noted in numerous
reports beginning with the early observations of Verrill (1898).
To some extent this resistance is afforded not by physiological compensation but by avoidance of environmental extremes (Hausman, 1919;
Miyawaki, 1951; Sassaman and Mangum, 1970; Ellington, 1976), but in
any case it represents the "somatic plasticity" typical of colonizing species (Lewontin, 1965).

Haliplanella luciae is a recent arrival in the Maine intertidal
zone, since it is not listed in the survey of Kingsley (1901) nor in
that of Procter (1933). It presently occurs at isolated sites along
the entire Maine coast (Perkins and Larsen, 1975), and its presence
provides the opportunity to study this colonizing actinian in an
extreme environment during a significant northward range extension,
the former limit to its distribution on the North American Atlantic
coast being Salem, Massachusetts (Parker, 1919).

MATERIALS AND METHODS

Most of the sea anemones used in this study were collected from
the intertidal zone at Blue Hill Falls, Maine (44°22'N, 68°34'W).
All animals in this population have 12 orange stripes on an olive
green or brown column (Type 1 of Uchida, 1932). The water temperature at the collecting site varies seasonally (-1° to 19°C), and the
animals are subjected to even greater extremes during intertidal
exposure, since I have recorded surface substrate temperatures from
-29° to 41°C. The salinity range is 25-32°/ooS. Additional specimens were collected at Narrow River, Rhode Island; all individuals
were Uchida's Type 3, having 48 paired yellowish-white stripes.
Specimens were also obtained from the population at Indian Field
Creek, Virginia, studied by Sassaman and Mangum (1970); the material
available contained representatives of Types 1 and 3, with the latter
predominant. Unless otherwise stated, all experiments described were
performed on Maine specimens.

Groups of 20 individuals were maintained for 2 weeks at 40 combinations of temperature and salinity, and per cent survival was
recorded. The criterion for survival was the expansion of a specimen after its return to the culture water at 30°/ooS.

The oxygen consumption by animals maintained for 10 days at 5°C
(cold acclimated) or at 15°C (warm acclimated) and 30°/ooS was determined polarographically at a series of common temperatures. From 8
to 15 specimens were used at each experimental condition, and weight

regression coefficients were used to correct by covariance the data for each group to the common dry weight of 30 mg.

After 2 weeks of maintenance at 17.5°C and 5, 12, 17.5, 22.5, or 30°/ooS, the water content of specimens was determined as in Pierce and Minasian (1974), and the lyophilized tissues were homogenized in 80% ethanol. Free amino acids (FAA) thus extracted were determined on a Technicon NC-1 amino acid analyzer. Specimens of H. luciae from Virginia were treated identically, except that no amino acid analyses were performed. In addition, the time course of change in tissue hydration and FAA concentration after acute transfer of Maine specimens from 30°/oo to 12°/ooS was followed.

Starch gel electrophoresis was used to separate enzymes in centrifuged homogenates of Maine specimens. Details of methodology will be presented elsewhere (Shick and Lamb, in preparation).

RESULTS AND DISCUSSION

The results of the survival experiments (Table I) provide quantitative confirmation of what has long been observed in the field,

TABLE I. Survival of H. luciae at different combinations of temperature and salinity. Figures represent per cent survival of individuals in groups of 20 after 2 weeks of exposure to the experimental conditions.

Temperature (°C)

Salinity (°/oo)	1.0	2.5	5.0	7.5	10.0	15.0	20.0	22.5	25.0	27.5
0.5	100		20			5	0	0	0	
1.0	100	100	80		100	50	5	5	0	
2.5						100		95	90	0
5.0	100			100		100		100	100	
10.0		100				100				
15.0						100		100		100
20.0			100		100		100			
30.0	100	100				100		100		100
35.0	100	100			100				100	

that <u>Haliplanella</u> <u>luciae</u> is exceedingly eurythermal and euryhaline.
This certainly was to be expected, but the very sharp temperature and
salinity limits to survival were surprising. In fact, I originally
intended to present a response surface analysis of these data, but
the per cent survival contours were so compressed as to be meaning-
less, except as a graphic illustration of the narrow range within
which extensive mortality occurs. For example, 90% of the experimental animals survive for 2 weeks at 2.5°/ooS and 25.0°C, although
survival is reduced to 0% at 2.5°/ooS and 27.5°C, and at 1.0°/ooS
and 25.0°C. The likely basis for this phenomenon is discussed below.

   The effects of temperature on the rate of oxygen consumption
($\dot{V}_{O2}$) in <u>H</u>. <u>luciae</u> are shown in Figure 1. The positions of the
curves for cold and warm acclimated Maine animals indicate that partial positive thermal acclimation occurs. $\dot{V}_{O2}$ in cold acclimated
individuals is significantly greater than that in warm acclimated
animals at 5° and 15°C (P < .05; P < .02, respectively), although the
reverse is true at 20°C (P < .001). A very different situation is
seen in Virginia <u>Haliplanella</u>, the pattern being the reverse of that
expected in poikilotherms showing positive acclimation to temperature (Sassaman and Mangum, 1970).

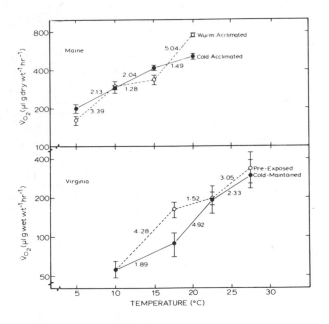

FIGURE 1. Temperature effects on oxygen consumption in <u>H</u>. <u>luciae</u>.
Vertical bars: ± 1 SE. Q_{10} values given for each temperature interval. Maine specimens: (o) maintained at 15°C; (●) maintained at 5°C.
Virginia specimens: (o) pre-exposed to each experimental temperature;
(●) maintained at 10°C. Curves for Virginia <u>H</u>. <u>luciae</u> redrawn with
permission from Sassaman and Mangum (<u>Mar</u>. <u>Biol</u>., 7: 123-130, 1970).

The dissimilar patterns of temperature acclimation in the Maine
and Virginia populations are of considerable interest, and are probably related to the very different thermal regimes experienced by
these populations. Sassaman and Mangum (1970) attribute the pattern
of reverse acclimation in Virginia Haliplanella to its more southern
distribution relative to Metridium senile, which shows positive
acclimation. Virginia Haliplanella utilizes avoidance of cold
extremes rather than physiological compensation, with a large percentage of laboratory and field specimens becoming encysted in mucus
at temperatures below 10°C. Records for the area reveal that the
water temperature falls and remains below 10°C only from mid-December
to mid-March (U.S. Dept. of Commerce, 1968; Mangum, personal communication), and a period of dormancy of this duration may reasonably be
expected. Conversely, the Maine population experiences water temperatures below 10°C during much of the year, from late October to early
June (Shick, unpublished). These animals show no evidence of encyst-
ment at these temperatures, remaining expanded and active even at 0°C
and below, and exhibit positive thermal compensation of metabolic
rate. According to the acclimation criteria of Prosser (1957), the
occurrence of dissimilar patterns of temperature compensation in the
two populations would indicate a genetic, rather than a latitudinal,
physiological racial, or reversible non-genetic basis for the difference.

As befits a euryhaline species, H. luciae is a good regulator of
cell volume over the non-lethal salinity range (Figure 2). Upon
transfer from water of 30°/ooS to salinities of 17.5°/ooS and below,
the initial response of all specimens examined is one of avoidance by
contraction and mucus secretion. Both Maine and Virginia animals
begin to re-expand after 10-15 h in water of 12°/ooS; this is in good
agreement with the results of Miyawaki (1951) for Japanese
Haliplanella (as Diadumene) luciae, which re-expands after 5-10 h in
"3/8 sea-water" (12.4°/ooS). Between 10°/oo and 12°/ooS, the re-
expansion of Maine and Virginia individuals is less predictable, and
below 10°/ooS it does not occur. This correlates well with the tissue hydration data in Figure 2, which indicate a decrease in volume
regulation ability below 12°/ooS in these animals. Thus it appears
that Miyawaki's (1951) statement regarding Japanese Haliplanella
that "the critical point for acclimatization lies between 2/8 seawater and 3/8" (8.3°/oo and 12.4°/ooS) is applicable to other populations as well. Further, the pattern of tissue hydration change as a
function of salinity is identical in Maine and Virginia specimens
(Figure 2).

As in another euryhaline sea anemone (Diadumene leucolena, studied by Pierce and Minasian, 1974), volume regulation in Maine
H. luciae is accomplished by the regulation of the intracellular FAA
pool, which increases in concentration with increasing salinity
(Table II). Most of this increase occurs as taurine and glycine,

TABLE II.   Intracellular free amino acids (μM/gram dry weight) of
H. luciae acclimated for 2 weeks to various salinities, and 16 hours
following transfer from 30°/ooS to 12°/ooS.   ND: Not detected.

Amino acid	Salinity (°/oo)					30 → 12 (16 h)
	5.0	12.0	17.5	22.5	30.0	
Taurine	3.61	46.92	80.43	85.68	177.22	115.11
Aspartic acid	0.02	0.50	1.23	1.09	3.81	1.01
Threonine	0.03	0.67	0.65	0.61	0.07	0.61
Serine	0.06	2.47	2.70	2.06	2.44	1.41
Glutamic acid	0.32	3.88	6.56	4.99	11.57	3.41
Proline	7.21	5.73	7.36	2.85	9.34	4.33
α-amino-butyric acid	ND	1.09	1.84	1.72	1.24	0.79
Glycine	1.10	24.13	50.91	51.99	55.33	21.19
Alanine	0.20	2.02	1.72	3.02	2.93	2.17
Citrulline	0.05	0.08	0.24	0.07	0.52	ND
Valine	1.86	1.16	0.90	0.65	0.07	1.13
Cysteine	2.91	2.60	3.03	1.75	2.15	1.94
Methionine	1.36	0.71	0.29	0.25	0.39	0.63
Isoleucine	1.18	0.81	0.18	0.27	0.33	0.48
Leucine	1.81	1.06	0.29	0.25	0.46	0.86
Tyrosine	3.05	1.46	1.14	1.13	2.00	2.00
Phenylalanine	1.13	0.38	0.18	0.23	0.54	0.65
Lysine	6.07	2.72	2.62	3.28	4.59	3.79
Tryptophan	0.16	0.08	0.09	0.09	0.16	0.07
Histidine	1.17	0.40	0.90	0.48	1.11	0.80
Arginine	0.34	0.05	0.61	0.52	0.23	0.16
Total	33.64	98.92	163.87	162.99	276.50	162.54

which together constitute 84.1% of the total FAA concentration at
30°/ooS, and which always account for more than 70% of the total over
the non-lethal range of 12-30°/ooS (Figure 3).   At 5°/ooS, the decline
in volume regulation ability (Figure 2) is correlated with the decline
in ability to regulate taurine and glycine concentrations (Figure 3).

The time course of tissue hydration following acute transfer of
anemones from 30°/oo to 12°/ooS is shown in Figure 4.   The increase
in water content is completed within 8-16 h following transfer, which
parallels the time involved in the contraction, mucus secretion, and
re-expansion response discussed above.   Tissue water then remains
constant at least to 48 h, after which physiological compensation
occurs, since there is a significant decrease in per cent water in
tissues after 344 h at 12°/ooS.   Likely involved in this compensation
is the long-term adjustment of taurine concentration, which begins by
16 h and is complete by 344 h after transfer (Table II).

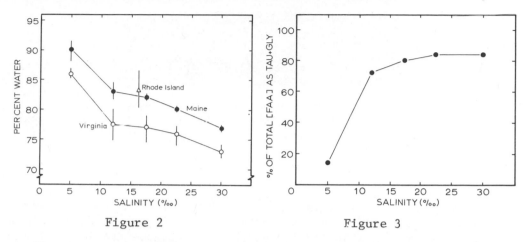

Figure 2   Figure 3

FIGURE 2.   Total water (as % wet weight) as a function of acclimation salinity in H. luciae. For Maine and Virginia specimens, each point represents the mean, and the vertical bars the range, of 5 observations.   Data for 10 Rhode Island specimens from W. R. Ellington (personal communication).

FIGURE 3.   Per cent of total intracellular free amino acid concentration comprised by taurine and glycine, as a function of acclimation salinity in H. luciae.

To summarize this line of research, Haliplanella luciae survives indefinitely over the range of 12-30°/ooS largely due to efficient physiological compensation entailing regulation of the FAA pool. Below 12°/ooS survival time is finite and the response of the actinian is one of resistance adaptation involving contraction and mucus secretion, which permits survival during periods of drastically reduced salinity caused by heavy rainfall and freshwater runoff.

The presence of only Type 1 H. luciae in the population at Blue Hill Falls, the abrupt rather than gradual increase in mortality

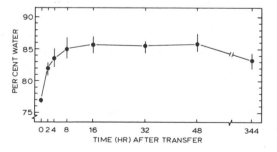

FIGURE 4.   Change in total water (as % wet weight) in H. luciae as a function of time following transfer from 30°/oo to 12°/ooS. Each point represents the mean, and the vertical bars the range, of 5 observations.

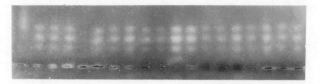

FIGURE 5. Fixed heterozygosity at a "tetrazolium oxidase" locus in
the population of H. luciae at Blue Hill Falls, Maine. For comparison, a Rhode Island individual homozygous for the fast allele is
included at the extreme left.

as temperature and salinity extremes were approached, the low variability in the results of the hydration experiments in Maine as
opposed to Virginia anemones, and the frequent observation of asexual
fission in Maine specimens, all suggested that the genetic variability in the population might be similarly low. Subsequently, of 213
specimens with recognizable gonads examined microscopically between
August 1975 and April 1976, all proved to be males. Further, electrophoretic studies revealed no variation among the specimens. Among
9 enzymes examined, 11 of the loci identified appear to be monomorphic
and therefore nothing could be determined with certainty regarding
the number of clones present. The demonstration of fixed heterozygosity at a "tetrazolium oxidase" locus (Figure 5), however, is
strong evidence that the 55 individuals examined after collection
throughout an area of approximately 1100 square meters (in which the
population density may be 4000 individuals/m^2) are members of a
single clone.

    A lack of genetic variability may be common in other local
populations of H. luciae as well, particularly those which have
resulted from single, isolated introductions or "colonizing episodes."
R. B. Williams (personal communication) reports that all of the
specimens that he sexed at Plymouth, England, were females, and that
those from Wells were exclusively males. Such a highly clonal local
population structure might explain the early observation that well-
established populations of H. luciae often disappear suddenly (Parker,
1919; Allee, 1923; Stephenson, 1935). Levinton and Fundiller (1975)
have demonstrated genotype-related selective mortality in the bivalve
Mya arenaria subjected to environmental stress. The present study
has established that local specimens of H. luciae are highly resistant
to environmental extremes, but that as the severity of conditions is
increased, there is an abrupt and massive incidence of mortality
rather than a gradual elimination of individuals whose genotypes confer less resistance, since only a single genotype is present. Genetic
and concomitant physiological uniformity thus has a profound impact
on geographic distributional phenomena, and other papers (Shick and
Lamb, in preparation) will consider the genetic structure of
geographically separated populations of Haliplanella luciae.

SUMMARY

Specimens of Haliplanella luciae from Maine are eurythermal, remaining expanded and active at 0°C, and exhibit positive temperature acclimation of the rate of oxygen consumption. The anemone is a good regulator of cell volume over the range of 12-30°/ooS, the volume regulation being effected by changes in the intracellular concentrations of taurine and glycine. Below 12°/ooS the response is one of contraction and mucus secretion. The electrophoretic demonstration of fixed heterozygosity at a "tetrazolium oxidase" locus is strong evidence that the large, all-male population under study is a single clone. These results are discussed in terms of the characteristics of colonizing species and observed geographic distributional phenomena.

ACKNOWLEDGMENTS

I thank C. P. Mangum for her provision of Virginia specimens of H. luciae, and W. R. Ellington for making available Rhode Island specimens and unpublished data. R. J. Hoffmann gave helpful suggestions regarding electrophoresis of cnidarian enzymes and critically read the manuscript. I am especially grateful to A. N. Lamb for his assistance in the performance and analysis of the electrophoretic studies. A grant from the University of Maine Faculty Research Fund provided research and travel support.

REFERENCES

ALLEE, W.C., 1923. Studies in marine ecology: IV. The effect of temperature in limiting the geographical range of invertebrates of the Woods Hole littoral. Ecology, 4: 341-354.

ATODA, K., 1973. Pedal laceration of the sea anemone, Haliplanella luciae. Pub. Seto Mar. Biol. Lab., 20: 299-313.

DAVIS, D.W., 1919. Asexual multiplication and regeneration in Sagartia luciae Verrill. J. Exp. Zool., 28: 161-263.

ELLINGTON, W.R., 1976. The early responses of a euryhaline anemone to hypo-osmotic stress. Amer. Zool., 16: (in press).

HAND, C., 1955. The sea anemones of Central California. Part III. The acontiarian anemones. Wasmann J. Biol., 13: 189-251.

HAUSMAN, L.A., 1919. The orange striped anemone (Sagartia luciae, Verrill). An ecological study. Biol. Bull., 37: 363-371.

KINGSLEY, J.S., 1901. Preliminary catalogue of the marine invertebrates of Casco Bay, Maine. Proc. Portland Soc. Nat. Hist., 2: 159-183.

LEVINTON, J.S., AND D.L. FUNDILLER, 1975. An ecological and physio-
    logical approach to the study of biochemical polymorphisms.
    Proc. 9th Eur. Mar. Biol. Symp., pp. 165-178.
LEWONTIN, R.C., 1965. Selection for colonizing ability. Pages 77-
    91 in H.G. Baker and G.L. Stebbins, Eds., The Genetics of Colon-
    izing Species. Academic Press, New York.
MIYAWAKI, M., 1951. Notes on the effect of low salinity on an actin-
    ian, Diadumene luciae. J. Fac. Sci. Hokkaido Univ., Ser. VI,
    Zool., 10: 123-126.
MIYAWAKI, M., 1952. Temperature as a factor influencing upon the
    fission of the orange-striped sea-anemone, Diadumene luciae.
    J. Fac. Sci. Hokkaido Univ., Ser. VI, Zool., 11: 77-80.
PARKER, G.H., 1919. The effects of the winter of 1917-1918 on the
    occurrence of Sagartia luciae Verrill. Amer. Nat., 53: 280-281.
PERKINS, L.F., AND P.F. LARSEN, 1975. A preliminary checklist of the
    marine and estuarine invertebrates of Maine. Pub. No. 10, The
    Research Institute of the Gulf of Maine, 37 pp.
PIERCE, S.K., JR., AND L.L. MINASIAN, JR., 1974. Water balance of a
    euryhaline sea anemone, Diadumene leucolena. Comp. Biochem.
    Physiol., 49A: 159-167.
PROCTER, W., 1933. Biological Survey of the Mount Desert Region.
    Part V. Marine Fauna. Wistar Institute, Philadelphia, 402 pp.
PROSSER, C.L., 1957. Proposal for the study of physiological vari-
    ation in marine animals. Ann. Biol., 33: 191-197.
SASSAMAN, C., AND C.P. MANGUM, 1970. Patterns of temperature adapta-
    tion in North American Atlantic coastal actinians. Mar. Biol.,
    7: 123-130.
STAUBER, L.A., 1950. The problem of physiological species with spe-
    cial reference to oysters and oyster drills. Ecology, 31: 109-
    118.
STEPHENSON, T.A., 1935. The British Sea Anemones, Vol. II. The Ray
    Society, London, 426 pp.
TORREY, H.B., AND J.R. MERY, 1904. Regeneration and non-sexual
    reproduction in Sagartia davisi. Univ. Calif. Pubs., Zool., 1:
    211-226.
UCHIDA, T., 1932. Occurrence in Japan of Diadumene luciae, a remark-
    able actinian of rapid dispersal. J. Fac. Sci. Hokkaido Imp.
    Univ., Ser. VI, 2: 69-82.
U.S. DEPARTMENT OF COMMERCE, 1968. Surface water temperature and
    density, Atlantic coast, North and South America. 3rd ed.
    C.&G.S. Pub. 31-1, 102 pp.
VERRILL, A.E., 1898. Descriptions of new American actinians, with
    critical notes on other species, I. Amer. J. Sci., 6: 493-498.
WILLIAMS, R.B., 1973a. The significance of saline lagoons as refuges
    for rare species. Trans. Norfolk Norwich Nat. Soc., 22: 387-
    392.
WILLIAMS, R.B., 1973b. Are there physiological races of the sea
    anemone Diadumene luciae? Mar. Biol., 21: 327-330.

NUTRITION EXPERIMENTALE CHEZ LES CERIANTHAIRES

Yves Tiffon

Laboratoire Maritime, Luc-sur-Mer (F) et Département

Anatomie-Biologie Cellulaire, C.H.U.Sherbrooke,P.Q. (CDN)

Les expériences qui vont être relatées ont été réalisées sur deux espèces de Cérianthes = C. Lloydi et C. membranaceus.

DIGESTION EXTRACELLULAIRE ET EXTRACORPORELLE

Lorsqu'une proie touche un tentacule marginal, elle est très rapidement transférée aux tentacules labiaux. Ceci a déjà été décrit par Arai et Walder (1973) pour l'espèce Pachycerianthus fimbriatus. Mais le fait est d'observation courante chez les deux espèces que nous avons utilisées. Si cette proie est relativement petite, elle disparait rapidement dans la cavité gastrique. Si elle est plus grosse, la proie est retenue plus longtemps au contact des tentacules labiaux. D'où l'hypothèse d'une digestion extracellulaire et extracorporelle possible au contact des sécrétions glandulaires des tentacules labiaux.

Les figures 1 et 2 illustrent la méthode utilisée. La proie expérimentale consiste en un abdomen de Crangon préalablement bouilli et fixé à un hameçon. La proie est amenée au moyen d'un fil au contact des tentacules labiaux et maintenue ainsi pendant une heure. Au bout d'une heure, les abdomens sont placés sur une plaque photographique (Photochemical plate, No. 60, Ilford) préalablement développée, fixée, séchée et imprégnée de tampon Tris 0, 1M, pH = 8,2. Au contact, la gélatine est digérée par les sécrétions retenues sur l'abdomen. Une enzyme protéolytique active à pH = 8,2 est donc sécrétée au niveau des tentacules labiaux (Tiffon, 1975).

Les tentacules labiaux diffèrent des tentacules marginaux par

Fig. 1 et 2. Nutrition expérimentale au moyen d'un abdomen de
Crangon fixé à un hameçon (PR) et amené au contact des tentacules
labiaux (TTL). Les tentacules marginaux (TTM) restent étalés
à la périphérie.

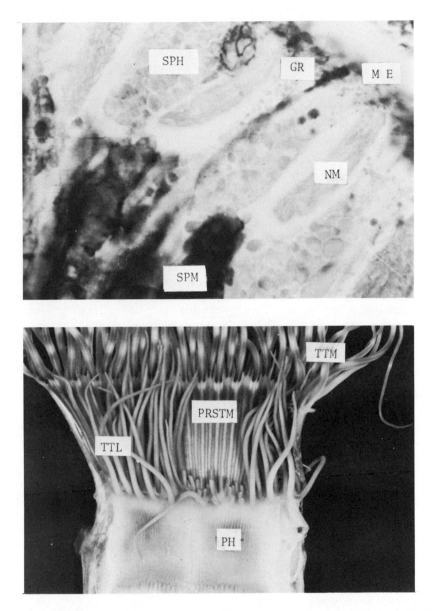

Fig. 3. Structure histologique d'un tentacule labial: cellules
glandulaires sphéruleuses (SPH), spumeuses (SPM), grains de trypsinogène (GR), mucus expulsé (ME).
Fig. 4. Disposition des tentacules labiaux (TTL), et des tentacules marginaux (TTM) par rapport au péristome (PRSTM) et au pharynx
(PH).

la nature des cellules glandulaires qui les constituent. (fig. 3).
Les tentacules labiaux contiennent en effet, outre des cellules
glandulaires de type muqueux, des cellules glandulaires dont les
sécrétions prennent la coloration au p-d.m.a.b. de Lison (1960).

Tout se passe donc comme si les tentacules labiaux avaient
un rôle digestif. Il convient cependant de noter qu'il n'est pas
évident que ce type de digestion ait un rôle de premier plan dans
la nutrition des Cérianthes. L'absence de publications concernant
des observations faites en plongée ne peuvent nous renseigner sur
ce point précis.

DIGESTION DANS LA CAVITE GASTRIQUE

La plupart des proies expérimentales macroscopiques sont im-
médiatement avalées. Elles sont digérées dans la cavité gastrique
grâce à l'équipement glandulaire des filaments largement pourvus
en cellules glandulaires à grains de trypsinogène (Tiffon et Bouillon, 1975). La digestion extracellulaire dans la cavité est immé-
diatement suivie d'une phagocytose par l'endoderme des cloisons.
Les fibres musculaires sont phagocytées dans l'endoderme des cloisons stériles et nulle part ailleurs. Ce type de nutrition par
des grosses proies a donc une efficacité restreinte. Chez un
animal non pourvu de milieu intérieur, il est évident que chaque
type de tissu se nourrit pour lui-même.

En dehors de toute considération sur la valeur énergétique
de la proie offerte, nous devons constater que la phagocytose des
fibres musculaires ne s'effectue que dans une zone restreinte.

NUTRITION PARTICULAIRE

Nous avons alors cherché une proie qui pourrait servir de
transition entre les muscles de Crangon et une nourriture particu-
laire. Un caillot de sang offre l'avantage d'être avalé comme
une proie macroscopique et de libérer ensuite les hématies au contact des tissus où aura lieu la phagocytose.

Cette technique nous a permis de mettre en évidence une phagocytose intense dans l'endoderme des cloisons stériles, mais aus-
si dans l'endoderme de la colonne et des tentacules. La taille
des hématies de poulet utilisées étant située entre 7,5 et 12,5 mμ
nous nous sommes alors adressé à des bactéries (Escherichia coli)
dont la taille est située entre 0,5 et 6 mμ. Les bactéries, préalablement marquées par de la thymidine tritiée ont été retrouvées
dans l'endoderme des cloisons stériles. Toutefois, cette expérience ne permet pas de préciser si les bactéries sont phagocytées

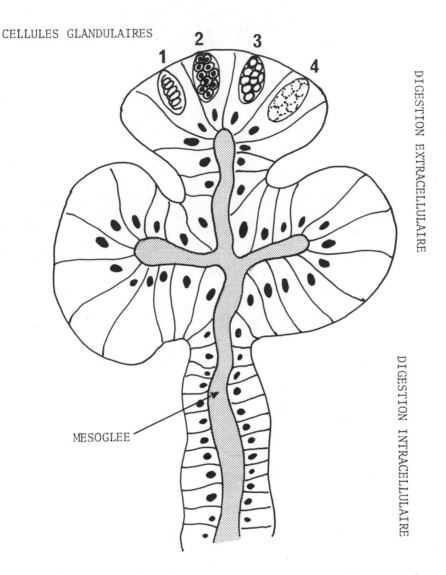

Fig. 5. Représentation schématique d'une cloison du gastroderme montrant les zones où s'opère la digestion de contact et la digestion intracellulaire.
Cellules glandulaires: 1 = hématocyte; 2 = cellule à grains de trypsinogène; 3 = cellule sphéruleuse; 4 = cellule spumeuse.

directement ou si elles sont fixées dans un premier temps sur un
substrat, éventuellement, le mucus de l'animal pour être ensuite
phagocytées en même temps que le mucus. Une expérience de nutrition avec des particules de latex a permis de mettre en évidence
une induction de la phagocytose des particules par une protéose-
peptone et par le tauroglycholate de sodium.

SUBSTANCES ORGANIQUES EN SOLUTION

Les travaux concernant la nutrition des Invertébrés par des
substances organiques ont surtout mis en évidence l'absorption des
acides aminés (Stephens, 1968; Schlichter, 1975) mais peu de travaux ont été réalisé en utilisant les protéines en solution.

En ce qui nous concerne (Tiffon et Daireaux, 1974) nous avons
pu démontrer une pinocytose intense dans l'endoderme de C. Lloydi
à tous les niveaux. Le phénomène est particulièrement intense dans
l'endoderme des cloisons fertiles. Rappelons à ce sujet que seules
les hématies de poulet étaient phagocytées à ce niveau.

Fait inattendu, l'ectoderme des tentacules labiaux est le siè-
ge d'une pinocytose intense contrairement à l'ectoderme des tentacules marginaux où aucune trace de ferritine ne peut être détectée.

Par contre, les acides aminés (hydrolysat de protéine de chlo-
relle) en solution sont absorbés intensément au niveau de l'ectoderme des tentacules marginaux alors qu'ils ne le sont pas au niveau des tentacules labiaux.

Il y a là une spécialisation dans la fonction des tentacules
dont il faudra chercher l'origine dans l'ultrastructure des tissus.

Des travaux expérimentaux dans d'autres groupes zoologiques
ont utilisé des protéines en solution, Schmidt (1970), chez les
éponges; Pasteels (1968), chez les moules; Little et Gupta (1968)
chez les pogonophores. Récemment, McCammon (1974) à partir d'observations réalisées à l'Aquarium de Vancouver a mis au point une
méthode de nutrition des brachiopodes à partir d'un extrait de
coeur de boeuf. Cette voie est peut être à approfondir pour la
nutrition en écloseries de larves destinées à l'aquaculture. La
pinocytose cellulaire pourrait être activée par des substances
adjuvantes du type de celles utilisées avec succès par Ricketts
chez les protistes.

En dehors de cet intérêt pratique, et sans nous prononcer
sur la valeur énergétique des protéines pinocytées, leur présence
en solution dans le milieu où vivent les animaux peut être un facteur écologique déterminant pour leur répartition.

Ceci a été souligné pour la première fois par Reiswig (1971) pour les Démosponges tropicales, mais il conviendrait d'étudier le rôle de ce facteur dans la répartition des espèces d'Anthozoaires littoraux et abyssaux.

BIBLIOGRAPHIE

Arai, M.N. et G.L. Walder, 1973. The feeding responses of Pachycerianthus fimbriatus. Comp. Biochem. Physiol., 44A: 1085-1092.

Lison, L., 1960. Histochimie et Cytochimie animales. Gauthier-Villars, Paris, 842 p.

Little, C. et B.L. Gupta, 1968. Pogonophora. Uptake of dissolved nutrients. Nature, 218: 873-874.

McCammon, H.M., 1974. Maintenance of some marine filter feeders on beef heart extract. Culture of marine invertebrate animals. W.L. Smith et M.H. Chanley, Editors, 15: 15-27. Plenum Pub. Comp. New York.

Pasteels, J.J., 1968. Pinocytose et athrocytose par l'épithélium branchial de Mytilus edulis. Analyse expérimentale au microscope électronique. Z. Zellf., 92: 339-359.

Reiswig, H.M., 1971. Particle feeding in natural population of three marine Demosponges. Biol. Bull., 141: 568-591.

Schlichter, D., 1975. The importance of dissolved organic compounds in sea water for the nutrition of Anemonia sulcata (Coelenterata). Proc. 9th European Mar. Biol. Symp., H. Barnes Editor, 395-405 p. Aberdeen University Press.

Schmidt, I., 1970. Phagocytose et pinocytose chez les Spongillidae. Z. Vergl. Physiol., 66: 398-420.

Stephens, G.C., 1968. Dissolved organic matter as a potential source of nutrition for marine animals. Am. Zool., 8: 95-106.

Tiffon, Y. et M. Daireaux, 1974. Phagocytose et pinocytose par l'ectoderme et l'endoderme de Cerianthus Lloydi. J. Exp. Mar. Biol. Ecol., 16: 155-165.

Tiffon, Y., 1975. Hydrolases dans l'ectoderme de C. Lloydi, C. membranaceus et Metridium Senile: Mise en évidence d'une digestion extracellulaire et extracorporelle. J. Exp. Mar. Biol. Ecol., 18: 243-254.

Tiffon, Y. et J. Bouillon, 1975. Digestion extracellulaire dans la cavité gastrique de C. Lloydi. Structure du gastroderme, localisation et propriétés des enzymes protéolytiques. J. Exp. Mar. Biol. Ecol., 18: 255-269.

FOULING COMMUNITY STRUCTURE: EFFECTS OF THE HYDROID, OBELIA DICHOTOMA, ON LARVAL RECRUITMENT

Jon D. Standing

Department of Zoology, University of California

Berkeley, California 94720

Hydroids are common, if not always conspicuous, animals in marine epibenthic communities, frequently growing on hard substrata, plants, and other animals. Yet the ecological roles of hydroids in these communities are only beginning to be understood. Hydroid stolons may overgrow and smother small sessile animals attached to the substratum or prevent the settlement of animals requiring the substratum for attachment (Coe, 1932; McDougall, 1943). Chemical suppressants produced by some hydroids can inhibit the growth of other hydroids (Katô, et al., 1967). Furthermore, some hydroid epizoites do not occur on hydranths, suggesting either predation on settling larvae or avoidance of the nematocyst-bearing regions by motile forms (Eggleston, 1972; Hughes, 1975).

Nevertheless, a rich variety of motile and sessile epizoites use the stems and stolons of hydroids as a substratum (Pyefinch and Downing, 1949; Round, et al., 1961; Hughes, 1975). Nudibranchs may be attracted to their hydroid prey by chemical cues (Tardy, 1962). However, most epizoites on hydroids probably respond to the physical conditions present there. Primary settlement of mussels onto hydroids and filamentous algae may involve tactile recognition of suitable substrata (De Blok and Geelen, 1958; Bayne, 1964; Seed, 1969), and similar behavior patterns may explain the host preferences of caprellid amphipods (Keith, 1971). Hughes (1975) explains the distributions of most of the epizoites on the hydroid, Nemertesia, by current velocities and the consequent changes in food supplies and sediment deposition within the colonies. McDougall (1943) observed changes in the sessile epibenthos on pilings that he attributed to sedimentation beneath Tubularia canopies and to sloughing of colonies from the substratum.

155

    Most of these explanations of hydroid roles in epibenthic
communities have relied on observational or correlative evidence.
In a general way they suggest that hydroids may influence the
recruitment and distribution of co-occurring animals. In this
study I have experimentally removed the hydroid, Obelia dichotoma
(Linnaeus, 1758), from settling plates and studied subsequent
larval recruitment onto the plates. This approach has revealed
additional information about the roles of hydroids in the communities of which they are a part.

                            METHODS

    This study took place in Bodega Harbor, California, located
about 62 miles northwest of San Francisco. Bodega Harbor is a
shallow, natural embayment having inshore, oceanic water
conditions. Water temperatures range from about 10°C in the
winter and spring to about 14°C in the late summer and fall. The
study site was located near the entrance channel at the U.S.
Coast Guard Station wharf.

    The settling plates used were 10 x 10 cm squares of 7 mm
masonite hardboard which were rough on one side. The plates were
attached to a rack which was suspended from a floating pier at a
depth of 1 m from the surface. The aspect of the plates was
horizontal, and to minimize sedimentation effects, the lower sides
(rough sides) served as the settling surfaces.

    The plates were first immersed on April 30, 1974. Obelia
dichotoma settled and grew on the lower surfaces, covering all of
the area within one month. On five experimental plates, the stems
of Obelia polyps were removed by clipping them with scissors close
to the surface. Obelia stolons and the other animals on the
plates remained intact. The clipping began on June 2 and was
repeated every 8 days because the stems and hydranths regenerated
rapidly. The Obelia stems on five control plates were not clipped.

    The plates remained in the water for nearly 2 months after
the clipping was initiated. They were then removed from the water
on July 27 and preserved in alcohol for later analysis. Two of
the most abundant sessile animals settling during this period were
the ascidian, Ascidia ceratodes, and the barnacle, Balanus
crenatus. The Ascidia from each of the plates were removed and
counted. The individuals on each plate were then pooled and
weighed, after drying to a constant weight at 60°C. Next, all of
the barnacles on each plate were counted, and the percentage
cover of the barnacles was determined using a random point method
with 100 points per plate.

    Experimental and control plates were analyzed statistically

with single classification analysis of variance tests (F). The
percentage cover data underwent an angular transformation prior to
testing. The relationship between the numbers of Ascidia and
percentage cover of barnacles was analyzed with a Model II
regression, using Bartlett's three-group method.

RESULTS

Obelia and Associated Fauna

Obelia dichotoma settled in May and June, but not in July, in
this study. Boyd (1971) recorded Obelia settlement during the
winter, spring, and early summer in Bodega Harbor and suggested
that settlement may be associated with low water temperatures. In
addition, my own observations suggest that there are two or three
major settlements, interspersed by periods of little or no settlement, during the period of low water temperatures. Obelia
dichotoma will settle onto barnacles, Ascidia, mussels, and
substrata that are free of other animals. Stems grew up to 12 cm
long in this study. Colonies regress after two or three months of
growth.

The fauna settling on the plates already occupied by Obelia
was typical of the fouling assemblage in Bodega Harbor (Boyd,
1971; Standing, et al., 1975). Balanus crenatus, Ascidia
ceratodes, and Bowerbankia gracilis were the most abundant sessile
animals. Balanus settled onto the plates on spring tides, occupying substratum between the Obelia stolons. The Ascidia settled
between the barnacles in late June and July, and the Bowerbankia
appeared in July, covering most of the other sessile animals with
a furry coat. Other sessile animals were represented by only a
few individuals or colonies. Various polychaetes and amphipods
became associated with the growing barnacles and especially the
Ascidia. Epizoites on the Obelia included the nudibranchs,
Dendronotus frondosus and Hermissenda crassicornis; the caprellid
amphipods, Caprella californica and C. gracilior; the pycnogonid,
Anoplodactylus viridintestinalis; and a number of smaller
unidentified animals.

Balanus Recruitment

The barnacle data are shown in Table 1. Significantly greater
numbers of Balanus settled on the experimental plates than on the
control plates having Obelia stems (F = 9.81; .01<P<.025). The
barnacles on the plates with reduced Obelia growth also occupied
significantly more space there than those with normal Obelia
growth (F = 17.34; .001<P<.005). These data suggest that Obelia

stems, or correlated factors, have negative effects on barnacle recruitment.

Initially, I thought that the Obelia were preying on the cyprid larvae of the barnacles. This interpretation would explain the lower barnacle settlements beneath hydroid canopies. However, two further observations failed to support this view. First, fresh Obelia dichotoma will not feed on intact B. crenatus cyprids in the laboratory. Cyprids do not appear to stick to the tentacles, and there is no mouth-opening response. However, if cyprid cuticles are experimentally damaged, they readily stick to the tentacles and elicit a mouth-opening response. Secondly, B. crenatus cyprids are slightly too large to be ingested by O. dichotoma hydranths. The cyprids are about 850 x 280 x 365 µm, but O. dichotoma hydrothecae are only about 260 µm in diameter. Consequently, an Obelia-cyprid predation interpretation does not appear to be correct.

Explaining the data on the basis of barnacle predators also appears untenable. There were no sea stars, snails, chitons, or nemerteans on the plates. Amphipods and a few small polyclad flatworms were present, but their numbers were not sufficiently different to explain the differences in barnacle numbers in the two conditions. There may have been fewer caprellid amphipods on the experimental plates than on the control plates. Some caprellids were inevitably lost when Obelia stems on the

Condition	Mean Density (number/10 cm^2)	Mean Percentage Cover
Stems Removed; Experimental Plates	137.8 (74.2-201.4)	31% (27%-35%)
Stems Present; Control Plates	55.6 (20.3-90.9)	8% (6%-11%)

Table 1. Effects of Obelia stems on density and percentage cover of Balanus crenatus. Data are expressed as means of five replicate settling plates; 95% confidence limits are given in parentheses.

experimental plates were clipped, although they reinvaded the plates as the Obelia regenerated. Keith (1969) and others have found crustacean (mainly amphipod) fragments in caprellid guts, but the food of many caprellids, including Caprella californica, consists largely of detritus and diatoms. It therefore appears unlikely that caprellid predation on cyprids or spat could explain the differences.

In addition, the present data can not be explained by differences in the amount of surface space available to the cyprids for settlement. Obelia stolons were present on both experimental and control plates; in fact, they may have occupied more space on the plates that were clipped than on the controls. Most of the barnacle recruitment preceded the settlement of Ascidia, and in addition, the large Ascidia on the experimental plates held more space than the smaller ones on the control plates.

Thus, it appears that the Obelia have an interference effect on B. crenatus cyprids. Dense canopies of branching hydrocauli and hydranths probably block contact between the cyprids and the settling surface. The physical barrier provided by these canopies is removed by the clipping operation, permitting barnacle cyprids to attach and metamorphose on the surfaces. Also, Obelia canopies probably decrease current velocity gradients in the understory. Crisp (1955) has shown that two species of intertidal barnacles tend not to attach to glass tubes in very low velocity gradients. It is not clear whether or not this phenomenon applies to fixation and metamorphosis of the subtidal barnacle, B. crenatus.

Ascidia Recruitment

The Ascidia data are presented in Table 2. There were significantly higher numbers of Ascidia on plates with unclipped Obelia stems than on experimental plates (F = 32.60; P<.001). Obelia stems, or related variables, appear to have positive effects on Ascidia numbers. A possible direct effect concerns physical conditions in the understory. Tadpole larvae of Ascidia may select dark, protected places for attachment and metamorphosis (personal observations). Currents are probably reduced beneath Obelia canopies, and light may also be attenuated there.

Most of the indirect effects involve interspecific interactions between Ascidia and barnacles. The numbers of Ascidia present on 10 replicate plates was inversely correlated with the percentage cover of barnacles (r_{xy} = -.82; F = 16.0; .001<P<.005). The regression for this relationship is shown in Figure 1. The numbers of Ascidia increase with decreasing barnacle coverage.

The data suggest three possible interpretations. First, the
barnacles could be preying on the tadpoles. B. crenatus will feed
on Ascidia tadpoles when presented with large numbers of them in
the laboratory. However, this interaction is probably relatively
unimportant in the field. Tadpoles that are settling tend to swim
rapidly towards dark surfaces and attach on contact. This
behavior directs them to substrata between light-reflecting
barnacle shells, thus avoiding capture by barnacles.

Secondly, Ascidia may be competing with barnacles for settling
sites. The barnacles exploited a considerable amount of potential
settling surface before the tadpoles arrived. Therefore, tadpoles
were precluded from the space occupied by barnacles, settling
between them instead. Recent observations suggest that barnacle
recruitment invariably precedes Ascidia settlement in the field.
Tadpoles settle directly on barnacles only rarely.

Finally, there was probably interference competition between
the Ascidia and the barnacles, and also intraspecifically, between
the Ascidia themselves. Comparing the numbers of barnacles with
the percentage covers in Table 1, the barnacles associated with
low numbers of Ascidia were larger than those on the control
plates (Table 2). The sizes of the Ascidia were also inversely
related to their numbers (Table 2) (r_{xy} = -.76; F = 11.2; .01<P<
.025). Ascidia grow very rapidly in Bodega Harbor, eventually

Condition	Mean Density (number/10 cm^2)	Mean Dry Weight (mg)
Stems Removed;	110.4	1.99
Experimental Plates	(73.8-147.0)	(1.71-2.27)
Stems Present;	196.6	1.01
Control Plates	(176.4-216.8)	(0.77-1.25)

Table 2. Density and size (dry weight) of Ascidia ceratodes on
settling plates having Obelia stems experimentally removed and
naturally present. Data are expressed as means of five replicate
plates; 95% confidence limits are given in parentheses.

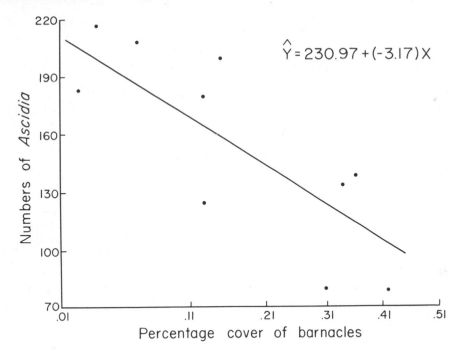

$$\hat{Y} = 230.97 + (-3.17)X$$

Figure 1. Relationship between the numbers of <u>Ascidia</u> <u>ceratodes</u>
and the percentage cover of barnacles. The data were analyzed
with a Model II regression, using Bartlett's procedure.

smothering most of the barnacles beneath them and usually dominating these fouling communities for 18 months or more. Additional
proof of these competition interactions must await further removal
experiments.

DISCUSSION

    Observations presented in this study suggest that the hydroid,
<u>Obelia</u> <u>dichotoma</u>, does not feed on <u>Balanus</u> <u>crenatus</u> cyprids.
Nevertheless, hydroids may be larval predators in some situations.
The absence of many epizoites from hydranths may provide some
evidence for larval predation (Eggleston, 1972; Hughes, 1975).
McDougall (1943) and R.H. Karlson (cited in Sutherland, 1974) have
reported the exclusion of many larval recruits from colonies of
<u>Hydractinia</u> <u>echinata</u>, and colonies of this species on hermit crab
shells will feed on polychaete, crab, snail, and bivalve larvae

(Christensen, 1967). McDougall (1943) also noted large numbers of
Balanus improvisus on surfaces kept free of Tubularia crocea, and
few barnacles in areas dominated by this hydroid. T. crocea will
feed on a variety of zooplankton, including portunid zoeae, in the
laboratory (Mackie, 1966). These studies suggest that some
hydroids, at least, may be predators on potentially settling
larvae. Further studies are necessary to evaluate the importance
of this phenomenon.

An Obelia interference effect best explains the barnacle
recruitment data recorded in this study. Hydroid canopies appear
to provide a physical barrier to cyprid settlement. This effect
may continue after the colonies regress because of the persistence
of stem perisarc. Hatton (1938), Southward (1956), and Luckens
(1970) have described similar barrier effects caused by algae.
Various fucoids interfere with barnacle recruitment by preventing
cyprid access to the substratum. Cyprids and spat may also be
dislodged by a whiplash effect of fucoid fronds (Lewis, 1964), but
this mechanism probably does not apply to hydroids. Filamentous
algae (Enteromorpha, Ectocarpus) may also exert a barrier effect
on cyprids (personal observations). Pyefinch (1948) believed his
barnacle numbers to be inversely related to the density of
Ectocarpus, Enteromorpha, and Obelia.

The effects of Obelia on Ascidia recruitment may be both
direct and indirect. Physical conditions underneath Obelia
canopies are direct effects which may promote the settlement of
tadpole larvae. The ascidians on Nemertesia cluster around the
holdfast, apparently responding to the physical conditions there
(Hughes, 1975). Obelia affects the recruitment of Ascidia
indirectly by its interference effect on barnacle numbers. Dense
hydroid or algal canopies reduce barnacle numbers. When barnacle
numbers are low, more space is available for the settlement and
growth of Ascidia, a dominant in this fouling community. Dense
aggregations of Ascidia, in turn, harbor a rich motile infauna of
polychaetes and amphipods. If the effects of interference canopies
are minimal, barnacle recruitment increases, promoting different
species mixes in this community. Less space is available for
Ascidia settlement in these situations, and mussels could replace
Ascidia as the community dominant because rough barnacle surfaces
promote the secondary settlement of mussels (De Blok and Geelen,
1958; Seed, 1969). Thus, some hydroids and algae play important
roles in their communities, roles which may be disproportionate to
their abundance and biomass.

This work was carried out at the Bodega Marine Laboratory,
Bodega Bay, California. I sincerely thank its director, Dr. Cadet
Hand, for the use of the facilities and for his helpful comments on
the manuscript.

SUMMARY

Effects of the thecate hydroid, Obelia dichotoma (Linnaeus, 1758), on larval recruitment were studied in a fouling community in Bodega Harbor, California. On experimental settling plates, stems of Obelia polyps were removed by clipping with scissors. Significantly greater numbers of the barnacle, Balanus crenatus, settled on these plates than on control plates having longer Obelia stems. The barnacles also occupied more space on the plates with reduced Obelia growth than on those with normal growth. The hydroid appears to interfere with barnacle recruitment by preventing cyprid access to the substratum.

By contrast, the solitary ascidian, Ascidia ceratodes, occurred in significantly higher numbers on the control plates than on the experimental plates. These and other data suggest that physical conditions beneath the hydroid canopy were favorable for ascidian settlement and that there were competitive inter-actions between the ascidians and the barnacles. By controlling barnacle numbers, Obelia can affect the numbers of ascidians and associated species, thus influencing the species compositions of these fouling communities for considerable periods of time.

REFERENCES

Bayne, B.L. 1964. Primary and secondary settlement in Mytilus edulis L. (Mollusca). J. Anim. Ecol. 33: 513-523.
Boyd, M.J. 1971. Fouling community structure and development in Bodega Harbor, California. Ph.D. Dissertation, University of California, Davis. 191 pp.
Christensen, H.E. 1967. Ecology of Hydractinia echinata (Fleming) (Hydroidea, Athecata) I. Feeding biology. Ophelia 4: 245-275.
Coe, W.R. 1932. Season of attachment and rate of growth of sedentary marine organisms at the pier of the Scripps Institution of Oceanography, La Jolla, California. Bull. Scripps Inst. Oceanogr., Univ. Calif. Tech. Ser. 3: 37-86.
Crisp, D.J. 1955. The behaviour of barnacle cyprids in relation to water movement over a surface. J. Exp. Biol. 32: 569-590.
De Blok, J.W. and H.J. Geelen. 1958. The substratum required for the settling of mussels (Mytilus edulis L.). Arch. Néerl. Zool., Vol. Jubilaire: 446-460.
Eggleston, D. 1972. Factors influencing the distribution of sub-littoral ectoprocts off the south coast of the Isle of Man (Irish Sea). J. Nat. Hist. 6: 247-260.
Hatton, H. 1938. Essais de bionomie explicative sur quelques espèces intercotidales d'algues et d'animaux. Ann. Inst. Oceanogr. Monaco, 17: 241-348.

Hughes, R.G. 1975. The distribution of epizoites on the hydroid
    Nemertesia antennina (L.). J. Mar. Biol. Ass. U.K. 55:
    275-294.
Katô, M., E. Hirai, and Y. Kakinuma. 1967. Experiments on the
    coaction among hydrozoan species in the colony formation.
    Sci. Rep. Tôhoku Univ. Ser. 4 (Biol.) 33: 359-373.
Keith, D.E. 1969. Aspects of feeding in Caprella californica
    Stimpson and Caprella equilibra Say (Amphipoda). Crustaceana
    16: 119-124.
_____. 1971. Substrate selection in caprellid amphipods of
    southern California, with emphasis on Caprella californica
    Stimpson and Caprella equilibra Say (Amphipoda). Pac. Sci.
    25: 387-394.
Lewis, J.R. 1964. The Ecology of Rocky Shores. English Univer-
    sity Press, London. 323 pp.
Luckens, P.A. 1970. Breeding, settlement and survival of
    barnacles at artificially modified shore levels at Leigh, New
    Zealand. N. Z. J. Mar. Freshwater Res. 4: 497-514.
McDougall, K.D. 1943. Sessile marine invertebrates of Beaufort,
    North Carolina. Ecol. Monogr. 13: 321-374.
Mackie, G.O. 1966. Growth of the hydroid Tubularia in culture.
    In "The Cnidaria and their Evolution" (W.J. Rees, ed.), pp.
    397-410. Academic Press, New York.
Pyefinch, K.A. 1948. Notes on the biology of cirripedes. J. Mar.
    Biol. Ass. U. K. 27: 464-503.
_____ and F.S. Downing. 1949. Notes on the general biology
    of Tubularia larynx Ellis & Solander. J. Mar. Biol. Ass. U.
    K. 28: 21-43.
Round, F.E., J.F. Sloane, F.J. Ebling, and J.A. Kitching. 1961.
    The ecology of Lough Ine. X. The hydroid Sertularia
    operculata (L.) and its associated flora and fauna: effects
    of transference to sheltered water. J. Ecol.. 49: 617-629.
Seed, R. 1969. The ecology of Mytilus edulis L.
    (Lamellibranchiata) on exposed rocky shores. I. Breeding
    and settlement. Oecologia 3: 277-316.
Southward, A.J. 1956. The population balance between limpets and
    seaweeds on wave-beaten rocky shores. Rep. Mar. Biol. Sta.
    Pt. Erin, No. 68: 20-29.
Standing, J.D., B. Browning, and J.W. Speth. 1975. The natural
    resources of Bodega Harbor. California Department of Fish
    and Game. 183 pp.
Sutherland, J.P. 1974. Multiple stable points in natural
    communities. Amer. Nat. 108: 859-873.
Tardy, J. 1962. Observations et expériences sur la métamorphose
    et la croissance de Capellinia exigua (A. & H.) (Mollusque,
    Nudibranche). C. R. Acad. Sci. Paris 254: 2242-2244.

THE ZONATION OF HYDROIDS ALONG SALINITY GRADIENTS IN SOUTH CAROLINA

ESTUARIES

Dale R. Calder

Marine Resources Research Institute
South Carolina Wildlife and Marine Resources Department
Charleston, South Carolina   29412

Only a small percentage of the water on earth is brackish, yet the biological productivity, ecological significance, and economic importance of estuarine areas is considerable. As a result, environments of reduced salinity have received increasing attention from marine biologists during this century, and particularly within the last two decades.

Estuaries are difficult to define because they vary widely in their origin, geology, size, shape, and hydrography. One of the most widely accepted definitions is that of Pritchard (1955), who described an estuary as "a semi-enclosed coastal body of water having a free connection with the open sea and within which the sea water is measurably diluted with fresh water runoff". Estuaries thus represent a transition zone from the sea to fresh water. However, an estuary is more than an ecotone between fresh water and the ocean, it is a biotope with its own unique set of factors (Carriker, 1967; Boesch, 1976). Estuaries differ hydrographically from both marine and fresh water areas in a number of important aspects, including a typically pronounced salinity stratification and oscillations in salt content. In addition, horizontal salinity gradients from the mouth to the head of an estuary have marked effects on indigenous organisms. The number of species in estuaries is low compared with the marine environment, and species diversity declines with decreasing salinity (Remane, 1971). The biota of estuarine areas is made up largely of euryhaline marine species (Percival, 1929; Gunter, 1950), and relatively few fresh water species occur even at low salinities (Kinne, 1971; Remane, 1971). Consequently, there is no corresponding increase in the number of fresh water species upestuary to compensate for the decrease in the number of marine species.

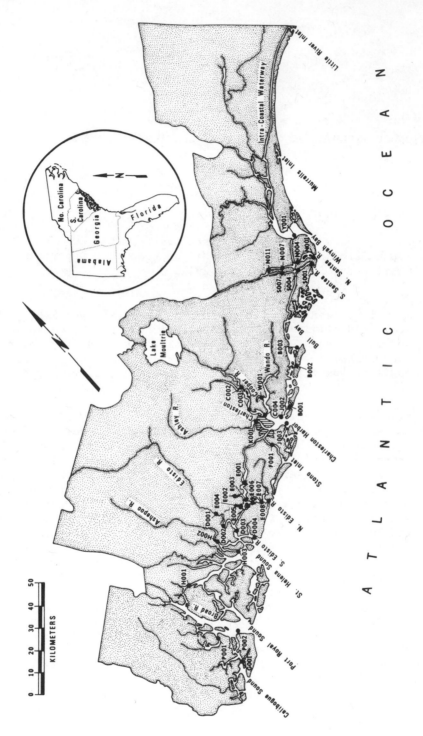

Fig. 1. Sampling stations in coastal South Carolina.

TABLE I

List of stations, with observed bottom salinity range and number of
hydroid species collected at each station. Stations with one asterisk
were sampled only in 1974; those with two asterisks were sampled only
in 1975.

Station	N	W	Depth (m)	Observed Salinity Range (^{o}/oo)	No. Species
Y001	33° 15.6'	79° 15.4'	4	0.2 − 13	4
**N001	33° 08.2'	79° 14.8'	3	0.2 − 31	3
**N004	33° 10.2'	79° 17.5'	3	0.2 − 10	4
**N007	33° 10.6'	79° 20.7'	5	0.1 − 0.2	1
**N011	33° 13.0'	79° 24.2'	7	0.1 − 0.3	3
**S001	33° 07.9'	79° 16.4'	3	10 − 30	3
S004	33° 08.8'	79° 19.2'	4	0.2 − 26	6
**S007	33° 09.8'	79° 22.3'	3	0.2 − 2	1
B003	32° 55.9'	79° 36.2'	7	28 − 34	12
B002	32° 54.2'	79° 40.7'	6	29 − 34	22
B001	32° 47.5'	79° 49.5'	5	23 − 33	14
J003	32° 44.9'	79° 51.6'	12	22 − 33	10
J002	32° 47.1'	79° 53.2'	4	16 − 28	8
W001	32° 53.1'	79° 52.6'	4	8 − 17	4
C004	32° 51.1'	79° 56.0'	7	4 − 23	5
C003	32° 53.8'	79° 57.6'	6	1 − 14	6
C002	32° 58.2'	79° 55.5'	6	0.1 − 5	1
K001	32° 49.0'	79° 58.1'	6	11 − 19	5
F001	32° 44.9'	80° 00.7'	4	10 − 21	5
E001	32° 41.2'	80° 13.4'	7	17 − 27	11
*E002	32° 41.1'	80° 17.3'	4	21 − 26	9
E003	32° 38.8'	80° 15.7'	6	19 − 31	13
*E004	32° 37.9'	80° 18.6'	6	13 − 28	1
*E005	32° 36.2'	80° 17.7'	8	23 − 29	9
*E006	32° 36.5'	80° 14.8'	6	25 − 30	2
E007	32° 35.0'	80° 13.5'	6	26 − 31	15
*E008	32° 33.6'	80° 10.7'	11	29 − 31	19
*D001	32° 39.7'	80° 24.8'	3	0.1 − 0.3	0
*D002	32° 36.3'	80° 25.4'	9	0.2 − 3	0
D003	32° 33.7'	80° 23.7'	5	5 − 17	6
*D004	32° 29.7'	80° 21.2'	5	24 − 33	15
H003	32° 30.9'	80° 27.9'	7	11 − 26	15
H002	32° 34.0'	80° 29.9'	5	2 − 12	5
H001	32° 32.1'	80° 43.7'	4	14 − 27	6
P002	32° 16.2'	80° 43.2'	7	27 − 32	13
P001	32° 17.2'	80° 49.0'	10	24 − 31	15
G001	32° 10.9'	80° 47.8'	5	25 − 30	12

Changes in species diversity are not uniform along the salinity gradient. Rapid changes occur in many estuaries around the 30 °/oo, 18-15 °/oo, and 8-5 °/oo isohalines (Kinne, 1971; Remane, 1971; Boesch, 1972, 1976). While other factors such as temperature, suitable substrate, currents, turbidity, and biotic interactions may play a role in these changes, salinity is generally the master factor involved (Gunter, 1961; Wells, 1961; Kinne, 1971; Remane, 1971).

Little is known about patterns of hydroid zonation under estuarine conditions. Salinity tolerances are known for a number of hydroids occurring in brackish waters, including such species as Cordylophora caspia, Garveia franciscana, and Protohydra leuckarti (Kinne, 1956; Crowell and Darnell, 1955; Clausen, 1971). Thiel (1970) and Meyer (1973) examined the distribution of hydroids in the Kiel Bay estuary. The goals of this study were to identify the more common species of hydroids found in estuarine areas of South Carolina, and to delineate their distribution with reference to salinity.

MATERIALS AND METHODS

The study area encompassed the region from the mouth to the head of various estuarine systems across the South Carolina coastal zone. Collections were made at 31 stations during January, April, August, and October of 1974, and at 29 stations during the same months in 1975 (Fig. 1, Table I). Tows of three to five minutes were made at each station using a 30 kg modified oyster dredge. Tows longer than five minutes at most stations usually resulted in clogged dredge bags and damaged specimens. After preliminary sorting of the catch on station, hydroids and a representative sample of firm substrates were preserved in 10% neutralized formaldehyde solution and returned to the laboratory for microscopic examination. Data for this study were recorded only on attached hydroids with hydranths present; colonies that were either dead or dormant were discarded. Hydroids on pelagic Sargassum and beach drift, as well as nearshore and offshore areas along the coast, were not included.

Bottom salinity samples were taken prior to each dredge tow with a Van Dorn bottle. These samples were analyzed in the laboratory using a Beckman RS7B induction salinometer. Depth data were recorded from a Raytheon DE-725B recording fathometer.

RESULTS AND DISCUSSION

Hydroids were heterogeneously distributed in estuaries of South Carolina, and the assemblage changed more or less continuously along the halocline (Fig. 2). Of the 40 species included, 27 were restricted to salinities of 18 °/oo or higher. When classified according to salinity-related ecological categories (Remane, 1971), the fauna

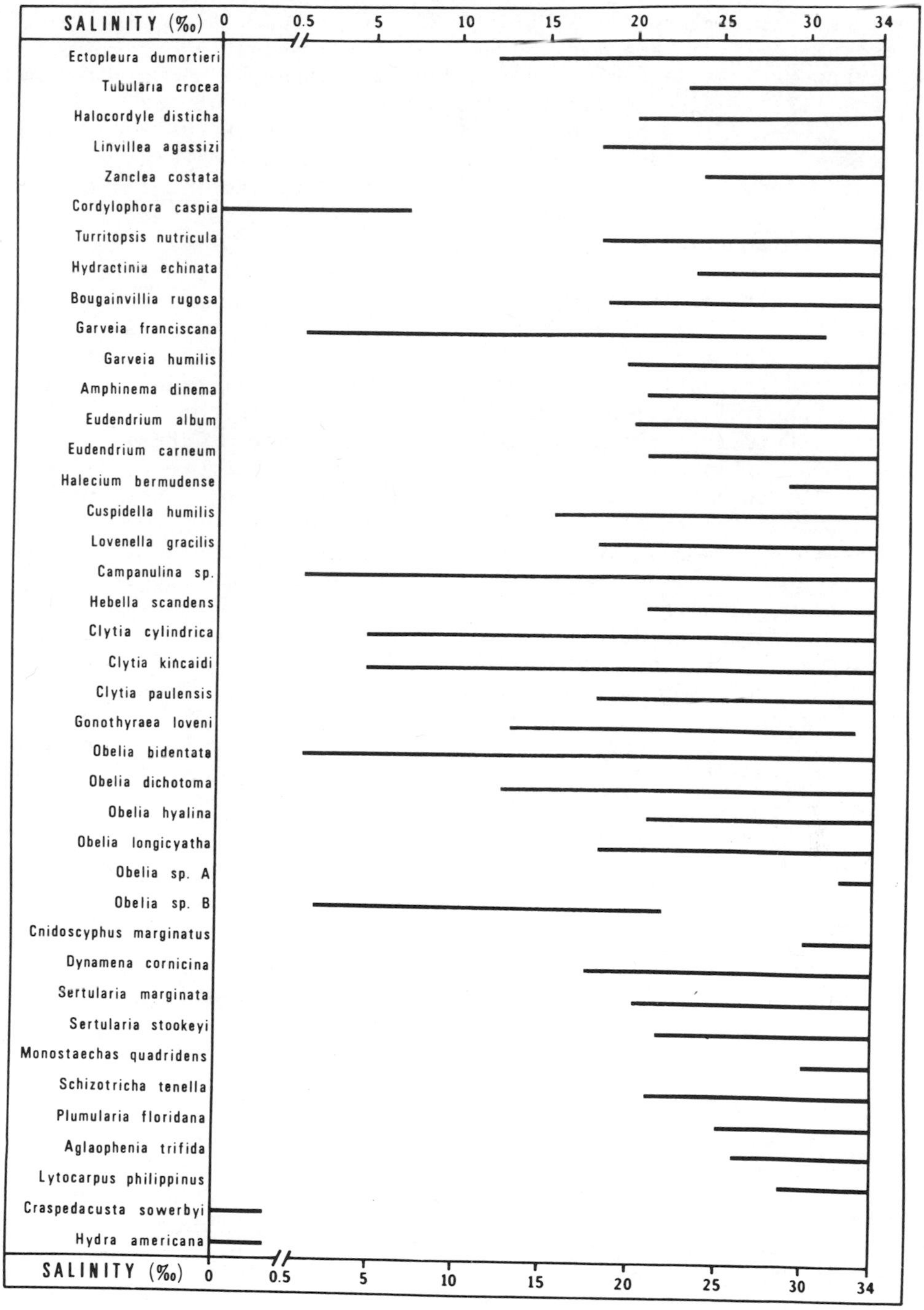

Fig. 2. The hydroids and their observed salinity distribution.

included three stenohaline marine, 32 euryhaline marine, and two
stenohaline limnic species, as well as three estuarine endemics.
Cordylophora caspia was categorized as an estuarine endemic following
Remane (1971), although it is also present in many strictly fresh
water areas. There were no hydroids common to both the euhaline and
limnetic zones in South Carolina waters. Species with the widest
distribution across the estuarine gradient were Campanulina sp.,
Obelia bidentata, and Garveia franciscana. All three were found from
miooligohaline waters to the euhaline zone.

The patterns of hydroid zonation in Fig. 2 do not necessarily
represent the absolute salinity tolerances of each species. The
interaction of many environmental factors may cause the field or
"ecological potential" of a species to differ substantially from its
actual "physiological potential" (Kinne, 1971; Schlieper, 1971). For
example, Cordylophora caspia is known to tolerate salinities of at
least 35 °/oo in the laboratory (Kinne, 1956), although it is normally
restricted in nature to salinities below 10-15 °/oo. In estuaries of
South Carolina during this study it was not observed in salinities
higher than 7 °/oo. Crowell and Darnell (1955) found that Garveia
franciscana grew under salinities ranging from 1-35 °/oo in the
laboratory, although it is normally most abundant under field conditions at salinities below 15 °/oo. The salinity tolerance of a
species may be further complicated by biological as well as physical
factors. Predation and interspecific competition may restrict the
distribution of a species to a narrower segment of an estuary than it
is physiologically adapted to occupy. Individuals of a given species
from different geographic areas may display different salinity tolerances, both in the field and in the laboratory (Kinne, 1971). Salinity tolerances are also known to vary from one stage to another in the
life history of some species, and dormant stages in particular are
characteristically more eurytopic than active phases (Kinne, 1971;
Schlieper, 1971).

A decline in number of species was evident along estuarine halo-
clines from a maximum in areas of high salinity to a minimum in the
limnetic zone (Fig. 3). While Remane (1934, 1971) and Kinne (1971)
have shown that overall species richness rises rapidly toward fresh
water from a minimum in the miomesohaline zone, few species of limnic
hydrozoans are known and no such trend was evident in this study.
With the addition of Cordylophora caspia to the fauna, an increase in
the number of species occurred in the lower mesohaline zone. However,
the remaining marine and estuarine species dropped out at progressively lower salinities, and the occurrence of Hydra americana and
Craspedacusta sowerbyi below 0.5 °/oo was not enough to compensate
for their disappearance.

In several respects the observed zonation corresponded rather
closely with the subdivisions of the Venice System (Symposium on the

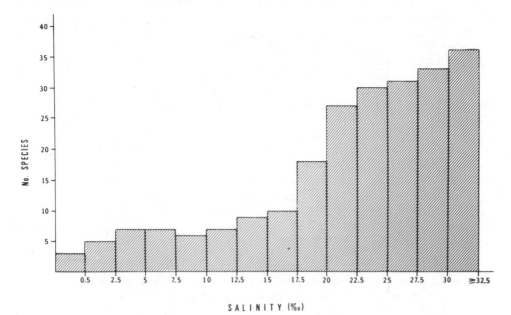

SALINITY (‰)

Fig. 3. Number of hydroid species collected in relation to salinity.

Classification of Brackish Waters, 1958). Faunal change appeared to
be most pronounced between the 15-20 o/oo isohalines, approximating
the mesohaline-polyhaline boundary. In addition, <u>Cordylophora</u> <u>caspia</u>
was the only species present in both the oligohaline and limnetic
zones, suggesting a faunal break near 0.5 o/oo. However, no note-
worthy discontinuity was observed at 5-8 o/oo, the border between the
oligohaline and mesohaline zones. Neither was there a marked change
in species composition at the lower limit of the euhaline zone. The
polyhaline-euhaline boundary would have been more distinct had the
number of stenohaline marine species been larger. Sampling at stations in permanently euhaline waters beyond the mouths of the estuaries would have increased the number of such species substantially.
However, part of any such increase in species richness could also be
attributed to the greater stability of water temperature in coastal
waters of South Carolina. Differences between the oligohaline and
mesohaline zones would have been accentuated if more euryhaline
limnic species had been represented, but few such hydrozoans exist.

The assemblage of hydroids collected at a given location depended primarily upon salinity, although presence of a suitable substrate
was also important. The number of species was typically maximal at
stations in areas of high and relatively stable salinity (Table I),
such as Price Creek (B002) and DeVeaux Bank (E008). The fewest

species were observed at stations in areas of low salinity or fresh
water, those which alternated between fresh and brackish conditions,
or those which underwent pronounced oscillations in salinity. While
salinity may be a master factor controlling animal distribution in
estuaries, its influence may be modified by a variety of factors and
factor combinations (Kinne, 1971). Relatively few species were collected at stations E004 and E006 during this study in the North
Edisto River, although salinity was comparable to adjacent areas of
the same river system having diverse hydroid assemblages. The scarcity of hydroids at these two stations was attributed to an apparent
lack of substrates suitable for hydroid attachment. In the case of
species undergoing seasonal activity-inactivity cycles, water temperature becomes an important limiting factor (Calder, 1971). _Garveia_
franciscana, a dominant member of brackish water epifaunal communities during summer and autumn in South Carolina estuaries, was largely dormant in winter across its entire salinity range.

The stress of highly variable salinity in reducing the number of
species present has been emphasized by Bacci (1954) and Dahl (1956),
among others. Boesch (1976) noted that the distribution of species
in a poikilohaline estuary is governed by conditions of low salinity,
and that species are displaced downestuary under poikilohaline conditions. His views are supported by the present study, particularly
from observations on the North and South Santee Rivers. Excluding
areas of strictly fresh water, only nine species of hydroids were
found in the entire Santee system during the study, although oyster
shells and other firm substrates were present in abundance at stations N001, N004, S001, and S004. The nine hydroids represented in
the samples included all three estuarine endemics (_Cordylophora_
caspia, _Garveia_ _franciscana_, and _Obelia_ sp. B) and six euryhaline
marine species (_Cuspidella_ _humilis_, _Campanulina_ sp., _Clytia_ _kincaidi_,
C. cylindrica, _Obelia_ _bidentata_, and _O. dichotoma_). Each of these
species may be found in the upper reaches of more homoiohaline estuaries. Although the salinity at stations near the mouths of the
North and South Santee Rivers occasionally approached or exceeded
30 °/oo, no stenohaline marine or euryhaline marine I species were
collected there during the study.

Short-term variations in salinity appear to have a greater impact on epifaunal organisms such as hydroids than on the infauna. In
the Pocasset River, Massachusetts, Sanders, Mangelsdorf, and Hampson
(1965) found that salinity was much more stable in the sediments than
in the overlying water column. As a result, the epifauna was subjected to greater physiological stress and was poorly represented compared with the infauna. For these reasons, infaunal and epifaunal
species of equal salinity tolerance may differ in their distribution
in estuaries (Carriker, 1967; Kinne, 1971).

Hydroids found in low salinity areas of South Carolina, as well

as those occurring in the poikilohaline Santee system, were also
present under relatively polluted conditions in the lower Cooper
River and Charleston Harbor. Seven of the nine species recorded in
estuarine regions of the Santee area were also collected at station
C003 on the Cooper River in an industrialized area, and six were
found at station C004 near a naval base. The remaining species were
present, along with several others, in polluted Charleston Harbor.
The species composition at station C003 was essentially indistinguishable from non-polluted or less-polluted areas of comparable salinity,
substrate, and geographic location elsewhere in the state (e.g. stations D003, H002, W001). This suggests that hydroids tolerant of both
low and widely variable salinity are also tolerant of at least certain
types of pollution stress. Boesch (1972, 1974) believed that benthic
organisms in mesohaline and oligohaline waters of the James River,
Virginia, were more resistant and resilient to perturbations, including salinity fluctuations and multiple pollution stress, than those
restricted to areas of higher salinity.

ACKNOWLEDGEMENTS

    I am indebted to J. M. Bishop, D. F. Boesch, R. J. Diaz, E. B.
Joseph, P. A. Sandifer, and M. H. Shealy, Jr., for their constructive
criticisms of the original manuscript. This study, Contribution No.
59 from the South Carolina Marine Resources Center, was supported in
part by funds from Coastal Plains Regional Commission Contract No.
10340031.

REFERENCES

Bacci, G., 1954. Gradienti di salinita e distributione di molluschi
    nel lago di Patria. Boll. Zool., 21.
Boesch, D. F., 1972. Species diversity of marine macrobenthos in the
    Virginia area. Chesapeake Sci., 13: 206-211.
Boesch, D. F., 1974. Diversity, stability and response to human dis-
    turbance in estuarine ecosystems. Proc. First Int. Congr. Zool.,
    The Hague, 109-114.
Boesch, D. F., 1976. A new look at the zonation of benthos along the
    estuarine gradient. In press in B. C. Coull, Ed., Ecology of
    Marine Benthos. Univ. South Carolina Press, Columbia.
Calder, D. R., 1971. Hydroids and Hydromedusae of southern Chesapeake
    Bay. Virginia Inst. Mar. Sci., Spec. Pap. Mar. Sci., 1: 1-125.
Carriker, M. R., 1967. Ecology of estuarine benthic invertebrates: A
    perspective. Pages 442-487 in G. H. Lauff, Ed., Estuaries.
    Amer. Ass. Adv. Sci. Pub. 83.
Clausen, C., 1971. Interstitial Cnidaria: Present status of their
    systematics and ecology. Pages 1-8 in N. C. Hulings, Ed., Pro-
    ceedings of the first international conference on meiofauna.
    Smithsonian Contrib. Zool. 76.

Crowell, S., and R. M. Darnell, 1955. Occurrence and ecology of the hydroid Bimeria franciscana in Lake Pontchartrain, Louisiana. Ecology, 36: 516-518.

Dahl, E., 1956. Ecological salinity boundaries in poikilohaline waters. Oikos, 7: 1-21.

Gunter, G., 1950. Seasonal population changes and distributions as related to salinity, of certain invertebrates on the Texas coast, including the commercial shrimp. Pub. Inst. Mar. Sci., 1: 7-51.

Gunter, G., 1961. Some relations of estuarine organisms to salinity. Limnol. Oceanogr., 6: 182-190.

Kinne, O., 1956. Zur Ökologie der Hydroidpolypen des Nordostseekanals (Laomedea loveni Allman, Cordylophora caspia Pallas, Perigonimus megas Kinne). Z. Morph. Ökol. Tiere, 45: 217-249.

Kinne, O., 1971. Salinity. Animals. Invertebrates. Pages 821-995 in O. Kinne, Ed., Marine ecology. A comprehensive integrated treatise on life in oceans and coastal waters. Wiley-Interscience, New York.

Meyer, H. U., 1973. Über den Einfluss von Milieufaktoren auf die Hydroidenfauna der Kieler Bucht. Kieler Meeresforsch., 29: 69-75.

Percival, E., 1929. A report on the fauna of the estuaries of the River Tamar and the River Lynher. J. Mar. Biol. Ass. U. K., 16: 81-108.

Pritchard, D. W., 1955. Estuarine circulation patterns. Proc. Amer. Soc. Civil Engineers, 81 (717): 1-11.

Remane, A., 1934. Die Brackwasserfauna. Zool. Anz., 7 (suppl.): 34-74.

Remane, A., 1971. Ecology of brackish water. Pages 1-210 in A. Remane and C. Schlieper, Biology of brackish water. Second edition. Wiley-Interscience, New York.

Sanders, H. L., P. C. Mangelsdorf, Jr., and G. R. Hampson, 1965. Salinity and faunal distribution in the Pocasset River, Massachusetts. Limnol. Oceanogr., 10 (suppl.): R216-R229.

Schlieper, C., 1971. Physiology of brackish water. Pages 211-350 in A. Remane and C. Schlieper, Biology of brackish water. Second edition. Wiley-Interscience, New York.

Symposium on the Classification of Brackish Waters, 1958. Symposium on the classification of brackish waters, Venice, 1958. Oikos, 9: 311-312.

Thiel, H., 1970. Beobachtungen an den Hydroiden der Kieler Bucht. Ber. Deutsch. Wiss. Komm. Meeresforsch., 21: 474-493.

Wells, H. W., 1961. The fauna of oyster beds, with special reference to the salinity factor. Ecol. Monogr., 31: 239-266.

RESPONSES OF HYDRA OLIGACTIS TO TEMPERATURE AND FEEDING RATE

Barbara Hecker and Lawrence B. Slobodkin

State University of New York at Stony Brook

Stony Brook, New York 11794

If one is concerned with the mechanisms of adaptation to environmental disturbances on the non-genetic level, hydra are particularly appropriate for several reasons. First, the problem is somewhat simplified since the behavioral repertoire of hydra is severely limited, the morphology at least on a gross and light microscopy level is reasonably simple and a spectrum of responses ranging from the cellular to the behavioral can be recorded on genetically identical individuals. As a consequence of this simplicity it ought to be possible to catalogue the response of hydra to change in such a way that we can distinguish between the rapid and slower responses and get some sense of what the implications of each of these responses might be for the further adaptive capacities of the hydra themselves. The rationale for this type of inquiry is presented in Slobodkin and Rapoport (1974) and papers cited therein.

MATERIALS AND METHODS

Hydra oligactis (Forrest, 1959) from Carolina Supply House, which have been maintained in our laboratory for the past four years were used for these experiments. Stock cultures at each of three temperatures had been maintained at that temperature for at least six months prior to the onset of all experiments. These stock cultures were fed unlimited amounts of Artemia nauplii several times a week. The animals were maintained in "M" solution as modified by Lenhoff and Brown (1970). All weighings were performed of freeze dried individuals on a Cahn electro balance. Cell counts were conducted by macerating hydra (David, 1973), and counting the resulting solution in a Levy counting chamber (Bode, et. al., 1973). Floating

175

was measured in Carolina culture dishes (3 cm. depth). The shape
of dish for this type of measurement is significant as indicated by
Ritte (1969).

RESULTS

Size and Temperature

The relation between size and temperature in Hydra oligactis
is indicated in Table I. The weight of hydra varies inversely with
temperature. The data in Table I was collected from freeze-dried
specimens by weighing adult hydra individually and buds in groups
of 10 at 13°C, 20°C and 27°C. Weights of immature hydra (non-
budding) at 15°C, 20°C and 25°C indicated the same inverse
relationship. Hydra were also disassociated and estimates of total
cell number were made (Table II). The probable error of these
estimates is rather small as indicated by Bode, et. al. (1973). It
will be seen that the ratios of weights at the different temperatures
are essentially identical with the ratios of the cell numbers at those
temperatures indicating that the weight difference is due to an
increment in cell number rather than an increment in individual cell
size.

Table I. Dry weights (mg. $\times 10^{-2}$) of adults and detached buds
maintained at three temperatures. Adults (10) were
weighed individually and buds (140) were weighed in
groups of ten.

Temperature	Buds	Adults
13°C	4.95 ± 0.60	50.39 ± 12.01
20°C	3.83 ± 0.67	38.50 ± 6.80
27°C	2.12 ± 0.69	22.51 ± 4.01

Table II. Dry weights and cell counts (total cells per hydra) of
non-budding hydra maintained at three temperatures.

Temperature	Weight (mg. $\times 10^{-2}$)	Number of cells ($\times 1000$)
15°C	6.06 ± 1.30	163 ± 15
20°C	3.94 ± 0.87	103 ± 16
25°C	1.72 ± 0.35	52 ± 8

Budding Rate as a Function of Temperature and Food Supply

In Fig. 1 we present the relation between budding rate, temperature, and the frequency of ad lib feeding. The animals for this study were taken from stock cultures at 13°C, 20°C and 27°C and subjected to feeding periodicities of every 1st, 2nd, 4th and 8th day.

It is apparent from these results that at 27° the budding rate can be considerably higher than at 13°, if the feeding frequency is sufficiently great. However, with a diminution of the frequency of feeding the number of buds/hydra/day declines much more rapidly at the higher temperature. At 27° the hydra population cannot in fact maintain itself with feedings less frequent than one every four days. Since these studies were done with ad lib feeding by flooding the animals with Artemia, there is every reason to believe that the same general result would occur on more restricted feedings.

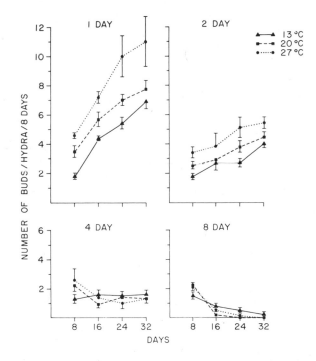

Fig. 1. Weekly budding rates of hydra maintained at three temperatures and on four feeding periodicities. Points are averages of 5 groups of 10 hydra each.

Studies on Response to Temperature Change

Hydra that had been maintained at each of the three
temperatures, 13°, 20°, and 27° were transferred to the other
possible temperatures or were kept at the initial temperature as a
control. This generated a 3 x 3 grid. For example, we had hydra
from 13° transferred to 20° or 27° or maintained at 13° as a control.
During the entire course of the experiment weekly measurements were
made of budding rate. The hydra were maintained at their new
temperatures, which will be called the stress temperatures, for
a period of three or four weeks and were then returned to their
initial temperature, which will be called the <u>acclimated</u> <u>temperature</u>.
Then after a period of three weeks they were once again transferred
to the stress temperature. As seen in Figs. 2 and 3 the budding
rate altered drastically on the temperature change.

In the case of animals that were moved from cooler to warmer
temperatures, the budding rate increased very rapidly becoming
higher than that of the controls. It then fell below that of the
controls and prior to the termination of the first stress period
had essentially approximated that of the controls. When these
animals were then returned to their acclimated temperatures they
did not revert to the budding rates of the controls. When after a
period of three weeks the animals were again subjected to stress
temperatures, the accommodation to the new temperature was
considerably more rapid, showing much less overshooting (Fig. 2).

Essentially the reverse occurred when the animals were moved
from the higher temperatures to lower temperatures. There was a
significant drop in budding rate, followed by an increase to well
above that of the controls. [Daily <u>ad</u> <u>lib</u> feeding reduces budding
rate at 13°C and 20°C (cf: Discussion)]. This again resulted in an
overshooting in the response of budding rate to temperature. On
reversion to their initial, warmer, temperatures there was a very
rapid return to the budding rate of the control. On the second
stress the response to the lowering of the temperatures was again
a drop in budding rate but considerably less than that on the
occasion of the first stress (Fig. 3).

Floating as a Function of Temperature and Food

In Table III are shown the results of measurements of floating
rate. The floating animals were predominantly buds that were
produced by adults that were maintained at low densities (1 hydra/
15 ml) in 3 cm. deep solution. At 27° there was essentially no
floating at all regardless of frequency of feeding. At 20° floating
was found when feeding was relatively infrequent and was not found
when the food supply was frequent and abundant. This corresponds

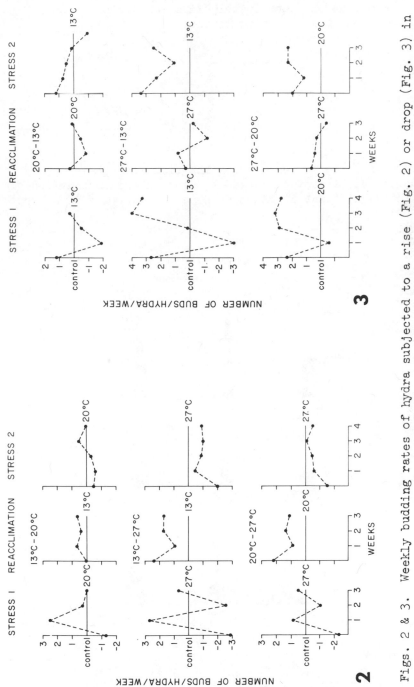

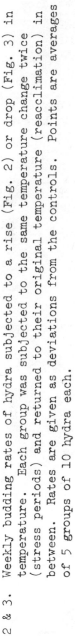

Figs. 2 & 3. Weekly budding rates of hydra subjected to a rise (Fig. 2) or drop (Fig. 3) in temperature. Each group was subjected to the same temperature change twice (stress periods) and returned to their original temperature (reacclimation) in between. Rates are given as deviations from the controls. Points are averages of 5 groups of 10 hydra each.

with the results of Łomnicki and Slobodkin (1966) who did their
studies on floating at approximately this temperature. At 13°
floating occurred at all feeding levels.

DISCUSSION

It will be noted that the buds produced at 13° are considerably
heavier than those produced at 27°. The ratio of adult body size
to bud size is essentially constant at approximately 10:1 at all
temperatures.

Maturation time is measured as the time from the dropping of
a new bud to the first appearance of a bud anlage on that individual.
There is no significant difference in maturation time as a function

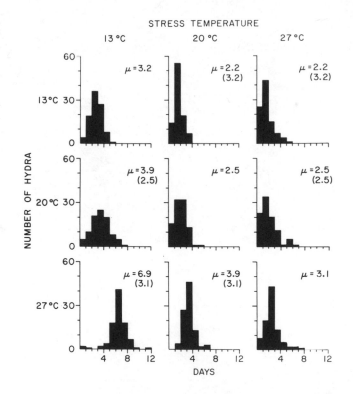

Fig. 4. Maturation time of hydra subjected to temperature changes.
         Maturation time is the time interval between detachment of
         a bud to the first appearance of a bud anlage on that
         individual. The graphs on the diagonals are the controls.

of temperature if temperature is held constant but maturation time
can be materially influenced by altering temperatures (Fig. 4).
Maturation time can be shortened by taking a bud produced at 13°
and transferring it to 27°. While at each temperature the weight
of a bud is approximately 1/10 that of its parent, the weight of
a newborn bud at 13° is approximately 1/2 the weight of an adult
27° animal. From this and the floating data we infer that in a
stream or pond in which several water temperatures are available,
those hydra in the cooler water temperatures will be sending out
emigrants to the warmer water temperatures. From the very warm
water no animals will be emerging. Animals from the cooler water
temperatures will begin reproducing quite rapidly in the warmer
water.

Table III.   Floating in detached buds of hydra maintained at three
             different temperatures and on four feeding schedules.
             The values are the percent of buds produced that were
             floating. Number in ( ) is the total number of buds
             produced.

| Feeding | Temperature | | |
Periodicity	13°C	20°C	27°C
1	62 (527)	6 (639)	1 (888)
2	57 (242)	49 (358)	3 (420)
4	54 (136)	64 (92)	3 (102)
8	61 (28)	100 (1)	0 (4)

Table IV.   Population increase, $d \frac{N_{16}}{N_1}$, over a 16 day interval.

             Each determination is based on five replicates, each
             starting with 5 adult hydra that had been maintained
             on the feeding schedules for 32 days.

| Feeding | Temperature | | |
Periodicity	13°C	20°C	27°C
1	9.60 ± 2.20	42.76 ± 5.62	81.12 ± 12.88
2	6.28 ± 0.63	10.16 ± 1.13	14.16 ± 0.85
4	3.04 ± 0.09	3.04 ± 0.48	3.00 ± 0.47
8	1.88 ± 0.58	1.24 ± 0.22	0.84 ± 0.17

The tolerance to feeding frequency differences is considerably less in warm water than in cold water (Table IV). For reasons indicated in Slobodkin (1964), the possible array of food particle sizes that can be consumed is considerably greater for large hydra than small ones.

We, therefore, conclude that while Hydra oligactis is physiologically capable of maintaining itself in warm water this is by no means an optimal ecological situation, despite the fact that the intrinsic rate of natural increase that can be achieved on the given high food level in warm water is considerably higher than that that can be achieved at the same feeding level in cold water. It is of critical significance that at 13° and at a slower rate at 20° ad lib feeding daily is not totally beneficial to the animal. In fact, continual ad lib feeding results in an eventual depression of the budding rate and the development of sexuality at 13° and morphological abnormalities at 20°. We suspect that this is related to the necessity that the hydra maintain correlated rates of feeding, budding and development and the incapacity to divert food that exceeds these legitimate rates to any storage organ (e.g. fat body).

We believe that these results are of material significance in defining and understanding such concepts as ecological niche optimality particularly since in this case the conditions associated with maximal fecundity are obviously not optimal.

Acknowledgements: This research was supported by NSF Grant 31-756A in General Ecology and by the General Electric Foundation. The data in this paper will be included in a Ph.D. dissertation by Barbara Hecker. This paper is contribution Number 178 in the Program of Ecology and Evolution at the State University of New York at Stony Brook.

SUMMARY

Floating, body size, budding rate and cell number were measured in Hydra oligactis as a function of temperature and food supply. At higher temperature body size was smaller, growth and budding rates were higher while floating rates and resistance to starvation were lower.

Therefore, in H. oligactis the temperature and feeding conditions that result in maximal growth and reproduction are clearly not optimal for population survival. This raises the theoretical issue of redefining optimality in terms other than reproductive rate.

REFERENCES

Bode, H., S. Berking, C.N. David, A. Gierer, H. Shaller, and
    E. Trenkner. 1973. Quantitative analysis of cell types
    during growth and morphogenesis in Hydra. Wilhelm Roux'
    Arch. Entwicklungsmech. Organismen, 171:269-285.

David, C.N. 1973. A quantitative method for maceration of hydra
    tissue. Wilhelm Roux' Arch. Entwicklungsmech. Organismen,
    171:259-268.

Forrest, H. 1959. Taxonomic studies on the hydras of North
    America. VII. Description of Chlorohydra hadleyi, new
    species, with a key to the North American species of hydras.
    Amer. Mid. Nat., 62:440-448.

Lenhoff, H. and R. Brown. 1970. Mass culture of hydra: an
    improved method and its application to other aquatic
    invertebrates. Lab. Animals, 4:139-154.

Łomnicki, A. and L.B. Slobodkin. 1966. Floating in Hydra
    littoralis. Ecology 47:881-889.

Ritte, Uzi. 1969. Floating and sexuality in laboratory populations
    of Hydra littoralis. Ph.D. Thesis, Univ. of Michigan.

Slobodkin, L.B. 1964. Experimental populations of hydrida.
    J. Anim. Ecol. 33 (Suppl.); 131-148.

Slobodkin, L.B. and A. Rapoport. 1974. An optimal strategy of
    evolution. Quart. Rev. 49,3:181-200.

Biology of Ctenophores,
Siphonophores, and Medusae

A LARGE-SCALE EXPERIMENT ON THE GROWTH AND PREDATION

POTENTIAL OF CTENOPHORE POPULATIONS

M. R. Reeve and M. A. Walter

Rosenstiel School of Marine and Atmospheric Science
University of Miami
4600 Rickenbacker Causeway, Miami, Florida, 33149

INTRODUCTION

In reviewing the ecology of planktonic ctenophores, Walter (1976) pointed out how laboratory estimations of such parameters as growth, feeding and fecundity had been utilized to interpret population dynamics of ctenophores in the natural environment. In organisms with short life-cycles, which inhabit the water column, it is usually impossible to follow the growth and mortality characteristics of cohorts by the use of such techniques as size/frequency histograms. The observer can never be sure he is sampling the same population, over a period of time, and even if this were so, continuous breeding and rapid growth tend to quickly obscure the identification of distinct cohorts. It is also not possible to experimentally manipulate the population. This report describes an experiment in which plankton populations were studied in three containers for six weeks following empoundment of a water column and its natural plankton assemblage.

METHODS

The design and operation of closed transparent polyethylene cylinders 15 m deep and holding 68 m^3 of water supported by flotation collars from the sea surface at Saanich Inlet, British Columbia, were described by Parsons (1974).

Twice weekly, paired 20-cm mouth diam. nets of 243-μm mesh on a "bongo" frame were vertically hauled up through the water column.

The procedure was repeated, yielding four samples from each container on each occasion. Each sample represented $0.45m^3$ of water
filtered. Formalin-preserved samples were counted for total zooplankton (excluding ctenophores) and standard deviations of replicates calculated. All preserved ctenophores (i.e. Pleurobrachia)
were measured (oral/aboral pole) except where they exceeded 30/sam-
ple, but in any case all animals exceeding 3.5 mm were measured in
every sample.

When it became clear that there was a population of the lobate
ctenophore Bolinopsis in the containers, a single vertical haul was
made with a 50-cm diam. 500-µm mesh net and 5-liter cod-end reservoir in each container on 4 occasions. Since these animals
disintegrate when preserved, and as a result of handling in the
unpreserved state, samples were returned to the laboratory unpreserved within a few minutes and all the animals in every sample
were measured immediately. Dry weight and organic carbon content of
Pleurobrachia over their entire size range were measured as described by Reeve and Baker (1975). In the case of Bolinopsis these
values were taken to be the same as those determined by those
authors for the related genus Mnemiopsis. The clearance rate, or
volume of water swept clear of food by Pleurobrachia of various
sizes (see Gauld, 1951; Kremer, 1975) was measured overnight in 48
separate experiments at various food concentrations. To three
containers (designated L, N and O) nutrient additions were made
frequently in an effort to stimulate primary production by maintaining ambient nitrogen and silicon concentrations of 2.5 µg-A/l
and phosphorus at 0.25 µg-A/l. A variety of other parameters were
regularly measured by the Controlled Ecosystem Pollution Experiment
group, including plant pigments and ^{14}C assimilation of phytoplankton.

Measurements of growth rate of Pleurobrachia were made in the
laboratory at $13^\circ C$ (near to the mean environmental temperature)
with abundant food in the form of live copepods, starting with
larvae hatched out from mature adults, according to methods of Baker
and Reeve (1974).

The containers were first sampled July 23, 1975. On August 6
and 14, 10 vertical hauls were made in container N with a 1-m mouth
diam. 500-µm mesh net in an effort to remove ctenophores. On the
assumption that the animals were immediately randomly distributed
preceding each tow, the procedure should have removed 82% of the
total population retained by the mesh on each occasion. Inspection
of the hauls showed that no other organisms were retained in the
nets. The collected ctenophores were transferred live to container
0. In additon, on August 18, some two liters of Pleurobrachia,
which had been collected with a seine net of mesh size 13 mm from a
surface of a separate 10-m diam. container, were added to container

0. This can be estimated to be equivalent to a final concentration of at least 50, 10-mm animals/m^3, using conversion factors provided by Hirota (1972) and Reeve and Baker (1975).

RESULTS AND DISCUSSION

Dry weight was related to length in Pleurobrachia by the expression log dry weight (mg) = -1.54 + 2.65 log length (mm), and bodily carbon content was 3.6-5.3% of dry weight. Figure 1 depicts size frequency histograms (numbers/m^3 against length) for 12 sampling days for containers L, N and 0 (except that container L was not sampled on the last two dates). Approximate periods over which ctenophores were removed are also indicated. An accident resulted in loss of some ctenophores from container 0 on August 8. In Figure 2 changes in the biomass of chlorophyll, copepods (numbers in the 243-μm mesh net) and Pleurobrachia and Bolinopsis (expressed as organic carbon) are presented, also with indications of periods of ctenophore removal and addition. Figure 3 presents growth rate curves of two laboratory populations of Pleurobrachia at 13°C and a comparison curve for 20°C on a logarithmic axis in terms of organic carbon. The inset contains the results of 48 experiments of water clearance rate expressed as ml/day for a range of sizes of Pleurobrachia. Although there appeared to be an indication that lower food concentrations produced higher clearance rates, the data were not significant statistically, and a single regression line was calculated for the pooled data.

The initially captured Pleurobrachia population was very low in all containers, consisting of a few animals between 3 and 7 mm over the first two sampling days, ranging from less than 1/m^3 in L to 3/m^3 in 0. In absolute terms, these data are highly unreliable because the 4 replicate tows from each container represented only 1.8 m^3 of water filtered. All that can be claimed is that there were a few animals present capable of sustained reproduction (see Walter, 1976). By August 1, a small population of young were in evidence. By the next date (August 5) more young had been produced and there were signs of growth of those hatched earlier. Sixteen days from the beginning (August 8) there were 11, 3 and 1 animals/m^3 in containers L, N and 0 respectively over 5 mm, which in L must have been produced in the containers because there were no large animals on prior sampling dates. Assuming, conservatively, that animals had reached only 5 mm from the beginning of the experiment (i.e. 16 days) their body carbon weight would be 106 μgC, a size which the laboratory-raised, abundantly-fed, animals required 20 days to reach.

Pleurobrachia is capable of very high growth rates, at least as fast as the small copepods which comprise its food (e.g. Paffenhöfer and Harris, in press) and considerably faster than those

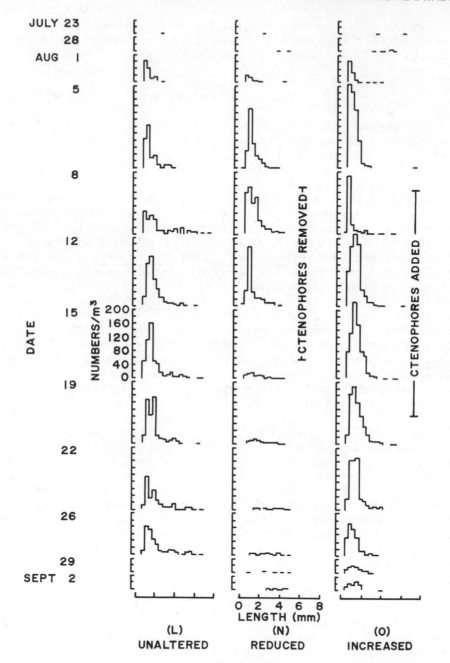

Figure 1.   Size/frequency histograms for populations of Pleuro-brachia in containers L (unaltered), N (ctenophores removed) and 0 (ctenophores added).   Units--numbers/m³ against length in mm.   No samples were taken from container L on the last two sampling dates.

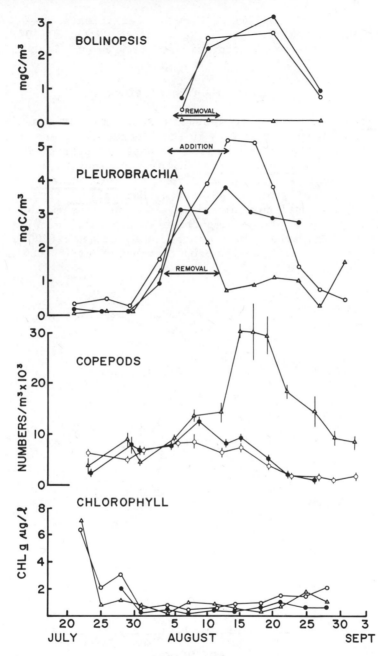

Figure 2. Changes in biomass of chlorophyll, copepods, Pleurobrachia and Bolinopsis in the units indicated for containers L (●)
N (▲ , ctenophores removed) and 0 (O , ctenophores added). Vertical bars (copepods)--standard deviation of replicates.

reported by Hirota (1974). Growth coefficients of 0.47 (over 50%/day) during the first three weeks and 0.09 (about 10%/day) for the next two weeks may be approximated from Figure 2 at 13°C.

Undisturbed Populations

Pleurobrachia grew at least as fast under the semi-natural conditions of the containers as in the laboratory, where an abundant food supply was always maintained. We have no estimate of the total food available to Pleurobrachia in the containers. In the 243-μm mesh net, older stages of the copepod Pseudocalanus dominated initially, gradually being replaced by Paracalanus (Koeller, Grice and Gibson, personal communications) with numbers in the range of 5,000/m³ (Figure 2). More importantly, probably, considering the requirement for smaller food of the newly-hatched (200-μm, cteno-phores) was a peak of naupliar production of about 150,000/m³ at the

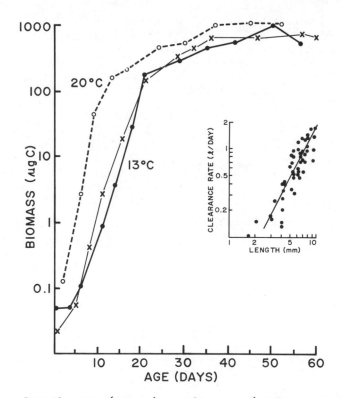

Figure 3. Growth rate (organic carbon, μgC) of two populations of Pleurobrachia at 13 and one at 20°C. Inset--Water clearance rate (liters/day) of Pleurobrachia at 13°C as a function of length (mm), with computed regression line.

end of July (Beers, personal communication). Over this period, the ctenophore biomass of container L, for instance, increased at a daily rate of 25%.

Other points of interest over this part of the experiment are the high fecundity of ctenophores (see review of Walter, 1976) which enabled the initial small seed population to rapidly produce a new cohort of young, and the high mortality rate in this growing cohort, although there were no known predators in the containers. Because of this pattern of mortality, and the continuous recruitment of young, the size/frequency histograms in container L attain a fairly constant pattern where interpretation of successive cohorts becomes very doubtful. This condition, as noted in the Introduction, is typical of short generation zooplankton populations.

The data for container L, in which the Pleurobrachia population was not manipulated at any time, can be used to calculate production of the population in a manner similar to that described by Mullin and Brooks (1970), Hirota (1974) and Reeve and Baker (1975). The data are summarized in Table 1. The size/frequency categories of Figure 1 were consolidated into five groups, selected to represent equal (7-day) development periods on the basis of Figure 3. Growth coefficients for each category were obtained as noted above, and a single overall mortality coefficient for the entire 5 categories (0.35) derived from the slope of the logarithm of numbers of each size category against mean size. Except for the first category, numbers for each category were inversely related to size, and could be approximated to a straight line. The first category was ignored in this computation because it was thought to have been inadequately sampled.

The third size category, representing the period immediately prior to the onset of maturity, was responsible for some three-quarters of the total production averaged over the whole period. Although production of category 1 was underestimated by using the observed numbers, if a theoretical number is computed by extrapolation of the survival line described in the preceding paragraph (which would increase the numbers by 1.5 orders of magnitude) production would approximate that of the second size category. Over the 10 sampling dates, total computed biomass and production in mg organic carbon/m^3 is shown in Table 2.

By August 13, the ctenophore population reached a biomass and production plateau which was maintained remarkably constant throughout the rest of the period. Taking the last 6 sampling dates, the mean population biomass was 6.01 mgC/m^3 + 1.28 standard deviation, and mean daily production was 2.68 mgC/m^3, yielding a daily production/biomass ratio of 0.45. This very high ratio is to some extent a reflection of the fact that the population was mostly composed of young, fast-growing animals. Reeve and Baker (1975) estimated a mean ratio of 0.12 for a whole year in Biscayne Bay, Florida, with a peak of 0.2 at the time of the seasonal rapid population increase, and Hirota (1974) found values of 0.02 off California.

Since the staff of the facility were also measuring primary production (^{14}C assimilation) during the experiment, it is possible

TABLE 1

SIZE CLASS	A	B	C	D	E
Length Range (mm)	<0.8	0.8–1.9	2.0–5.1	5.2–7.1	7.2–8.3
Duration (days)	7	7	7	7	7
Mean Biomass/Animal (μgC)	0.1	4.1	63	209	364
Mean Biomass/m^3 (mgC)	0.003	0.72	2.63	0.34	0.12
Mean Production/m^3 (mgC)	0.002	0.40	1.32	0.027	0.010
Mean Water Clearance Rate (ml/animal/day)	5	36	240	710	1,150

TABLE 2

	July		August							
	23	28	1	5	8	12	15	19	22	26
Total Biomass mgC/m^3	0.07	–	0.31	1.75	4.40	7.55	7.57	5.63	5.50	5.41
Total Production mgC/m^3	0.04	–	0.15	0.87	1.55	3.68	3.45	2.76	2.37	2.25
Total Water Clearance Liters/m^3/day	0.26	–	2.59	10.8	17.7	35.7	34.6	28.5	24.3	24.1

to compute that the ratio of Pleurobrachia to primary production over the same period was 4.3%. If Bolinopsis had been taken into account, this ratio would be somewhat higher. This ratio may be compared with that derived by Mullin and Evans (1974) of 3% for a shore-based tank of comparable size. It is tempting to speculate in the light of the apparent stability of population size once the initial build-up was achieved, and the high mortality rate in the absence of predators, that the population was in some way regulating itself, with the effect of conserving its food supply in a manner analogous to that suggested by Steele (1974) and others for her-bivorous copepods which may cease feeding at low "threshold" levels of phytoplankton. The ctenophores appeared to be limiting their numbers rather than their feeding activity. Walter (1976) reviewed laboratory experimental data which suggested that although cteno-phores might exert some food intake regulation, ingestion rates nevertheless continued to increase up to extremely high food con-centrations. It is also interesting that such speculation does not confirm the impressions of field workers who associate ctenophore blooms with eradication of other zooplankton. It is not profitable to extend such speculation further at this point because there are a number of interrelated factors which could affect the situation on which we have less adequate data.

When it became clear that there were significant numbers of Bolinopsis present we made an attempt to assess their biomass. Their extreme fragility to all forms of handling, including manipu-lations in the net and laboratory, made them extremely difficult to quantify and we regard the accuracy of our biomass estimations (Figure 2) as only within an order of magnitude of the true value. It is clear, however, that they were of the same order of magnitude of biomass on two occasions as the Pleurobrachia populations and were competing for the same food source.

Following Kremer (1975) an estimation of the total Pleuro-brachia water clearance rate capacity for container L, expressed in liters of water/m^3, or a percentage of the total water column, can be made from the relationship of Figure 3 (inset). Table 1 contains the estimated water clearance rate for the average individual of each size class, and Table 2 records total water clearance for each sampling day in liters/m^3. Taking the peak value on August 13, and assuming that the Bolinopsis population was responsible for roughly the same intensity of predation, it may be calculated that the total peak clearance rate of ctenophores was of the order of 70 liters/m^3, or some 7% of the total container volume. Kremer (1975) provided some similar total clearance estimations from her own data and the data of other workers for natural ctenophore populations, which suggested that such indices rarely exceeded 10%/day.

Manipulated Populations

From our experience on the survival of net-collected ctenophores in the laboratory, we would expect that most of the Bolinopsis collected and transferred to container O would not survive, but
a significant fraction of the Pleurobrachia would survive. Inspection of the upper two graphs of Figure 2 tends to confirm this,
because, during the period of removal and addition, container O
increased its Pleurobrachia population in comparison with the unmanipulated container L, but there were no changes in the relative
Bolinopsis population biomass. Bolinopsis was virtually eliminated
from container N, the process quite possibly damaging animals which
were not actually collected. The Pleurobrachia population was
reduced to less than a third of its former level.

The clearest effect of these manipulations (in the sense that
it would have been predicted) is the very rapid increase in the
older stages of the copepod Paracalanus over a period of 10 days, to
concentrations some four times higher than those characteristic of
all the containers on August 5, while the numbers in L and O
remained relatively constant. This dramatic increase is presumed
attributable by ourselves and Parsons (personal communication) to a
developing cohort of copepods resulting from the naupliar peak
referred to earlier, the bulk of which were cropped by the ctenophores of containers L and O before they could grow to a size large
enough to appear in the 243-μm mesh of the sampling net. There are,
however, several phenomena which may not be accounted for so
readily.

The Pleurobrachia population in N, although much reduced, was
significantly larger both in biomass and size of individuals, than
at the beginning of the experiment, but failed to make the same
rapid recovery (unless the last sampling date is an indication of
this). Beers (personal communication) indicated that naupliar
populations remained low throughout the period. Since we seriously
underestimated the numbers of newly-hatched ctenophores, we do not
know if they were produced but failed to grow, or were not produced.
Even more surprising was how little increase in ctenophore biomass
could be observed in container O relative to L, particularly
following the large addition of large Pleurobrachia estimated as at
least 50/m^3. Traces of them or any increase of offspring were not
evident on subsequent sampling dates (Figure 2), and instead there
occurred a rapid population decline to an ultimate level below that
of N from which the animals had been removed. As noted above, it is
unlikely that handling could account for the death of all of them,
and in any case higher mortalities (relative to L) must have been
occurring in the fraction of the population which originated in O.
Superficially, at least, it appears as if the sudden addition of
ctenophores could have produced an over-compensation in the mor-

tality pattern which we speculated above might be responsible for the stability of biomass in the undisturbed container L.

Ctenophores did not rapidly eradicate the zooplankton population in container L once their population had stabilized (as noted earlier) although a gradual decline to about $1,000/m^3$ began after August 15. There was a small difference in the copepod standing stock between L and 0, reflecting the increased numbers of ctenophores in the latter, but this was eliminated towards the end of the experiment as ctenophores in 0 became rapidly reduced. The most striking phenomenon is the rapid decline in copepods from container N following their peak. Predators had been much reduced, but food supply also appears to have been limiting. Reference to Figure 2 (bottom) indicates that phytoplankton standing crop, following the demise of a bloom when the water column was captured, remained below 1 μg/l in all containers for most of the period. This is in agreement with the observation of Reeve et al. (in press) that copepod populations in the range of $2,000-5,000/m^3$ (as estimated from a 200 μm-mesh net) can control the accumulation of phytoplankton biomass in such containers, and suggests that copepods were food-limited. The decline of copepod standing stock, even in containers L and 0, then, cannot be attributed with certainty solely, or even largely to predation.

ACKNOWLEDGEMENTS

This work was supported by National Science Foundation grants DES75-17761 (Biological Oceanography) and ID073-09759 A01 (IDOE). The authors wish to express their grateful thanks to all Principal Investigators and staff of the Controlled Ecosystem Pollution Experiment at Saanich Inlet, British Columbia, for use of their data and assistance.

SUMMARY

An experiment is described in which ctenophore populations in a group of three large containers floating at the sea surface were monitored over 41 days. The growth of a cohort of Pleurobrachia from a small seed population occurred within 2-3 weeks, which was faster than the growth of laboratory reared populations at the same temperature. The population reached a production plateau of 2.7mgC/day with a daily production/biomass ratio of 0.45, and was calculated to clear 3.5% of the total container water volume daily. About 4% of primary production was converted to Pleurobrachia biomass. In a container from which 70% of the ctenophores were removed the copepod population increased 4 times compared to an undisturbed container. There were high mortalities of both copepods and ctenophores which could not be related to predation.

LITERATURE CITED

Baker, L. D. and M. R. Reeve, 1974. Laboratory culture of the lobate ctenophore Mnemiopsis mccradyi with notes on feeding and fecundity. Mar. Biol. 26: 57-62.

Gauld, D. T., 1951. The grazing rate of planktonic copepods. J. Mar. Biol. Ass. U. K. 29: 695-706.

Hirota, J., 1972. Laboratory culture and metabolism of the planktonic ctenophore, Pleurobrachia bachei A. Agassiz. Pages 465-484 in A. Y. Takenouti et al., Eds., Biological Oceanography of the Northern North Pacific Ocean. Idemitsu Shoten, Tokyo, 626 pp.

Hirota, J., 1974. Quantitative natural history of Pleurobrachia bachei in La Jolla Bight. Fish. Bull. U. S. 72: 295-335.

Kremer, P. M., 1975. The ecology of the ctenophore, Mnemiopsis leidyi in Narragansett Bay. Ph.D. dissertation, University of Rhode Island.

Mullin, M. M. and E. R. Brooks, 1970. The effect of concentration of food on body weight, cumulative ingestion and rate of growth of the marine copepod Calanus helgolandicus. Limnol. Oceanogr. 15: 748-755.

Mullin, M. M. and P. M. Evans, 1974. The use of a deep tank in plankton ecology. II. Efficiency of a planktonic food chain. Limnol. Oceanogr. 19: 902-911.

Paffenhöfer, G-A. and R. P. Harris, 1976. Feeding, growth and reproduction of the planktonic copepod Pseudocalanus elongatus Boeck. J. Mar. Biol. Ass. U. K. (In press).

Parsons, T. R., 1974. Controlled ecosystem pollution experiment (CEPEX). Environ. Cons. I: 224.

Reeve, M. R. and L. D. Baker, 1975. Production of two planktonic carnivores (chaetognath and ctenophore) in south Florida inshore waters. Fish. Bull. U. S. 73: 238-248.

Reeve, M. R., J. C. Gamble and M. A. Walter, 1976. The behavior of copepods and other zooplankton in copper contaminated enclosed water columns. Bull. Mar. Sci. 26: (In press).

Steele, J. H., 1974. Spatial heterogeneity and population sta-
    bility. Nature 248: 83.

Takahashi, M., W. H. Thomas, D. L. R. Seibert, J. Beers, P. Koeller
    and T. R. Parsons, 1975. The replication of events in
    enclosed water columns. Arch. Hydrobiol. 76: 5-23.

NUTRITIONAL ECOLOGY OF <u>AGALMA</u> <u>OKENI</u> (SIPHONOPHORA: PHYSONECTAE)

Douglas C. Biggs

Woods Hole Oceanographic Institution

Woods Hole, Massachusetts   02543   U.S.A.

Siphonophores are among the 10 major groups of zooplankton collected by plankton tows in the upper 200 m of the Sargasso Sea (Grice and Hart, 1962; Deevey and Brooks, 1971) and they comprise 45 - 67% of the macroplankton (i.e., siphonophores, medusae, molluscs, chaetognaths, and thaliaceans) in areas of the Mediterranean (Boucher and Thiriot, 1972). Their widespread occurrence in the open ocean suggests that siphonophores are important components of oceanic ecosystems. However, most are extremely delicate and are rarely seen alive. None have been studied in their natural environment, and because of their fragility little information has been available on their metabolism and production.

METHODS

During 1973-1975, I participated in a program of 171 SCUBA dives in subtropical oceanic regions of the western North Atlantic Ocean. Most dives were made during the day in the upper 30 meters, where temperatures ranged from 23 - 29°C. <u>Agalma</u> <u>okeni</u> (Figure 1) was the most common physonect siphonophore encountered by divers, and its behavior and physiological ecology were studied in the laboratory aboard ship and in the field.

Colonies of <u>A. okeni</u> were individually collected in 130-980 ml glass jars and incubated at surface temperature in the laboratory aboard ship for 1 - 6 hours after enclosure. Oxygen consumption and ammonia excretion were estimated by difference from control jars of sea water which were collected at the same time as the siphonophores. The tension of dissolved oxygen was measured

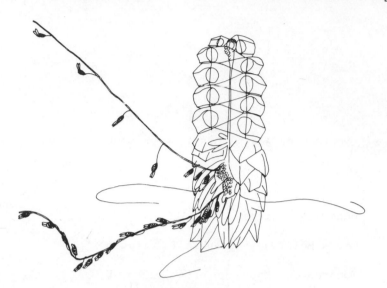

Figure 1. Colony of _Agalma_ _okeni_ with ten nectophores and two
gastrozooids, drawn from in situ photograph.   Tentacles are partially extended in fishing posture.

with a polarographic oxygen electrode (Kanwisher, 1959), and ammonia was measured by the phenol-hypochlorite method (Solorzano, 1969).
Methodology has been detailed in a previous report (Biggs, in prep).
Total dissolved nitrogenous excretion was determined by Kjeldahl
digestion ashore.   All metabolic measurements were standardized to
body protein, as determined by the Lowry method (Lowry, et al.,
1951), using bovine serum albumin as a reference standard.

     Several colonies of _A._ _okeni_ were maintained in 3.8-liter and
20-liter cylindrical aquaria for determination of short-term growth.
I estimated growth rates of _A._ _okeni_ by counting the increase in
number of nectophores and gastrozooids in 12 colonies in captivity
for 1 - 4 days.   Aquaria were kept in the dark; laboratory temperatures ranged from 24 - 26°C.   Water was changed every second day.
Most colonies were allowed to feed on stage-2 _Artemia_ nauplii or
on _Acartia_ and _Pleuromamma_ spp. copepods, though three captured
and ingested shrimp (_Leander_ _tenuicornis_) 15 - 20 mm long.   _Artemia_
nauplii were provided at densities greater than 100/liter and
copepods at greater than 20/liter.

RESULTS

Feeding and Digestion

A small colony of A. okeni which had been fed carmine-dyed Artemia nauplii egested the carmine and other undigested material within 2 - 3 hours after capture. Larger prey required longer times for digestion. For example, three colonies which had captured Leander tenuicornis and another which had captured a crab megalops in situ required 7 - 18 hours before egestion. Palpons and gastrozooids remained swollen for 18 - 48 hours.

Agalma okeni is primarily a nocturnal feeder. Tentacles were contracted in over 90 of 114 colonies observed in situ during daylight hours, while 7 colonies observed at night each had tentacles extended in fishing posture. In the laboratory, colonies would extend their tentacles only in darkness or reduced light. Since A. okeni captured prey 15 - 20 mm long on several occasions in the laboratory, vertically-migrating crustacea and fish may be an important fraction of its diet.

Respiration and Excretion

Oxygen consumption and ammonia excretion (Figure 2) can be expressed by regression equations fitted by the method of least squares:

$$\log Y = 0.87 \log X + 1.07; \ r^2 = 0.77 \qquad (1)$$

$$\log Z = 0.82 \log X - 1.21; \ r^2 = 0.79 \qquad (2)$$

where Y = oxygen consumption (μl O_2/hr); Z = ammonia excretion (μg-at NH_4^+-N/hr); X = body protein (mg BSA); r^2= coefficient of determination.

In 8 specimens of A.okeni, ranging in size from 1.3 mg to 10.1 mg body protein, ammonia excretion averaged 69% of total dissolved nitrogenous excretion (Table 1). Most other groups of planktonic invertebrates are also primarily ammonotelic (e.g., Corner and Cowey, 1968; Jawed, 1973; Mayzaud and Dallot, 1973). For colonies of A.okeni, the atomic ratio of oxygen consumed to nitrogen excreted (O:N) was 13 ± 5.5. Metabolism is apparently based on catabolism of protein, since protein catabolism (16% N and 1.04 liters O_2 needed for complete combustion of 1 g) yields an O:N ratio of about 8, while catabolism of a 50:50 mixture by weight of protein and lipid (2.02 liters O_2 needed for complete combustion of 1 g) yields an O:N ratio of about 24 (Ikeda, 1974).

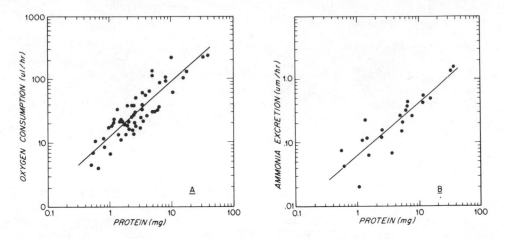

Figure 2. Oxygen consumption (A) and ammonia excretion (B) of colonies of <u>Agalma okeni</u> collected in situ and incubated at environmental temperatures of 23 - 29°C.

COLONY SIZE (mg protein)	AMMONIA EXCRETION (µg NH_4^+/hr)	TOTAL NITROGENOUS EXCRETION (µg NH_4^+ equivalents/hr)	$\dfrac{NH_4^+}{TOTAL\ N}$
1.3	0.4 ± 0.1	0.5 ± 0.2	0.80
2.4	1.5 ± 0.1	2.2 ± 0.2	0.68
2.5	2.6 ± 0.1	4.7	0.56
8.6	3.7 ± 0.1	5.0 ± 0.2	0.74
8.6	5.9	5.9	1.00
8.6	4.4 ± 0.2	7.7 ± 0.1	0.58
8.6	11.7 ± 0.1	13.2 ± 0.6	0.89
10.1	8.2 ± 0.2	23.6 ± 3.9	0.35

Table 1. Nitrogenous excretion by colonies of <u>Agalma okeni</u> collected in situ and incubated at environmental temperatures of 23 - 29°C.

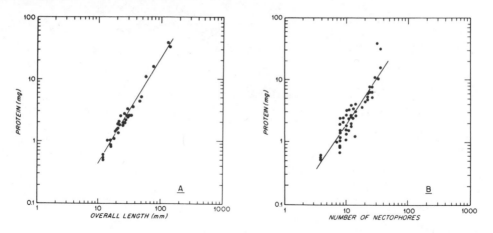

Figure 3. Protein content of colonies of <u>Agalma okeni</u> as a function of overall length (A) and number of nectophores (B).

Growth and Reproduction

As colonies of <u>A</u>. <u>okeni</u> grow, individual nectophores increase
in size and protein content, though at any time all except 2 - 4
apical nectophores (buds) are roughly equivalent. Colonies having
the same number of nectophores are similar in size, and colony biomass can be estimated from either the number of nectophores or from
overall colony length (Figure 3). For example, the mean difference
in size between colonies with 2 and 3 pairs of nectophores is 480
μg protein, while the mean difference in size between colonies with
6 and 7 pairs of nectophores is 600 μg protein (Figure 3).

Most colonies of <u>A</u>. <u>okeni</u> added 1 - 2 pairs of nectophores
within 1 1/2 - 2 1/2 days. Five colonies maintained for 3 1/2 -
4 1/2 days added 2 - 3 pairs of nectophores (Table 2). All colonies
budded a new gastrozooid and tentacle within 1 1/2 days and showed
nectosome growth. Small colonies (1 or 2 pairs of nectophores)
doubled in protein in 2 days, while larger colonies (4 - 6 pairs
of nectophores) added 33 - 36% more protein (Table 2).

This steady increase in size was offset by accidental loss of
gelatinous parts. Although the effects of laboratory confinement
are difficult to assess, the faculty for autotomy is so well developed in most physonect siphonophores that shedding of nectophores and bracts probably occurs in situ as well as in the laboratory.

Colonies of <u>A</u>. <u>okeni</u> first show well-developed gonophores at
about the 14±2 nectophore stage (32±4 mm overall length).

INITIAL SIZE NECTOPHORES PLUS BUDS	*PROTEIN (µg)	DIET	NECTOPHORES + BUDS ADDED A	B	C	FINAL SIZE *PROTEIN (µg)
1+0	100	nauplii	-	2+1	-	300
3+1	500	nauplii; shrimp	-	3	5+1	1600
4+1	600	nauplii; shrimp	-	-	5+1	1900
5+0	700	nauplii	-	2+1	-	1100
6+0	900	nauplii	0+2	-	-	1100
6+1	1100	copepods	-	4+2	6+2	3500
6+1	1100	nauplii	-	4+1	-	2100
7+0	1100	nauplii	1	-	-	1300
7+0	1100	nauplii	1+1	-	-	1400
8+0	1400	nauplii	-	1+2	-	1900
10+1	2100	copepods	1+1	-	5	3500
13+0	2500	nauplii; shrimp	-	2+2	-	3500

*Calculated from Figure 3

Table 2.   Increase in size of colonies of Agalma okeni maintained
for 1 - 4 days in the laboratory (A = after $1\pm1/2$ days; B = after
$2\pm1/2$ days; C = after $4\pm1/2$ days).

DISCUSSION

Mackie and Boag (1963) maintained colonies of Nanomia cara
(Family Agalmidae) in the laboratory at 12 - 14°C on a diet of
fresh crab meat or small crustacea. They found that small colonies
of N. cara budded a pair of nectophores and a gastrozooid about
every three days, which approximates rates of budding observed in
A. okeni.

I can now estimate the energy requirements of colonies of A.
okeni. A small colony with 3 pairs of nectophores has about
1.0 mg of protein (Figure 3B) and consumes about 12 µl O_2/hr
(Figure 2). Assuming an oxycaloric equivalent of 4.9 cal/ml, 3.5
calories would be consumed in respiration over a period of
2 1/2 days.

The caloric value of a Candacia sp. copepod is about 0.5
calories (Shushkina and Sokolova, 1972). If assimilation by siphonophores ranges between 70 - 90% of ingestion, a colony with 1.0
mg protein would have to ingest 8 - 10 such copepods in 2 1/2
days to balance its metabolism. Assimilation efficiencies of 70 -
90% are not unrealistic for aquatic carnivores (Welsh, 1968);

values of 80% and 88% have been reported for <u>Sagitta</u> <u>hispida</u>
(Cosper and Reeve, 1975) and <u>Euphausia</u> <u>pacifica</u> (Lasker, 1966),
respectively. The ctenophore <u>Pleurobrachia</u> <u>pileus</u> may have an assimilation efficiency as great as 90% (Hirota, 1972).

If siphonophores living under natural conditions increase in
size at the rates I have measured in the laboratory, a colony of
<u>A.okeni</u> with 3 pairs of nectophores would grow to the 8 or 10
nectophore stage in 2 1/2 days. This increase in size corresponds
to mean production of 480 or 960 µg protein, respectively (Figure
3<u>B</u>). Since the caloric value of protein is about 5.5 kcal/gram
(Morowitz, 1968), production represents 2.6 - 5.3 calories. Additional consumption of 6 - 15 copepods of <u>Candacia</u> size should support this increase in size, for a total ingestion of 14 - 25 copepods over a 2 1/2 day period. A colony of <u>A</u>. <u>okeni</u> initially three
times larger, with 3.0 mg protein and 14 nectophores, would have to
consume 29 - 46 copepods of similar size to balance its respiratory
energy losses and increase in size to the 16 or 18 nectophore stage.

The preceding calculations suggest that growth in siphono-
phores like <u>A</u>. <u>okeni</u> may be quite efficient. In fish and euphau-
siids, the greatest fraction of ingestion goes to support respiration (Table 3). Physonect siphonophores able to grow to a colony
size 2 pairs of nectophores larger in 2 1/2 days should have a
higher ratio of growth to respiration,or more like chaetognaths
in production efficiency (Table 3). Hirota (1972) estimated that
68% of food ingested by the ctenophore <u>Pleurobrachia</u> <u>bachei</u> was
incorporated into growth and egg production. However, difficulties
with organic weight determinations caused Hirota to overestimate

SPECIES	PRODUCTION	RESPIRATION	EGESTION	SOURCE
<u>Agalma</u> <u>okeni</u>				
3.0 mg protein	33%	47%	20%	
1.0 mg protein	48%	32%	20%	
Carnivorous fish	20%	60%	20%	Welsh (1968)
<u>Euphausia</u> <u>pacifica</u>	29%	59%	12%	Lasker (1966)
<u>Sagitta</u> <u>elegans</u>	35%	37%	28%	calculated from Sameoto (1972)

Table 3. A comparison of production, respiration, and egestion
estimated for <u>Agalma</u> <u>okeni</u> with other marine carnivores.

production, and ctenophore growth efficiency is now believed to be less than 20% (M. R. Reeve, personal communication).

Most siphonophores are monoecious, and colonies of Agalmidae have clusters of male and female gonophores associated with each stem group. Female gonophores have 1 - 4 eggs, which measure about 0.3 - 0.7 mm in diameter (Totton, 1965). When fertilized in the laboratory at 14°C, eggs of Agalmidae require about 2 - 3 weeks to develop to the postlarva (e.g., Carré, 1969, 1971 , 1973). In tropical and subtropical environments, development is probably more rapid and might proceed in 1 - 2 weeks. If A.okeni can grow from the postlarva to the 14-nectophore adult in 1 1/2 - 2 weeks, as the data in Table 3 suggest, this physonect, like other gelatinous carnivores living in warm-water oceanic regions (Baker and Reeve, 1974) probably has a generation time of only 2 1/2 - 4 weeks. Since A. okeni is not restricted to feeding on copepod-size prey, a colony with only 1 mg protein which captured a mysid or euphausiid less than an inch long (with 5 mg protein) should gain enough energy for 1 - 2 weeks maintenance or to increase in size to within the range of reproductive capacity.

This research was supported in part by an NSF Graduate Fellowship and Grant Nos. GA39976 and GA31893 from the National Science Foundation. I am grateful to G. Woodwell for allowing me to use Kjeldahl facilities in his laboratory, and thank J. Teal, R. Harbison, L. Madin, and P. Wiebe for helpful discussions. Contribution No. 3765 of the Woods Hole Oceanographic Institution.

SUMMARY

Agalma okeni was the most common physonect siphonophore encountered by SCUBA divers during the day in the upper 30 m of subtropical oceanic regions of the Western North Atlantic Ocean. In situ observations indicated that A. okeni is primarily a nocturnal feeder. A large portion of its diet may be vertically-migrating crustaceans and fish, although epipelagic copepods can be captured as well. Digestion of shrimp and megalops larvae required 7 - 18 hours.

Colonies with 1 - 10 mg body protein consumed 12 ± 5.5 μl O_2/mg protein-hr and excreted 1.0 ± 0.6 μg NH_4^+/mg protein-hr. The ratio of oxygen atoms consumed to nitrogen atoms excreted was 13 ± 5.5. A small colony with 6 nectophores requires 2.8 - 5.0 calories to balance daily rates of oxygen consumption and short-term growth, while a medium-size colony with 14 nectophores requires 5.8 - 9.2 calories. Generation time in tropical and subtropical environments is probably 2 1/2 - 4 weeks.

REFERENCES

Baker, L. D., and M. R. Reeve, 1974. Laboratory culture of the lobate ctenophore Mnemiopsis mccradyi, with notes on feeding and fecundity. Mar. Biol., 26: 57-62.

Biggs, D. C., in prep. Respiration and ammonia excretion by open-ocean gelatinous zooplankton. Submitted to Limnol. Oceanogr.

Boucher, J., and A. Thiriot, 1972. Zooplancton et micronecton estivaux des deux cents premiers mètres en Mediterranée occidentale. Mar. Biol., 15: 47-56.

Carré, D., 1969. Étude histologique du développement de Nanomia bijuga (Chiaje, 1841), Siphonophore Physonecte Agalmidae. Cah. Biol. Mar., 10: 325-341.

Carré, D., 1971. Étude du développement d'Halistemma rubrum (Vogt, 1852) Siphonophore Physonecte Agalmidae. Cah. Biol. Mar., 12: 77-93.

Carré, D., 1973. Étude du développement de Cordagalma cordiformis Totton, 1932, Siphonophore Physonecte Agalmidae. Bijd. tot Dierkunde, 43: 113-118.

Corner, E. D. S., and C. B. Cowey, 1968. Biochemical studies on the production of marine zooplankton. Biol. Rev., 43: 393-426.

Cosper, T. C., and M. R. Reeve, 1975. Digestive efficiency of the chaetognath Sagitta hispida Conant. J. Exp. Mar. Biol. Ecol., 17: 33-38.

Deevey, G. B., and A. L. Brooks, 1971. The annual cycle in quantity and composition of the zooplankton of the Sargasso Sea off Bermuda. II. The surface to 2000 m. Limnol. Oceanogr. 16: 927-943.

Grice, G. D ., and A. D. Hart, 1962. The abundance, seasonal occurrence, and distribution of the epizooplankton between New York and Bermuda. Ecol. Monogr., 32: 287-309.

Hirota, J., 1972. Laboratory culture and metabolism of the planktonic ctenophore, Pleurobrachia bachei A. Agassiz. pp. 465-484, In Takenouti, et al. (eds.), Biological Oceanography of the Northern North Pacific Ocean. Idemitsu Shoten, Tokyo.

Ikeda, T., 1974. Nutritional ecology of marine zooplankton. Mem. Fac. Fish. Hokkaido Univ., 22: 1-97.

Jawed, M., 1973. Ammonia excretion by zooplankton and its signifi-
    cance to primary production during summer. Mar. Biol., 23:
    115-120.

Kanwisher, J., 1959. Polarographic oxygen electrode. Limnol.
    Oceanogr., 4: 210-217.

Lasker, R., 1966. Feeding, growth, respiration, and carbon utili-
    zation of a euphausiid crustacean. J. Fish. Res. Bd. Canada,
    23: 1291-1317.

Lowry, O. H., N. J. Rosebrough, A. L. Farr, and R. J. Randall,
    1951. Protein measurement with the Folin Phenol reagent. J.
    Biol. Chem., 193: 265-275.

Mackie, G. O., and D. M. Boag, 1963. Fishing, feeding, and diges-
    tion in siphonophores. Publ. Staz. Zool. Napoli, 33: 178-196.

Mayzaud, P., and S. Dallot, 1973. Respiration et excretion azotée
    du zooplancton. I. Evaluation des niveaux métaboliques de
    quelques espèces de Mediterranée occidentale. Mar. Biol.,
    19: 307-314.

Morowitz, H. J., 1968. Energy Flow in Biology. Academic Press,
    New York. 179 pp.

Sameoto, D. D., 1972. Yearly respiration rate and estimated energy
    budget for Sagitta elegans. J. Fish. Res. Bd. Canada, 29:
    987-996.

Solorzano, L., 1969. Determination of ammonia in natural waters by
    the phenol hypochlorite method. Limnol. Oceanogr., 4: 799-
    801.

Shushkina, E. A., and I. A. Sokolova, 1972. Caloric equivalents of
    the body mass of the tropical organisms from the pelagic part
    of the ocean. Oceanology, 12: 860-867.

Totton, A. K., 1965. A Synopsis of the Siphonophora. British Mus-
    eum, Natural History. 230 pp.

Welsh, H. E., 1968. Relationships between assimilation efficien-
    cies and growth efficiencies for aquatic consumers. Ecology,
    49: 755-759.

BEHAVIOUR OF PLANKTONIC COELENTERATES IN TEMPERATURE AND SALINITY

DISCONTINUITY LAYERS

Mary Needler Arai

University of Calgary

Calgary, Alberta, Canada

In an earlier paper (Arai 1973) I described the response of
Sarsia tubulosa and Pleurobrachia pileus to discontinuities between
upper layers of lower salinity water and lower layers of high
salinity water. Both species aggregated actively to salinity
discontinuities of as little as 2‰. Under fjord conditions where
such a response might be ecologically significant the halocline
is usually accompanied by a thermocline. The present paper
describes further experiments investigating responses to
discontinuity layers, particularly those involving temperature
discontinuities.

MATERIALS AND METHODS

Sarsia tubulosa (M. Sars) and Pleurobrachia pileus (O.F. Muller)
var bachei Agassiz were collected from Departure Bay, Ladysmith
Harbour, or Sooke Harbour, B.C. Animals were dipped from the
surface layers of water using small beakers or glass jars. They
were transferred as quickly as possible to jars or cylinders
containing sea water of the desired salinity and placed in a
laboratory cooler. Sarsia tubulosa was fed euphausids broken
into small pieces. Pleurobrachia pileus was fed young stages of
Artemia salina.

Surface salinities and temperatures in which animals were
caught varied from 27.0 to 28.7‰ and 15-17°C for S. tubulosa
from Departure Bay or Ladysmith Harbour, 31.4‰ and 16°C for
S. tubulosa from Sooke Harbour, and 23.5 to 27‰ and 17-18°C for
Pleurobrachia pileus from Departure Bay or Ladysmith Harbour.

211

As it was not known what portion of the water column the animals
caught at the surface might have previously occupied each animal
was held at an arbitrary temperature and salinity for at least
3 days. Departure Bay and Ladysmith Harbour animals were held at
25‰ and 10 or 13°C, those from Sooke Harbour at 30‰ and 10°C.

Filtered and U. V. sterilized sea water was diluted as
required by distilled water. For routine mixing, salinity
measurements were made with previously calibrated hydrometers
and were found to be accurate to better than 0.1‰ of salinity.

The chamber used consisted of a 2-liter, 41.28-cm high
cylinder enclosed in a black painted wooden box in which a
horizontal plexiglass baffle allowed application of different
temperatures of running cooled water in the upper and lower
sections. The box had a glass front through which the cylinder
could be observed and a hinged lid including a fluorescent light
above frosted glass. During experiments the room was darkened
with black curtains, and the overhead light of the chamber turned
on.

In establishing a discontinuity layer 1,000 cm^3 of the higher
density water was first poured into the cylinder. A disc which
supported the end of some flexible plastic tubing attached to a
separatory funnel was then floated on the surface. Water was
delivered at a controlled rate by the separatory funnel through
pin holes at the end of the tubing to form the upper layer of
lower density water with a minimum of turbulence. Discontinuities
formed by water of two salinities gave very sharp interfaces
which were shown by methylene blue dye tests to be stable for
several hours. Animals passing through the interface occasionally
produced some mixing, the maximum being approximately 20 cm^3,
i.e. a vertical distance of approximately 4 mm. In experiments
with water of two temperatures, some heat transfer occurred so
temperatures were constantly monitored by thermistors at the 900
and 1,100 cm^3 levels. The maximum deviation at each thermistor
from the temperature expected in that half of the cylinder was
less than .15°C.

In each experiment, the cylinder was filled close to the
1,000 cm^3 line. Five animals were placed in each cylinder, being
transferred with as little accompanying water as possible in a
5 cm^3 beaker attached to a glass rod. The higher density water
level was then adjusted precisely to the 1,000 cm^3 line, and
the upper portion of the cylinder filled to the 1,900 or 2,000 cm^3
level. The position of the animals (top of the bell in Sarsia,
top of the body in Pleurobrachia) was recorded at 5 min intervals
for 1 h. To allow the animals to recover from the transfer, the
data presented is that collected in the second half-hour of
recording, each graph representing the results from two experiments.

RESULTS

Sarsia tubulosa

Arai (1973) showed that S. tubulosa is strongly attracted to
light. When placed in a homogeneous column of the temperature and
salinity to which they were acclimated the animals aggregate at
the top of the column, i.e. toward the overhead light. This was
confirmed in the controls for the present series of temperature
experiments (see Fig. 1). When temperature discontinuities are
added to the column the animals become distributed between the
discontinuity layer and the top of the column, increasing numbers
being present at the discontinuity layer with increasingly large
temperature differentials (Fig. 1). However in none of the
temperature experiments were most animals aggregated around the
discontinuity level as occurred at some salinity discontinuities
(Arai 1973). As shown in Fig. 2 if temperature and salinity
discontinuities were combined the effect was slightly greater
than for the corresponding temperature or salinity discontinuity
present separately, but again the result was not striking.

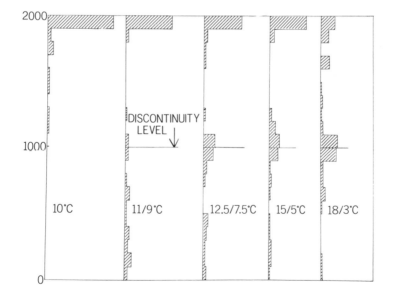

Fig. 1. Distribution in chamber of Sarsia tubulosa (previously held
in 25‰ 10°C water) in 25‰ water with temperature as labelled; to
show effect of temperature discontinuities.

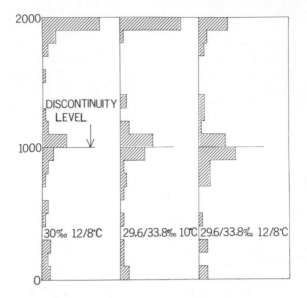

Fig. 2. Distribution in chamber of _Sarsia tubulosa_ (previously held in 30‰ 10°C water); to show the effect of combining temperature and salinity discontinuities.

Records were kept of the behaviour of each animal as well as its position. Each animal was rated as swimming, fishing or floating (see Arai 1973 for definitions). The results of this analysis are shown in Table 1. As noted in the salinity experiments fishing activity was markedly increased in any column including a discontinuity layer.

Table 1. Percentage of _Sarsia tubulosa_ involved in various activities (second half-hour each experiment).

Temperature	Swimming	Fishing	Floating
10°C	52%	46%	2%
11/9°C	25%	75%	
12.5/7.5°C	27%	73%	
15/5°C	33%	67%	
18/3°C	20%	78%	2%

It seemed possible that animals would actively avoid very low
salinity water. A series of experiments were therefore run in which
animals acclimated to 25‰ 10°C water were placed in columns in which
the lower half contained 25‰ water and the upper half various
dilutions of this water with distilled water. No avoidance reactions
were found. At dilutions above 15‰ the animals aggregated around
the discontinuity as expected. As the salt concentration decreased
further (to approximately 10‰ and less) the animals became concentrated just below the discontinuity level.

Pleurobrachia pileus

It was previously shown (Arai 1973) that in a homogeneous
column P. pileus tend to aggregate at the top and bottom of the
column (see also Fig. 3). When temperature discontinuities were
introduced the animals partially aggregated at the interface
(see Fig. 3). Moreover in this species when temperature was
combined with a salinity discontinuity small enough to cause only
partial aggregation at the interface by itself, the two factors
supplemented one another so that all animals were aggregated
around the discontinuity level (Fig. 4).

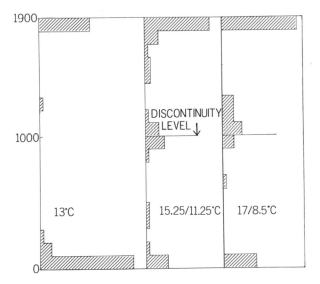

Fig. 3. Distribution in chamber of Pleurobrachia pileus (previously
held in 25‰ 13°C water) in 25‰ water with temperature as labelled;
to show effect of temperature discontinuities.

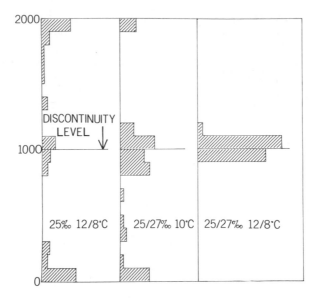

Fig. 4. Distribution in chamber of <u>Pleurobrachia</u> <u>pileus</u> (previously
held in 25‰ 10°C water); to show the effect of combining temperature
and salinity discontinuities.

DISCUSSION

The results show that <u>S</u>. <u>tubulosa</u> and <u>P</u>. <u>pileus</u> aggregate at
temperature discontinuity layers. This aggregation is due to an
active response rather than passive flotation in a density barrier.
It was not possible for technical reasons, to run as extensive
controls as was done with the salinity series. However, comparison
with the salinity controls shows that the relatively low density
changes associated with temperature changes would not produce a
passive barrier to animal movement up and down the column. The
mechanism of such an active response is not known. Since both
salinity and temperature changes produce density changes, it is
tempting to speculate that the animals are sensitive to modification
of the rate of pressure increase with depth.

The ecological advantage of aggregation in or around a
discontinuity layer, especially when associated with increased
fishing behaviour as demonstrated for <u>S</u>. <u>tubulosa</u>, may be the
concentration of food organisms in the layer. Fraser (1970),
Greve (1972), and Hirota (1974) have summarized their own and
previous work showing that <u>Pleurobrachia</u> are miscellaneous feeders
but that crustacea, particularly copepods are the dominant food.

Lebour (1922) stated that <u>Sarsia</u> <u>tubulosa</u> ate only copepods. While our own observations indicate a wider range of prey, copepods are again a main item of the diet. A variety of field and laboratory studies have shown that copepods become concentrated in or near discontinuity layers. It is particularly pertinent that Hansen (1951) found <u>Sarsia</u> <u>tubulosa</u> occurred only in the discontinuity layer of Oslo Fjord and that it was accompanied by three species of copepods.

This research was supported by a grant from the National Research Council of Canada. I am grateful for the able assistance of Mr. J. Dempsey, Mr. W. Roland, and Mr. B. Smith. I am also grateful for facilities and assistance of members of the staff of the Pacific Biological Station, Nanaimo where most of this work was carried out.

SUMMARY

The behaviour of <u>Sarsia</u> <u>tubulosa</u> (M. Sars) and <u>Pleurobrachia</u> <u>pileus</u> (O.F. Muller) var <u>bachei</u> Agassiz was investigated in chambers where salinity or temperature discontinuity layers could be established. It was shown, as previously shown for salinity, that these animals will aggregate at temperature discontinuity layers. When simultaneous discontinuities of temperature and salinity were established the two environmental parameters were found to be supplementary in eliciting aggregation.

REFERENCES

Arai, M. N., 1973. Behaviour of the planktonic coelenterates, <u>Sarsia</u> <u>tubulosa</u>, <u>Phialidium</u> <u>gregarium</u>, and <u>Pleurobrachia</u> <u>pileus</u> in salinity discontinuity layers. <u>J</u>. <u>Fish</u>. <u>Res</u>. <u>Board</u> <u>Can</u>, 30: 1105-1110.

Fraser, J. H., 1970. The ecology of the ctenophore <u>Pleurobrachia</u> <u>pileus</u> in Scottish waters. <u>J</u>. <u>Cons</u>. <u>Cons</u>. <u>Int</u>. <u>Explor</u>. <u>Mer</u>, 33: 149-168.

Greve, W., 1972. Okologische Untersuchungen an <u>Pleurobrachia</u> <u>pileus</u> 2. Laboratoriumsuntersuchungen. <u>Helgol</u>. <u>Wiss</u>. <u>Meeresunters</u>, 23: 141-164.

Hansen, K.V., 1951. On the diurnal migration of zooplankton in relation to the discontinuity layer. <u>J</u>. <u>Cons</u>. <u>Cons</u>. <u>Int</u>. <u>Explor</u>. <u>Mer</u>, 17: 231-241.

Hirota, J., 1974. Quantitative natural history of <u>Pleurobrachia</u>
    <u>bachei</u> in La Jolla Bight. <u>Fishery</u> <u>Bull</u>., 72: 295-335.

Lebour, M.V., 1922. The food of plankton organisms. <u>J</u>. <u>Mar</u>. <u>Biol</u>.
    <u>Assoc</u>. <u>U</u>. <u>K</u>., 12: 644-677.

CRASPEDACUSTA SOWERBII: AN ANALYSIS OF AN INTRODUCED SPECIES

Thomas S. Acker

University of San Francisco

San Francisco, California 94117   U.S.A.

The ecology of *Craspedacusta sowerbii* Lankester, a fresh-water cnidarian, is a source of interest and confusion for many biologists. The medusoid stage draws special attention because in common parlance at least, "jellyfish do not belong in fresh-water." For the investigator, on the other hand, this same stage causes puzzlement since the medusae occur only infrequently. Both factors have led to an expanded literature. Russell (1953) in his first volume of *The Medusae of the British Isles,* compiled a separate but not exhaustive bibliography containing 150 entries on this species. In the light of this rather large and at times conflicting literature, Acker and Muscat (1976) reviewed the information with the purpose of defining more clearly the questions that need to be asked of its ecology. This paper continues that approach by considering the implications of *C. sowerbii's* original habitat, suggesting the factors that control medusoid formation, and explaining the probable reasons for the sporadic appearance of medusa populations.

THE ORIGINAL HABITAT AND PRESENT DISTRIBUTION

Kramp (1950) presented convincing evidence that the Yang-tse-kiang region in China is the original habitat for the genus *Craspedacusta.* In the upper valley (103-109° E. longitude) the two extant species, *C. sowerbii* and *C. sinensis,* are found together. From 109° E. longitude to the mouth of the Yang-tse river, only *C. sowerbii* occurs. In this latter area *C. sowerbii* is generally located in the tributaries and lakes that feed or border this major river. *C. sowerbii* has also been reported south of this watershed area but only in artificial tanks in Canton, Amoy, and Hong Kong. The natural distribution along the Yang-tse-kiang is roughly 103-120° E. longitude by 24-33° N. latitude.

The reported world-wide distribtuion today is far more extensive; it is generally circumterrestrial in the temperate zones of
both hemispheres. Since its first discovery in England, it has been
found in the northern hemisphere in numerous European countries,
Canada, the United States, Panama Canal Zone, Hawaii, the Phillipine
Islands, Japan, China, and Soviet Central Asia (Tashkent and the Sea
of Aral). In the southern hemisphere it has occurred in South
America, Australia, and New Zealand. In Africa and India its place
appears to be taken by the genus *Limnocnida*. Limnomedusans are not
reported in the Arabian peninsula.

How did *C. sowerbii* get such a wide distribution? There is
general agreement that man is the principal agent for its transportation across major land and sea barriers. This is an accidental
distribution done in conjunction with importation of water plants
for conservatories or commercial purposes. Hence we account for its
initial discovery in Regents Park, London -- a legacy of the well-
known Chinese-English associations. Kramp (1950) suggests that
Craspedacusta was transported from China to other parts of the
world with the water-hyacinth, *Eichornia*. Within any general locality,
however, other agents, both physical and biological, assist its dispersal. *C. sowerbii* makes this transporation easier by forming resistent bodies under two conditions at least. Payne (1924) noted
that in winter the polyp contracted into a solid mass and secreted
a chitinous-like covering around itself. Reisinger (1957) indicated
that in late summer, after a polyp forms a second medusa, the remaining polyp dedifferentiates and becomes covered with a perisarc. He
called this body a podocyst and noted that it remained viable for
days after drying and can resist temperatures up to 50° C. The fact
that the polyp can form a drought resistent body allows for considerable distribution and its establishment over wide areas.

This distribution pattern, however, is probably not complete.
Most records of *C. sowerbii* are for the active and relatively visible medusoid stage. The sessile and microscopic polyp stage is
easily overlooked and less often reported. Yet it is the polyp
stage that is more tolerant to most ecological factors. The polyp
can exist in lentic and lotic situations, can withstand long periods
of food shortage, tolerates wide variations of temperature and light,
can reproduce itself asexually, and is persistent throughout the
year except when it contracts into a resting stage. The medusa is
a transient stage; it requires a relatively lentic environment, needs
abundant plankton to sustain its more active life and support its
growth to maturity, is less tolerant of temperature extremes and
strong light, and can only recur in dependence upon the polyp.
Furthermore, there are more narrow environmental conditions (a
bottle-neck) that control medusoid budding. In a word, the polyp
can live without the medusa; the medusa cannot live without the
polyp. These relationships can be summarized in the following
diagram.

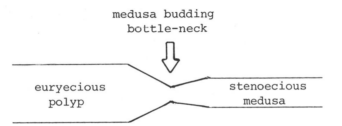

It is probably reasonable to say, then, that the distribution of *C. sowerbii* in the microscopic but hardy polyp form is wider than the present distribution pattern (based mainly on the medusa) suggests.

The original habitat and the adopted habitat show some important differences. Most of the localities in North America where *C sowerbii* is reported lie between 33-43° N. latitude. In Europe they are found principally from 43-53° N. latitude. The presumed original habitat in China lies between 24-33° N. latitude. Kramp (1950) noted that this Yang-tse-kiang region in China is temperate and moist, that the contrast between summer and winter is not great, that there are no severe frosts, and that June, July and August are the rainy months. This description does not equally describe the climate of Europe or most of North America.

One result of these latitudinal and climatic differences is an earlier seasonal appearance of the medusa in China. .Kramp (1950) in several places indicated that in the Yang-tse-kiang region, *Craspedacusta* first occurs in spring and early summer and that it does this regularly. An additional charming but also informative note is contributed by Uchida and Kimura (1933) who reported that the jellyfish is apparently well known by the population around the great river Yang-tse. The people call it *Tao-hwa-yu* (peach blossom fish) or *Tao-hwa-shan* (peach blossom fan) because it makes its appearance when the peach blossoms are in full bloom. These local names tell us of its spring occurrence and also imply a certain regularity of appearance. This contrast with virtually all of the records for Europe and North America. In these areas *Craspedacusta* first appears in late summer and only sporadically from year to year. Even in the most southern parts of the United States, parts below 33° N. latitude, its earliest seasonal record is the 1st of July (Lytle, 1962).

CONDITIONS FOR MEDUSA BUDDING

Reisinger (1957), McClary (1959), and Lytle (1959) have all experimented in the laboratory with *C. sowerbii*. Reisinger obtained medusa budding only with a sudden elevation in temperature from 20 to 25-27° C. McClary performing his experiments in the dark got medusa budding only at temperatures above 26° C. Lytle working in

the light got medusa budding at constant temperatures of 20±1° C by
feeding each colony (4.5 polyps/colony) approximately two *Aeolosoma*
worms every two days. At week eight the budding peaked at 0.4
medusa buds/colony/week. However, Lytle got significantly better
budding results (3 medusa buds/colony/week) when under the same feeding regimen the water in his experimental tanks rose from 19 to 23°
C over a period of ten weeks and then gradually fell to 19° C by the
sixteenth week. In this design the medusa budding began in week nine,
peaked at week twelve and fifteen, and then dropped at week sixteen.
Lytle also reported that daily feeding as opposed to alternate day
feeding triggered increased frustule formation (asexual reproduction)
but only minimal changes in the number of medusa buds. Experiments
performed in our lab likewise show strong frustule production with
satiation feeding, but not medusa budding (Acker and Muscat, unpublished data).

    The only quantitative field study for *Craspedacusta* which in-
volved both temperature and food was done by Dunham (1941) in
McKeever Pond, Columbus, Ohio. This concrete reservoir built for
fire protection measured 18 by 9 meters with a maximum depth of 2
meters. The data redrawn in a graph is shown in *Figure 1*.

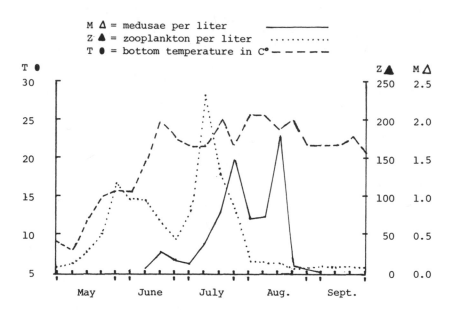

Fig. 1.   Variation in number of medusae in relation to temperature
          and zooplankton.

The graph shows three important elements: medusae appear
when temperatures reach 20° C and are rising; medusa abundance lags
2 or 3 weeks behind zooplankton peaks; and, medusae disappear with
diminished food supplies and falling temperatures.

Summarizing, then, from the laboratory and field data, I can
provisionally suggest that the conditions for medusa budding include
minimal temperatures approximating 20° C, a rising temperature profile, and adequate but not satiation food levels. Light-dark rhythms, food quality, CO_2/O_2 levels, and other environmental conditions remain untested. Likewise, the synergism of all of these
factors as well as the oscillation effects are unknown.

It is especially important here to note the role of oscillating
or changing factors. Literature from both fresh-water and marine
environments support the fact that these changes clue the organism
for specific actions that are adaptive. Such changes may induce
sexuality. Goetsch (1927) and Stolte (1928) noted that when
Chlorohydra viridissima and *Hydra attenuata* are transferred from a
low temperature to a higher temperature, they proceed toward sexual
reproduction. They suggested that the "shock" which occurs during
this transfer influences the physiology of the animal. Goetsch
generalized that all kinds of shocks produced by changes in the
environment might provoke the sexual process if they are not too
damaging. He suggested that changes of water, of feeding routine,
or even jolting while cultures are being transported, might have this
effect.

In this same vein, Goy (1973) reported an interesting experience with a hydroid related to *Craspedacusta*. She had accidentally
left the hydroid of *Gonionemus suavensis* in a vessel and the unchanged
sea water rose over a period of time to $53\%_0$ salinity. When the
water in the glass vessel was renewed to a normal $38\%_0$ and the hydroid fed, it developed a medusa bud. She noted that this scene is
partially duplicated in the Bay of Villefranche-sur-mer (France)
when the spring rainfalls lower the salinity in the bay (although
not so dramatically) and the temperatures rise. Shortly thereafter,
the medusa buds of *Gonionemus suavensis* appear.

In brief, then, it may be possible to induce budding by many
artificial shocks, ones not found generally in nature. But the three
factors of 20° C, rising temperatures, and intermediate food levels
appear to be interconnected natural phenomena that form at least
part of the bottle-neck conditions controlling medusa budding in
Craspedacusta.

THE SPORADIC APPEARANCE OF MEDUSA POPULATIONS

One factor stands out in most European and North American

literature regarding *C. sowerbii*. The medusa appears irregularly;
it is present in a body of water one year (usually in late summer
or fall) and then absent in the same body of water for the next or
several following years. Why does this happen? One level of answer
probably lies in the fact that *C. sowerbii* is a recently introduced
species in most localities where it occurs. In these introduced
areas, the ecological conditions are significantly different from
its original habitat. Besides the changed latitude with its climatic
differences, Kramp (1950) mentioned the unusual microhabitat that
spawns medusae in the Yang-ste-kiang region. He noted that the
medusae are very abundant in the flood pools along the river bed;
the lakes in which the medusae occur are shallow and rise with the
summer rains. Such localities are notable for their stress conditions. They are subject to warming from the sun and cooling from
new flows of water. They give rise to fluxes of plankton in response to nutrient additions and depletions and temperature changes.
In these shifting conditions (shocks) the medusae occur first in
spring and then recur at other times even into January and February
of the following year. I might note that many of the medusan occurrences in Europe and North America take place in shallow concrete
ponds or aquaria where similar stress conditions often prevail.
The previously mentioned McKeever Pond in Ohio is simply one example.
The medusan appearances in Europe and America take place generally
later in the year partly because the water temperatures in these
more northern latitudes seldom rise to 20° C before late summer.
Furthermore, positive temperature shifts will not induce medusa
budding unless food supply is adequate but not superabundant. Temperature and food levels must be in coordination. At this time
we know nothing of the other factors that may singly or together
affect medusa budding. But we can assume that in the Yangtse-
kiang region, *Craspedacusta* has evolved in response to its particular climatic clues within its microhabitat. Therefore, it
occurs regularly.   In the introduced areas the narrow conditions
for medusa budding are less often met and medusae irregularly occur.
When they do occur, we have already noted a shifted season for initial
budding from the Chinese spring-early summer to the European/American
late summer-fall.

     A second factor then enters in. Polyps exist in both lentic and
lotic situations; medusae, however, can only mature in lentic situations. If budding takes place in a lotic environment, it will
probably go unnoticed unless the 1 mm bud reaches a quiet pool or
gets quickly into a lake or pond and can mature. The Yang-ste-kiang
region offers regularly such a situation. It is bordered by numerous sluices, irrigation canals, shallow lakes, and flood pools.
These are the suitable microhabitats for medusoid growth. The usual
situation in Europe or North America is not similar. In these introduced areas, then, we have lowered probability for two events (budding and maturation) which are sequentially dependent. The probability that two sequentially dependent events (E and F) will both

happen is the product of the probability that E alone will happen *times* the conditional probability that F will happen given that E has happened. In this formulation it becomes easier to understand the sporadic occurrence of *C. sowerbii*. The decreased probabilities of the budding event and the maturation event (E and F) in the introduced areas are multiplied together. The product (a lower probability) suggests why medusae are irregularly found in areas outside the original habitat. Only in the Yang-tse-kiang region where *Craspedacusta* has evolved in response to the climate and microhabitat, does it occur regularly and in spring. Here only can *Craspedacusta* be called *Tao-hwa-yu*, the peach blossom fish.

ACKNOWLEDGEMENTS

The author wishes to express appreciation to Triad Metal Products, Cleveland, Ohio, for research support, and to Mr. William Gammie, Berlin Heights, Ohio, for permission to collect and do research on his family's private land.

LITERATURE CITED

Acker, T.S. and A.M. Muscat, 1976. Ecology of *Craspedacusta sowerbii* Lankester, a freshwater Hydrozoan. *Am. Midl. Nat.*, 95: 323-336.

Dunham, D.W., 1941. Studies on the ecology and physiology of the freshwater jellyfish, *Craspedacusta sowerbii*. Ph.D. thesis. Ohio State University, 121 pp. University Microfilms, Ann Arbor, Mi. (*Diss. Abstr.* 36: 57-62).

Goetsch, W. 1927. Die Geschlechtsverhältnisse der Susswasserhydroiden und ihre experimentelle Beeinflussung. *Roux Arch. Entw-Mech.*, 111: 173-249.

Goy, J., 1973. *Gonionemus suvaensis*. Structural Characters, Developmental Stages and Ecology. *Publ. Seto Marine Biol. Lab.*, 20: 525-536.

Kramp, P.L., 1950. Freshwater Medusae in China, *Proc. Zool. Soc. London*, 120: 165-184, 2 plates.

Lytle, C.F., 1959. Studies on the Developmental Biology of *Craspedacusta*. Ph.D. thesis. Indiana University, 116 pp. University Microfilms, Ann Arbor, Mi. (*Diss. Abstr.* 20: 1491).

_____ 1962. *Craspedacusta* in Southeastern United States, *Tulane Stud. Zool.*, 9: 309-314.

McClary, A., 1959. Growth and differentiation in *Craspedacusta sowerbii*. Ph.D. thesis. Univ. of Michigan, 101 pp. University Microfilms, Ann Arbor, Mi. (*Diss. Abstr.* 21: 381).

Payne, F., 1924. A study of the freshwater medusae, *Craspedacusta ryderi*. *J. Morph.*, 38: 387-430.

Reisinger, E., 1957.   Zur Entwicklungsgeschichte und Entwicklungs-
    mechanik von *Craspedacusta* (Hydrozoa, Limnotrachylina). *Z.
    Morph. u. Okol. Tiere*, 45:   656-698.
Russell, F.S., 1953. The Medusae of the British Isles I: Antho-
    medusae, Leptomedusae, Limnomedusae, Trachymedusae, and Narco-
    medusae. Cambridge University Press, Cambridge, 530 pp.
Stolte, H.A., 1928.   Analyse der Bedingungen für Knospung und
    Sexualitat bei *Hydra attenuata* Pallas.   Biol. Zbl. 48(5):
    273-302.
Uchida, T. and S. Kimura, 1933.   Notes on fresh-water medusae in
    Asia.   *Annot. zool. jap.*, 14: 123-126.

LIFE HISTORY OF <u>CARYBDEA ALATA</u> REYNAUD, 1830 (CUBOMEDUSAE)

A. Charles Arneson and Charles E. Cutress

Department of Marine Sciences
University of Puerto Rico
Mayaguez, Puerto Rico 00708

INTRODUCTION

Until 1971, Cubomedusae polyps had never been raised beyond the very earliest stages. Although most workers on jellyfish until that time grouped the Cubomedusae with the Scyphozoa, it was realized that knowledge of complete life histories of Cubomedusae was essential to understanding the phylogeny of this group.

In 1971, in connection with a project to investigate the jelly-fishes of Puerto Rico, attempts were made to rear Cubomedusae. Four genera and six species occur about the shores of this island. Five of the species are abundant at least seasonally. The first species reared was the larviparous <u>Tripedalia cystophora</u> Conant (Werner, Cutress and Studebaker, 1971; Werner, 1975). Next, <u>Carybdea marsupialis</u> (L.) was reared, first from polyps collected in the field and subsequently from gametes (Studebaker, 1972; Cutress and Studebaker, 1973).

These studies revealed that cubopolyps differ in several important respects from scyphopolyps and there are marked differences between developmental stages of <u>T</u>. <u>cystophora</u> and <u>C</u>. <u>marsupialis</u>, both members of family Carybdeidae. Additional Cubomedusae life history studies were considered desirable.

On July 23, 1973 and again on August 12, 1974, large numbers of spawning <u>Carybdea alata</u> Reynaud were collected at a pier near Aguadilla, Puerto Rico. From these, blastulae were obtained and reared through the polyp generation to young medusae. The developmental stages of this species, comparisons with other cubopolyps and comments on the biology and ecology of the medusae covered here are a part of a thesis (Arneson, 1976).

227

METHODS AND MATERIALS

Adult Carybdea alata, attracted to light from a 12-volt seal-beam auto head lamp, were taken by dip net. Very young medusae were collected with a meter plankton net. Laboratory cultures were initiated using approximately 100 planulae per 4.5 inch glass stacking dish containing clean but unfiltered sea water. Cultures were maintained in these dishes at ambient temperatures (range 26-29°C) in running sea water baths in a partially darkened room. Initially the polyps were fed fragments of Isognomon alatus hepatopancreas. Later they were fed every second day with Artemia first nauplii. Cultures were cleaned and water changed after each feeding. Subcultures of detached buds were made. Newly liberated medusae were also fed Artemia nauplii, but later this diet was augmented with living plankton of suitable size and bits of fresh fish.

ADULT MEDUSAE

Carybdea alata is a four-tentacled medusa with bell height approximately twice the diameter. The species is circumtropical and neritic but is confined largely to cleaner offshore waters. The medusae are strong swimmers yet avoid choppy surface conditions. Unless the sea is calm, they remain almost motionless near the bottom. With the usual abatement of wind at night, the medusae rise to the surface to feed. They come readily to a light source, probably attracted primarily by the cloud of plankton that reaches the illuminated area first.

A spawning aggregation of several hundred C. alata was first observed about 2000 hours on July 23, 1973 during night lighting from the pier near El Faro de Aguadilla. The medusae collected and preserved ranged in size from 35 x 61 to 65 x 118 mm. The gonads of the females, normally transparent, were opaque white, indicating fertilization had taken place. Female medusae placed in pails of sea water soon released blastulae and males still released quantities of sperm. In about 10 hours the gonads of both sexes were depleted. The following night many medusae were still present at the pier, but except for one or two, their gonads were empty and transparent. On the third night noticeably fewer medusae were present, and a week later none were seen although several visits were made. C. alata were not seen again at the pier until early May of the following year at which time five adult size but non-ripe specimens were collected. Although the pier was visited frequently during July 1974 in anticipation of another epidemic spawning, only five medusae with near ripe gonads were collected. It was not until August 12 that C. alata spawned in the same manner as the previous year. Except that the water was calm on both spawning nights, correlation of spawning with environmental factors was obscure.

Direct observations of a single population of C. alata for more than a few days proved impossible. Nevertheless, some insight into the life history of the species can be obtained when field observations, plankton samples and data on cultures, over a period of three years, are compared. Laboratory cultures started at the time of spawning began to metamorphose in about ten weeks with the largest number of medusae being produced in November and December. That this represented normal development time was verified by finding in two consecutive years newly released C. alata in plankton samples in December. The oldest medusa raised was 44 days old, and specimens recovered from plankton collected at that time could not be distinguished from it except for slightly larger pedalia.

In the laboratory cultures, after the first wave of metamorphosis, of primary polyps, the secondary polyps continue to metamorphose in small numbers the year round.

Carybdea alata with bell heights of 60 to 118 mm have been collected every month of the year at one or another site about western Puerto Rico. In late September, the gonad is but a thin line on either side of the septa. Little change in size of the gonads is noticed until March or April when they begin to swell rapidly. By late July or early August, the gonads are full sized and ripe.

On one occasion, October 3, 1974 at Mona Island, eight C. alata medusae with bell heights of 19-21 mm were collected. These had ripe gonads although many large medusae of the same species also present had virtually none.

The evidence, that medusae do not attain full size in one year, gonads require a full year to mature, and some medusae spawn at about half size, indicates that C. alata may live for more than a year.

POLYP

The eggs of Carybdea alata are fertilized while in the ovary, and zygotes attain the blastula stage before being shed. Collected blastulae become swimming planulae in 24 hours. The planulae are slightly pyriform and have a midbody annulus of approximately 16 reddish-brown pigment spots. Development times and dimensions for certain polyp and metamorphic stages are given in Table 1. The planulae begin to settle in four days and exhibit about equal affinity for clean Isognomon shell and glass. They attach pointed end down and become small featureless, flagellated spheres. Two days later, four small swellings appear on the distal end of the polyp, and within 15 hours these become short tentacles. The tentacles have a core of large, vacuolated, entodermal cells. The ectodermal epithelium consists of sparsely flagellated, almost squamous cells. Two to three small eurytele nematocysts (20 x 8 μ), resembling closely stenoteles,

Table 1. Summarized developmental data for <u>Carybdea</u> <u>alata</u>.

Age in days	Event	Length or height in mm	Diameter in mm
1	Blastulae released.	-	-
2	Planulae.	0.17	0.1
6	Planulae settle.	-	-
8	Polyps with 4 tentacles.	0.36	0.1
31	Polyps with 6 tentacles. Four small nematocysts at each tentacle tip replaced with a single large stenotele.	0.6	0.2
54	Polyps with 12 tentacles. Budding commences.	1.2-1.5	0.3-0.5
68	Polyps with 16 tentacles. Metamorphosis commences.	1.3-2.0	0.5
75	Medusae liberated.	2.3-2.5	1.5-2.0
119	Oldest medusae raised.	6.0	4.0
	Youngest medusae obtained in plankton. (Polyp remnant present.)	2.5	2.0

migrate from the polyp stalk to the tip of each tentacle. The pigmented spots, present on the planula, have now migrated to the stalk
margin. The polyps are photonegative but in darkness expand and move
in inchworm fashion over the bottom. Twelve days later a small round
mouth, encircled by euryteles, is formed.

A month after settlement, two more tentacles appear. As soon
as they are formed, a single large stenotele nematocyst (65 x 23 u),
one of several that arise just prior to this event on the midstalk,
migrates beneath the tentacle ectoderm to a cavity at the tip of each
of two new tentacles. Similar large stenoteles replace the two or
three small euryteles at the tips of the original four tentacles as
soon as they are expended. As each of these nematocysts is discharged, a replacement migrates to the vacant cavity from the stalk.
The cavity at the tip of each polyp tentacle is surrounded by a cone
of fused flagella (<u>sic</u> "ciliary cones," Picken and Skaer, 1966) which
possibly function as multiple cnidocils. Subsequent tentacles are

added, two at a time, at variable rates depending on the amount of
food consumed.

When the polyp has 12 tentacles, it begins to form buds on the
midstalk. These arise as a concentration of several large stenoteles in a slight swelling. In a day the swelling may reach a length
of 0.5 mm and gain a mouth and four tentacles. In about six days,
there are six to eight tentacles at which time the bud detaches. Up
to three buds at different stages of development may accumulate on a
polyp. Occasionally smaller buds with only four tentacles detach and
seem to fare as well as the larger ones. Budding may continue well
after onset of metamorphosis, often with the last bud detaching only
a day or so before the medusa releases.

Detached buds usually settle near the parent polyp where they
grow, produce more buds and build up a clone of secondary polyps of
one sex. Secondary polyps are indistinguishable from primary polyps
with the exception that the former are noticeably more hardy.

METAMORPHOSIS

Conversion from polyp to medusa in all reared species of Cubomedusae differs from the process of strobilation in other Scyphozoa
in that one polyp completely metamorphoses into one free swimming
medusa. This event in the development of C. alata can be completed
as early as 75 days after settlement of the planula but in most cases
takes around 90 days. The time interval seems to depend largely on
the amount of food the polyps receive.

Metamorphosis usually begins two to three days after the polyp
develops the definitive number of 16 tentacles. It is first evidenced by a clumping together of the tentacles into four equal groups.
This is followed by a coalescing of tentacle bases, resorption of the
tentacles, and appearance of four pigmented eye spots and a statocyst
at the base of each remnant of resorbed tentacles. The polyp now
acquires a tetramerous symmetry. At this time the epithelium on the
upper portion of the polyp becomes suffused with ochre colored pigment. This epithelium is that of the medusa and is more densely
flagellated and contains holotrich and eurytele nematocysts. Four
medusa tentacles arise between the four remnants of the polyp tentacles. These differ from polyp tentacles in that they are hollow
and pigmented and have annular batteries of microbasic euryteles.
The polyp hypostome persists well into metamorphosis and can be distinguished from the developing medusa manubrium, the proximal nema-
tocyst-free portion, by its numerous glistening, small euryteles.

As pigmentation migrates down the stalk, remaining buds detach,
the distal end flares and becomes medusiform, and the last traces of
the hypostome disappear. At this time the velarium is obvious. Four

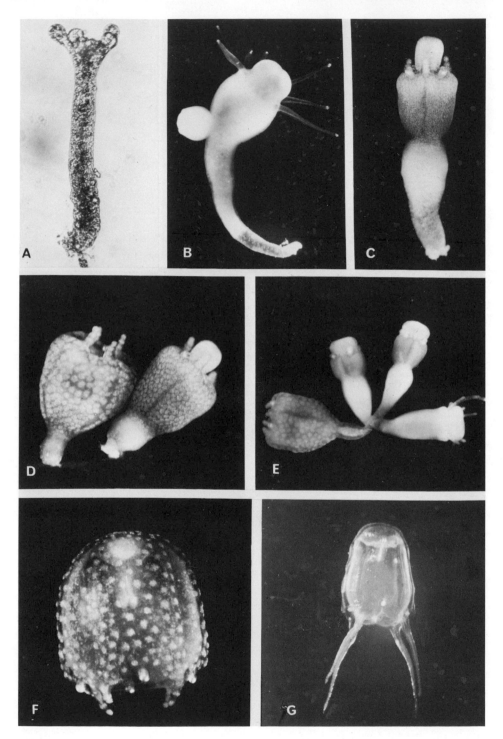

EXPLANATION OF PLATE FIGURES

A. Five day polyp with four tentacles.

B. Fifty day, 12 tentacled polyp with two day old tentacled bud.
    Note single stenotele at tip of each polyp tentacle.

C. Sixty eight day old polyp, three days into metamorphosis. Note
    pigment on oral end, polyp tentacles, sensory organs.

D. Seventy two day old polyp and 70 day old polyp, medusa form
    apparent in 72 day polyp.

E. Cluster of metamorphosing polyps. From right to left: polyp one
    day into metamorphosis, tentacles grouping; two middle polyps
    four and five days into metamorphosis; polyp on left pulsing and
    ready to swim away.

F. One day old medusa, tentacles and sensory clubs visible.

G. Six and a half to seven week old medusa taken in field. Note
    loss of pigment and beginning of pedalia.

interradial canals and the ring canal are present, but no canals have
as yet penetrated the velarium or the still sessile rhopalia. With
conversion completed except for the small remnant of stalk, the medusa pulsates intermittently and in an hour or two detaches and swims
away. The last remaining trace of the polyp is carried away by the
medusa as a small stub on the apex of the bell. In 10 to 13 hours,
this is absorbed by the medusa. Newly detached medusae of C. alata
are 2.5 x 2 mm, have a darkly pigmented exumbrella and four short,
hollow tentacles. Nematocysts of the exumbrella are microbasic euryteles (20 x 8 μ) and spherical holotrichs (13 x 13 μ). The latter
and some of the former are disposed in warts on the exumbrellar surface. Nematocysts of the tentacles are mostly euryteles (30 x 7 u).
Four single gastric cirri are present, and these contain small euryteles. Additional gastric cirri and velar canals appear when the
medusa's bell height is between 10 and 15 mm, and at this stage the
pigment is dispersed so well the animal is transparent.

Small C. alata medusae are distinctive and easily recognized.
Those reared in the laboratory compare in all respects with those
taken in plankton samples and with three described species of juvenile medusae. Carybdea aurifera Mayer (1900, p. 70, pl. 25), Carybdea verrucosa Hargitt (1904, p. 70, pl. 6) and Tamoya punctata Fewkes
(1883, p. 84, fig. 4) can all be considered synonyms of C. alata.
All these were collected in areas where large C. alata were commonly
taken, and the authors considered them juveniles when describing them.

                                    DISCUSSION

The development of Carybdea alata is similar to that described
by Studebaker (1972) for C. marsupialis. Polyps of the two species
differ in number of tentacles (maximum 16 in C. alata, 24 in C. marsupialis) and maximum length (2 mm in C. alata, 1.5 mm in C. marsupialis). The cnidoms are identical although the stenoteles of C.
marsupialis are more numerous and somewhat smaller than those in C.
alata. Polyps of both species lose their stenoteles when they metamorphose, and these are replaced by euryteles and holotrichs. Newly
released medusae of the two species differ in pigmentation (denser in
C. alata) and initial number of tentacles (four in C. alata, two in
C. marsupialis).

There are more marked differences between developmental stages
of these two species of Carybdea and those described for Tripedalia
cystophora (Werner et al., 1971; Werner, 1973, 1975). The thin bean
shaped euryteles of Tripedalia are not found in either species of
Carybdea. Both species of Carybdea have but a single large stenotele
at the tentacle tip whereas Tripedalia has a single large stenotele
surrounded by a number of bean shaped euryteles. The smaller number
of polyp tentacles (9) and size (1.4 mm) in Tripedalia are less important. The periderm which Werner reports for Tripedalia is but an

accumulation of mucus and detritus and, although sometimes present
also about the pedal disc of Carybdea, must be discounted for comparative purposes.

SUMMARY AND CONCLUSIONS

Carybdea alata most commonly occurs near the edge of the shelf.
Young medusae taken in the plankton near the shelf edge, comparable
to day old laboratory medusae, indicate that polyps of the species
are in the immediate vicinity.

Growth rates of medusae and maturation time of gonads indicate
that C. alata medusae live more than a year.

The polyps of the species have 16 solid tentacles, produce tentaculate buds and convert totally to medusae.

Early growth stages of C. alata medusae correspond in all respects with C. aurifera Mayer (1900), C. verrucosa Hargitt (1904),
and Tamoya punctata Fewkes (1883), and these may be considered
synonyms.

The presence of typical stenoteles, previously known only from
the Hydrozoa, and multiple cnidocil-like structures of the polyp tentacle tips strongly indicate hydrozoan affinities as do the spherical
holotrichs of the medusae.

LITERATURE CITED

Arneson, A. C., 1976. Life history of Carybdea alata Reynaud, 1830,
    with notes on its biology. M.S. thesis, Dept. Mar. Sci., Univ.
    Puerto Rico, Mayaguez, in preparation.

Cutress, C. E. and J. P. Studebaker, 1973. Development of the Cubo-
    medusae, Carybdea marsupialis. Proc. Assoc. Is. Mar. Labs. of
    Carib., 9: 25.

Fewkes, J. W., 1883. On a few medusae from the Bermudas. Bull. Mus.
    Comp. Zool. Harv., 11 (3): 81-89, pls. 1.

Hargitt, C., 1904. The medusae of the Woods Hole region. Bull. U.S.
    Bur. Fish. for 1904, 24: 21-79, pls. 1-7.

Mayer, A. G., 1900. Some medusae from the Tortugas, Florida. Bull.
    Mus. Comp. Zool. Harv., 37 (2): 13-82, pls. 1-44.

Picken, L. E. and R. J. Skaer, 1963. A review of researches of nem-
    atocysts. In The Cnidaria and their Evolution, Ed. by W. J.
    Rees, Academic Press, London (Symp. Zool. Soc. Lond. 16: 19-49).

Reynaud, --, 1830. Carybdea alata n. sp. In Lesson's Centurie
    Zoologique, 244 pp., 80 pls.

Studebaker, J. P., 1972. Development of the Cubomedusae, Carybdea
    marsupialis. M.S. thesis, Dept. Mar. Sci., Univ. Puerto Rico,
    Mayaguez, 52 pp., 4 pls.

Werner, B., C. Cutress and J. Studebaker, 1971. Development of the
    Cubomedusae, Tripedalia cystophora. Nature, London, 232: 582-
    583.

Werner, B., 1973. Spermatozeugmen und Paarungsverhalten bei Tripe-
    dalia cystophora (Cubomedusae). Mar. Biol., 18: 212-217.

Werner, B., 1975. Bau und Lebensgeschichte des von Tripedalia cyst-
    ophora (Cubozoa, class. nov., Carybdeidea) und seine Bedeutung
    für die Evolution der Cnidaria. Helgoländ wiss. Meeresunters.,
    27: 461-504.

CUBOMEDUSAE: FEEDING - FUNCTIONAL MORPHOLOGY, BEHAVIOR AND PHYLOGENETIC POSITION

R. J. Larson

Department of Marine Science
University of Puerto Rico
Mayaguez, Puerto Rico

Introduction

The cubomedusae, well known as the "sea-wasps", are also notorious because of their voracious feeding habits. F. Conant (1898), who made the first notable study of the biology of cubomedusae, was amazed that Carybdea could capture and swallow relatively large fish. Barnes (1966) who worked on Australian cubomedusae was greatly impressed by the ability of Chiropsalmus and Chironex to consume numerous shrimp of the genus Acetes. Detailed studies of prey capture and transfer, and other aspects of feeding are nonexistent. In fact, except for a few published accounts, such as Southward's (1955) paper on Aurelia and Smith's (1936) paper on Cassiopea, no other extensive work has been reported concerning feeding in medusae. In this paper, a comprehensive account of feeding behavior and functional morphology of feeding structures are provided for the cubomedusae, with particular emphasis on Carybdea marsupialis (L.).

Feeding in Carybdea marsupialis

Throughout the year, medusae were attracted to a night light at La Parguera, Puerto Rico. They swam mostly near the surface and pulsated continuously except for short, infrequent "rest" periods and during prey transfer. The swimming velocity ranged from 3-6 m per min and the pulsation rate was 75-150 pulses per min. Medusae neither swam toward prey nor avoided obstacles, indicating that the ocelli show only directional sensitivity to light.

Prey capture takes place on the tentacles which can reach
lengths of 30 cm or more (about 10 times the bell height).
Annular nematocyst batteries occur at intervals along their
length. Prey contacting a tentacle causes discharge of these
batteries. Consequent adhesive forces may be so strong that a
large fish escapes only by breaking the tentacle. Evidently
adhesion results from penetration of nematocyst threads into the
prey, and from retention of the capsules within the cnidoblast.
The tentacle then contracts to several centimeters in length,
and stiffens (Fig. 1 A). The pedalium bends inward by flexing
at the margin and forces the tentacle holding the prey into the
bell cavity so that the short manubrium can grasp the food
(Fig. 1 B). If prey has adhered to several tentacles, multiple
pedalia action occurs.

Contraction of the tentacles, pedalium flexing, and probably
nematocyst discharge, occur in response to amino acids of the
prey. To test this, plain paper or paper soaked in 10^{-4} gm/ml
proline solution were touched to the tentacle. There was no
perceptible response. Then small pieces of paper soaked in fish
homogenate or in a dilute (10^{-4} gm/ml) reduced glutathione solution was used. Adhesion, tentacle contraction and pedalia
bending resulted. A pedalium-tentacle attached to a small portion
of the bell margin was cut from the medusa. When food was touched

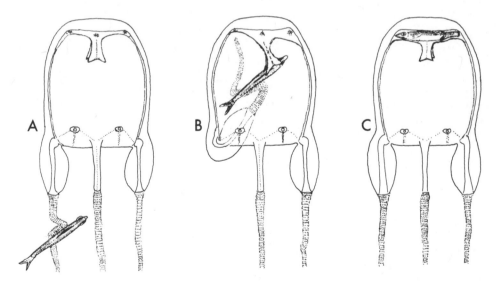

Fig. 1. Prey transfer in <u>Carybdea marsupialis</u>. A. Fish attached
         to contracted tentacle. B. Pedalium flexed, prey grasped
         by lips. C. Fish coiled within stomach.

to the tentacle, bending occurred. Therefore the flexing response
can act independently. Activity of the pedalia is however also
under neurological control as indicated by the crumpling response
which takes place due to severe chemical or physical stimulii or
ablation of rhopalia (Berger, 1900; Yatsu, 1917; and Larson,
unpublished). Pedalia flexing does not occur in all cases. For
example, isopods caught on the tentacles are usually released
after several minutes. The thick exoskeleton is not easily
penetrated by nematocysts, so there was probably insufficient
leakage of amino acids to illicit the bending response.

Once the prey is within the bell cavity, the medusa's
swimming is inhibited and stops within several seconds. When
forceps were used to touch food to the lips, swimming also
ceased. Possibly amino acids act as the inhibitory stimulus.
Cessation of swimming during feeding has also been observed in
other medusae e.g. Aurelia ephyrae (by Horridge, 1956 a),
Nausithoe (by Horridge, 1956 b), Sarsia (by Passano, 1973) and
Phacellophora ephyrae (by Larson, unpublished). Apparently, this
response results from the inability of some medusae to transfer
food while swimming. When food is near the manubrium, searching
activity occurs involving manubrial bending and twitching, and
lip flaring. Dilute glutathione (10^{-4} gm/ml) or fish homogenate
pipetted into the bell cavity produced a considerable increase
in this behavior. Again, proline had no noticeable effect.

The lips are very prehensile so that once food is contacted,
it is promptly engulfed. By a combination of muscular activity
and ciliary creeping, a 20 mm fish can be swallowed in less than
a minute. Apparently, amino acids initiate this feeding activity,
since paper soaked in either glutathione solution or fish homogenate was swallowed. But untreated paper and proline-soaked paper
provoked no obvious response unless the medusa had been starved
for several days.

Medusae often have concentrations of nematocysts along the
lip margin. These probably act to secure the prey to the lips
during food transfer. But since this is not the case in cubomedusae, the method of removal of prey by the lips from the
tentacles is a dilemma. Probably during food transfer the nematocyst capsules are released by the cnidoblast. In one experiment,
food was touched to an isolated pedalium-tentacle-bell margin
preparation. Adhesion occurred and after the pedalium flexed,
the food was dropped.

Primarily, cilia and muscular contractions transport the
mucus-covered food up the central, interradial portion of the
highly elastic manubrium and into the stomach (Fig. 2 B). Small
prey e.g. copepods, that become attached to or near the gastric

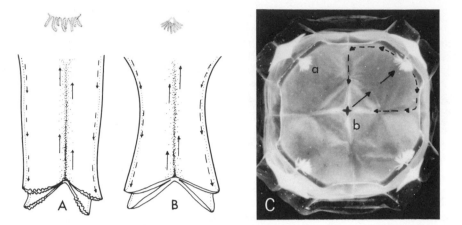

Fig. 2. Manubrium and stomach ciliary tracts. A. Aurelia aurita
ephyra; manubrium (after Southward, 1955). B. Carybdea
marsupialis, manubrium. C. C. marsupialis, aboral view of
stomach- a. gastric cirri, b. gastric furrow. (Solid arrows
show path of food; dashed arrows show path of wastes).

cirri, are possibly held there by nematocysts. The gastric cirri
often cling to large prey, e.g. fish which are usually coiled
within the stomach. Extracellular proteases are produced by the
cirri and stomach wall (Ishida, 1931; and Larson, unpublished).
Digestion of a 20 mm fish takes only 3-4 hours, during which time
wastes are continuously expelled. They move peripherally from the
cirri and into the perradial stomach furrow (Fig. 2 C), then down
the corners of the manubrium and eventually are expelled from the
lips (Fig. 2 B). Food and wastes generally can move simultaneously along their respective ciliary tracts. However, to expel large
indigestibles like a fish skeleton requires reversal of the interradial ciliary tracts. These tracts were first noted by Conant
(1898) who observed that cells of the perradial furrows differed
from the other gastrodermal cells, and that they "probably represent specialized ciliated courses". These same ciliary tracts
also occur in Tripedalia and probably in the other cubomedusae
as well, since the cruciform shape of the manubrium is characteristic of these medusae.

                          Diet of Cubomedusae

     Diets have been reported for a few cubomedusae (Table 2);
and for C. marsupialis (Table 1), (Larson, unpublished). Although
fish were not numerically abundant in the diet, they were

Table 1. Summary of gut contents of <u>Carybdea</u> <u>marsupialis</u> collected
          at a night light at La Parguera, Puerto Rico (100 + spec.)

Prey	% of Total No. of prey	% of Total Biomass of prey	Mean No. of prey/medusa
Polychaetes (Mostly <u>Ceratonereis</u>)	5	5	1
Misc. crustaceans (Mostly <u>Acartia</u>)	80	10	6
Fish (Mostly <u>Jenkinsia</u>)	15	85	1

most important due to their high biomass. Polychaetes were
probably a seasonal food during the surface swarming of their
reproductive stages. Crustaceans are eaten by all sizes of
<u>Carybdea</u>. However, small medusae are more dependent on them than
larger ones, because of the difficulty of capturing fish. The
above data might be considered biased because medusae attracted
to the light fed on animals that were also phototropic, but as
Carybdea would normally be attracted to the same areas as positively phototropic animals, these same prey would probably be
captured. Crustaceans often formed plankton clouds below the
night light, while fish were not nearly as abundant. Some form
of active selection for fish is indicated since numerically, an
expected proportion of copepods were not found in the gut contents.
Results from Table 1 are congruous with the data from Table 2.
Both disclose the importance of crustacea and fish in the diet
of medusae.

Phylogenetic Position of the Cubomedusae

Hydromedusae and scyphomedusae have different methods of
feeding, digestion and egestion, according to their respective
morphology. Scyphomedusae typically have a cruciform manubrium
for swallowing and transporting prey to a spacious stomach where
gastric cirri aid in food digestion. They also have ciliated
tracts which move food proximally in the interradii and waste
distally in the perradii. These tracts have been observed in
<u>Aurelia</u> (by Southward, 1955), <u>Cassiopea</u> (by Smith, 1936), and I
have noted them in <u>Chrysaora</u>, <u>Pelagia</u>, <u>Phacellophora</u> and
<u>Phyllorhiza</u> as well as the cubomedusae <u>Carybdea</u> and <u>Tripedalia</u>.

Table 2. Summary of Cubomedusae Diets.

Species	Coelenteron Contents	Authors *
Carybdea alata	Polychaetes, mysids, crab megalopae	5
Carybdea marsupialis	Polychaetes, misc. crustaceans (copepods, isopods, amphipods, stomatopod larvae, mysids, caridean shrimp and larvae, crab zoeae), chaetognaths, fish	2, 5
Carybdea rastoni	Polychaetes, mysids, fish	3, 4, 5, 8
Chiropsalmus quadrumanus	Misc. crustaceans (amphipods, cumaceans, stomatopod larvae, Lucifer, caridean shrimp, crab larvae), fish	5, 6, 7
Chiropsalmus quadrigatus and Chironex fleckeri	Caridean shrimp (Acetes), other small crustaceans, fish	1
Tripedalia cystophora	Copepods (Oithona)	5

* 1. Barnes (1966) 2. Berger (1900) 3. Gladfelter (1973) 4. Ishida (1936)
  5. Larson (unpublished) 6. Phillips and Burke (1970) 7. Phillips et. al. (1969)
  8. Uchida (1929)

In the hydromedusae which have been studied, i.e. <u>Phialidium</u>
(by Roosen-Runge, 1967) and <u>Polyorchis</u>, <u>Sarsia</u> and <u>Olindias</u> (by
Larson, unpublished), food and wastes are moved by muscular peristalsis, ciliary activity and distortion of the canals during
swimming. Unlike the scyphomedusae, most hydromedusae lack
distinct ciliary tracts (but see Mackie and Mackie, 1963). Wastes
are rejected in toto as a bolus rather than being continuously
egested as in the scyphomedusae (Larson, unpublished). Because
hydrozoans have a tubular coelenteron, they can most effectively
use peristaltic action. The voluminous gastric cavity of the
scyphomedusae necessitates the use of cilia for circulation of
fluids and particles. Additionally, the cruciform shape of the
manubrium provides the needed segregation for the two opposing
tracts used in prey transport and waste removal.

There are two distinct groups of medusae which differ in the
morphology and function of the coelenteron. Is there a third
group? Werner (1973) has re-examined the cubomedusae and has
suggested that they be separated from the scyphozoans. Present
studies clearly show that certain structures associated with the
coelenteron of cubomedusae and scyphomedusae are similar in both
morphology and function. These structures may well be homologous.
(See Thiel, 1966).

Summary

Cubomedusae have a group of anatomical structures which are
sequentially co-ordinated during feeding, i.e. elongate, contractile,
nematocyst laden tentacles for paralyzing and securing prey,
pedalia which bend inwards for bringing the tentacle and prey
into contact with the lips, an active, muscular elastic manubrium
for grasping the food and transporting it to the stomach, and
gastric cirri for extracellular digestion of the prey. Each of
these behavioral responses is autonomous and is probably initiated
by the presence of amino acids from the prey as it is moved toward
the next structure. But physiological nervous mediation is also
involved. Ciliated tracts can move food and wastes continuously
to and from the stomach. Wastes do not accumulate in the gastric
cavity as they do in the hydromedusae, where wastes are rejected
as a bolus. The diet of cubomedusae consists mostly of crustaceans
and fish. But polychaetes are seasonally important when they swarm.
The phylogenetic position of the cubomedusae is debatable. The
morphology and function of the feeding and digestive structures
are scyphozoan in nature and are possibly homologous in these
two groups.

I wish to express appreciation to the staff at the marine laboratory at the University of Puerto Rico, La Parguera, Puerto Rico where this study was conducted. I am indebted to Professor Charles E. Cutress for introducing me to cubomedusae. I also wish to thank the staff at the marine station at Bamfield, British Columbia for use of their facilities during my study of hydromedusae there. I kindly acknowledge the assistance of Ms. Maureen E. Downey, from the U.S.N.M., Smithsonian Institution, Wash., D.C., for her careful proof-reading. Finally I wish to express my sincere thanks to Ms. Kathleen S. Stemler, from the U.S.N.M., who not only typed and proof-read, but helped in other ways as well.

Literature Cited

Barnes, J. H., 1966. Studies on three venomous Cubomedusae. In: The Cnidaria and their evolution. Ed. by W. J. Rees, Symp. Zool. Soc. Lond., 16: 305-332.

Berger, E. W., 1900. Physiology and histology of the Cubomedusae, including Dr. F. S. Conant's notes on the physiology. Mem. Biol. Lab. Johns Hopkins Univ., 4 (4): 1-84.

Conant, F. S., 1898. The Cubomedusae. Mem. Biol. Lab. Johns Hopkins Univ., 4 (1): 1-61.

Gladfelter, W. B., 1973. A comparative analysis of the locomotory systems of Medusoid Cnidaria. Helgolander Wiss. Meeresunters, 25: 228-272.

Horridge, G. A., 1956 a. The nervous system of the ephyra larva of Aurellia aurita. Quart. J. Microscop. Sci., 97: 59-74.

Horridge, G. A., 1956 b. The nerves and muscles of Medusae. V. Double innervation in Scyphozoa. J. Expt. Biol., 33: 366-383.

Ishida, J., 1936. Note on the digestion of Charybdea rastonii. Annot. Zool. Jap., 15: 449-452.

Mackie, G. O. and G. V. Mackie, 1963. Systematic and biological notes on living Hydromedusae from Puget Sound. Nat. Mus. Canada Bull., 199: 63-84.

Passano, L. M., 1973. Behavioral control systems in medusae: a comparison between hydro- and scyphomedusae. Publ. Seto. Mar. Biol. Lab., 20: 615-645.

Phillips, P. J. and W. D. Burke, 1970. The occurrence of sea wasps (Cubomedusae) in Mississippi Sound and the northern Gulf of Mexico. Bull. Mar. Sci., 20: 853-859.

Phillips, P. J., Burke, W.D. and E. J. Keener, 1969. Observations on the trophic significance of jellyfishes in Mississippi Sound with quantitative data on the associative behavior of small fishes with medusae. Trans. Amer. Fish. Soc., 98 (4): 703-712.

Roosen-Runge, E. C., 1967. Gastrovascular systems of small hydromedusae: mechanisms of circulation. Science, 156: 74-76.

Smith, H. G., 1936. Contribution to the anatomy and physiology
        of Cassiopea frondosa. Pap. Tortugas Lab., 31: 17-52.
Southward, A. J., 1955. Observations on the ciliary currents of
        the jelly-fish Aurelia aurita L.   J. Mar. Biol. Ass. U.K.,
        34: 201-216.
Thiel, H., 1966. The evolution of the Scyphozoa. A review. In:
        The Cnidaria and their evolution. Ed. by W. J. Rees,
        Symp. Zool. Soc. Lond., 16: 77-117.
Uchida, T., 1929. Studies on the Stauromedusae and Cubomedusae,
        with special reference to their metamorphosis. Jap. J.
        Zool., 2: 103-193.
Werner, B., 1973. New investigations on the systematics and
        evolution of the class Scyphozoa and the phylum Cnidaria.
        Publ. Seto. Mar. Biol. Lab., 20: 35-61.
Yatsu, N., 1917. Notes on the physiology of Charybdea rastonii.
        J. Coll. Sci. Tokyo, 40 Art. 3: 1-12.

AN EDWARDSIID LARVA PARASITIC IN MNEMIOPSIS

Sears Crowell

Marine Biological Laboratory, Woods Hole
and Department of Zoology
Indiana University, Bloomington, In. 47401

In some years a vermiform larval sea anemone is abundant inside the ctenophore Mnemiopsis leidyi in the Woods Hole, Massachusetts region. This report deals with the identification of its adult form, its occurrence over the years and the extent of infection of the host, observations on penetration of the host, and data on the time required for development to the adult stage.

My own informal observations over many years have been made only from June to late August. A fall sabbatical in 1964 allowed collecting through the month of October; and having begun this study then, I continued collections and observations during the following summer. With an assistant, Susan Oates, the development of larval anemones was followed with specimens transported back to Indiana University at the end of the summer of 1965.

IDENTIFICATION

The parasitic form was named Edwardsia leidyi by Verrill (1898) but we now know, because we have raised the larvae to the adult form, that it transforms to the adult anemone earlier described by Verrill (1873) as Edwardsia lineata. When Carlgren (1893) subdivided the genus Edwardsia he placed E. lineata in genus Milne-Edwardsia; but in a later publication (Carlgren, 1949) he accepted the designation of Fagesia, a name applied by Delphy (1938) because Milne-Edwardsia was preempted. But so is Fagesia! A new name or an older synonym needs to be designated, but I am not yet prepared to do this.

OCCURRENCE

From year to year the numbers of Mnemiopsis at Woods Hole vary
considerably. So, too, does the incidence of infection by the larval
anemone. Collectors at the Supply Department of the Marine Biological Laboratory say that Mnemiopsis makes a brief appearance in April
or May, then becomes rare, then appears in great abundance from late
July through September; not, however, in every year. Verrill (1873)
reports that it is often abundant in Long Island Sound and New York
Harbor in February, also, in Woods Hole waters from July to September. In the collecting records of the summer invertebrate classes
at Woods Hole from 1922 through 1942, the ctenophore was reported in
twelve of the twenty-one years. The larval anemone was reported
only in 1932, 1933, and 1934. Specimens given to me by Milton Gray,
a collector at Woods Hole, are labelled "7-27-30" and "Fall, 1943".
Hargitt (1912) writes: "The occurrence of the larvae is most
erratic. During some seasons it abounds to such an extent that
hardly a specimen of Mnemiopsis can be found without from one to
a half dozen or more of the parasites--and during other summers
scarcely a single specimen can be found at all."

Both the larvae and Mnemiopsis were abundant at Woods Hole in
the late summer of 1964 and 1965, but none were seen by me nor
reported in 1966. Sometimes the Mnemiopsis are found in masses
against rocks or walls at the shore, and in such numbers that a
single sweep with a bucket or net may pick up as many as fifty.
In the middle of Buzzards Bay, September 10, 1964, I estimated
that there was at least one ctenophore per square meter, and nearly
all, of those that could be seen clearly, had parasites.

A collection of 73 Mnemiopsis, October 1, 1964 showed 25 free
of parasites; 48 with from 1 to 7; total number of anemones: 116.
The level of infection dropped during October: October 19, 3 infected of 50 Mnemiopsis; October 27, 3 in over 100; October 28, 1 in
50; October 19, none in 30; October 30, none in 8.

In 1965 no Mnemiopsis were seen in June; some--not parasitized
--were taken in July; by early August some were found with larval
anemones. On August 10 an estimate was made that one in ten Mnemiopsis had well-developed larvae. A collection of 253 ctenophores
was made on September 3. Of these, 56 lacked parasites, 197 had from
1 to 10. The total number of parasites was 540.

OBSERVATIONS

It is easy to observe the pinkish-brown larvae within the
ctenophore since they are opaque and the host is transparent. Most
of the ctenophores were from 4 to 8 cm long. Within the ctenophore
the larvae are elongated and thin; large specimens are over 30 mm

long but barely 1 mm thick. Most of the body is in the mesoglea,
but commonly the oral end is at the expansion of the enteron between the flattened stomodaeum and the funnel. Hargitt (1915) refers
to "parasites within the canals," but I have not seen any so located.

In the laboratory the ctenophores disintegrate in a few days
and the larvae are released. In a few cases we observed a larva
work its way out of an apparently healthy ctenophore. After leaving
the host the larvae shorten so that they become about three times as
thick. They swim freely, aboral end forward, using cilia.

When liberated larvae were put in dishes with uninfected Mnemiopsis some entered the new host. After contacting the surface of
a comb jelly, a larva is often moved about by the comb plates and
cilia of the host, but eventually it may stick at a comparatively
undisturbed spot. It then bends itself so that its oral end makes
contact with the host's epidermis. From this position it penetrates
the epidermis and mesoglea and makes its way to the favored position.
I'm uncertain of the relative importance of cytolysis, muscular
contractions, and ciliary action during penetration and movement
within the ctenophore. The cilia would not be effective unless the
direction of their beat were to be reversed. Some larvae wandered
for as long as two days without reaching the favored position; others
reached it in a few hours.

In 1965 we kept 15 larvae under observation and watched for their
metamorphosis. Nine completed it: the first on the 31st day; seven
on days 31 to 63; and two after a free-swimming life of over 150 days.
There was no correlation between size of the larvae and the time of
transformation. Those which remained larval continued to swim about
for several months; the extreme case was a specimen which already
showed pharynx and eight mesenteries on November 9. It developed no
further, grew smaller, and died the following July.

Description of the metamorphosis is beyond the scope of this
report. A 16-tentacled stage is reached in just a few days once
the change begins. Additional tentacles are often added later.
Mark (1884) followed the process and obtained an adult form sufficiently developed so that he conjectured that it was Edwardsia
lineata, a proposal which Hargitt (1912) thought unlikely.

DISCUSSION AND CONCLUSIONS

We do not know that a parasitic phase is essential in the life
cycle of this anemone. The ability to penetrate the host shows an
adaptation of behavior which one must presume has a selective advantage. Stephenson (1935) describes a single instance of infection
of a ctenophore by a larva of a European anemone in the same genus

as ours (Fagesia? carnea). This larva developed to the adult form.
Stephenson doubts, however, that this is usual since larvae are more
commonly found living freely. Also, we do not know what effect para-
sitism may have on the Mnemiopsis. Infected specimens appear to be
just as vigorous as uninfected ones.

Penetration of the host is oral end forward through the epidermis and mesoglea. I have not seen a larva enter the host by a
gastrovascular route. The larvae can leave the Mnemiopsis while
the latter is still intact as well as when it disintegrates. Peak
abundance of the larvae seems to correspond with peak abundance of
Mnemiopsis in August and September.

I am grateful to Susan Oates who assisted me in the day-to-day
care and observations of the larvae and adults at Indiana.

Contribution No. 1036 of the Department of Zoology, Indiana
University.

REFERENCES

Carlgren, O. 1892. Beiträge zur Kenntniss der Edwardsien. Över-
sigt af K. Vetenskaps-Acad. Förhandl., 49: 451-461.
Carlgren, O. 1949. A survey of the Ptychodactiaria, Corallimorpharia,
and Actinaria. K. Sven. Vetenskaps-Acad. Handl., 4th ser. 1
(1): 1-121.
Delphy, J. 1938. Les actinies athénaires (Actiniaria, Athenaria)
de la faune française. Bull. du Mus. Hist. Nat. Paris., (2)
10: 619-622.
Hargitt, C.W. 1912. The Anthozoa of the Woods Hole Region. Bull.
U.S. Bureau of Fisheries, 32: 225-254.
Mark, E.L. 1884. Polyps. In selections from embryological mono-
graphs compiled by A. Agassiz, W. Faxon, and E.L. Mark. Memoirs
of the Museum of Comparative Zoology at Harvard College. Vol IX,
No. 3: 1-52 + 13 plates.
Stephenson, T.A. 1938. The British sea anemones. II. The Ray Society, London. 426 pp.
Verrill, A.E. 1873. Report upon the invertebrate animals of Vineyard Sound and the adjacent waters, with an account of the
physical characters of the region. U.S. Commission of Fish
and Fisheries, Part 1, report. 1871-1872: 295-778. (Also
published separately in 1874: Verrill and Smith, Invertebrate
animals of Vineyard Sound.....).
Verrill, A.E. 1898. Description of new American actinians with
critical notes on other species. I. Amer. J. Sci., 6: 493-498.

TOWARDS A THEORY OF SPECIATION IN BEROE*

Wulf Greve, John Stockner, and John Fulton

Biologische Anstalt Helgoland, Helgoland, Federal
    Republic of Germany
Pacific Environment Institute, West Vancouver, Canada
Pacific Biological Station, Nanaimo, Canada

INTRODUCTION

Evolutionary processes in ctenophores cannot be investigated
on the basis of fossils. Circumstantial evidence only can provide
us with an insight into the history of this phylum. Such evidence
may be the presence or absence of true larvae; they are missing in
the orders Cydippida, and Beroida while the other three orders
Lobata, Cestida, and Platyctenida have cydippid larvae at least as
an embryonic stage (Dunlap Pianka, 1974). Therefore Cydippida and
Beroida are assumed to be more primitive. Other circumstantial
evidence is the missing tentacles in the order Beroida, which in
combination with the above mentioned direct development, suggests
that this taxon originates from common ancestors with early
cydippid forms. From these were derived the order Cydippida and
the three orders with cydippid larvae.

A common ecological character of the members of the species
Beroe which has been experimentally investigated is their specialization in feeding upon tentaculate ctenophores (Kamshilov 1960;
Greve 1969; 1970; Swanberg 1974). This prey specialization does
not occur to the same degree in Beroe forskali and B. ovata in the
Mediterranean, which were fed by M.L. Hernandez-Nicaise with salps
and other soft bodied plankton (personal communication).

The genus Beroe consists of at least five genuine species:
B. abyssicola Mortensen 1927, B. cucumis Fabricius 1780, B. forskali
Milne Edwards 1841, B. gracilis Kuenne 1939, B. ovata Bosc

---

* supported by the German-Canadian Programme for scientific and
  technological cooperation.

1802, and a list of further species including B. clarki Agassiz
1860, B. hyalina Moser 1908, B. roseola Agassiz 1860 and others
which have not been sufficiently investigated for there to be
certainty about their taxonomic position.

MATERIALS AND METHODS

Of the five species mentioned above only three are known to
be part of the temperate zone fauna where our investigations were
carried out (Figure 1).   These are B. abyssicola which so far seems
to be endemic to the Strait of Georgia and the surrounding waters,
(Kozloff 1974; Shih et al. 1971; Fulton 1968), B. cucumis, which
is a cosmopolitan species (Moser 1909), and B. gracilis, endemic
to the German Bight (south east North Sea) (Liley 1958).

Investigations were carried out in the southern North Sea,
the English Channel, the Irish Sea, and in Scottish waters where
ctenophores were known to be abundant (Scott 1914; Fraser 1970).
Long oblique plankton hauls, each providing sample sizes up to
500 m^3, were repeated at 10 mile intervals.   The ctenophores were
sorted and counted while still alive aboard the vessel.   This is
at present the only way to count Bolinopsis infundibulum, which
disintegrate in formalin (Greve 1970).

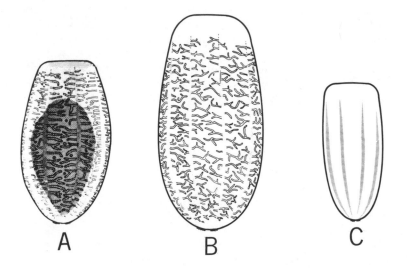

Figure 1.   Moderate zone Beroe: A.   B. abyssicola with violet pharynx,
solid mesogloea with anastomoses.   B.   B. cucumis with unconnected
anastomoses.   C.   B. gracilis without anastomoses, small.

Specimens from the Strait of Georgia were taken in samples
of approximately 6 m³ or 1 m³. They were obtained during survey
cruises in the years 1966, 1967, 1968 and in 1973, 1974,
1975. The latter samples were sorted especially for this investigation, the earlier ones were sorted for a previous study (Parsons
et al. 1970).

RESULTS

Beroe gracilis proved to be a predator of P. pileus not only
in the German Bight but also in Bristol Channel and Morecambe Bay
(both Irish Sea) (figure 2). B. cucumis and Bolinopsis infundibulum
were detected only in a few specimens usually occurring together.
In the north west of the North Sea P. pileus was extremely abundant
but B. gracilis could not be detected. Although B. cucumis was
found it was not abundant. From the sizes of the P. pileus it could
be concluded that the population was at least two months old which
would have given any Beroe species sufficient time for population
growth.

The Strait of Georgia investigations suffered from small sample
size which only indicated the extent of prey populations of Pleurobrachia. Small specimens (up to 15 mm) of Pleurobrachia from the
Strait of Georgia could be identified as P. bachei and were differentiated from P. pileus by the position of the tentacle pouches
relative to the pharynx (figure 3). Larger specimens (greater than

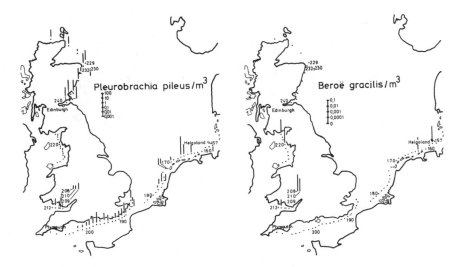

Figure 2. Distribution of Pleurobrachia pileus and Beroe gracilis
around the British Isles in early August 1974.

15 mm) could not be distinguished from P. pileus on the basis of published descriptions and further work is necessary for a final description of the species.

As for the Strait of Georgia Beroida, only three specimens were found in a total of 66 samples pre-selected to coincide with the time of highest probability of their occurrence. The total volume of water filtered in samples examined (about 400 m^3) was less than the volume of water filtered in single hauls in the North Sea--British Isles study so that the total of three specimens of Beroe represents a high abundance of the predator. The Beroe were found in September and October samples from waters of the Strait of Georgia at a location where Pleurobrachia had been most abundant in July to August (10 specimens or more per m^3).

One of the Beroe was identified on the basis of solid body structure and anastomoses as B. abyssicola. The other two, 10 and 16 mm respectively, showed no sign of anastomoses indicating an affinity to B. gracilis; possibly B. roseola Agassiz 1860. However, anastomosing is a valid criteria only for Beroe larger than 10 mm and the number and size of specimens available for study was insufficient for species determination.

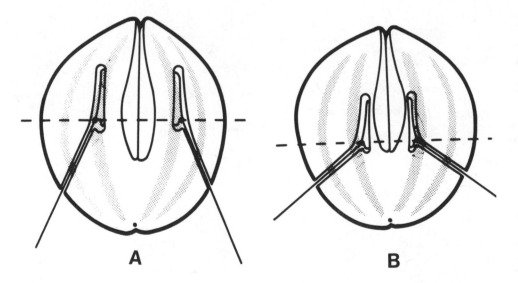

Figure 3. Pleurobrachia: A. P. pileus, tentacle pouches distant from and on the same level as the pharynx. B. P. bachei, tentacle pouches close to the pharynx and on the same level as the oesophagus.

DISCUSSION

The three temperate zone species of Beroe can be distinguished
according to size, the degree and kind of branching of their meri-
dional and pharyngeal vessels, the tint of their pharynx and, their
specialization for certain prey. The latter criterion is of special
importance for the ecology of each species. As speciation processes
depend upon the ecological adaptations of a species towards the
peculiar conditions of its environment this nutritional aspect was
used for the description of a possible pathway of speciation in Beroe.

Beroe gracilis is the Beroe with the highest degree of specialization known so far. It is connected with Pleurobachia in an
ecological feedback system not only in the German Bight but in other
regions where P. pileus is abundant. Though B. cucumis occasionally
feeds on Pleurobrachia too, its population dynamics do not respond
to the cydippid even if it is the only Beroe present as shown by the
results from Scottish waters. This indicates that B. cucumis really
depends upon Bolinopsis infundibulum for prey even if both prey
species are available at the same time. It must be added, that
many reports about Pleurobrachia and its "constant companion" Beroe

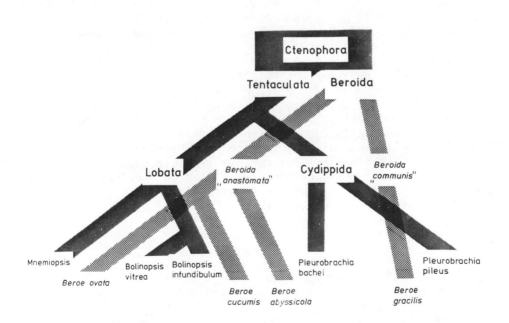

Figure 4. Possible pathways of speciation in Beroe. Recent
species with their assumed prey in the lower levels. Lines indicate the hypothetical evolution of Beroe following the evolution
of Tentaculata.

cucumis (Moser 1909) have to be re-examined owing to the disappearance of Bolinopsis from the preserved samples and to the confusion
of young B. cucumis and B. gracilis partially due to an erroneous
translation of Kuenne's (1939) species description by Liley (1958)
(Greve 1975).

Figure 4 summarizes a hypothesis of speciation due to prey
specialization for a few of the recent species. The early Beroe
(having neither true larvae nor tentacles) have split from early
Tentaculata (having no true larvae but tentacles) which resemble
recent Cydippida. With the development of a new type of predation
by oral lobes some Tentaculata increased in size. The Beroida
accordingly developed larger species, which in order to gain
proper nutrition for their increased body size, developed branched
gastral vessels. The new species of Beroe maintained the ability
to catch and digest Cydippida while the more primitive form never
gained the faculty of eating Lobata. While the common species
developed into the recent B. gracilis, the anastomosed Beroe or
"Beroida anastomata" again split into different species.

This speciation process includes the widening of the food
spectrum up to the feeding upon soft bodied plankton other than
ctenophores and, the spatial separation of species. B. ovata
prey upon warm water species and B. abyssicola can be regularly
found in the Strait of Georgia only below 100 m (W.A. Heath,
personal communication). No information about the feeding of this
species is available, but P. bachei may be the prey of the deep
water species, as it is also found in deep water part of the year
(P.V. Fankboner, personal communication).

The decreasing degree of prey specialization from moderate
zone Beroe to warm water species provides an example of the change
in optimal foraging in two different systems. In a community in
which all populations are synchronized in their oscillations by
spring phytoplankton bloom and trophic level, a high degree of
adaptation of the predator to the prey enables survival at minimal
prey abundance. In less tuned communities a broad prey spectrum
enables survival by prey switching.

REFERENCES

Agassiz, L. 1860. Ctenophorae. In: Contributions to the natural
       history of the United States. Little Brown, Boston 3: 153-301.

Dunlap Pianka, H. 1974. Ctenophora. In: Reproduction of marine
       invertebrates. (Ed. Giese, A.C. and J.S. Pearse) Acad. Press
       Inc.: 201-265.

Fabricius, O. 1780. Fauna Groenlandica. Hafniae et Lipsiae.
       16 + 452 pp.

Fraser, J.H.   1970.   The ecology of the ctenophore Pleurobrachia
    pileus in Scottish waters.   J. Cons. Int. Explor. Mer.   33
    (2): 141-168.

Fulton, J.   1968.   A laboratory manual for the identification of
    British Columbia marine zooplankton. Fish. Res. Board Can.
    Tech. Rep. No. 55: 141 pp.

Greve, W.   1969.   Zur Oekologie der Ctenophore Pleurobrachia
    pileus. Math.-nat. Dissertation, Kiel, 138 pp.

          1970.   Cultivation experiments on North Sea ctenophores.
    Helgoländer Wiss. Meeresunters.   20: 304-317.

          1975.   Ctenophora. Fiches d'Ident. Zooplankton 146:
    6 pp.

Kamshilov, M.M.   1960.   Feeding of a ctenophore Beroe cucumis.
    Dokl. Akad. Nauk S.S.S.R. 130: 1138-1140.

Kozloff, E.N.   1974.   Keys to the marine invertebrates of Puget
    Sound, the San Juan Archipelago, and adjacent regions. Univ.
    of Washington Press, Seattle and London.   226 pp.

Kuenne, C.   1939.   Die Beroe der suedlichen Nordsee, Beroe gracilis
    n. sp. Zool. Anz.   127: 172-174.

Liley, R.   1958.   Ctenophora. Fiches d'Ident. Zooplankton 82: 5 pp.

Mayer, A.G.   1912.   Ctenophores of the Atlantic coast of North
    America.   Publ. Carnegie Inst. Washington 162: 1-58.

Mortensen, T.   1927.   Two new ctenophores. Vidensk. Meddel, Foren.,
    83: 277-288.

Moser, F.   1908.   Japanische Ctenophoren. (Beiträge zur Naturges-
    chichte Ostasiens. Hrsg. von F. Doflein. Abh. 4.) München,
    Abh. Ak. Wiss. Suppl. 1: 1-78.

          1909.   Die Ctenophoren der Deutschen Südpol-Exped.,
    2, Zool. 3: 115-192.

Parsons, T.R., R.J. LeBrasseur, and W.E. Barraclough.   1970.
    Levels of production in the pelagic environment of the Strait
    of Georgia, British Columbia: A review. J. Fish. Res. Board
    Can. 27(7): 1251-1264.

Shih, C.T., A.J.G. Figueira, and E.H. Grainger.   1971. A synopsis
    of Canadian marine zooplankton. Fish. Res. Board Can. Bull.
    176: 127-128.

Swanberg, N.   1974.   The feeding behaviour of _Beroe ovata_.   Mar.
     Biol.   24: 69-76.

Scott, A.   1914.   The mackerel fishery off Walney in 1913.   Rep.
     Lakes Sea Fish. Labs.   22: 19-25.

Reproductive Biology

SEA ANEMONE REPRODUCTION: PATTERNS AND ADAPTIVE RADIATIONS

Fu-Shiang Chia

Department of Zoology, University of Alberta

Edmonton, Alberta, Canada

It is known that sea anemones (Anthozoa: Actinaria) are capable of reproducing in a variety of ways both sexually and asexually (Stephenson, 1928; Hyman, 1940; Uchida and Yamada, 1968). Even in the same species the mode of reproduction may differ from population to population (Schmidt, 1967; Rossi, 1975; personal observations). The versatility of reproduction can, perhaps, be best illustrated by Stephenson's comment (1928): ". . . . breeding, from our imperfect knowledge of it, would appear to be erratic, the individual being a law unto itself."

There have been several recent reviews on the development and reproduction of coelenterates (Campbell, 1974; Spaulding, 1974; Mergner, 1971) but none of these dealt with reproductive patterns in sea anemones in sufficient detail. The purpose of this paper is to categorize the reproductive patterns and to comment on the adaptive radiations of these patterns.

ASEXUAL REPRODUCTION

There are four patterns of asexual reproduction: pedal laceration, longitudinal fission, transverse fission and budding; species known to reproduce asexually in each of the patterns are listed in Table I.

Pedal laceration. There are two ways the pedal disk can lacerate: constriction or tearing. In the first case, the animal attaches itself stationary while the margins of the pedal disk constrict into small pieces and eventually separate from the disk. In the second case, some marginal areas of the pedal disk become

firmly attached to the substrate while the animal moves away
slowly and in doing so the attached areas are torn apart from the
main disk. In both situations the separated pieces of the disk
are usually small (1-10 mm diameter) but always contain both
ectoderm and endoderm and some mesoglea. These lacerated pieces
soon roll up and heal, and the subsequent development is essentially a process of regeneration, replacing the wanting parts of
an individual. The advantage of pedal laceration is that one
adult can produce a large number of small new individuals but the
disadvantage of being small is the high mortality rate before it
reaches adult size.

     Longitudinal fission. The initial fission begins either at
the aboral end (aboral-oral) or at the sides of the column
(side-side), but the plane of the fission is always parallel with
the long axis of the mouth. The advantage of longitudinal fission
is that the daughter individuals are large; they can perform the
functions of an adult soon after fission. It is possible that
they can feed during the process or immediately after fission.
In all considerations, longitudinal fission appears to be the most
efficient method of asexual reproduction; it permits colonization
of new habitats more rapidly than any other reproductive methods.

     Transverse fission. This method was known for years in only
two species, Gonactinia prolifera and Aiptasia couchii. More
recently Schmidt (1970) described the transverse fission of a
third species, Anthopleura stellula, from the Red Sea. The situation in A. stellula is limited to laboratory observations, induced
by lowering the salinity from its natural Red Sea environment.

     Budding. Budding is known in the family Boloceroideae in
which there are three genera: Boloceroides, Boloceractis and
Bunodeopsis (Carlgren, 1949; Robson, 1966). There is apparently
a sphincter at the base of each tentacle and the contraction of
the sphincter can cause the tentacle to autotomize. The shed
tentacle will regenerate a new animal from the broken end.
Budding may occur in situ, and in this case the new individual
always faces the adult by its oral end. Although all three genera
can regenerate from tentacles both in situ or in vitro, Bunodeopsis
can also bud from the basal part of the column.

                         SEXUAL REPRODUCTION

     Classifications of sexual reproduction in this section are
based on three critera: the mode of fertilization, larval
nutrition and larval habits. Fertilization can be internal or
external. If it is internal, the development can be either
larviparous (release at larval stage) or viviparous (release at

juvenile stage); if it is external, the development is oviparous.
When considering larval nutrition, the larvae can be feeding
(planktotrophic, parasitic, or detritotrophic) or non-feeding
(lecithotrophic). Larval habits can be pelagic, benthic, brooding (external) or viviparous. After a close examination of the
known sexual developing forms, the development of actinarians
can be categorized into the following seven patterns: oviparous-
pelagic-planktotrophic, oviparous-pelagic-lecithotrophic,
oviparous-pelagic-parasitic, oviparous-benthic-detritotrophic,
oviparous-brooding-lecithotrophic, larviparous-pelagic-
lecithotrophic, and viviparous (Table I).

    Oviparous-pelagic-planktotrophic. This type of development
usually starts from small, but more numerous eggs (less than
300 μ in diameter); the pelagic life can be as long as many weeks.
The advantage of this mode of development is clearly the ability
to disperse, but, on the other hand, the mortality rate, resulting
from starvation, predation and adverse environmental conditions
during the long larval life is great.

    Oviparous-pelagic-parasitic. The difference between this
pattern and the previous one is the larval diet; it changes from
free living and feeding on plankton to parasitic and feeding on
the tissue of medusae (medusophillous) or ctenophores. It has
been shown that the parasitic larval life is obligatory
(Spaulding, 1972) which undoubtly increases the larval mortality
rate but in the meantime has gained the assurance of nutrition
and protection once a host is secured.

    Oviparous-benthic-detritotrophic. There is only one species,
Halcampa duodecimcirrata, known for certain to have benthic and
detritus feeding larvae (Nyholm, 1949). The larva of this species
is not ciliated; it burrows in the sand and feeds on detritus.
Two other species, Halcampa decementaculata and Halcampa crypta,
have been inferred to have a similar mode of reproduction
(Strathmann, 1969; Siebert & Hand, 1974). The omission of the
pelagic larval phase in this type of development will certainly
reduce the mortality rate but, in the meantime, the animal is
handicapped for dispersal.

    Oviparous-pelagic-lecithotrophic. This type of reproduction
differs from the previous three types in the changing of larval
nutrition, from feeding to non-feeding. This is achieved by
increasing yolk content in the egg; hence, an increase in the
size of the egg. The diameter of the lecithotrophic eggs is in
the range of 300 μ to 1,100 μ. The increase of size naturally
results in a reduction of number of eggs produced. This type of
development does have the advantage of dispersal although the
pelagic life is short; it also removes the possibility of larval

death by starvation. Its disadvantage lies in the reduced number
of eggs produced.

     Oviparous-brooding-lecithotrophic. Six species in the genus
Epiactis are known to brood externally on the column. Dunn (1972)
has studied the reproduction of E. prolifera from central
California extensively and found that the larvae are not ciliated.
The advantage of external brooding is again in the reduction of
larval mortality; its disadvantage is the lack of a dispersal
phase. Brooding in E. prolifera is obligatory; small juveniles
when less than 4 mm in base diameter cannot survive when dislodged
from the parent.

     Larviparous-pelagic-lecithotrophic. This type of development
differs from the previous ones in that fertilization has become
internal and, indeed, the first phase of larval development takes
place inside the coelenteron. This mode of reproduction has
clearly the advantage of preventing fertilization waste and the
reduction of mortality during the earlier stages of larval development; it also has the advantage in dispersal, however short the
pelagic larval life may be. Thus, in all considerations, this
appears to be a very successful method of reproduction.

     Viviparous. In this type of reproduction, there is essen-
tially no waste of eggs and the juveniles, when released, are
able to live an independent life. The disadvantage is again the
lack of a dispersal larval stage.

<hr>

Table I

     Species list of asexual and sexual reproductive patterns.
References and other methods of reproduction are given in parenthe-
sis.

<hr>

ASEXUAL REPRODUCTION
  Pedal laceration: Actinothoë lacerata (Stephenson, 1935),
    Aiptasia annulata (Cary, 1911), A. couchii (Stephenson, 1938),
    A. pallida (Cary, 1911), A. tagetes (Cary, 1911), Amphianthus
    dohrnii (Stephenson, 1935), Cylista leucolena (Cary, 1911),
    Diadumene cincta (Stephenson, 1928), D. leucolena (Hand, 1955,
    sexual), D. luciae (Stephenson, 1928), Gephyropsis dohrnii
    (Stephenson, 1928), Metridium senile (Stephenson, 1928; ovi-
    parous), Phellia gausapta (Stephenson, 1928), Sagartia
    anguicoma (Stephenson, 1928), S. elegans (Stephenson, 1928;
    viviparous), S. lacerata (Stephenson, 1928)
  Longitudinal fission: Actinia bermudensis (Walton, 1918;
    viviparous), A. cavernosa (Walton, 1918), Actinothoë sphyrodeta
    (Stephenson, 1935; viviparous), Anemonia sulcata (Stephenson,

1928), Anthopleura aureoradiata (Parry, 1951), A. elegantissima
   (Hand, 1955; oviparous), A. thallia (Stephenson, 1928),
   Diadumene lighti (Hand, 1955; sexual), Haliplanella luciae
   (Stephenson, 1928), Metridium senile (Stephenson, 1935;
   oviparous, pedal laceration), Protanthea simplex (Walton,
   1918; oviparous), Sagartia davisi (Torry & Merry, 1904), S.
   sphyrodeta (Stephenson, 1935), Syanthina smithi (Walton,
   1918)
Transverse fission: Aiptasia couchii (Stephenson, 1935),
   Anthopleura stellula (Schmidt, 1970; oviparous), Gonactinia
   prolifera (Blockman & Hilger, 1888)
Budding: Boloceratis (Carlgren, 1949; Robson, 1966),
   Boloceroides (Carlgren, 1949; Robson, 1966), Bunodeopsis
   (Carlgren, 1949; Robson, 1966)
SEXUAL REPRODUCTION
   Oviparous-pelagic-planktotrophic: Anthopleura elegantissima
   (Siebert, 1974; longitudinal fission), A. xanthogrammica
   (Siebert, 1974), Metridium senile (Gemmill, 1920; pedal
   laceration), Protanthea simplex (Nyholm, 1959)
   Oviparous-pelagic-planktotrophic or lecithotrophic: Actinothoë
   anguicoma (Stephenson, 1935), A. spherodeta (Stephenson, 1935),
   Aiptasia mutabilis (Gemmill, 1921), Anemonia sulcata
   (Stephenson, 1935), Anthopleura aureoradiata (Parry, 1951),
   A. ballii (Stephenson, 1935), A. stellula (Schmidt, 1970;
   transverse fission), Calliactis parasitica (Stephenson, 1935;
   viviparous), Cereus pedunculatus (Rossi, 1975; viviparous,
   parthenogenesis), Diadumene lighti (Hand, 1955), Edwardsia
   beautempsii (Stephenson, 1935), Eloactis mazeli (Stephenson,
   1935), Halcampa chrysanthellum (Stephenson, 1935), Illyanthus
   mitchelli (Stephenson, 1935), Metridium canum (Parry, 1951),
   Paraphillia expensa (Stephenson, 1935), Sagartia anguicoma
   (Stephenson, 1928), S. elegans (Stephenson, 1935), S.
   troglodytes (Nyholm, 1949; viviparous), S. undata (Stephenson,
   1928), S. viduata (Stephenson, 1928; viviparous), Sagartiogeton
   sp. (Widersten, 1968), Stomphia coccinea (Stephenson, 1935),
   Tealia felina (Stephenson, 1935), Zaolutus actius (Hand, 1955)
   Oviparous-pelagic-parasitic: Edwardsia callimorpha (Stephenson,
   1935), Milne-Edwardsia carnea (Stephenson, 1935), Peachia
   carnea (Widersten, 1968), P. clava (Parry, 1951), P. hastata
   (Stephenson, 1935), P. hili (Stephenson, 1935), P. parasitica
   (Calgren, 1949), P. quinquecapitata (Spaulding, 1972)
   Oviparous-benthic-detritotrophic: Halcampa crypta (Siebert &
   Hand, 1974), Halcampa decmentaculata (Strathmann, 1969),
   Halcampa duodecimcirrata (Nyholm, 1949)
   Oviparous-pelagic-lecithotrophic: Adamsia palliata (Gemmill,
   1920), Bolocera tueidae (Gemmill, 1921), Stomphia didemon
   (Siebert, 1973), Tealia coriacea (Gemmill, 1921), Tealia
   crassicornis (Chia & Spaulding, 1971)
   Oviparous-brooding-lecithotrophic: Epiactis arctica (Calgren,

1949), E. fecunda (Calgren, 1949), E. japonica (Uchida &
Iwata, 1954), E. lewisi (Calgren, 1949), E. marsupialis
(Calgren, 1949), E. prolifera (Dunn, 1972)
Larviparous-pelagic-lecithotrophic: Cribrinopsis fernaldi
(Siebert & Spaulding, 1976), Tealia crassicornis (Nyholm, 1949)
Viviparous: Actinia bermudensis (Walton, 1918; longitudinal
fission), A. equina (Stephenson, 1935), A. tenebrosa (Parry,
1951), Aiptasia couchii (Stephenson, 1935; transverse fission,
pedal laceration), Bunodactis chrysobathys (Parry, 1951),
B. verrucosa (Stephenson, 1935; oviparous), Calliactis
parasitica (Stephenson, 1935; oviparous), Cataphellia
brodricii (Stephenson, 1935), Cereus pedunculatus (Stephenson,
1935; oviparous, parthenogenesis), Cnidopus ritteri (Carlgren,
1934), Hormathia coronata (Stephenson, 1935), H. digitata
(Stephenson, 1935), Sagartia troglodytes (Stephenson, 1935),
Tealia crassicornis (Stephenson, 1935)

ON THE ORIGIN OF ASEXUAL REPRODUCTION

From a review of sea anemone reproduction, a general trend
emerges and it indicates that asexual reproduction is associated
with either one or all three of the following factors: stressful
environment, small body size, and poor nutrition. The three
factors may be interrelated, that is, small body size may be the
result of poor nutrition and poor nutrition may be an indication
of a stressful environment. As a specific example, the reproduction of Anthopleura elegantissima of the Pacific sea shore of
North America supports this hypothesis well. At the upper
intertidal zone this species reproduces predominantly by the
asexual method (longitudinal fission) resulting in numerous
clusters (clones) of small individuals. At the lower intertidal
zone or the deeper pools, one finds the solitary and large ones
which do not indulge in asexual reproduction. Both populations
are known to reproduce sexually (Siebert, 1974). In this case we
are comparing a stressful (greater fluctuation of temperature,
salinity, desiccation, etc.) and nutrient-poor upper intertidal
environment with a less stressful and nutrient-rich lower intertidal environment. Nutrient poverty of the upper intertidal
should be understood from the standpoint of both food resources
as well as the limitation of feeding time (they feed only when
submerged). Along the same line of reasoning, one should also
note Metridium senile. It has been my observation that small
individuals of this species of the fouling community undergo active
pedal lacerations while the large individuals of the subtidal
environment do not. Another example is found in Cereus
pedunculatus. In this species, the asexual reproduction is in the
form of parthenogenesis which occurs only in the populations of
upper intertidal where the water is heavily polluted, and the

animal is 2 to 3 times smaller than those of subtidal populations
whose reproduction is sexual and oviparous (Rossi, 1975).   Under
the above-mentioned selective pressures, a population is favored
to reproduce asexually and the sexuality may eventually be lost.

EVOLUTION OF SEXUAL REPRODUCING PATTERNS

Of the seven patterns of sexual reproduction, the oviparous-
pelagic-planktotrophic is considered the primitive one.   This is
based on the common belief that a pelagic and feeding larva is a
primitive form of development of marine benthos (Jägersten, 1970).
If this is so, the origin of other modes of reproduction can be
traced to some simple alteration of either larval nutrition,
larval diet, larval habit or the mode of fertilization.   My interpretation of the adaptive radiation of the seven patterns is
presented in Figure 1.

From an analysis of the relative rate of dispersal and survival of the seven patterns, it is concluded that the viviparous
form has the highest rate of survival but the lowest rate of
dispersal; on the other hand, the oviparous-pelagic-planktotrophic
pattern has the highest rate of dispersal but the lowest rate of

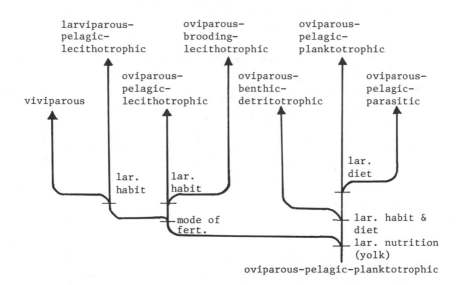

Figure 1.   Possible evolutionary relationships of the seven
sexual reproduction patterns.   Marks of the separation points of
the lines indicate the major alterations which have occurred in
either larval nutrition, larval diet, larval habit, or mode of
fertilization.

survival. The rest of the five patterns fall in between the two.
Since all patterns are in existence today, we should assume that
they are equally successful in maintaining a steady state of the
population. If so, how do the oviparous-pelagic-planktotrophic
animals and those of other patterns close to it cope with the low
rate of survival? The compensation factor here is the increase in
the number of eggs produced. In the final analysis, therefore,
the oviparous-pelagic-planktotrophic pattern has the net advantage
of dispersal. In the long run, this is the most favored pattern
of reproduction in the survival of a species. If so, why, then,
do the viviparous or brooding strategies still exist? I have
argued elsewhere that this is the "poor-man's game" in coping with
an adverse situation (Chia, 1976). The essential point is that
when nutrients for gamete production are limited it is more
advantageous for the animal to adopt a strategy of high survival
rate, giving up the "luxury" of dispersal. This strategy is good
for short term success, and in the long run species reproducing
in such a way will have a greater chance of extinction; but as
long as the environmental pressure of limited nutrient supply for
gamete production exists, there will be more species to adopt the
brooding or viviparous modes of reproduction. In other words,
these methods of reproduction will continue to exist although they
are not the best strategy.

I thank Drs. D. P. Abbott, L. G. Harris, and N. R. Howe for
useful discussions during my stay at Hopkins Marine Station of
Stanford University, where this paper was written. I thank also
Dr. C. Hand for reading the manuscript. The preparation of the
paper is supported from a grant and a travel fellowship from the
National Research Council of Canada.

REFERENCES

Blockmann, F. and C. Hilger. 1888. Über <u>Gonactinia</u> <u>prolifera</u>
      Sars, eine durch Quertheilung sich vermechrende Actinie.
      Morph. Jahrb. Leipzig, 13: 385-401.
Campbell, R. D. 1974. Cnidaria <u>In</u> Reproduction of Marine Inverte-
      brates I. Acoelomate and Pseudocoelomate Metazoans. (Giese,
      A. C. and J. S. Pearse, ds.) Academic Press, New York.
      pp. 133-199.
Carlgren, O. 1934. Some Actinaria from Bering Sea and Arctic
      waters. J. Washington Acad. Sci. 24: 348-353.
-------- 1949. A survey of the Ptychodactiaria, Corallimorpharia,
      and Actinaria. K. Svenska Vetenskap. Handl. 1: 1-120.
Cary, L. 1911. Pedal laceration in actinians. Biol. Bull. 20:
      81-108.

Chia, F. S. 1976. Classification and adaptive significance of
    developmental patterns in marine invertebrates. Thalassia
    Jugoslavica (in press).
-------- and J. G. Spaulding. 1972. Development and juvenile
    growth of the sea anemone Tealia crassicornis. Biol. Bull.
    142 (2): 206-218.
Dunn, D. F. 1972. Natural history of the sea anemone, Epiactis
    prolifera Verrill, 1869, with special reference to its repro-
    ductive biology. Ph.D. Thesis, Univ. Calif. Berkeley. 187
    pp.
Gemmill, J. F. 1920. The development of the sea anemones
    Metridium dianthus (Ellis) and Adamsia palliata (Bohadsch).
    Phil. Trans. Roy. Soc. London 209: 352-375.
-------- 1921. The development of the sea anemone Bolocera
    tueidae (Johnst.). Q. J. Microsc. Sci. 65: 577-587.
Hammatt, M. L. 1906. Reproduction in Metridium by fragmental
    fission. Amer. Nat. 40: 583-591.
Hand, C. 1955a. The sea anemones of central California. Part
    II. The endomyarian and mesomyarian anemones. Wasmann.
    J. Biol. 13 (1): 37-99.
-------- 1955b. The sea anemones of central California. Part
    III. The acontiarian anemones. Wasmann J. Biol. 13 (2):
    189-251.
Hyman, L. H. 1940. The Invertebrates. I. Protozoa through Cteno-
    phora. McGraw-Hill, New York. pp. 365-661.
Jägersten, G. 1972. Evolution of the metazoan life cycle. Aca-
    demic Press, New York. pp. 280.
Mergner, H. 1971. Cnidaria. In Experimental Embryology of
    Marine and Freshwater Invertebrates. (Reverberi, G., ed.).
    American Elsevier Co., Inc., New York. pp. 1-84.
Nyholm, K. G. 1949. On the development and disperal of Athenaria
    with special reference to Halcampa duodecimcirrata. Zool.
    Bidr. Fran Uppsala 27: 467-505.
-------- 1959. On the development of the primitive actinian
    Protanthea simplex Carlgren. Zool. Bidr. Fran Uppsala 33:
    69-77.
Parry, G. 1951. Actinaria of New Zealand. Part I. Rec. Cant.
    Mus. 6: 83-119.
Robson, E. A. 1966. Swimming in Actinaria. In The Cnidaria and
    their Evolution (Rees, W. J., ed.). Academic Press, New
    York. pp. 333-360.
Rossi, L. 1975. Sexual races in Cereus pedunculatus (Boad).
    Pubbl. Staz. Zool. Napol 39 (suppl.): 462-470.
Schmidt, H. 1967. A note on the sea anemone, Bunodactis verrucosa
    Pennant. Pubbl. Staz. Zool. Napol. 35: 252-253.
-------- 1970. Anthopleura stellula (Actiniaria, Actiniidae) and
    its reproduction by transferse fission. Mar. Biol. 5 (3):
    245-255.

Schmidt, H.   1972.   Prodromus zu einer Monographie der Mediterranen
    Aktinien.   Zoologica 42: 1-146.   (This important reference
    came to my attention only after the manuscript was completed.)

Siebert, A. E., Jr.   1973.   A description of the sea anemone
    Stomphia didemon sp. nov. and its development.   Pac. Sci. 27
    (4): 363-373.

-------- 1974.   A description of the embryology, larval develop-
    ment and feeding of the sea anemones Anthopleura elegantissima
    and Anthopleura xanthogrammica.   Can. Jour. Zool. 52: 1383-
    1388.

-------- and C. Hand.   1974.   A description of the sea anemone
    Halcampa crypta, new species.   Wasmann Jour. Biol. 32 (2):
    327-336.

-------- and J. G. Spaulding.   1976.   The taxonomy, development and
    brooding behavior of the anemone, Cribrinopsis fernaldi sp.
    nov. Biol. Bull. 150: 128-138.

Spaulding, J. G.   1972.   The life cycle of Peachia quinquecapitata,
    an anemone parasitic on medusae during its larval development.
    Biol. Bull. 143: 440-453.

-------- 1974.   Embryonic and larval development in sea anemones
    (Anthozoa: Actiniaria).   Am. Zool. 14 (2): 511-520.

Stephenson, T. S.   1928 and 1935.   The British Sea Anemones.   Vols.
    I and II.   London, Ray Society.

Strathmann, M.   1969.   Cnidaria.   Methods in developmental biology.
    Laboratory manual, Friday Harbor Laboratories, Univ. of Wash.

Torrey, H. B. and J. R. Mery.   1904.   Regeneration and non-sexual
    reproduction in Sagartia davisi.   Univ. Calif. Publ. Zool.
    Berkeley 3: 211-226.

Uchida, T. and F. Iwata.   1954.   On the development of a brood-
    caring actinian.   J. Fac. Sci. Hokkaido Univ.   Series VI.
    Zool. 12.

-------- and M. Yamada.   1968.   Cnidaria In Invertebrate Embryology
    (Kume, M. and Dan, K., eds.).   J. C. Dan, transl.   Nolit Publ.
    House, Belgrade.   pp. 86-116.

Walton, A. C.   1918.   Longitudinal fission in Actinia: J. Morph.
    31: 43-52.

Widersten, B.   1968.   On the morphology and development in some
    cnidarian larvae.   Zool. Bidr. Uppsala 37: 139-182.

REPRODUCTION OF SOME COMMON HAWAIIAN REEF CORALS

John S. Stimson

Zoology Department

University of Hawaii, Honolulu, Hawaii 96822

INTRODUCTION

The study of coral ecology has primarily been one of
distribution and zonation and little attention has been paid to
the adaptations of particular species, aside from the obvious one
of morphology. Now workers are beginning to study such questions
as the relative ability of different species to capture zooplankton
food (Porter, 1974; Johannes & Tepley, 1974) and how coral species
rank in terms of predator-preferences (Goreau, et al, 1972). Such
studies can be useful in interpreting coral community structure.
In this work I have asked whether different coral species have
different modes and temporal patterns of reproduction, and if they
do, how these might represent adaptations to the environments of
these species.

At this point in time, knowledge of coral reproduction is
based on observations from relatively few species and can be
summarized only in a very general way; Vaughan and Wells (1943)
stated that: "Scleractinian polyps are hermaphroditic, unisexual
or sterile."

The mode of sexual reproduction is known for only two of the
fifteen common Hawaiian species. Harrigan (1972) has described
the timing of planulation in one of these, Pocillopora damicornis,
in great detail, and Edmondson (1946) has described the timing of
planulation in both P. damicornis and Cyphastrea ocellina. The
mode and timing of reproduction of the most important Hawaiian
species in terms of cover and number of colonies, has not yet been
determined.

METHODS

To assess the lunar phase and season of planulation, individual colonies or representative parts of colonies were broken off the substrate and returned to the lab in buckets containing sea water. Colonies were housed individually in glass or plastic containers in still or aerated seawater for 1 day. The number of planulae released over 24 hours by each colony was recorded.

Because many species failed to release planulae, histological sections were made of 1-2 cm^2 pieces of colonies in order to determine when gonads developed and whether planulae were present in polyps. Pieces of coral were fixed in formalin or Bouins solution and decalcified in Bouins or 2.5% nitric acid. Frozen sections were made by staining tissue with oil red 0 and methylene blue, sectioning by hand and mounting in glycerine jelly. Corals with thicker tissue such as Porites species and Montipora species were paraffin sectioned and stained with Mallory's triple stain.

The common species of Hawaiian corals have been examined for planulation (Table 1); of these; Porites compressa, Porites lobata and Pocillopora meandrina are by far the most common, comprising together approximately 95% of the cover on Hawaiian reefs.

TABLE 1

Characteristics of Hawaiian corals examined for planulation and presence of gonads

Species	Habitat	Approx. Depth Range	Growth Form
Cyphastrea ocellina	reef flat	1m	encrusting
Montipora species	calm water	1-5m	encrusting & erect
Pavona explanulata	reef fronts	2-5m	encrusting & erect
Pavona varians	calm water	1-3m	encrusting
Pocillopora damicornis	reef flat	1m	branching
Pocillopora meandrina	reef front	1-10m	branching
Porites compressa	calm water	1-25m	branching
Porites lobata	reef front	1-35m	massive & encrusting

The colonies examined for the release of planulae, were of at least average size for these species, except in the case of Porites colonies which were generally < 60cm in diameter. It is usually difficult to determine the size of Montipora and Porites colonies because what appears to be an individual colony may only be part of a once larger colony which fragmented or divided as it grew.

RESULTS

Only two species of Hawaiian corals released planulae while held in the laboratory: Pocillopora damicornis and Cyphastrea ocellina, and they planulated in all seasons (Table 2). Colonies of each species have been examined in each lunar phase, but not in every lunar phase of every month; P. damicornis planulates primarily on the full moon in Hawaii, as shown by Harrigan (1972), and C. ocellina planulates primarily on the new moon (χ^2 test of independence, $\chi^2=9.22$, df=3,P<.05), despite the fact that they live in the same habitat. The distributions of both are almost entirely restricted to the reef flat.

TABLE 2

Number of coral colonies observed for the release of planulae and the number of Cyphastrea ocellina and Pocillopora damicornis which released planulae.

Species	Jan–Mar	Apr–Jun	Jul–Aug	Sep–Dec
Cyphastrea ocellina	7	32	7	5
number of colonies which released planulae	1	8	4	4
Montipora species	4	11	7	8
Pavona varians	2	2	7	5
Pocillopora damicornis	9	35	31	13
number of colonies which released planulae	1	3	5	3
Pocillopora meandrina	6	15	10	10
Porites compressa	3	13	7	3
Porites lobata	4	25	4	9

The number of planulae released by individual colonies was very variable and showed little relation to the size of the colony, values ranged from 1 to 47 for P. damicornis and from 1 to 9 for Cyphastrea ocellina.

Since most species did not release planulae I examined coral tissue from additional colonies for the presence of planulae in the polyps, and for signs of gonadal development. No planulae were found in the colonies examined (Table 3), but ovaries were found in the polyps of some colonies of all species examined. If ovaries were present they were generally found in all polyps of the sample. Pocillopora meandrina colonies contained ovaries and testes only in April and May; I do not know whether both ovaries and testes mature at the same time. The smallest fertile colony of P. meandrina was ∿22 cm in diameter. Only ovaries were seen in the Porites polyps, except in one instance in February in which polyps contained only testes. Polyps of P. damicornis and Montipora species were also hermaphroditic but may or may not be simultaneous hermaphrodites.

TABLE 3

Timing of gonadal development in Hawaiian corals which have not liberated planulae. Figures refer to the number of colonies examined by histological sections and the number of examined colonies whose polyps contained gonads.

Species	Jan	Feb	Mar	Apr	May	Jun	Jul	Aug	Sep	Oct	Nov	Dec
Pocillopora meandrina	7	2	12	14	10	4	6	0	24	23	9	0
colonies with eggs and testes	0	0	0	7	3	0	0	–	0	0	0	–
Porites compressa	4	1	2	2	1	1	4					
colonies with eggs	0	(testes)	0	0	0	0	1					
Porites lobata	4	0	2	3	2	0	4					
colonies with eggs	0	–	0	0	0	0	1					
Montipora species	1	3	2	5	5	1	1	0		0		5
colonies with eggs and testes	0	2	1	2	1	1	0	–		–		1

By counting eggs and ovaries in cross and longitudinal
sections it is possible to estimate the fecundity of the Hawaiian
corals. Table 4 shows the number of eggs in each polyp and the
number of ovaries per polyp. In the case of Pocillopora meandrina
the eggs became fewer as they increased in size. This probably
occurs in other species too, but I've not observed it. Despite
the presence of up to 24 eggs per polyp in Pocillopora damicornis
and Cyphastrea ocellina, no more than three planulae were ever
seen in polyps of these species. The species which planulated
have the smallest number of eggs per polyp except for Montipora
species, which had the largest eggs of any species.

TABLE 4

The fecundity of polyps of Hawaiian reef corals.

Species	Approximate Number of eggs per polyp	Number of Ovaries per polyp
Cyphastrea ocellina	24	–
Montipora species	12–16	4
Pavona explanulata	240	12
Pocillopora damicornis	18–24	6
Pocillopora meandrina	108–120	6
Porites compressa	36–48	12
Porites lobata	36–48	12

The size of eggs varied between species, being largest in
Montipora species, $\sim$ 280μ in diameter, and as small as 30μ in
diameter in the species of Pocillopora. Eggs of Porites species
were up to 120μ in greatest diameter.

Polyps of some Pocillopora colonies contained up to three
elongate bodies of fat in their coelenteron in all months. The
bodies are more tapered at one end than the other and grow to about
0.4mm in length. They are almost entirely fat, but contain
zooxanthellae and mesogleal tissue and show very limited development of a surface layer of cells. They never possessed a columnar
ectoderm or the structures of the larval coelenteron. Because the
bodies appeared independently of eggs in P. meandrina, that is in
all months and in all lunar phases, I don't think they are early
stages in the development of planulae.

DISCUSSION

The only Hawaiian species which planulated in the course of
this study or other studies of Hawaiian corals (Harrigan, 1972;
Edmondson, 1946) are Cyphastrea ocellina and P. damicornis, both
are reef flat species. Marshall and Stephenson (1933) and Atoda
(1947) have also commented on their failure to detect planulation
in many species; in both their studies, P. damicornis (called
P. bulbosa by Marshall and Stephenson (1933) was one of the
species which planulated.

The species of quieter or deeper water such as P. meandrina
and Porites are the most abundant corals in Hawaii, yet they did
not planulate in the course of the study. The data collected here
suggests that these species are seasonal in their production of
gametes. If P. meandrina and Porites species planulate, they
probably do so only in spring or summer, in contrast to the year
round planulation of P. damicornis and C. ocellina. It seems
unlikely that P. meandrina does not planulate, i.e., that it
reproduces by external fertilization, because all five pocilloporid
species I have examined at Enewetak atoll have planulated, both in
January, and in June and July (unpublished data).

A number of arguments lead to the hypothesis that Hawaiian
reef flat species have high rates of increase relative to corals
of other habitats. The first is the ease or frequency with which
planulae can be obtained from reef flat species compared to other
corals. Non reef-flat species may not reproduce by planulation
but this alternative seems unlikely, at least in the case of
P. meandrina, because I have observed planulation in all five
Pocilloporid species I examined at Enewetak. Also planulation is
regarded as the usual mode of sexual reproduction in corals,
(Hyman, 1940; Vaughan and Wells, 1943). Secondly, P. damicornis
and C. ocellina released planulae in every month of the year
(Edmondson, 1946; Harrigan, 1972), and according to Harrigan's data
it is likely that individual colonies can planulate in consecutive
months, thus a colony of P. damicornis may be able to reproduce
12 times a year. Colonies of P. meandrina have gonads in only two
months of the year; assuming this represents the time of planulation, then a colony of P. meandrina probably only planulates twice
a year and thus has a lower annual fecundity than an equivalent
size P. damicornis. A low annual fecundity suggests a lower potential rate of increase for the reef front species compared to its
reef flat congener.

Finally reef flat species may also reproduce at a younger age
than those from other habitats, and this too would result in a
greater potential rate of population increase. Loya (1976) has
presented an estimate of 2-3 years to sexual maturity for the
shallow water species of Stylophora pistillata. Harrigan (1972)

found that P. damicornis colonies as small as 5 cm in diameter can produce planulae; I would estimate these colonies to be about 1-2 years old using the growth rates of Edmondson (1929) and Tamura and Hada (1932). Colonies of P. meandrina, greater than ∿20cm in diameter, were the only ones whose polyps were fertile. Using Edmondson's (1929) figures and those of Tamura and Hada (1932) for Pocillopora molokensis, I estimate 2-4 cm growth in diameter in P. meandrina per year, which suggests at least 5 years to reach sexual maturity in this one deep water species. Connell (1974) has estimated the age at first reproduction of the solitary coral Fungia actiniformis at 10 years and of the massive coral Favia doreyensis at 8 years; both of these are also found in shallow water such as reef flats, but have very different growth forms than the 2 species of Pocillopora and Stylophora pistillata.

Mortality rates of reef flat species may also be higher than those of corals from other habitats because corals of reef flats are susceptible to biological disturbances such as algal over-growth (Edmondson, 1928) and physical disturbances such as sedimentation, extremes of salinity, temperature, and exposure (Loya, 1972; Glynn, 1973). Under such conditions selection would favor early reproduction and high fecundity; the ease with which planulae can be obtained from reef flat species year round in Hawaii is probably a simple measure of the influence of this selection.

SUMMARY

In previous studies and in the present one, many coral species have not planulated while under laboratory observation. Two Hawaiian hermatypic corals whose distributions are restricted almost entirely to reef flats, liberate planulae year round on lunar cyles. Polyps of one of these two, Pocillopora damicornis, contain as many as 4 fully developed planulae at a time. Colonies as small as 5 cm in diameter have planulated. The number of planulae released by colonies of these reef flat species is very variable.

The most abundant and wide spread Hawaiian species have not planulated, but eggs have been seen in histological sections of polyps of Pocillopora meandrina in the spring and summer, and in species of Porites in the summer. Polyps of P. meandrina contain ∿ 100 eggs in 6 ovaries, more than were seen in ovaries of P. damicornis; mesenteries with ovaries alternate with those containing testes. Only colonies greater than ∿ 20cm in diameter have contained eggs and testes. Porites spp. contain eggs in 12 mesenteries, a total of ∿ 36 eggs per polyp. Testes have been seen only in spring.

REFERENCES

Atoda, K.   1947.   The larva and postlarval development of some
    reef-building corals.   I. *Pocillopora damicornis* (Dana).
    Sci. Rep. of the Tohoku Uni., Ser. 4 (Biology), 18:24-47.

Connell, J.   1974.   Population ecology of reef corals.   In: The
    Geology and biology of coral reefs.   Academic Press, New York.

Edmondson, C. H.   1928.   The ecology of an Hawaiian coral reef.
    B. P. Bishop Mus. Bull. 45:1-79.

Edmondson, C. H.   1929.   Growth of Hawaiian corals.   Bernice P.
    Bishop Mus. Bull. 58:38pp.

Edmondson, C. H.   1946.   Behavior of coral planulae under altered
    saline and thermal conditions.   Occasional papers of B. P.
    Bishop Mus. 18(9) 283-304.

Glynn, P. W.   1973.   Ecology of a Caribbean coral reef.   The
    Porites reef flat biotope: Pt. 1 Meteorology & Hydrography.
    Marine Biology 20:297-318.

Goreau, T., J. Lang, E. Graham, P. Goreau.   1972.   Structure and
    ecology of the Saipan reefs in relation to predation by
    Acanthaster planci (Linn.) Bull. Mar. Sci. 22:113-

Harrigan, J.   1972.   The planula and larva of *Pocillopora
    damicornis*; lunar periodicity of swarming and substratum
    selection behavior.   Ph.D. Thesis University of Hawaii 213p.

Hyman, L.   1940.   The invertebrates: Protozoa through ctenophora.
    Vol. I McGraw-Hill, New York.   726 p.

Johannes, R. E. and L. Tepley.   1974.   Examination of feeding of
    the reef coral *Porites lobata* in situ using time lapse
    photography.   Proc. Second Int. Symp. on Coral Reefs 1:127-131.

Loya, Y.   1972.   Community structure and species diversity of
    hermatypic corals at Eilat, Red Sea. Mar. Biol. 13:100-123.

Loya, Y.   1976.   The Red Sea coral *Stylophora pistillata* is an
    r strategist.   Nature 259:478-480.

Marshall, S. & T. Stephenson.   1933.   The breeding of reef animals
    Part 1.   The Corals.   Sci. Repts. of the Great Barr. Reef
    Exped. III. 8:219-245.

Porter, J. W. 1974. Zooplankton feeding by the Caribbean reef-
    building coral _Monastrea cavernosa_. Proc. Second Int. Symp.
    on Coral Reefs. 1:111-125.

Tamura, T. and Y. Hada. 1932. Growth rate of reef building
    corals inhabiting in the South Sea Island. Sci. Rep. of
    the Tohoku Uni. Rep. 4, Vol. 7.

Vaughan, T. and J. Wells. 1943. Revision of the suborders,
    families and genera of the Scleractinia. Geol. Soc. Am.
    Spec. Paps. 44:363-

REPRODUCTION, GROWTH, AND SOME TOLERANCES OF *ZOANTHUS*

PACIFICUS AND *PALYTHOA VESTITUS* IN KANEOHE BAY, HAWAII

William J. Cooke

Dept. of Zoology and Hawaii Institute of Marine

Biology, Univ. of Hawaii, Honolulu, Hi., 96822

INTRODUCTION

Kaneohe Bay, on the island of Oahu, has for some time been recognized as a tropical reef ecosystem under considerable stress from such insults as sewage pollution from outfalls in the southern sector, and increased freshwater and terrigenous sediment input from urbanization of the surrounding watershed (Smith et al., 1973). In the southern sector, one of the most striking faunal changes has been the replacement of scleractinian corals on shallow patch and fringing reefs by extensive beds of *Zoanthus pacificus* Walsh and Bowers,1971. This replacement has apparently taken place rather rapidly, within the last decade and a half, as has been noted before, (Banner, 1968). No less striking are the dense (up to 12,000 polyps/m^2) beds of *Palythoa vestitus* Verrill, 1928, on the sand flats of these reefs. *P. vestitus* is reasonably common on other shallow reef flats in Hawaii, but at much lower densities. It was believed that some clues to the success of these species may be found by comparison of their reproductive patterns with those of the species they replaced.

This question is of some general significance in several other ways. Members of the genus *Palythoa* are known to produce a highly toxic chemical, palytoxin (Moore and Scheuer, 1971) which is of considerable interest for its physiological and pharmacological properties. Kimura *et al.*, (1971) reported that the degree of toxicity of *P. tuberculosa* at Naha, Okinawa is a function of the presence of eggs in the polyps. Obviously, cyclicity of reproduction and hence of toxin production is of great interest to those seeking to obtain palytoxin. In addition, Hashimoto *et al.*, (1969) reported that the filefish *Alutera scripta* was rendered toxic by ingestion of some *P.*

tuberculosa. Thus at certain times of the year a method for the introduction of palytoxin into the marine food chain is possible.

This inital study focused on the reproductive biology of the two species in an effort to understand how they could so quickly become dominant in the reef community. The study is part of continuing research on the comparative ecology of these two species in Kaneohe Bay which is seeking, as a general goal, to understand the adaptations which allow zoanthids to exploit this disturbed reef community. The initial hypothesis was that both species would demonstrate high biotic potential, demonstrated by year around reproduction.

MATERIALS AND METHODS

Collections were made at two week intervals beginning in October 1973 at a site on the reef on the northeast side of Coconut island, site of the Hawaii Institute of Marine Biology. This island is located in the southern sector of the bay, and extensive beds of *Palythoa* and *Zoanthus* are present on the reefs here, as they are on most patch and fringing reef in this portion of the bay. The collections were made along a limited portion of the reef which was representative of conditions along the reef as a whole.

Palythoa polyps were collected from a contiguous bed 200 meters long, and 5 to 10 meters wide, located along the seawall of the island. The intertidal height of the bed varied irregularly between +15cm above 0.00 level (MLLW), to -5cm. At the collection site, the density of the bed was between 500 to 1500 polyps/m^2, and was representative of the average density. *Palythoa* are loosely colonial, and thus at each date 20-30 individuals (of undetermined clonal ancestry) were collected. Collections of *Zoanthus* were made from the reef approximately 30-80 meters off the seawall in 10 to 40 cm of water. In this zone, the colonies of *Zoanthus* are contiguous over many square meters on dead heads of *Porites* and *Montipora* corals, and coralline algae pavements. Several colonies were collected at each date, the exact distribution of the polyps reserved for gonad studies depended on other experimental demands.

For live examinations, the colonies were maintained in seawater from the site, and were examined within three hours of collection. The majority of the collection was fixed in Bouin's, and stored in ethanol. Macroscopic examination of the polyps was done under a dissecting scope on longitudinally sliced individuals pinned out in a wax dish. For the initial assessment reported here, the polyps were examined and scored on a scale of 0 - 3, without reference to sex. A score of "0" indicated no gonads; "1" indicated some gonad tissue present in the body of the mesentery; "2" indicated swelling of the

gonadal mass; and "3" indicated that numerous large folded bulges
protruded from the mesenteries. The score was usually a composite
for any polyp, as within a polyp, both "0" and "3" mesenteries could
some times be found. For a few dates, information on the sex of the
gonads was obtained by whole mounts of stained and cleared mesenteries. This was necessary because the differences between egg masses
and spermaries were not visible macroscopically, as they are in *P.
tuberculosa*.

     Growth studies were carried out in seawater tables at HIMB
supplied with unfiltered running seawater drawn from the reef close
to the collecting site. The tables were exposed to direct sun only
during the morning hours. *Palythoa* and *Zoanthus* have been maintained in these tables for over two years. 20 polyps were separated
from their substrate and placed in individual glass petri dishes on
the bottom of the seawater table. Photographic records were taken
every two weeks (*Zoanthus)* or every week (*Palythoa*), and in this
way, the number of buds (scored .5), newly opened polyps (1), as
well as size, tentacle number and so forth were easily obtained. The
salinity tolerances were established by exposing polyps (under
constant light and temperature) to experimental salinities in
aerated noncirculating seawater (diluted with distilled water) in
the laboratory.   Test salinities ranged from $0^o/oo$ to $30^o/oo$ and
exposure times were 1 day and 1 week. Survival was determined by
behavioral observation in the test beakers, and by recovery on
return to full ($34^o/oo$) seawater in the seawater tables. Polyps
were maintained in $34^o/oo$ control beakers in the laboratory for over
6 weeks.

                              RESULTS

     The eggs, spermaries, and sperm are very similar in both
species. Eggs range from 75μm to 280μm with a large nucleus approximately 40μm in diameter, with a distinct nucleolus 13μm in diameter. Spermaries range from 50μm to 390μm. Live sperm in both are
small bell shaped cells with a rounded apical end and a bulge basally with two refractile bodies, presumably mitochondria. They both
averaged 3μm by 2μm, with tails 50μm long.

     The quantitative results are summarized in figures 1 and 2.
Figure 1 is a plot of the percentage of fertile polyps from January
through December, combining the 60 collection dates through the
several years. Note that the period from December through April has
no fertile *Palythoa* polyps, and that the number of fertile *Palythoa*
polyps rises sharply in late April, and drops drastically in the end
of September.

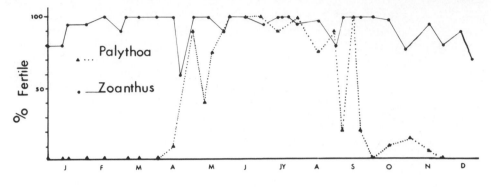

Figure 1. Number of fertile polyps by months, percentage.

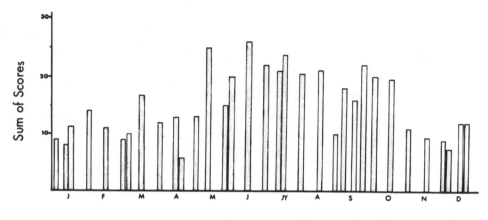

Figure 2. Sum of gonad rating scores by months for *Zoanthus*.

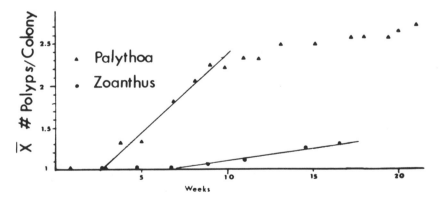

Figure 3. Mean number of polyps/colony vs time.

Although the pattern of fertile polyps for *Zoanthus* throughout the year did not reveal a seasonal cycle, and was generally consistently high, inspection of the scores, which rated the degree of involvement in gonad production, indicated that there was a peak of activity during the summer months. Figure 2 is a histogram of the sums of scores for *Zoanthus* collections throughout the year, combining years. Sample dates within four days of each other are combined and averaged in the graph, but in the analysis, no dates were lumped. As can be seen graphically the same general peak of activity is present in the period from May through September. When the dates are apportioned into four three month quarters and tested by the Kruskal-Wallace non-parametric analysis of variance, a significant difference between the quarters is obtained. As suggested by the histogram, the summer, early fall quarters are significantly ($p < 0.005$ $H' = 29.0976$ χ^2 exp.$= 12.838$) higher than the winter and spring quarters. It is thus concluded that *Zoanthus* is following the same basic seasonal cycle as *Palythoa,* but is much less drastically depressed in the low part of its cycle.

The results of the growth and budding experiment are shown on figure 3, which is a plot of the mean number of polyps/colony over a twenty week period. Each colony begins with one polyp, and over time, first buds, then fully formed polyps with column, oral disc and tentacles arise (but do not separate) from the basal part of the original polyp. Although both *Palythoa* and *Zoanthus* can divide by longitudinal fission, this was not encountered in the course of this experiment. Linear regression lines were drawn through the curves for the period of linear increase in numbers, when the active budding was not impeded by the geometry of the test situation, and the presence of already formed buds. The regression coefficient b (slope) for *Palythoa* is 0.19, while for *Zoanthus,* it is 0.03. Thus the rate of increase of new individuals by asexual production is about six times higher in *Palythoa* than in *Zoanthus* under these conditions.

As would be expected from its higher intertidal distribution, *Palythoa* demonstrated a significantly greater tolerance to reduced salinities than *Zoanthus* did. For *Palythoa,* the LD 50 for 1 day's exposure was approximately $10^o/oo$, while for *Zoanthus,* it was $20^o/oo$. For *Palythoa,* LD 50 for 1 week's exposure was $27^o/oo$.

DISCUSSION

Works on the controlling factors in seasonal reproduction have considered light, temperature, salinity, plankton factors, and unique endogenous rhythms cued, but not controlled by, one or some of the above factors. The analysis here concentrated on isolating controlling factors which allowed similar reproduction (in time) to

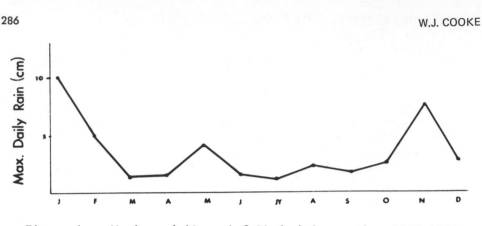

Figure 4.   Maximum daily rainfall (cm) by months, 1973-1975.

be expressed by both *Palythoa* and *Zoanthus,* yet which completely
depressed the activity in *Palythoa,* while modulating the cycle to a
much lesser extent in *Zoanthus.*   Of the above factors, only salinity
variation was thought to have a significantly different effect on
Palythoa populations, vs *Zoanthus* populations.   Located within 25
meters of each other, they are subject to essentially the same
variation in water temperature, light (discounting the shallow depth
difference) and plankton factors.   While the *Palythoa* will face
higher air temperatures in summer than in winter, this is more
likely to be a stressful rather than beneficial situation.

However, in this reef enviroment, the variation in salinity
which an animal is exposed to is a function largely of its exposure
to direct rain, as well to a lesser extent, to lower salinity lenses
of bay water caused by stream discharge.   Observations of rains at
rates between .8mm/hr and 144mm/hr have demonstrated insignificant
(approximately 2°/oo) reductions of the surface waters of the bay
during the duration of the rains.   However, on exposed sand surfaces
a rain of .8mm/hr will depress surface salinity in the water standing on the surface to 25°/oo, and -1cm interstitial water salinities
to 22°/oo.   Rains on the sand flats in the *Palythoa* beds at the rate
of 48mm/hr decreased salinities on the surface to 5°/oo, and -1cm
salinities to 16°/oo.

Figure 4 is a graph of the mean of the maximum daily rainfall
in each month combining the years 1973-75. Total rainfall during
months usually follows this same curve, but this data was considered
more informative of the short term stress the organisms might be
exposed to.   As can be seen, the period of heaviest precipitation is
in the months from October through April.   It is believed that
during periods of winter rains, the surface and interstitial salinities around the *Palythoa* colonies often becomes significantly reduced, sometimes to the extent of causing mass mortalities in the cases

when spring tides occur during exceptionally heavy rains (Banner, 1968), but almost certainly to the extent of causing physiological stress which would hinder normal gonad production, or cause resorption.

The effect of stream discharge is somewhat harder to assess, as the period of maximum discharge and hence lens formation is separated by periods of hours to a day or two at most, from the period of maximum rainfall. Five days after the storm recorded by Banner (1968), the lowest salinity of the surface waters of the south sector of the bay was 7°/oo. Additionally, corals were killed to a depth of 1 meter around Coconut Island, indicating that substantial freshening occured even at that depth. Certainly, *Palythoa* in its higher intertidal situation is exposed to reductions in salinities when rained upon during winter storms. In addition, when the tide returns, the lower salinity lens which is likely to overlie the beds may effectively retard the return of normal salinities. *Zoanthus* colonies, being generally submerged, are rarely exposed to direct rain. They may be affected by the freshened lens, if it extends deep enough to wash over them when the tide drops. Perhaps this influence is enough to depress gonad maturation during the rainy season, giving a peak of sexual production during the summer months.

Questions as to whether the observed seasonal patterns are also a function of the proposed ecological differences in heterotrophy between *Palythoa* and *Zoanthus* (Reimer, 1971) are premature at the present time, as no conclusive demonstration of the trophic budget in terms of plankton capture, photosynthetate transfer, and so forth has yet been made. If indeed, *Zoanthus* were deriving most of its energy from its zooxanthellae, then perhaps seasonal variation in solar radiation would be important in the cycle.

Unfortunately, there is little comparative evidence in the zoanthids on which to judge the hypothesis of salinity control of gonad development. Kimura *et al.*, (1971) did not record the exact ecological situation of their colonies, nor did Yamazato *et al.*, (1973) in the subsequent investigation, and neither suggested any controlling environmental factors. Herberts (1971) believed that *Zoanthus* and *Palythoa* reproduced in the warm months in Madagascar, but her primary concern was not with the reproductive cycles.

The high asexual production of *Palythoa* offers some insight into its ability to colonize the sand flats and maintain high populations there. *Palythoa* polyps were often seen loose on the surface of the sand, and can easily be induced to reattach to glass dishes, small rocks, etc, by the stolon-like basal part of the column. Budded and detached polyps thus are available for dispersal over the sand flats.

There is not yet sufficient evidence from the microscopical

studies to determine whether patterns of hermaphroditism as seen in
P. tuberculosa are present in these species. However, some *Zoanthus
pacificus* are hermaphroditic, with mesenteries present in mid-summer
with both sperm and eggs. Certainly, the control of sex determination within the group deserves much more study. Perhaps most of the
dioecious species are the result of incomplete sampling.

SUMMARY

The sperm and eggs of these two species are similar to each
other, and to those described for the group before. Both species
exhibit a peak of gonad production in the summer months. In *Palythoa,* reproduction is completely halted in the winter months, while
at the same period, it is only reduced in *Zoanthus*. Evidence indicates that *Palythoa* is exposed to significant salinity reduction in
winter rains, probably making gametogenesis difficult. *Zoanthus* is
likely less affected by the winter rains in its more submerged
habitat. *Palythoa,* has a higher rate of asexual production, which
allows it to disperse and colonize new areas around the colonies on
the sand flats.

LITERATURE CITED

Banner, A. H., 1968. A freshwater "kill" on the coral reef of
        Hawaii, *HIMB Tech. Rept. 15:* 1-29
Hashimoto, Y., N. Fusetani, and S. Kimura. 1969: Aluterin:
        a toxin of filefish *Alutera scripta,* probably origi-
        nating from a Zoantharian, *Palythoa tuberculosa.Bull.
        Jap. Soc. Sci. Fish.,* 35: 1086-1093.
Herberts, C., 1971. Étude systématique de quelques zoanthaires
        tempérés et tropicaux. *Tethys Supple.,* 3: 69-156.
Kimura, S., Y. Hashimoto, and K. Yamazato, 1971. Toxicity of the
        zoanthid *Palythoa tuberculosa. Toxicon,* 10: 611-617.
Moore, R. E., and R. J. Scheuer, 1971. Palytoxin: a new marine
        toxin from a coelenterate. *Science,* 172: 495-498.
Reimer, A. A., 1971. Chemical control of feeding behavior and role
        of glycine in the nutrition of *Zoanthus* (Coelenterata, Zoan-
        thidea). *Comp. Biochem. Physiol.,* 39: 743-759.
Smith, S. V., K. E. Chave, and D. Kam, 1973. *Atlas of Kaneohe Bay
        a Reef Ecosystem under stress.* 1-128.Univ. of Hawaii.
Walsh, G. W., and R. L. Bowers, 1971. A review of the Hawaiian
        zoanthids with descriptions of three new species.*Zool.
        Journ. Linn. Soc.,* 50: 161-180.
Yamazato, K., F. Yoshimoto, and N. Yoshihara. 1973. Reproductive
        cycle of a Zoanthid, *Palythoa tuberculosa* Esper. *Publ.Seto
        Mar. Biol. Lab.,* 20: 275-283

CHARACTERISTICS OF ASEXUAL REPRODUCTION IN THE SEA ANEMONE, HALIPLANELLA LUCIAE (VERRILL), REARED IN THE LABORATORY

Leo L. Minasian Jr.

Department of Biological Science

Florida State University, Tallahassee, Fla.   32306

INTRODUCTION

Longitudinal fission and pedal laceration are the most common asexual reproductive modes employed by sea anemones. In longitudinal fission, the sea anemone stretches transversely to generate two fragments of similar size. Tissue tears along a longitudinal plane parallel, and usually close to the oral-aboral axis (Torrey and Mery, 1904; Davis, 1919). In pedal laceration, many small tissue fragments separate from the pedal disk and regenerate into new individuals (Cary, 1911; Atoda, 1973).

Haliplanella luciae is a small (basal diameter about 5 mm), acontiate sea anemone which occurs intertidally in many different regions in the northern hemisphere (Stephenson, 1935; Hand, 1955). Although longitudinal fission has been described in most populations of H. luciae (Torrey and Mery, 1904; Davis, 1919; Stephenson, 1929; Field, 1949), pedal laceration has been found to occur only in Japanese populations (Atoda, 1973). These two different asexual processes do not appear to be exhibited simultaneously by H. luciae at any one locality.

This study describes some characteristics of asexual reproduction in Florida H. luciae, and shows that longitudinal fission can bring about rapid population growth. Since it can be cultured in the laboratory like hydrozoan (e.g. Loomis and Lenhoff, 1956) and scyphozoan (e.g. Spangenberg, 1965) polyps, H. luciae may be a useful research tool for studies investigating the simple systems of Anthozoa.

MATERIALS AND METHODS

Haliplanella luciae was collected from an oyster-shell rocky
intertidal zone at the Florida State University Marine Laboratory
near Turkey Point, Fla. in the northern Gulf of Mexico during autumn,
1974 and winter, 1975. In the laboratory, these anemones were placed
in 4" glass bowls, each with 10 anemones and about 150 ml of natural
sea water (28-30°/oo), obtained from the collection site. Anemones
were maintained at a temperature of 21-23°C and under a 14 hr:10 hr
light:dark cycle. The experiment consisted of nine culture bowls.
Feeding was as follows: three bowls were fed Artermia nauplii every
second day; three bowls were fed Artemia nauplii every fourth day;
three bowls were not fed. Anemones were fed freshly hatched Artemia
nauplii for 30 minutes, after which each bowl was rinsed with three
brief (about 3 seconds) jets of tap water to prevent microbial contamination. All bowls received this rinse every second day, at which
time the number of anemones in each culture bowl was counted, and
each bowl was refilled with fresh sea water.

Logarithmic growth rates (k) were calculated according to the
method of Loomis (1954), using the equation for logarithmic growth,
having been simplified for calculation purposes:

$$k = 0.693/T$$

Here T is the doubling time of the number of anemones in the culture
bowls. This was estimated to the nearest 0.1 days from the semi-log
plots shown in Figure 2. The day on the abscissa corresponding to
the point on the graph which indicated an exact doubling in the number of anemones was used to obtain this estimate. Where the graph
of number of anemones for a given treatment included more than one
doubling over the culture period, an average of each of the successive doubling times constituted T.

To determine the percentage of tissue contributed to daughter
individuals at the time of fission, anemones were placed in separate
plastic petri dishes (60 x 15 mm) with covers, which had been perforated to allow fluid exchange with the external medium. These
were kept within filtered sea water aquaria at room temperature (21-
23°C) under a 14 hr:10 hr light:dark cycle. Every second day the
anemones were fed to repletion with Artemia nauplii, after which the
petri-dish enclosures were rinsed briefly in tap water. At this
time, each dish was examined for the occurrence of fission. When
fission had occurred, the specimens were placed in vials containing
10% formalin in sea water to preserve the tissue intact and aid in
handling. After 48 hrs the fixative was poured off, and the samples
were frozen. Individual anemones were dried at 105°C for 24 hrs.
and the dry weights determined to the nearest 0.1 mg. A daughter:
parent tissue ratio was calculated, in which the dry weight of the
smallest fission product was divided by the total dry weight of both

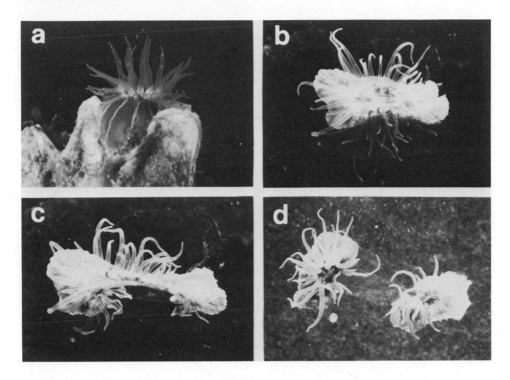

Figure 1. (a) <u>Haliplanella</u> <u>luciae</u> on natural oyster-shell substrate.
The diameter of this specimen is about 5 mm. (b) <u>H.</u> <u>luciae</u> in a culture bowl after initiation of longitudinal fission. (c) Same animal
as in (b) after several minutes of continued fission. Tissue on one
side of the oral disk has torn longitudinally. (d) "Daughter" anemones seconds after having completed longitudinal fission. A mucous
tract extending between the two new individuals is barely visible.

fission products (i.e. dry weight of the "parent" anemone). Thus,
the maximal obtainable value of 50% denotes a binary fission in
which both daughter anemones were of equal mass.

        This determination was performed on two separate samples. The
first sample was collected near Turkey Pt., Fla. on 18 Jan. 1975,
and starved in a holding tank for six days prior to the first feeding. The second group was collected at the same site on 30 Mar.
1975 and starved for only three days prior to the first feeding.
The Wilcoxon two-sample test (Sokal and Rohlf, 1969) was used to
evaluate differences between these two groups.

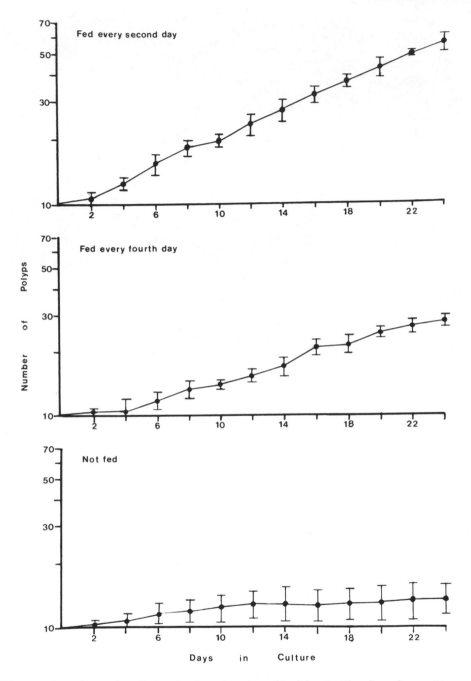

Figure 2. Growth of H. luciae by longitudinal fission in cultures
fed different quantities of Artemia nauplii. Each point represents
six culture bowls; error bars are standard deviations.

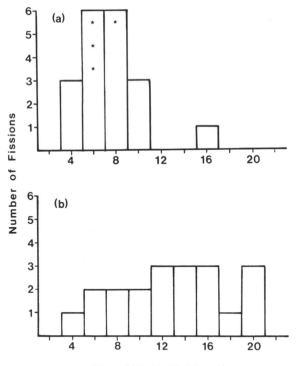

Figure 3. Temporal pattern of fission in H. luciae after transfer
to laboratory aquaria. The anemones in (a) were kept in the laboratory for six days prior to initial feeding. The anemones in (b)
were starved for only three days prior to the first feeding. Temporal distributions of the two groups are significantly different
(p<.0005; group (a): n = 19; group (b): n = 20). Stars indicate
the occurrence of ternary fissions in which three daughter anemones
were generated from a single parent.

RESULTS

     Haliplanella luciae from Turkey Pt., Fla. are dark greenish in
color with several longitudinal yellow-orange to orange stripes
(Fig. 1a). The specimens used in this study agreed closely with
descriptions of H. luciae from New England (Hargitt, 1912) and the
Texas coast (Carlgren and Hedgpeth, 1952; Hedgpeth, 1954). All
reproduction was by means of longitudinal fission (Figs. 1b-d).

     H. luciae reared in the laboratory showed sustained logarithmic
growth for the duration of the period over which cultures were maintained as indicated in Fig. 2. Feeding acts as a stimulus to growth:

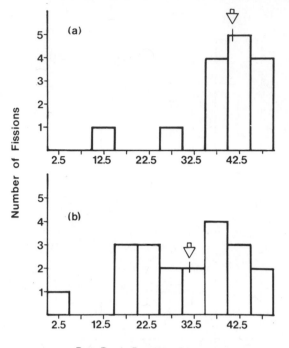

Per Cent Dry Weight of Parent

Figure 4. The percentage dry tissue weight which was contributed to the daughter anemone by the original parent anemone. The anemones in (a) were kept in the laboratory for six days prior to the initial feeding. Those in (b) were starved for only three days. Distributions of daughter:parent tissue ratios of the two groups are significantly different (.005>p>.0005; group (a): n = 15; group (b): n = 20). Median values are indicated by arrows. Ternary fissions were not included.

those cultures which were fed most frequently showed the fastest rates of growth, increasing their numbers by 550% over a period of 24 days. Population growth occurred at a lower, less uniform rate in cultures which were fed only once every four days, and showed an increase of 280% over the 24-day culture period. Thus, the anemones which were fed half as often as the most frequently fed group underwent fission approximately half as often. Unfed anemones underwent a limited increase of about 30% over the 24-day culture period.

H. luciae from Florida, fed every second day and reared under the conditions described, had a growth rate of k = 0.071, which corresponds to a doubling time of 9.6 days. Anemones reared under

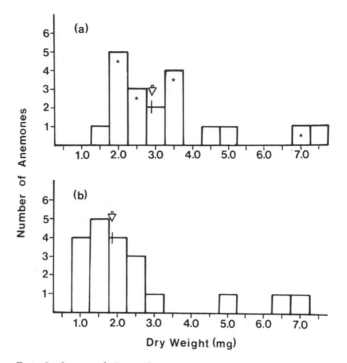

Figure 5. Total dry weight of anemones at the time of fission. Differences between the two groups are not significant (.45>p>.25; group (a): n = 19; group (b): n = 20). Sample medians are indicated by arrows. Each star indicates the occurrence of one ternary fission.

identical conditions, but fed only once every fourth day, had a growth rate of k = 0.044, and a doubling time of 15.6 days. Unfed H. luciae showed only a slight increase in numbers, as most fission activity stopped after the first 12 days of culture (Fig. 2).

H. luciae underwent longitudinal fission in as little as four days following the initial feeding as shown in Fig. 3. Fission occurred significantly sooner (p <.0005) and more synchronously among animals which were starved for six days than in those starved for three days prior to feeding. Therefore, anemones which were starved for a longer period underwent fission more promptly once feeding was renewed. There was also a significant (.005>p>.0005) difference between the two groups where anemones starved for six days (Fig. 4a) showed less variability in the proportion of tissue contributed to daughter individuals, and tended to divide more evenly (i.e. into equal halves). Those anemones starved for only three days showed greater variability in the proportion of tissue redistributed during longitudinal fission (Fig. 4b). In the majority of fissions (77.1%), more than 25% of the tissue present in the original anemone was con-

tributed to the daughter anemone (Figs. 3,5). Of the 35 binary fissions observed, 14 anemones contributed 40% or more of their original
tissue to daughter anemones (Fig. 4). Of the 39 cases of longitudinal fission, four fissions were ternary, where two fissions occurred
during the same two-day period (Figs. 3, 5). In these cases, the
anemone arbitrarily designated as the original parent underwent two
successive divisions. Whether or not it was the larger product of
the first division which again divides is not yet known. All ternary fissions occurred in the more synchronous group which was
starved for six days (Figs. 3a, 5a). The size of the anemones at
the time of fission was quite variable, but centered around 1.5-3.5
mg dry weight (Fig. 5). This size distribution probably reflected
that of anemones in the field. Differences between these two samples
with regard to size were not significant (.45>p>.25). There was
no correlation between the size of the anemones and the time between
first feeding and fission.

DISCUSSION

Longitudinal fission in H. luciae can sustain logarithmic growth
in laboratory culture. The rate of logarithmic growth (k = 0.071)
is lower than the same values for Hydra littoralis (Loomis, 1954)
and colonial hydroids (reviewed in Davis, 1971), which generally have
growth rates ranging from 0.20-0.35, and doubling times of two to
three days. However, the longitudinal fission of H. luciae is not
directly comparable to the asexual budding exhibited by some Hydrozoa. Rates of increase in hydranth number in colonial hydroids represent growth in a single, compound organism, whereas the asexual increase in numbers of H. luciae more closely resembles Hydra: both
types of polyps relinquish tissue at each asexual generation. Like
budding in Hydra (Campbell, 1967), longitudinal fission in H. luciae
may be associated with steady-state growth or maintenance of a certain size, since rapidly dividing anemones are probably smaller than
those which divide less frequently. However, there is little indication that the largest anemones sampled were more predisposed to
divide than anemones which were close to the median size in the sample. Thus, size per se does not appear to determine when fission
will occur.

One exogenous factor which regulates the rate of asexual proliferation is the frequency at which the anemones are fed. My data
confirms the observations of E. Evans that "feeding" stimulated division in H. luciae (Stephenson, 1935). Although the rate of asexual
growth is a function of feeding frequency, feeding in itself is not
an absolute requirement for the occurrence of fission, since a very
limited number of fissions also occurred among unfed anemones. In
addition, this study indicates that a period of starvation prior to
the initiation of regular feeding may produce greater synchrony among
anemones starved for longer periods. This may be analogous to star-

vation-induced synchrony of cytokinesis as is known to occur in both plant and animal cells. David and Campbell (1972) have found that the epithelial cells of Hydra, which have a relatively long but variable premitotic (G_2) phase, may be induced to divide more synchronously in response to feeding. Although feeding is of primary importance in regulating fission rate, the observed differences in fission synchrony may also reflect the fact that the two samples were collected three months apart, and thus may represent different, environmentally imposed physiological states. Metabolic events associated with fission may be more active in the more synchronous anemone group (Figs. 3a, 4a, 5a), since these anemones contributed more tissue to daughter anemones. Moreover, ternary fissions occurred only among the more synchronous anemones.

Stephenson (1929) also succeeded in culturing several species of asexually reproductive actinarians using small glass containers as were used in the present study. However, he reported that H. luciae showed high mortality and could not be satisfactorily maintained under such conditions. This work demonstrates that sustained proliferation of H. luciae in mass culture can be maintained for at least several weeks. Studies dealing with actinarian polyps rather than fresh-water Hydra are sometimes difficult since sea anemones are generally much larger, and not so easily maintained in quantity. Thus, Haliplanella luciae and perhaps other anthozoan species can provide new alternative tools for a variety of investigations into simple systems, for which formerly only Hydra was suited. Comparative studies could then provide insights into how the increased tissue complexity of the Anthozoa relates to modification of developmental and physiological processes in lower metazoans.

I wish to thank Dr. Richard N. Mariscal for providing advice during the course of this study, and for his criticism of this manuscript. This work was supported in part by N.S.F. Grant No. GB-40547 to Richard N. Mariscal.

LITERATURE CITED

Atoda, L., 1973. Pedal laceration in the sea anemone, Haliplanella luciae. Publ. Seto Mar. Lab., 20: 299-313.

Campbell, R.D., 1967. Tissue dynamics of steady state growth in Hydra littoralis. II. Patterns of tissue movement. J. Morphol., 121: 19-28.

Carlgren, O. and Hedgpeth, J.W., 1952. Actinaria, Zoantharia and Ceriantharia from shallow water in the Northwestern Gulf of Mexico. Publ. Inst. Mar. Sci. Univ. Texas, 2: 141-172.

Cary, L.R., 1911. A study of pedal laceration in actinarians. Biol. Bull., 20: 81-108.

David, C.N. and Campbell, R.D., 1972. Cell cycle kinetics and development of Hydra attenuata. I. Epithelial cells. J. Cell Sci., 11: 557-568.

Davis, D.W., 1919. Asexual multiplication and regeneration in Sagartia luciae (Verrill). J. Exp. Zool., 28: 161-263.

Davis, L.V., 1971. Growth and development of colonial hydroids. Pages 16-36 in H.M. Lenhoff et. al., Eds., Experimental Coelenterate Biology. University of Hawaii Press, Honolulu.

Field, L.R., 1949. Sea anemones and corals of Beaufort, North Carolina. Bull. Duke Univ. Mar. Stat., 5: 1-39.

Hand, C., 1955. The sea anemones of central California. Part III. The acontiarian anemones. Wasmann J. Biol., 13: 189-215.

Hargitt, C.W., 1912. The Anthozoa of the Woods Hole region. Bull. U. S. Bureau Fish., 32: 223-254.

Hedgpeth, J.W., 1954. Anthozoa: the anemones. Fish. Bull., 89: 285-290.

Loomis, W.F., 1954. Environmental factors controlling growth in Hydra. J. Exp. Zool., 126: 223-234.

Loomis, W.F. and H.M. Lenhoff, 1956. Growth and sexual differentiation of Hydra in mass culture. J. Exp. Zool., 132: 555-568.

Sokal, R.R. and F.J. Rohlf, 1969. Biometry. W.H. Freeman and Co., San Francisco, 776 pp.

Spangenberg, D.B., 1965. Cultivation of the life stages of Aurelia aurita under controlled conditions. J. Exp. Zool., 159: 308-318.

Stephenson, T.A., 1929. On methods of reproduction as specific characters. J. Mar. Biol. Ass. U. K., 16: 131-172.

Stephenson, T.A., 1935. The British Sea Anemones, Vol. II. Ray Society, London, 426 pp.

Torrey, H.B., and Mery, J., 1904. Regeneration and non-sexual reproduction in Sagartia davisi. Univ. Calif. Publ. Zool., 1: 211-226.

SOME OBSERVATIONS ON SEXUAL REPRODUCTION IN TUBULARIA

Richard L. Miller

Temple University
Department of Biology
Philadelphia, Pennsylvania   19122

Observations of the sexual reproduction of the hydroid
Tubularia have been made over a period of several years in connection
with my work on animal sperm chemotaxis prior to fertilization. What
follows is a summary of my findings, in an attempt to add this data
to the published literature on an organism which has been used almost
solely for studies of asexual reproduction (Rose, 1974).

I. The organism: Tubularia, a large athecate hydroid which may be
solitary or colonial depending on the species, is found in temperate
oceans in areas of high current. Fig. 1A and 1B are photographs of
male and female polyps of T. crocea. The two whorls of tentacles,
grape-like clusters of permanently sessile gonomedusae, turret-like
body, and generally unbranched stems are characteristic. The sexes
are separate and commonly clusters of polyps are grouped by sex.
Probably this is due to asexual origin of the cluster by basal
branching (T. crocea) or branching above the base (T. larynx).
Hermaphroditic polyps have been observed by others (Berrill, 1961).
Hermaphroditic gonomedusae are apparently rare. I have found none
after several years' work with the animals. Fertilization is internal, the sperm aggregating at and penetrating through a pore in
the apex of the female gonomedusa. Development of the embryo up to
the actinula larva takes place within the female gonomedusa which
is often quite distorted by the elongation of the tentacles of the
developing actinula. The mechanism for larval release is not known.

II. Spawning: Over 100 gonomedusae may be produced by large mature
polyps, but they do not mature simultaneously (Fig. 1A). In the
male, maturity is indicated by whitening of the contents of the
gonomedusa, and often the white color has a streaked or patchy distribution. White gonomedusae can be seen in figure 1A, and these

Figure 1. A. T. crocea, male. Note the white gonomedusae (arrows).
B. T. crocea, female. Scale = 0.5 cm.

almost surely contain mature sperm. In older males, swollen white
gonomedusae may be filled with parasitic ciliates. Gonomedusae
which do not show the obvious white color may produce motile sperm
in small quantities, however. Figure 2A shows four relatively immature male gonomedusae which have been cut from the polyp. Each
has a dark band surrounding the gonomedusa midline which is composed
of refractile bodies, some of which may be in motion. Figure 2B
shows one gonomedusa shedding sperm and it may be noted that the
dark area has moved upward to the apex of the gonomedusa.

     In the laboratory, spawning is induced by clipping male polyps
bearing mature gonomedusae from their stems, and adding them to 10
ml cold sea water (10°C). The preparation is placed on the stage
of a dissecting microscope at room temperature and observed until
sperm are seen issuing from one or more gonomedusae, as in Fig 2B.
Often sperm extrusion occurs simultaneously with polyp transfer,
but more usually occurs after some 5 to 20 minutes on the microscope
stage. It is not known whether handling, light, or increased water
temperature is responsible for spawning initiation under these conditions. Since males in the holding tanks spawn under a constant
temperature-continuous light regime, it seems likely that a combination of handling and increasing temperature induces the spawning
responses observed. Spawning is usually intermittent, and ripe gonomedusae on the same polyp do not necessarily spawn together. In
nature, it is likely that sperm are being emitted in smaller or
larger amounts more or less continuously from one to several gono-

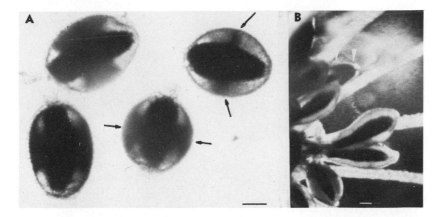

Figure 2.   A. Four male T. larynx gonomedusae photographed under
transmitted illumination. Note the dark bands at the level of the
equator (arrows). These would appear white under reflected light.
B. Male gonomedusa spawning. Note the apical dark mass (arrows)
compared to the other gonomedusae shown. Scale = 0.5 mm.

medusae at any one time for each male.

     Almost as soon as the isolated male polyp of T. larynx has ex-
panded in the sea water brief pulsations can be observed in several
of the gonomedusae. These pulsations are about 1-2 seconds in duration and may be as frequent as 1 per second. The pulses travel
apically. Both ripe and unripe gonomedusae show pulsations. They
have not been seen in male gonomedusae of T. crocea. At some point
the white streaks or patches become mobile within the gonomedusae
and can be seen to be motile small particles (presumably sperm)
moving between the inner wall of the bell of the gonomedusa and the
germinal mass within. Figure 3 shows this process in sequence.
Figure 3A shows a rounded gonomedusa with a dark band of mature
sperm basally (arrow). The succeeding frames are taken over a period
of 10 minutes while the pulsations continue, and by 3E it can be
seen that the dark mass has moved apically and begins to collect
beneath the apical tentacle zone (Fig. 3J). By this time the contractions become sustained and the sperm are squeezed through the
narrow opening at the center of the tentacle mass (Fig. 3L).

     Figure 3 does not show individual pulsations in any detail,
though the changes in shape from round to oval to tapered are quite
evident. Figure 4 shows pictures taken 2 seconds apart over somewhat more than five contraction cycles beginning with a contraction
reaching the apex, (4A), followed by relaxation (4B), contraction
below the apical sperm mass (4C), contraction at the center of the
apical sperm mass (4D) and relaxation (4E). Normally, sperm would

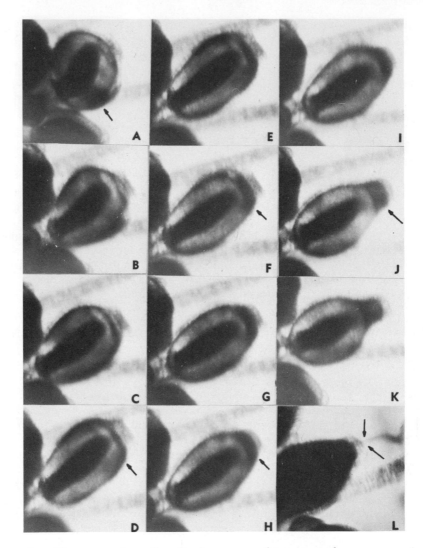

Figure 3. Photographs of a T. larynx male gonomedusa preparing to spawn. See text for details. Scale = 0.5 mm.

be emitted at 4D and 4E. This can be seen in Figure 2B. Two male gonomedusae spawning together are shown in Figure 5.

    Roosen-Runge and Szollosi (1965) describe spawning in the hydro-medusa Phialidium, where the mature sperm are released by a light induced rupture of adhesive attachments between the epithelial cells enclosing the testis. When shedding is completed, the cells re-establish the connections, restoring testis surface integrity. The

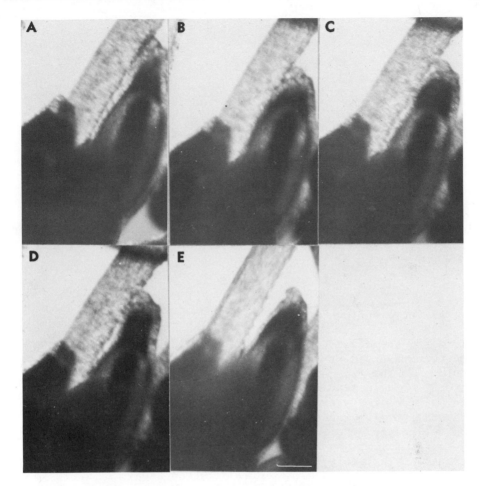

Figure 4. Photographs of one complete pulsation cycle in a male
gonomedusa of T. larynx. See text for details. Scale = 0.5 mm.

process is repeated 12 or 24 hours later, under normal daylength
conditions. It seems likely that the male gonomedusa of Tubularia
is behaving much like this, though the importance of a light stimulus
in the process has not yet been determined. The epithelium surrounding the spermatogenic mass must open in some manner, permitting
the ripe sperm to pass into the thin space between the bell wall and
the spermatogenic mass. The bell pulsations then progressively
drive the loose cells upward toward the apex, as the bell presses on
the central blastostyle-spermatogenic area complex. Presumably the
epithelium surrounding the spermatogenic mass heals up even before
the sperm reach the gonomedusa apex. Under suitable conditions, a
polyp which has spawned can be induced to spawn a second time after

Figure 5.   Two male gonomedusae (T. larynx) spawning. Enough force
is exerted to push the sperm out to 0.5 mm. from the apex.   Scale
= 0.7 mm.

several hours storage at 10°C. in the dark.

III. Sperm Chemotaxis: As mentioned earlier, sperm aggregations
have been observed at the apical tentacle zone of the female gonomedusa.   In the one case where more than one species of Tubularia
was available in sexual condition it was observed that no
species-specificity exists between the live sperm and the female
gonomedusa of T. crocea and T. marina.   Using extracts obtained
from the polyps of all three species studied (T. crocea, T. larynx,
and T. marina) it was found that sperm from any one species are
attracted by extracts from all of them.   Since measurement of extract
activity is based on a half-serial dilution assay, it would be interesting to know if this cross reactivity is uniform for each species
or concentration dependent.   One might expect, for example, that
though the sperm of T. crocea are attracted to the extracts of T.
larynx, they would be attracted much more by T. crocea extracts
under the same concentrations and conditions.   It has not been possible to perform these experiments, however, because of the seasonal
and geographical isolation of the species under study.

     Only female polyps produce extracts which possess attracting
activity.   The active agent is heat stable, acid-base stable, positively charged, of less than 700 molecular weight and is sensitive
to pronase (Miller and Tseng, 1974).   The molecular weight estimate
was derived from gel-filtration studies which also indicated similar
elution patterns of the sperm attractants obtained from all three
of the Tubularia species so far used (Fig 6 and Miller, 1972).   Work
is currently in progress to identify the substance responsible for
the sperm behavior.

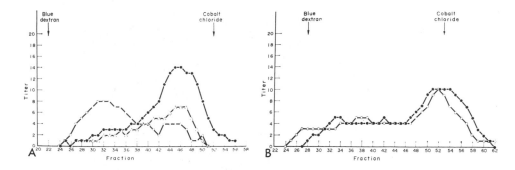

Figure 6.   Gel filtration, using Bio-gel P-2 and eluting with distilled water, of sperm attracting activity from A. Campanularia
calceolifera (O), T. marina (Δ), and T. crocea (●); B. T. crocea
(O) and T. larynx (●). From Miller, 1972).

IV. Sperm Behavior: The sperm clustering about the apical tentacles
of the female gonomedusa of Tubularia are freely motile and have
never been seen to adhere or migrate along the tentacles as reported
for Gonothyrea (Miller, 1970).   The ability of a particular female
gonomedusa to induce the clustering of sperm cannot yet be determined
by visual inspection of the contents, except that those containing
larvae in later stages of development are far less likely to induce
sperm aggregations than those containing early stages.   Unfortunately,
it seems impossible to identify unfertilized egg stages within the
gonomedusa without sectioning.   This prevents strict correlation of
embryonic stages with the ability of the gonomedusa to attract sperm.
A further complication is the fact that early oogenetic stages and
well-developed larvae may be present simultaneously in the same gonomedusa.

     Dark-field observations of the sperm behavior in the aggregations
is difficult because of the light scatter from the massive gonangium.
Therefore, extracts of female and male polyps were prepared by soaking
them in ethanol for several hours.   The alcohol was then evaporated
and the residue dissolved in sea water.   If this sea water is injected
(Miller, 1966) into a Tubularia sperm suspension an aggregation
rapidly develops at the site of the injection (Fig 7).   The sperm make
directed turns to approach the source of the attractant.   Preliminary
studies indicate that they turn less frequently both when moving up
the gradient and in the presence of uniform concentrations of attractant.   The most striking aspect of the behavior is that although
loops occur regularly in the tracks of sperm moving up the gradient,
the direction taken after the loops are completed is always up the
gradient.   The path directions never appear random, as has been
described for bacteria (Berg and Brown, 1972).

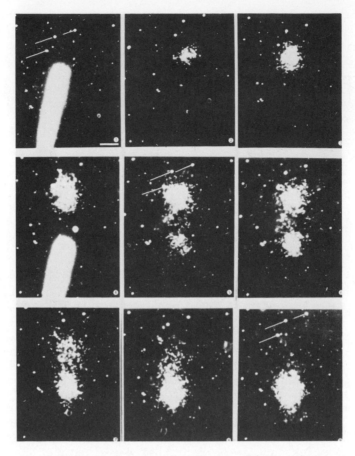

Figure 7. A series of photographs from a 16mm motion picture of
Tubularia sperm aggregating to sperm attractant introduced by a
micro-pipette. The pipette is inserted twice (7.1 and 7.4) and
the sperm in the older aggregation move into the newer one. Time
intervals are: 7.1, 0 sec.; 7.2,12 sec.; 7.3, 24 sec.; 7.4, 31 sec.
-- pipette in for 1.2 sec.; 7.5, 42 sec.; 7.6, 45 sec.; 7.7, 50 sec.;
7.8, 54 sec.; 7.9, 60 sec.. Scale = 100 μ

The sperm flagellum undergoes a change in wave morphology during
turning, which is characterized by suppression of beating over the
posterior one-third to one-half of the flagellum. Coupled with this
is a marked asymmetry of beat over the anterior half, such that the
flagellum forms a large wave of high amplitude and low radius of
curvature which is concave in the direction of the turn (Miller and
Brokaw, 1970). Figure 8 is one example of this behavior. It is not
yet clear if the amount of wave suppression or the angle of the major
bend can be related to the change in direction made by the sperm

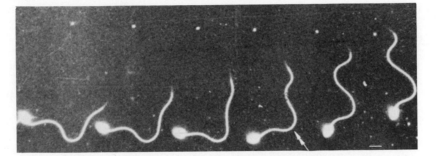

Figure 8. Multiple flash exposure of a <u>Tubularia</u> <u>crocea</u> sperm during
the transition from straight-line swimming to turning. The suppres-
ion of the distal waves in the tail begins with the image indicated
by the arrow and continues for 8 beat cycles. Scale = 2 μm.

during the turn. What is clear is that other hydroid sperm behave
in the same way during chemotaxis (Miller, unpublished) while the
sperm of other phyla which show chemotactic behavior do not (Miller,
1975).

V. <u>Sperm</u> <u>Chemotaxis</u> <u>as</u> <u>a</u> <u>Mechanism</u> <u>for</u> <u>Reproductive</u> <u>Isolation</u> <u>in</u>
<u>Tubularia</u>: There appears to be no species-specificity of sperm
chemotaxis within the genus <u>Tubularia</u>. It is still possible that
some specificity will be found based on careful tests using pure
attractants, but for the moment it seems that sperm chemotaxis would
actually serve to increase the chances of hybridization rather than
prevent it. In a situation like this, one has to explain why there
has been no divergence of molecular specificity, and the simplest
explanation which comes to mind is that of climatic or geographical
isolation.

     After many years of field observations I have yet to discover
a locality where more than one species of <u>Tubularia</u> is sexually
active simultaneously. The closest situation is on the New Jersey
Coast, where in one estuary <u>Tubularia</u> <u>crocea</u> and <u>T</u>. <u>larynx</u> overlap
in both time and location for periods of one to two months in the
late Fall and early Summer. Despite the physical presence of both
species, hybridization is unlikely, because one or the other is under
physiological stress due to the effects of water temperature at the
end of each season (Miller and O'Rand, 1974). For example, <u>T</u>. <u>larynx</u>
appears first in December when the sea water temperature is falling.
<u>T</u>. <u>crocea</u>, which has been flourishing all summer and fall, is in
reproductive decline - as evidenced by empty female gonomedusae,
massive infestations of male gonomedusae by ciliates (Geens, <u>et</u> <u>al</u>.,
1966) and almost complete absence of larvae. So it appears that even
if <u>T</u>. <u>crocea</u> sperm are attracted to <u>T</u>. <u>larynx</u> sperm attractant, at

the time when both species are present there are no T. crocea sperm
available to be attracted. If such climatic isolation is general in
Tubularia, there would be no reason to develop species-specific sperm
attractants. Whether foreign sperm exclusion mechanisms exist within
the gonomedusa, like those described by O'Rand,(1972) for Campanularia,
remains to be seen.

Bibliography:

Berg, H. C. and D. A. Brown 1972 Chemotaxis in Escherichia coli
    analyzed by three-dimensional tracking. Nature 239: 500-504.
Berrill, M. J. 1961 Growth, Development and Pattern. Freeman,
    San Francisco Figs 7-9, p. 207.
Geens, M., M. James and G. G. Holz, Jr. 1966 Destruction of the
    male gonophores of Tubularia larynx by a hymenostome ciliate of
    the genus Paranophrys. Biol. Bull. 131: 391.
Miller, R. L. 1966 Chemotaxis during fertilization in the hydroid
    Campanularia. J. Exp. Zool., 162: 23-44.
Miller, R. L. 1970 Sperm migration prior to fertilization in the
    hydroid Gonothyrea loveni. J. Exp. Zool. 175: 493-504.
Miller, R. L. 1972 Gel filtration of the sperm attractants of
    some marine hydrozoa. J. Exp. Zool. 182: 281-298.
Miller, R. L. 1975 Chemotaxis of the spermatozoa of Ciona
    intestinalis. Nature 254: 244-245.
Miller, R. L. and Brokaw, C. J. 1970 Chemotactic turning behaviour
    of Tubularia spermatozoa. J. Exp. Biol 52: 609-706.
Miller, R. L. and O'Rand, M. G. 1974 Utilization of chemical
    specificity during fertilization in the hydrozoa. In: The
    Functional Anatomy of the Spermatozoon. Edit. by B. Afzelius.
    Pergamon, New York pp 15-26.
Miller, R. L. and Tseng, C. 1974 Properties and partial purifica-
    tion of the sperm attractant of Tubularia. Amer. Zool. 14:
    467-486.
O'Rand, M. G. 1972 In vitro fertilization and capacitation-like
    interaction in the hydroid Campanularia flexuosa. J. Exp.
    Zool 182: 299-306.
Roosen-Runge, E. C. and D. Szollosi 1965 On the biology and
    structure of the testis of Phialidium Leukart (Leptomedusae)
    Z. Zellforsch 68: 597-610.
Rose, S. M. 1974 Bioelectric control of regeneration in Tubularia.
    Amer. Zool. 14: 797-803.

THE STRUCTURE AND FUNCTION OF CENTRIOLAR SATELLITES AND

PERICENTRIOLAR PROCESSES IN CNIDARIAN SPERM

Maurice G. Kleve and Wallis H. Clark, Jr.

Department of Biology, University of Houston, Houston,
Texas 77004 and National Oceanic and Atmospheric
Administration, National Marine Fisheries Service,
4700 Avenue U, Galveston, Texas 77550

INTRODUCTION

During spermiogenesis cnidarian sperm centrioles associate
with a number of accessory structures including microtubular nucleating satellites and pericentriolar processes. These centriolar
specializations are particularly well developed in cnidarian sperm
and appear to be highly involved in sperm development and function.

Microtubular nucleating satellites are thought to be involved
in the assembly and disassembly of microtubules (De The, 1964;
Boisson et al., 1969 and Tilney and Goddard, 1970). The microtubules that often radiate from satellites have been implicated in
cytoskeletal phenomena seen in a number of cell types (Porter, 1966;
Gibbins et al., 1969 and Tilney and Gibbins, 1969). With respect
to vertebrate sperm they are involved in nuclear shaping and formation of the manchette (Burgos and Fawcett, 1955; Tilney and Gibbins,
1969 and Rattner, 1972). The function of cytoplasmic microtubules
during spermiogenesis in invertebrate sperm is unclear.

Pericentriolar processes are unusual structures associated with
the centrioles of many animal cells, especially the distal centrioles
of invertebrate sperm including the Cnidaria (Dewel and Clark, 1972;
Summers, 1972; Hinsch and Clark, 1973 and Hinsch, 1974). In many
instances pericentriolar processes have been confused with microtubular nucleating satellites. Satellites associate with the anterior half of sperm centrioles in early spermiogenesis. As spermiogenesis advances pericentriolar processes appear at the posterior
end of the distal centriole existing simultaneously with the anterior
satellites for a short period. This period of coexistence, typical
of most cnidarian sperm centrioles, has undoubtedly contributed to

much of the confusion surrounding satellites and pericentriolar
processes.

In the present study we compared the structure and function of
satellites and pericentriolar processes in the hydrozoan, Hydractinia
echinata. Our preliminary studies have included a careful ultrastructural examination of the morphological relationships between
sperm centrioles and their associated specializations in both thin
section and whole mounts of disrupted sperm.

METHODS

Male colonies of H. echinata were collected from shells occupied
by the hermit crab Pagurus sp. in Galveston Bay, Texas or Woods Hole,
Massachusetts. Synchronous gonadal development was induced by exposure to continuous photoperiods of one to seven days (Ballard,
1942). Sperm or spermatids of uniform maturity were obtained from
excised gonophores with several strokes of a loose teflon homo-
genizer.

Isolated early spermatids or mature sperm were washed (3x) in
sterile seawater by sedimentation at 1000 xg to remove gonophore
cells and debris released as the gonophores were broken. Clean
preparations were sedimented at 5000 xg, resuspended in a hypotonic
saline and allowed to swell for 10 minutes. The swollen sperm were
disrupted by brief ultrasonication. These whole cell homogenates
were applied in a drop-wise fashion to 0.4% formvar coated, carbon
reinforced grids. The excess fluid was immediately drawn off with
filter paper. Whole mount preparations were stained by the addition of a drop of 0.2% aqueous uranyl acetate which was allowed to
stand for 60 seconds before being drawn off with filter paper.

Tissue for thin section electron microscopy was fixed in a
glutaraldehyde-paraformaldehyde mixture (Karnovsky, 1965) buffered
in 0.1M sodium phosphate (pH 7.2). Following a buffer wash, the
tissue was post-fixed in 1.0% osmium tetroxide buffered as above,
rapidly dehydrated in an acetone series and embedded in low viscosity
epoxy resin (Spurr, 1969). Both thin section and whole mounts of
disrupted sperm were examined with an AEI EM6B or Hitachi HS-8
electron microscope.

RESULTS

Microtubular nucleating satellites are the first centriolar
specializations to associate with Hydractinia sperm centrioles
during spermiogenesis. These satellites appear as spheres of electron dense material that seem to diffuse into the cytoplasm leaving
no well defined margin. At times a distinct connection exists

between a satellite and the outer surface of the centriolar matrix. Satellites are attached to the matrix by a stalk of electron dense material (Fig. 1). The stalk tapers, being widest at the point of matrix attachment (80 to 100 mµ) and narrowest at the point of satellite attachment (40 to 60 mµ). Individual variations in size and the diffuse periphery of the satellites do not allow precise measurement; however, the approximate value of 100 mµ is consistent with previous reports (Tilney and Goddard, 1970).

Satellites that surround and associate with Hydractinia sperm centrioles are the site of extensive microtubular radiation which persists through most of spermiogenesis (Fig. 2). These microtubules concentrate in the developing spermatid midpiece region. Micro- tubules extend from the distal and proximal centrioles to the peri- phery of the cell and are found anchored on the plasma membrane (Fig. 3). Microtubules also run between the sperm plasma membrane and the mitochondria forming a band that girdles the circumference of the sperm midpiece (Fig. 4).

Pericentriolar processes associate with the distal centrioles of Hydractinia sperm about mid-way through spermiogenesis. The pro- cess complex is composed of nine pericentriolar processes which ema- nate from the centriolar matrix between the triplets (Fig. 5). Each of the nine members is composed of a primary process which extends out from the matrix for 200 mµ terminating in a thickened tip and three secondary processes which radiate from that tip. The secondary processes extend into the cytoplasm between the mitochondria and plasma membrane where they terminate in close apposition to the mem- brane (Fig. 6). Inter-primary processes are additional components observed in thin sections. The structures interconnect adjacent primary processes by extending from the base of one primary process, near the point of matrix attachment, to a point near the thickened tip of the next primary process. Interprimary processes parallel the plane of the centriolar triplet blades while the primary pro- cesses are perpendicular to the triplet blades.

Disrupted whole mounts of spermatids from the early stages of spermiogenesis, before pericentriolar process formation is complete, yield preparations which contain centrioles with attached satellites (Fig. 7). The satellites in whole mount preparations demonstrate a considerably different morphology from those seen in thin section. They are generally ovoid in shape rather than spherical. Their peripheral margins are well defined and the outer edge of the ellipse (distal to the centriole) is often drawn to a point. Satellites seen in whole mount demonstrate a core and cortex. Often the core has a crystalloid appearance in these air dried preparations (Fig. 7). The tapered stalk seen in thin sections is readily apparent. Satellites viewed in whole mount preparations are no longer as- sociated with the numerous microtubules seen in thin section.

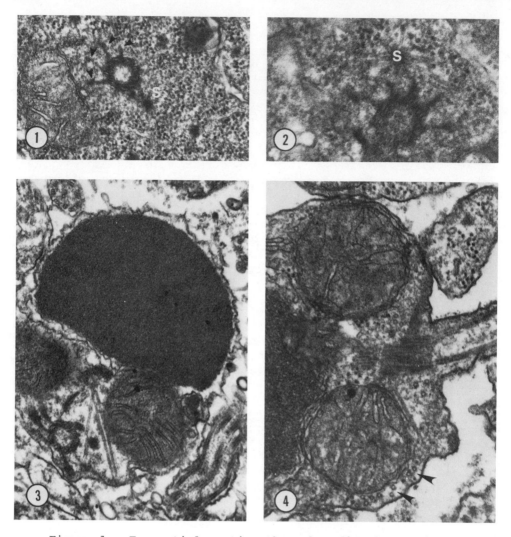

Figure 1.   Tangential section through a distal centriole re-
veals an anterior satellite (S) attached to the centriole by a
tapered stalk and the posterior pericentriolar processes (arrows)
which are starting of form.   X38,000.

Figure 2.   Distal centriole with closely positioned satellite
(S) which is the nucleation site for numerous microtubules.   Pericentriolar processes are also apparent.   X54,000.

Figure 3.   Longitudinal section of a late spermatid showing
several microtubules which run from the proximal centriole to the
posterior plasma membrane.   Pericentriolar processes can be seen
emanating from the distal centriole.   X32,000.

Figure 4.   Late spermatid showing microtubules that run in a
band between the plasma membrane and mitochondria of the midpiece
(arrows).   X45,000.

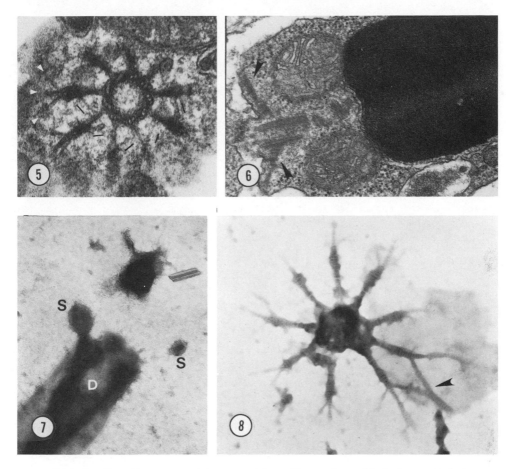

Figure 5. Cross section passing through a distal centriole
at the level of the pericentriolar processes. Nine primary processes can be seen emanating from the centriolar matrix. The primary processes terminate in thickened tips and three secondary processes emanate from these tips (arrows). Interprimary processes
which connect adjacent primary processes are marked on the micrograph by lines drawn parallel to them. X50,000.

Figure 6. Longitudinal section passing through the mid-plane
of a late spermatid. The pericentriolar processes (arrows) emanate
from the posterior end of the distal centriole and extend out into
the cytoplasm between the plasma membrane and mitochondria. X25,000.

Figure 7. Whole mount of free distal centriole (D) from early
spermatid. Two satellites (S) can be seen. One attached to the
anterior portion of the centriole by the tapered stalk, and the
other lying close to the centriole. X56,000.

Figure 8. Whole mount of a free distal centriole from a mature
sperm. The asymmetry of the complex is demonstrated by the two long
secondary processes. The point of anastamosis between these processes is marked by an arrow. X27,000.

Pericentriolar processes liberated from the disrupted sperm are well preserved (Fig. 8). They retain their primary processes ending in thickened tips and secondary processes which also appear to terminate in thickened tips. In whole mount the process complex appears asymmetrical, an aspect poorly discernible in thin section because of the angle of process projection. This asymmetry is demonstrated by the secondary extensions of two adjacent primary processes. These are much longer than those extending from the other seven (Fig. 8). Previous thin section observations of Hydractinia sperm (Hinsch and Clark, 1973) suggested that fusion took place between the outermost members of each set of three secondary processes. In whole mount, it is clear that this anastomosis takes place only between the long and not between the seven short secondary processes. Centriolar complexes viewed in whole mount and thin sections from early spermatids do not demonstrate this asymmetry and fusion. It is only observed in late spermatids and mature sperm.

DISCUSSION

In Hydractinia spermatids microtubules appear to play a significant role in the morphological changes taking place during spermiogenesis. The intimate association of microtubules with the spermatid plasma membrane (Fig. 3) and the high degree of organization they demonstrate in the girdle that surrounds the midpiece (Fig. 4) suggest extensive microtubular involvement in cytoskeletal shaping of the sperm midpiece. Microtubules have been implicated in a wide variety of cytoskeletal phenomena (Porter, 1966; Gibbins et al., 1969 and Tilney and Gibbins, 1969). Because of this involvement of microtubules in cytoskeletal shaping, it seems reasonable to assume that centrioles carrying microtubular nucleating satellites may play a regulative role in the structural changes taking place during sperm differentiation.

Although many researchers have failed to make a distinction between pericentriolar processes and satellites, Dewel and Clark (1972) have clearly distinguished these as separate structures. The data presented in this study confirm their view that pericentriolar processes are distinct from satellites in both form and location. The processes are situated at the posterior end of the distal centriole while satellites are more anteriorly located. Satellites are ovoid or spherically shaped structures that have a somewhat amorphous appearance in thin section. Satellites are apparently centers for the nucleation of microtubules which are found associated with them the entire time they are present in the developing spermatid. Pericentriolar processes are long, branching and striated structures always numbering nine, one for each centriolar triplet. They have not been observed to act as centers for the convergence of microtubules or to be involved in microtubular nucleation.

Since pericentriolar processes are distinct from satellites, the question remains, what is their function? Previous investigators have suggested two possible roles; one structural, i.e. an anchoring mechanism (Szollosi, 1964), and another contractile involved in locomotor activity (Summers, 1972; Dewel and Clark, 1972 and Kleve and Clark, 1974). The contractile role is very appealing when one considers the chemotactic responses demonstrated by many sperm (for review see Miller, 1973). The most obvious and clearly demonstrated chemotactic sperm behavior is seen in various cnidarian species including Hydractinia (Miller, 1974 and Miller and Tseng, 1974). It is interesting that sperm of species with elaborate pericentriolar processes often demonstrate extensive chemotactic behavior. Such responses necessitate directional selectivity by sperm. Response to chemotactic stimuli by alterations in the asymmetry of the sperm midpiece would provide the "rudder" necessary for directional select- ivity. Hydractinia sperm are radially symmetrical with exception of the asymmetric pericentriolar process array. This asymmetry may in some way give the sperm the orientation it requires to react to and move toward released chemotactins.

When sperm react chemotactically they are doing so as a one- cell sensory unit. In such an instance the sperm flagellum may well act in a manner similar to sensory cilia. It is interesting to note that a number of investigators have observed extensive pericentriolar process complexes associated with the basal bodies of a wide variety of sensory cilia. A partial list would include sensory cilia of the vertebrate inner ear and lateral line organs (Flock and Duvall, 1965; Wesall et al., 1965 and Flock and Jorgenson, 1974); olfactory cilia (Reese, 1965); and photoreceptive cilia (Tokuyosu and Yamada, 1959 and Horridge, 1964). Sensory cilia with complex centriolar struc- tures are also found in neurosensory cells of hydra (David, 1969 and 1974), cnidocils and distal flagella of cnidarian nematocytes (Westfall, 1965 and 1970) as well as the pedicellaria of echinoderms (Cobb, 1967). The role pericentriolar processes play in sensory cilia function is evasive. One cannot readily see the need for a contractile unit at the base of seemingly non-locomotor cilia unless they facilitate the orientation of the sensory cilia toward a stim- ulus. Their frequent association with sensory cilia would imply a functional connection. In the case of cnidarian sperm pericentriolar processes may serve in both a sensory and locomotor capacity.

(Supported in part by the National Oceanic and Atmospheric Administration Sea Grant #04-3-158-18 and the Link Foundation.)

SUMMARY

Microtubular nucleating satellites have been distinguished from pericentriolar processes in structure, location, chronology of appearance and function. Satellites associate with the anterior half of <u>Hydractinia</u> sperm centrioles early in spermiogenesis, are spherical in shape and involved in the assembly of microtubules responsible for spermatid shaping. Pericentriolar processes associate with the posterior end of distal centrioles late in spermiogenesis. They are long trifurcated structures always numbering nine, one for each centriolar triplet blade. Pericentriolar processes never associate with microtubules; however, they may be contractile and involved with the chemotactic behavior of sperm.

REFERENCES

Ballard, W. W. 1942. The mechanism for synchronous spawning in <u>Hydractinia</u> and <u>Pennaria</u>. <u>Biol. Bull.</u> 82(3):329.

Boisson, C., X. Mattei and C. Mattei. 1969. Mise en place et evolution du complexe centriolaire au course de la spermiogenèse d'<u>Upeneus</u> <u>prayensis</u> C. V. (Poisson Mullidae). <u>J. Microscopie.</u> 8:103.

Burgos, M. H. and D. W. Fawcett. 1955. Studies on the fine structure of the mammalian testis. I. Differentiation of the spermatids in the cat (<u>Felis</u> <u>domestica</u>). <u>J. Biphys. Biochem. Cytol.</u> 1(4):287.

Cobb, L. S. 1967. The fine structure of the pedicellaria of <u>Echinus</u> <u>esculentus</u> (L). II. The sensory system. <u>J. Roy. Microscop. Soc.</u> 88:223.

Davis, L. E. 1969. Differentiation of neurosensory cells in <u>Hydra</u>. <u>J. Cell Sci.</u> 5:699.

Davis, L. E. 1974. Ultrastructural studies of the development of nerves in <u>Hydra</u>. <u>Am. Zool.</u> 14:575.

De The, G. 1964. Cytoplasmic microtubules in different animal cells. <u>J. Cell Biol.</u> 23:265.

Dewel, W. C. and W. H. Clark, Jr. 1972. An ultrastructural investigation of spermiogenesis and the mature sperm in the anthozoan <u>Bundosoma</u> <u>cavernata</u> (Cnidaria). <u>J. Ultrastruct. Res.</u> 40:417.

Flock, A. and A. J. Duvall. 1965. The ultrastructure of the kinocilium of the sensory cells in the inner ear and lateral line organs. <u>J. Cell. Biol.</u> 25:1.

Flock, A. and J. M. Jorgenson. 1974. The ultrastructure of lateral line sense organs in the juvenile salamander <u>Ambystoma</u> <u>mexicanum</u>. <u>Cell Tiss. Res.</u> 152:283.

Gibbins, I. R., L. G. Tilney and K. R. Porter. 1969. Microtubules in the formation and development of the primary mesenchyme in <u>Arbacia</u> <u>punctulata</u>. I. The distribution of microtubules. <u>J. Cell Biol.</u> 41:201.

Hinsch, G. W. and W. H. Clark, Jr. 1973. Comparative fine structure of cnidarian spermatozoa. <u>Biol. Reprod.</u> 8:62.

Hinsch, G. W. 1974. Comparative ultrastructure of cnidarian sperm. <u>Am. Zool.</u> 14:457.

Horridge, G. A. 1964. Presumed photoreceptive cilia in a ctenophore. <u>Quart. J. Micr. Sci.</u> 105(3):311.

Karnovsky, M. J. 1965. A formaldehyde glutaraldehyde fixative of
    high osmolarity for use in electron microscopy. J. Cell Biol.
    27:137A.

Kleve, M. G. and W. H. Clark, Jr. 1974. Fine structure of centri-
    olar specilizations isolated from Hydractinia (Cnidaria) sperm.
    J. Cell Biol. 63:171A. (Abstract.)

Miller, R. L. 1973. Chemotaxis of animal sperm. In Behavior of
    microorganisms. A. P. Miravete, Editor. Plenum Press, London.
    31.

Miller, R. L. 1974. Sperm behavior close to Hydractinia and Ciona
    eggs. Am. Zool. 14(4):1250. (Abstract.)

Miller, R. L. and C. Y. Tseng. 1974. Properties and partial puri-
    fication of the sperm attractant of Tubularia. Am. Zool.
    14(2):467.

Porter, K. R. 1966. Cytoplasmic microtubules and their function.
    In Principles of Biomolecular Organization. G. E. W.
    Walstenholme and M. O. Conner, Editors. J. and A. Churchill
    Ltd., London. 308.

Rattner, J. B. 1972. Nuclear shaping in marsupial spermatids.
    J. Ultrastruct. Res. 40:498.

Reese, T. S. 1965. Olfactory cilia in the frog. J. Cell Biol.
    25:209.

Spurr, A. K. 1969. A low viscosity epoxy resin embedding medium
    for electron microscopy. J. Ultrastruct. Res. 26:31.

Summers, R. G. 1972. A new model for the structure of the centri-
    olar satellite complex in spermatozoa. J. Morphol. 137:229.

Szollosi, D. 1964. The structure and function of centrioles and
    their satellites in the jellyfish Phialidium gregarium. J.
    Cell Biol. 21:465.

Tilney, L. G. and J. R. Gibbins. 1969. Microtubules in the forma-
    tion and development of the primary mesenchyme in Arbacia
    punctulata. II. An experimental analysis of their role in
    development and maintenance of cell shape. J. Cell Biol.
    41:227.

Tilney, L. G. and J. Goddard. 1970. Nucleating sites for the
    assembly of cytoplasmic microtubules in the ectodermal cells
    of blastulae of Arbacia punctulata. J. Cell Biol. 46:564.

Tokuyosu, U. and E. Yamada. 1959. The fine structure of the re-
    tina. Studies with the electron microscope. IV. Morpho-
    genesis of outer segments of retinal rods. J. Biophys.
    Biochem. Cytol. 6:225.

Wesall, J., A. Flock and P. G. Lundquist. 1965. Structural basis
    for directional sensitivity in cochlear and vestibular sensory
    receptors. Cold Spring Harbor Symp. Quant. Biol. 30:115.

Westfall. J. A. 1965. Nematocysts of the sea anemone Metridium.
    Am. Zool. 5:377.

Westfall, J. A. 1970. The nematocyte complex in a hydromedusan,
    Gonionemus vertens Z. Zellforsch. 110:457.

EARLY DEVELOPMENT AND PLANULA MOVEMENT IN <u>HALICLYSTUS</u> (SCYPHOZOA:

STAUROMEDUSAE)

Joann J. Otto

Friday Harbor Laboratories, Friday Harbor, WA 98250 and

Develop. and Cell Biol., Univ. Calif., Irvine, CA 92717

Stauromedusae are unusual scyphozoans in that they are sessile and live attached by a stalk to eel grass or algae. Their planula larva is also unusual since it creeps about rather than using ciliary locomotion and since it has a constant number of endodermal cells (Kowalevsky, 1884; Wietrzykowski, 1910, 1912; Hanaoka, 1934).

These unusual features provoke interest in the development of the Stauromedusae. The embryogenesis of three species has been studied: <u>Lucernaria campanulata</u> (Kowalevsky, 1884), <u>Haliclystus octoradius</u> (Wietrzykowski, 1910, 1912) and <u>Thaumatoscyphus distinctus</u> (Hanaoka, 1934). Cleavage is irregular and gastrulation occurs by unipolar ingression. The planula larva then develops. After creeping about for a day or two, the planula settles, grows, and undergoes metamorphosis to form a juvenile medusa.

In this paper I examine the development of two additional species and confirm the generality of stauromedusan development. Also, I present an analysis of the planula's creeping movement and the cell behaviors which underlie it.

MATERIALS AND METHODS

Two species of <u>Haliclystus</u> were collected on San Juan Island, Washington, from June through August, 1974. One, tentatively identified as <u>H. stejnegeri</u> (P. Corbin, personal communication), was collected at False Bay and Cattle Point on eel grass and a variety of algae. <u>H. salpinx</u> was obtained from Smallpox Bay, also on eel grass and algae. Mature animals were maintained in the lab in slowly running sea water with a daily feeding of amphipods, caprellids, or crab zooeas.

Spawning was stimulated by placing the animals in the dark for at least 8 hr and then exposing them to light. The embryos were kept in dishes in filtered sea water. The temperature ranged from 12C to 15C. After the planulae settled in the bowls, some bowls were filled with unfiltered sea water with small crustaceans or a suspension of hard-boiled chicken egg yolk added as potential food for the embryos.

To study planula movement, time lapse movies on Plus X Reversal movie film were made of the planula with Nomarski differential interference and phase optics. A cooling stage maintained the temperature at 12C while the planulae were being observed.

For histological examination, planulae were fixed with 2.5% glutaraldehyde and 2% O_sO_4 in 0.2M phosphate buffer (pH 7.6) and embedded in Epon (Cloney and Florey, 1968). Thin sections were made and observed on a Zeiss electron microscope.

RESULTS

Adult Sexual Characteristics

The Stauromedusae are dioecious. An animal has 8 gonads, each consisting of a group of gonadal sacs arranged along the interradial septa. In sexually mature animals all the gonadal sacs contain gametes ready to be released. The gametes are released into the gastric cavity and are then expelled through the animal's mouth.

In Haliclystus salpinx the sexes can be distinguished. Viewed from the exumbrellar side, the females have irregularly shaped, bumpy gonadal sacs, and the males have spherical and smooth sacs. From the subumbrellar side, the center of the female sac appears granular and yellow, and the male sac has a white mass in the middle. Only sexually mature specimens of H. salpinx were found. These ranged from 1.5 to 3 cm wide at the bell margin (without tentacles). No correlation of size with the sex of the animal was observed. There was very little color variation in H. salpinx with both males and females being yellow-tan.

Males and females of H. stejnegeri cannot be distinguished from external appearance because the opacity of the tissue obscures the gonadal sacs. A variety of different colored animals was found. The color did not correlate with the sex of the animal which was determined by dissection or by placing it alone in a

dish to spawn. The diameter of the bell margin ranged in size
from 0.5 mm to 3.5 cm. Only animals above 1.7 cm in bell diameter
spawned. No correlation of size with the sex of the animal was
found.

H. stejnegeri began to spawn within 15-20 minutes of exposure
to light after being in the dark for at least 8 hours. H. salpinx
did not require a dark period for spawning to occur, but spawning
was more profuse when the animals were placed in the dark overnight. Specimens of both species spawned even when alone in a
dish indicating that the spawning of one sex is not necessary to
stimulate spawning of the other sex.

Early Development

Embryonic development is similar in the two species; therefore, they will be described together.

Oocytes (Fig. 1a) are released at the germinal vesicle stage
and after fertilization undergo germinal vesicle breakdown. An
extremely sticky fertilization membrane is produced which causes
the eggs to adhere to the substrate. In H. stejnegeri the eggs
are 35 mμ in diameter and in H. salpinx, 40 mμ. Eggs of both
species are yolky. The first cleavage (Fig. 1b) is total and
equal and occurs about 2 hours after sperm are added. The second
cleavage is perpendicular to the first. After the second cleavage, the cell divisions become asynchronous. After the embryo
has 15-20 cells, a stereoblastula is formed (Fig. 1c). The proliferation of endodermal cells continues until there are 16 cells.
As the planula develops, a rearrangement of endodermal cells
occurs and they line up in a single row (Fig. 1d).

About 24 hours after fertilization the planula hatched from
the fertilization membrane. The planula (Fig. 1d) is solid and has
16 endodermal cells in both species. The highly vacuolate endodermal cells are arranged like coins in a stack. In what I define
as the planula's resting state, all the endodermal cells except
the most anterior and posterior have a flat coin-like shape. The
most anterior and posterior cells have a dome shape. The planula
moves by creeping over the substrate as described in the section
on planula movement.

In my cultures, the planula stage lasted 1-3 days after which
the larvae settled and ceased movement. They usually aggregated
in clumps which contained variable numbers of larvae. After the
larvae settled they secreted a thin, clear covering. About 1 week
after fertilization nematocysts developed. I kept some settled
planulae for over 2 months but did not observe any further development although potential food was placed in the dishes as well as

322 J.J. OTTO

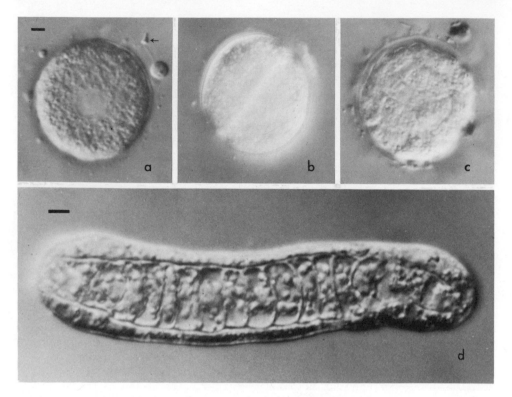

Figure 1. Haliclystus salpinx development. a) Oocyte in germinal vesicle stage with sperm nearby (arrow). b) Two cell stage. c) Stereoblastula. d) Planula. Bars in a (for a-c) and d represent 5mµ.

eel grass and the algae from which the animals had been collected in the hope that metamorphosis could be stimulated. The settled planulae maintained their form during this time. The proper stimulus for further development probably was not provided.

Although further development was not observed in laboratory cultures, I did find one H. stejnegeri in the field which was 0.5 mm in bell diameter and had 1 tentacle per arm plus developing anchors between arms. Since the planula is 100 mµ in length and 20 mµ in diameter, this represents a substantial increase in size from the planula stage. Several juveniles were found with three tentacles per arm. Aside from these very tiny animals, a number of immature H. stejnegeri were found which were of various sizes up to 1.7 cm in bell diameter.

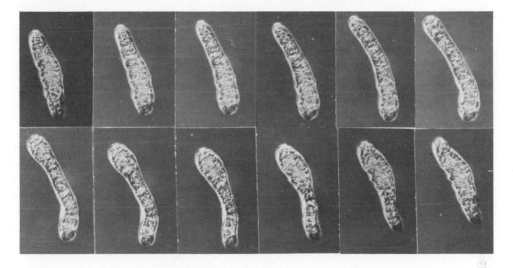

Figure 2. A cycle of <u>Haliclystus</u> <u>salpinx</u> planula's elongation
and retraction beginning at top left.   Interval between frames is
12 sec.

Planula Movement

     The planula moves by a series of extensions and retractions
(Fig. 2).   One complete movement cycle takes about 2 minutes.
The endodermal cells elongate in an anterior to posterior sequence
and cause the entire planula to lengthen.   The endodermal cells
then regain their flat coin-like shape in an anterior to
posterior sequence.   The planula is quite sticky and when the
anterior end protrudes, it probably forms some connections with
the substrate.   When retraction occurs, some of these anterior
connections hold and the planula moves slightly forward by breaking posterior connections.   During the elongation phase of movement, the posterior end of the planula does actually extend
slightly back beyond its relaxed position when the posterior cells
elongate; this is probably due to the planula not being firmly
attached to the substrate.   A trail of a thin substance was left
behind as the planula moved.   This secretion may be the actual
substrate with which the planula interacts.   Planulae were never
observed to use ciliary activity to move and do not have cilia on
the ectodermal surface.   Movement only occurred through the extension and retraction sequence.

     As the planula elongates and retracts, the shape of the endo-
dermal cells changes systematically.   Consider one cell (Fig. 3)
starting from the position in which it has a flat coin-like shape
(Fig. 3a).   When cells anterior to it elongate, its membranes

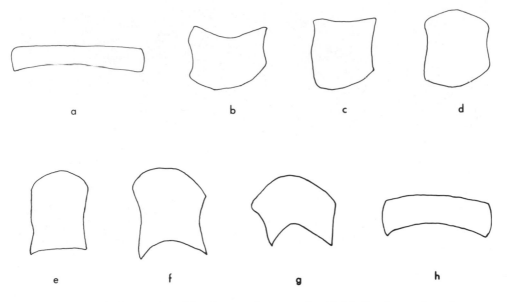

Figure 3. Endodermal cell shape changes in <u>Haliclystus</u> planula.
These drawings were traced from a 16 mm film and represent longitudinal sections through the fifth cell from the anterior end of
the planula.

become concave anteriorly and convex posteriorly. The cell then
elongates (Fig. 3b). As the cell reaches its maximum length, its
anterior membrane flattens (Fig. 3c) and then becomes convex
(Fig 3d). The posterior membrane moves from its convex position
to a flatter one (Fig. 3e); this occurs slightly later than the
anterior membrane shift such that at one point in the cycle both
the anterior and posterior membranes are convex (Fig. 3d). As the
cell shortens or the cell behind it contracts, the posterior membrane becomes concave (Fig. 3f, g). Then as shortening is completed (Fig. 3h), the flat coin-like shape is regained (Fig. 3a).

   Ultrastructural examination (Fig. 4) of the planula revealed
a lack of cilia in the ectoderm. The endodermal cytoplasm adjacent to the mesoglea has a meshwork appearance and excludes
organelles. This resembles other investigators' descriptions of
regions containing microfilaments, although no microfilaments could
be resolved. The histology of the planula also shows a large
amount of rough endoplasmic reticulum and Golgi in the ectoderm.
These organelles may be involved in the secretion of the material
through which the planula moves or perhaps the secretion of the
covering of the settled planula.

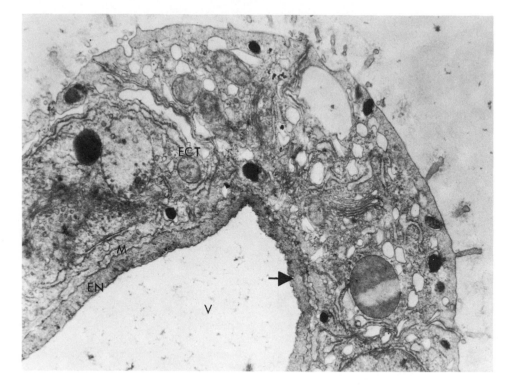

Figure 4. Cross section through a Haliclystus planula. Mesoglea
(M) separates ectoderm (ECT) from endoderm (EN). Endodermal cell
has large vacuole (V) and has a dense meshwork (arrow) which
excludes organelles (6300X).

DISCUSSION

Haliclystus embryogenesis, as that of other coelenterates, is
at first irregular in that the cleavages are asynchronous. However, Haliclystus development contains at least two striking
departures from normal coelenterate embryogenesis. The development
of a planula with a constant endodermal cell number is quite
different from most other coelenterates, where cell constancy is
not a usual feature. The other peculiarity of Haliclystus development is the lack of ciliary movement of the planula.

Although Wietrzykowski (1910, 1912) and Hanaoka (1934) described some aspects of the movement of the planula, they did not
determine the mechanism of movement. The cyclic changes in the
shapes of the planula's cells and thus of the planula are most
easily explained by the following model. The state of the planula

at its shortest length is defined as the resting state. In the
resting state the endodermal cells except for the most anterior
and posterior cells have a flat coin-like shape. The extreme
anterior and posterior cells have a dome shape. Consider one cell
in the middle of the planula. As the anterior part of the planula
elongates, the anterior membrane of the cell becomes concave and
the posterior membrane becomes convex. Since the planula is a
constant volume system, the lengthening of the anterior cells
results in increased pressure on the cell in question and the cell
bends posteriorly. The cell itself then begins to lengthen. The
best candidate for the force for this increase in length is the
activity of contractile elements, such as microfilaments, which
are most likely located at the periphery of the endodermal cells.
Although I could not resolve microfilaments in this area, it is
the only region devoid of other organelles, and it stretches
around the entire periphery of every endodermal cell. I assume
it contains the contractile apparatus. As the cell is reaching
its maximum length and smallest diameter, the anterior cell membrane flattens and then becomes convex while the posterior membrane remains convex. This shape is probably a result of several
forces: the pressure generated within the cell itself at maximal
circumferential contraction, the beginning of shortening and
relaxation in the immediately adjacent anterior cell, and the
adjacent posterior cell not having contracted yet. The posterior
cell membrane becomes concave when the cell posterior to the cell
in question lengthens and thus increases the pressure from the
posterior direction. The cell then begins to shorten with the
anterior membrane convex and posterior membrane concave. As
shortening proceeds, the cell membranes become flatter with the
anterior membrane again slightly ahead of the posterior. The
shortening of the cell increases the diameter of the planula
which would tend to flatten the end membranes, and since the cells
anterior to the cell are also shortening, they press the anterior
membrane to its flat position. Since the cell itself is shortening to maintain a constant volume, the posterior cell membrane is
also forced posteriorly to its flat position. The shortening and
relaxation of the adjacent posterior cell probably also contributes to this flattening of the posterior cell membrane. The cell
has then regained its flat coin-like shape. Each endodermal cell
except the most anterior and posterior ones go through the same
cell shape changes which result in the elongation and retraction
of the planula.

     The extreme anterior and posterior endodermal cells go
through similar cell shape changes that the central cells exhibit.
However, the anterior cell membrane of the anterior cell and the
posterior membrane of the posterior cell are not adjacent to
another endodermal cell; rather they abut directly on the mesoglea.
These membranes do not exhibit the concave curvature in the anterior
cell or the concave curvature in the posterior cell. This supports

the hypothesis that these membrane movements in the central cells
are due to pressure changes from adjacent endodermal cells. The
lengthening of the anterior cell initiates planula extension. The
increased pressure in this anterior cell causes not only the
anterior tip of the planula to extend, but also causes the cell's
posterior membrane to become convex which intitiates the cell
shape changes in the central cells analyzed above.

The exact mechanism of cell shortening is not clear. The
planula is of a constant volume so that non-contractile, elastic
elements of the planula (for example, mesogleal components) may
be the direct cause of retraction, or there may be an antagonistic
set of contractile elements connecting the anterior and posterior
endodermal cell membranes or existing in the ectoderm. Whatever
the cause of retraction, it is probably accompanied by a relaxation of the contractile agents that lengthen the cell.

Mere contraction and relaxation of cells will not result in
displacement of the planula. In addition some mechanism must
exist for the planula to interact with the substrate so that
elongation results in net forward movement. As Wietrzykowski
(1910, 1912) noted, Haliclystus secretes a sheathing substance as
it moves. This substance may not only act to stick the planula
to the substrate, but it may also function as a substrate itself
all around the planula. The planula has microvilli which insert
into this sheath and which probably provide a greater surface
area and thus greater friction between the sheath and the planula.

The movement of the planula is efficient since the larva is
subdivided into somewhat independent contractile units, namely
the endodermal cells. The planula is thus segmented as far as
movement is concerned. The "segmentation" of the planula allows
extension events to go on at one end of the planula while shortening events occur at the other end. For example, the anterior end
often begins elongating before the wave of shortening has passed
through the posterior cells. This is reminiscent of movement in
earthworms which has been similarly analyzed by Gray and Lissman
(1938).

The movement of the planula bears some resemblance to non-
ciliary movement in other coelenterates. The colonial sessile
hydrozoans expand their colonies by the stolon and upright tips
moving along the substrate and pulling out tissue behind them.
As the tips move, they go through extension and retraction cycles
just as the Haliclystus planula does. The mechanism of this movement is not yet known (for review, see Campbell, 1974) although
in the case of the solitary hydrozoan Corymorpha palma, an analagous
movement in the holdfast has been shown to be caused by peristalsis (Campbell, 1968). However, the movement of the Haliclystus
planula is different from the stolon tips in that stolon tips

contain a coelenteron in which fluid moves back and forth. In the
Haliclystus planula fluid is confined to individual cells which
together are probably responsible for the movement of the entire
planula.

SUMMARY

The early development and planula movement of Haliclystus
stejnegeri and H. salpinx are described. Both species have small
(35-40 mµ), yolky eggs. After the first two cell divisions, cleavage becomes irregular and a stereoblastula develops. At the end
of gastrulation there are 16 endodermal cells. No other case of
cell constancy in coelenterates is known. The endodermal cells
become arranged end to end as the planula forms. In culture the
planula creeps about for 1-3 days and then settles.

The planula moves by alternately elongating and contracting.
The elongation and contraction are caused by endodermal cell shape
changes which are probably due to a circular contractile apparatus
girdling the cylindrical cells.

ACKNOWLEDGMENTS

I thank Drs. Richard Campbell, Robert Fernald, Paul Illg,
and Thomas Schroeder for their advice and encouragement during
the course of this work. Much of this study was carried out at
the Friday Harbor Laboratories. The use of their facilities is
gratefully acknowledged.

REFERENCES

Campbell, R. 1968. Holdfast movement in the hydroid Corymorpha
    palma: Mechanism of elongation. Biol. Bull. 134: 26-34.

Campbell, R. 1974. Development. Pages 179-210 in L. Muscatine
    and H. Lenhoff, Eds., Coelenterate Biology: Reviews and New
    Perspectives. Academic Press. New York.

Cloney, R. and E. Florey. 1968. Ultrastructure of cephalopod
    chromatophore organs. Z. Zellforsch. Microsk. Anat. 89:
    250-280.

Gray, J. and H. Lissmann. 1938. Studies in animal locomotion.
    VII. The earthworm. J. Exp. Biol. 15: 506-517.

Hanaoka, K.   1934.   Notes on the early development of a stalked
    medusa.   Proc. Imp. Acad. Japan. 10: 117–120.

Kowalevsky, A.   1884.   Zur Entwicklungsgeschichte der Lucernaria.
    Zoologischer Anzeiger 7: 712–717.

Wietrzykowski, W.   1910.   Recherches sur le dévelopment des
    Lucernaridés.   Archives de Zoologie Expérimentale et Générale
    (Paris) 5: x–xxvii.

Wietrzykowski, W.   1912.   Recherches sur le dévelopment des
    Lucernaires.   Archives de Zoologie Expérimentale et Générale,
    5e Serie.   10: 1–95.

LARVAL DISPERSAL IN THE SUBTIDAL HYDROCORAL Allopora californica VERRILL (1866)

Georgiandra Little Ostarello

College of Notre Dame

Belmont, California 94002

The larval stages of benthic marine organisms are critical periods in the life cycles, for both recruitment of new individuals into the local population and dispersal to exploit new areas must take place at this time. Various strategies to achieve these ends have evolved in different organisms.

In the subtidal hydrocoral Allopora californica the fertilized eggs are retained within the female colony during larval development, then the planulae push out of their ampullae to leave the parent colony (Ostarello, 1973). In a related species, A. petrograpta, it was found in laboratory studies that immediately after emergence, the planulae actively moved downwards seeking the bottom of the dish. When placed in a gentle current, these planulae quickly attached themselves to the bottom or sides of the dish (Fritchman, 1974).

The present investigation was carried out to determine if similar behavior is found in situ in Allopora californica. The larval dispersal patterns were investigated by studying the patterns of settlement of new colonies around existing mature Allopora colonies,

MATERIALS AND METHODS

This study of larval dispersal was done off Carmel River State Beach and Point Lobos State Reserve at Carmel, California. Previous work on the life history of A. californica, done at the same locations, showed that the reproductive cycle culminates with release of planulae during late October through November (Ostarello, 1973). Field data on settlement of new colonies were collected during Octo-

ber and November 1975 before known high attrition rates could reduce
the number of new colonies. Newly settled <u>Allopora</u> <u>californica</u>
colonies appear first as round, flat, slightly pink plates measuring
1-2 mm in diameter. Within a few days the first cyclosystem is
formed, complete with a gastrozooid and several dactylozooids.
Thirteen divers were trained to recognize these stages and to
collect reliable data.   Forty underwater man-hours over eight diving
days were necessary for collection of the data.

     The divers, working in pairs, located isolated, mature colonies
of <u>Allopora</u> <u>californica</u>. A colony was considered an isolate if it
was separated by at least 100 cm. from any other mature <u>Allopora</u>
colony. Each isolate was tagged and a small piece was collected
for later histological examination to determine the sex of the colony
and to access the development of the male and female gonophores.

     Actual counts of newly settled <u>Allopora</u> colonies were made in
a rectangular area, 20 cm. x 100 cm., along a line starting at the
base of the isolate and extending out 100 cm. The rectangle was
divided into twenty equal counting areas, numbered consecutively
from the base of the isolate, each area measuring 5 cm. x 20 cm.
This counting area was drawn on flexible clear plastic which could
be placed on the substrate to delineate the area to be counted.
The direction of the counting was east or west from the base of the
isolate, whichever direction led furthest away from the nearest
neighboring mature <u>Allopora</u> colony (Fig. 1). Data were recorded
on opaque plastic sheets and later transcribed.

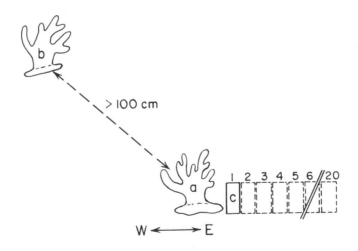

Figure 1. Counting method used to determine distribution of new
colonies around mature isolated <u>Allopora</u> <u>californica</u> colonies.
(a) isolate (b) nearest neighboring mature <u>Allopora</u> colony
(c) first 5 cm. x 20 cm. counting area

Upon returning to the beach, the collected pieces were immediately relaxed in sea water with a few menthol crystals sprinkled
on the surface. After about 2 hours they were killed with 1-2%
formalin in sea water (10 minutes), then fixed in Bouin's picroforma-
lin. Following decalcification the coral was processed for paraffin
sectioning using cedarwood oil as a clearing agent to soften the
yolky tissue. Sections were cut at 8 μ and stained with Heidenhain's
hematoxylin so results would be comparable with previous histological
data (Little, 1971; Ostarello, 1973).

RESULTS

Data were obtained on a total of 58 isolated colonies, 27
females and 31 males. Comparing the number of isolates which had
no new colonies in the twenty counting areas with those that had
at least one new colony shows that the occurence around female isolates was much greater than around male isolates (Table 1). The
difference between female and male isolates is statistically
significant (p < 0.01) using a 2 x 2 contingency table to obtain
a chi-square value.

Table 2 gives the actual number of colonies found in each of
the 20 counting areas examined around each isolate. The total
number of new colonies near the 3 male isolates was small and showed
no distinct pattern. The female colonies with adjacent juveniles
tended to have more new colonies close to the female isolate and
fewer as the distance from the isolate increased. In most cases
the total number of new colonies was not large, though the data indicate a definite pattern of settlement with most new colonies settled close to the female colonies.

Histological examination of the male and female gonophores in
the isolates showed several unusual conditions. In previous work

TABLE 1
Occurence of newly settled Allopora colonies
around mature male and female isolates

	Number of Isolates with New Colonies	Number of Isolates without New Colonies
Females	20	7
Males	3	28

TABLE 2

Distribution in the counting areas around those
isolates which had newly settled colonies

Sex	Counting Area from Base of Isolate														
	1	2	3	4	5	6	7	8	9	10	11	12	13	14	15–20
Male	2	0	0	0	0	0	0	0	0	0	0	0	0	0	0
Male	0	0	0	1	0	0	0	0	0	0	1	0	0	0	0
Male	0	0	0	1	0	0	0	0	0	0	0	0	0	0	0
Female	67	6	3	0	0	0	0	0	0	0	0	0	0	0	0
Female	20	5	14	8	3	1	2	3	1	1	0	0	0	0	0
Female	6	3	0	0	0	0	0	0	0	0	0	0	0	0	0
Female	5	3	3	4	3	3	3	1	1	0	0	0	0	0	0
Female	3	3	1	0	0	0	0	0	0	0	0	0	0	0	0
Female	2	0	0	0	0	0	0	0	0	0	0	0	0	0	0
Female	1	1	2	1	0	0	1	0	0	0	0	0	0	0	0
Female	1	1	1	0	0	1	0	0	0	0	1	0	0	1	0
Female	1	1	1	0	0	0	0	0	0	0	0	0	0	0	0
Female	1	1	0	0	0	0	0	0	2	0	0	0	0	0	0
Female	1	0	0	0	0	0	0	0	0	0	0	0	0	0	0
Female	1	0	0	0	0	0	0	0	0	0	0	0	0	0	0
Female	0*	2	3	0*	0*	0*	1	0	0	0	0	1	0	0	0
Female	0	2	2	0	1	0	0	0	0	0	0	0	0	0	0
Female	0	1	1	0	1	1	0	0	0	0	0	0	0	0	0
Female	0	1	0	1	0	0	0	0	0	0	0	0	0	0	0
Female	0*	0*	1	3	1	0	0	0	2	0	0	0	0	0	0
Female	0	0	1	0	1	0	0	0	0	0	0	0	0	0	0
Female	0*	0*	1	0	0	0	0	0	0	0	0	0	0	0	0
Female	0	0	1	0	0	0	0	0	0	0	0	0	0	0	0

*obstructing sponge prevented settlement

on typical closely spaced colonies, it was found that close to 100%
of the eggs in a female colony were fertilized, and only rarely did
an unfertilized egg degenerate and the surrounding ampulla fill with
tissue and calcareous material (Ostarello, 1973). In the isolates,
however, numerous ampullae were filled in, indicating a much greater
number of unfertilized eggs. Considering the distance from the nearest male colony (at least 100 cm.), this is not unexpected. It does,
perhaps, account for the relatively low total number of new colonies
counted. It was also found that development of gonophores within
both male and female isolates was not synchronized either within
a given colony or between colonies. For example, gonophores in a
female isolate showed stages of development ranging from very small
eggs with no yolk deposition to eggs with moderate yolk deposition
and a greatly enlarged trophodisc. Such wide intracolony and intercolony variation is not typical of what has been found in colonies

that are clumped together (Ostarello, 1973). Whatever mechanism
serves to synchronize the clumped members of the population is
apparently missing or ineffective in isolated colonies.

DISCUSSION

The distribution of new colonies around existing mature Allopora
colonies showed a strong tendancy towards settlement near a
female colony. If dispersal and settlement were random, the new
colonies would have been randomly or irregularly distributed and
settlement near male and female isolates would have been similar.
If dispersal were to culminate with gregarious settlement, where
planulae settled near existing colonies of the same species, the
new colonies would have been clustered close to both male and female
isolates. The fact that settlement showed a clustering around
female colonies but not around male colonies suggests that planulae
leaving the female parent quickly seek the ocean bottom and generally
move only a short distance before settling. Some new colonies did
settle well away from the female, demonstrating that some dispersal
prior to permanent settlement is possible. This hypothesis agrees
with the laboratory observations on A. petrograpta, where planulae
moved by means of cilia and/or muscular movements along the bottom
of the dish and did not stay in the water column to be moved by
currents over considerable distances (Fritchman, 1974).

It is surprising that almost equal numbers of male and female
isolates were found. If settlement tends to be close to the female
parent, one would think that female isolates would soon have their
offspring surrounding them and they would no longer be separated
by 100 cm. from the nearest colony. The fact that this did not
occur may be attributed to the low fertilization rate in isolated
colonies and the known low survival of newly settled colonies in
this species (Ostarello, 1973). The low fertilization rate seen
in isolates was probably due to the considerable water motion in
these areas. Effective sperm transport to isolates must be very
difficult and luck alone accounts for most of the success. Chemotaxis, as reported in other hydroids (Miller, 1966), is effective
only over a very short range, and if it occurs at all in Allopora,
it could only account for directing the sperm the last few millimeters.

The role of larval dispersal in species survival in A. californica has two distinct parts. First, the recruitment of new
colonies into the local population with a reasonable chance of survival is insured since most of the planulae settle close to the
female parent. By not entering the plankton they reduce mortality
due to being carried to unsuitable areas or to being devoured by
other plankters. Clustering of colonies is certainly an advantage
in the reproductive cycle of Allopora californica.

The second role of the larval stage in <u>Allopora</u> is species
dispersal. A few planulae are carried some distance from the female
parent. Almost all of these die, but a small percentage may land
on suitable sites for attachment. Such sparse dispersal succeeds
in colonizing new areas only because of the longevity of the individual colonies. Since a colony 25–30 cm. wide is probably 50–100
years old (Little, 1971), recruitment of colonies into an area
can occur over many years in order to establish a new group of
colonies.

This strategy of larval dispersal in <u>Allopora californica</u> has
allowed successful distribution of the species over a 600 mile
range, from central California to Baja, California (Durham, 1947).
At the same time the recruitment of new colonies into the local
population is insured.

I am deeply indebted to the 12 University of California,
Berkeley certified divers who aided in collecting the data. Mr.
Lloyd F. Austin both helped with the underwater work and allowed
me to use his laboratory for the histological preparation of
material. Mrs. Emily Reid prepared the figure and tables.
Special thanks are due to my husband, John, who helped in collecting data and critically reviewing the manuscript.

SUMMARY

In the subtidal hydrocoral <u>Allopora californica</u> the developing
planulae are retained within the female colony and are released in
October and November. Data were collected on settlement of new
colonies around mature male and female colonies that were isolated
from other hydrocorals by at least 100 cm. The distribution of
new colonies showed that released planulae settle very close to the
female colony; settlement around male isolates is rare. Planktonic
dispersal is not the general rule in this species.

Histological data revealed that a much lower percentage of the
eggs in the isolated female colonies was fertilized as compared
with non-isolated members of the population. The development of
male and female gonophores in the isolated colonies was not synchronized either within a colony or between colonies. Previous data
showed a high degree of intra- and intercolony synchronization
of gonophore development in clustered colonies.

LITERATURE CITED

DURHAM, W., 1947. Corals from the Gulf of California and the North Pacific Coast of America. Geol. Soc. of Am. Mem., 20: 1-68.

FRITCHMAN, H.K., 1974. The planula of the stylasterine hydrocoral Allopora petrograpta Fisher: Its structure, metamorphosis and development of the primary cyclosystem. Proc. Second Intl. Coral Reef Symp., 2: 245-258.

LITTLE, G., 1971. Aspects of the biology of the subtidal hydrocoral Allopora californica Verrill (1866) with emphasis on reproduction, recruitment and mortality of new colonies, and regeneration. Ph.D. thesis, University of California, Berkeley.

MILLER, R.M., 1966. Chemotaxis during fertilization in the hydroid Campanularia. J. Exp. Zool., 161: 23-44.

OSTARELLO, G.L., 1973. Natural history of the hydrocoral Allopora californica Verrill (1866). Biol. Bull., 145: 548-564.

VERRILL, A.E., 1866. Synopsis of the polyps and corals of the North Pacific Exploring Expedition under Commodore C. Ringgold and Captain John Rodgers, U.S.N., from 1853 to 1856; collected by Wm. Stimpson, naturalist to the expedition; with descriptions of some additional new species from the west coast of North America. Pt. 3: Madreporaria. Proc. Essex Inst., 5: 17-50.

LARVAL ADHESION, RELEASING STIMULI AND METAMORPHOSIS

W.A. Müller, F. Wieker and R. Eiben

Zoological Institute, Technical University

Braunschweig, West Germany

Free living stages of sedentary organisms can be considered adaptations to enable immobile species to exploit a scattered or transient ecological niche. The task to prospect for and to identify a congenial habitat is consigned, as a rule, to larvae or larva-like buds, that is to those stages which actually transform into the sessile phase. This is the case even if a metagenetic life cycle has provided the species with a dominant swimming phase capable of brooding the eggs throughout development until the settling stage. Eventually, the larvae are set free and they have to find a suitable substratum themselves. But how can larvae comply with such a task? Their sensory equipment is very limited and does not qualify them to locate an appropriate habitat from a distance. This applies especially to coelenterate larvae. They depend, therefore, on a hierarchy of key or sign stimuli indicative for their adult environment. Viewed in terms of behaviour, the larva displays appetitive or searching activity which continues until the larva is presented with a specific releasing stimulus triggering fixation and metamorphosis. The effective stimulus must be derived from characteristic substrate properties which can be explored by mechanoreceptors and/or a chemical contact sense (Crisp, 1974). Recognition of an adequate substratum, therefore, depends on physical contact or, at least, on a close range approach. The word 'close' can be defined in this context in relation to the various forces of adhesion.

For tiny larvae it is not an easy task to establish intimate contact with a solid surface. They have to overcome several physical barriers such as forces of surface tension and repulsive electrostatic potentials. Thus, the physical conditions affecting adhesion

can not be left out of consideration when evaluating factors determining settlement. We aim to illustrate the significance of such
forces and the nature of releasing stimuli by comparing three examples: the settlement of the planulae of Hydractinia echinata, the
settlement of the swimming buds of Cassiopea xamachana, and the
adhesion of Bowerbankia gracilis. This bryozoan may be included in
our comparison because it demonstrates an unconcealed dependence upon
physical surface properties.

Adhesion of particles, e.g. larvae, to submerged substrata is
influenced by several physical conditions including electrostatic
and ion exchange properties, cohesiveness, surface energy, and tension of wetting (e.g. Weiss, 1970; Baier, 1970). The significance
of tension of wetting is clearly documented by our investigations on
larvae of Bowerbankia (Eiben, in press). These tiny, rounded larvae
cannot establish contact on surfaces with high wettability properties
due to capillary forces which impede dislodgement of the water film at
the larva-substrate interface.

Wettability of solids can be characterized by the hydrophobic
contact angle δ which is determined by the relative tensile strength
of the liquid to the solid and to itself. In the classical equation
of Young (1) the angle is treated as the result of the mechanical
equilibrium of a drop resting on a plane solid surface under the action of three surface tensions:

$\gamma_{L/V}$ at the interface of the liquid and vapor phases,
$\gamma_{S/L}$ at the interface of the solid and the liquid,
$\gamma_{S/V}$ at the interface of the solid and vapor.

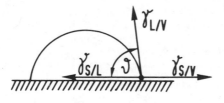

Fig. 1. Contact angle
of a sessile
drop.

Young's equation: $\gamma_{S/V} - \gamma_{S/V} = \gamma_{L/V} (\cos \delta)$ (1)
Tension of wetting: $(\sigma_{S/L})$ is determined by:
$$-\sigma_{S/L} = \gamma_{S/V} - \gamma_{S/L} \qquad (2)$$
Hence: $\sigma_{S/L} = -\gamma_{L/V} (\cos \delta)$ (3)

Thus, the contact angle is a useful inverse measure of spread-
ability and wettability, since the tendency for the liquid to spread
increases as δ decreases. The cosine of δ is a direct measure, in
the sense that, as the value of cos δ increases towards unity, the
wettability also increases.

Larvae of Bowerbankia can establish contact with hydrophobic
surfaces exhibiting contact angles >17°. An explanation for this

observation can be given on the basis of a qualitative model taking
the larva as a drop of a second liquid (as for example an oil drop)
and considering its adhesion in terms of competition between larva
and water for wetting the solid surface. Then adhesion can be
accounted for by assuming that the tension of wetting solid to water
is smaller than of the solid to the larval surface: $\sigma_{solid/water}$
$< \sigma_{solid/larva}$. The ratio between these two wetting tensions
determines whether the larva has to invest energy in order to dislodge the water film from the solid surface or whether the larva
is attracted to the solid.

Thermodynamic work of displacement:

$$\left|\sigma_{S/La/W}\right| = \left|\sigma_{S/W}\right| - \left|\sigma_{S/La}\right| \qquad (4)$$

where the indices are: S = solid, La = larva, W = water. When a
critical ratio $\left|\sigma_{S/W}\right| \gtrless \left|\sigma_{S/La}\right|$ is surpassed the left handed
term of the equation (4), that is the thermodynamic work of dis-
placement, becomes negative. That means, adhesion of the larva turns
into an exergonic event and the larva is passively attracted to the
solid. The attractive force must be strong enough to overcome the
electrostatic repulsion barrier which is built up by electrical
double-layer energies between the larva and the solid. In Bowerbankia, an over-all decrease in free energy sufficient to overcome
such barriers is achieved at contact angles >17°. Substrata
exhibiting lower angles will never be settled on, those with angles
above this critical value will always be colonized. If the percentage of settling larvae is plotted against the contact angle an
all-or-none response becomes manifest: At the critical point the
rate swings suddenly up from 0% to 100%.

The primary adhesion to which the larva is subjected provides
shearing forces acting upon the larval cilia. Shearing of cilia,
in turn, constitutes the releasing stimulus triggering ultimate
fixation by a particular adhesive organ, and metamorphosis.

Some coelenterate larvae are believed to settle also as a
result of simple contact (Hawes, 1958; cited by Crisp, 1974).
However, for most sessile organisms it would be uneconomical to
produce larvae which fix permanently as soon as they touch a surface by chance. At least in Hydractinia mechanical contact is
insufficient to initiate metamorphosis. The planulae of Hydractinia
can attach to a substratum in three different ways: (1) by means of
a mucous adhesive which is secreted by gland cells mainly at the
anterior pole, (2) by means of sticky threads discharged by nematocysts which are concentrated in the posterior region. These two
methods allow a reversible attachment. (3) A third method of
anchorage is employed during metamorphosis: ectodermal cells
provide a chitinous cement for a permanent fixation.

Again, reversible contact is facilitated on hydrophobic

substrata. The planulae readily attach with their anterior pole
to surfaces showing low wettability, and take a vertical position.
Upon contact, again, shearing forces work on the cilia but this
stimulus per se never induces metamorphosis. The vertical
position of the larvae benefits their transfer to another, mobile
substratum.

Hydractinia echinata is most abundantly found on disused
gastropod shells inhabited by hermit crabs. If the searching
larvae have not been exposed to a specific sign stimulus leading
to metamorphosis the planulae remain standing vertically poised
on the substratum until an object is dragged past. The movement
stimulates the planulae to stick to the passing object by dis-
charging atrichous isorhizas. This transfer is an event which is
strongly promoted by the velocity gradient in the fluid between the
larva and the passing surface. The drag of the moving solid is
transmitted to the liquid bringing the liquid layer close to the
surface into motion. Thus, a velocity gradient is set up in the
shear plane surrounding the object. The movement of the solid
against the liquid phase leads to boundary potentials across the
shear plane often referred to as zeta-potentials or streaming potentials (Weiss, 1970). Probably such potentials are implicated in
evoking the larval response. The change of velocity along the larva,
in addition to the impulse of collision, constitutes the condition
that stimulates the larva to fire and maintain its holdfasts. Thus,
when planulae are drawn into a pipette, they immediately attach
to the sides, but when the relative movement stops they eventually
expel their nematocysts and resume crawling about.

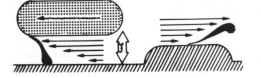

Fig. 2.   Transfer of planu-
          lae to a second
          substratum by
          velocity gradients.

Transfer to a mobile substratum (or, conversely, to a substratum over which water is flowing) again does not initiate metamorphosis. The larvae must receive a particular releasing stimulus. We
have identified two natural sign stimuli which trigger the
irreversible onset of metamorphosis and, hence, of permanent
fixation.

The first releasing stimulus consists of a pheromone-like
lipid which can be emitted by certain marine bacteria (Müller, 1973).
These bacteria themselves can sorb to surfaces. They are regularly
found on gastropod shells as well as on the cuticle of crabs, but
are also found on other surfaces. The presence of these bacteria
may be the cause for the participation of Hydractinia in fouling
communities. Evidence exists for the localization of the inducing

lipid in the outer layer of the bacterial envelope from which it
may diffuse in very minute quantities to become adsorbed to the
larval cell membrane.   We have succeeded in isolating and purify-
ing this labile agent by chromatographic methods up to a degree
that now allows the determination of its chemical nature.

The second sign stimulus is less well defined.   It appears to
derive from the peculiar vibrations produced by crabs.   If shells
which had been previously sterilized and deprived of organic matter
by boiling with KOH are reoccupied by hermit crabs, they will be
colonized by metamorphosing planulae within two hours.   The planulae
are whirled up by the migrating crab, attach to the shell, prefer-
entially in pits and grooves, and undergo metamorphosis.   Immobile,
vacant shells are ineffective, though they are accepted as
support for reversible adhesion.   In the short time available no
effective bacterial film could have been built up on the mobile
shell.   However, we have not yet been able under sterile conditions
to produce stimulating oscillations with mechanical devices.   Such
work should be done in order to determine the decisive physical
parameter.

Bacteria-borne stimuli are the dominant, if not the sole factor
determining settlement and development in Cassiopea (Wieker, 1975).
The ciliated and rather large swimming buds of Cassiopea are endowed
with sufficient powers of propulsion to make contact with every
substratum.   However, the buds never adhere to a solid as a result
of a mere collision.   The surface of the solid must be covered with
a bacterial film in order to be attractive.   The stimulus promoting
settlement issues only from certain living bacteria.   Even subtle
methods of killing the bacteria such as drying in air or irradiation
with UV removes their attractive quality.   Moreover, the bacteria
are effective only in their sedentary state.   In contrast to
Hydractinia, Cassiopea cannot be induced to convert into the polyp
state by applying bacterial suspensions.   Accordingly, for isolating
an effective strain we had to develop a particular assay that presents
the microbes in the form of a stable film grown on glass plates.
The plates are held in a horizontal position because the success of
settlement is also a function of surface angle.   Cassiopea favours
an angle of zero, that is the under surface.

The isolated bacterial strain can be offered, with moderate
effectiveness, to larvae of Hydractinia too.   It appears, therefore,
that we are dealing with related phenomena.   However, attempts to
analyze the physiological response disclose unexpected differences.

In Hydractinia, the bacterial lipid can be substituted by
exposing the larvae for 3 hours to particular ionic environments
(Müller and Buchal, 1973).   Two procedures are successful: (1)
The first may be referred to as 'ion exchange' mechanism by which

K^+-ions dislodge Ca^{++} ions from anionic receptor sites P at the larval cell membrane:

$$P_2Ca + 2 K^+ \rightleftarrows 2 PK + Ca^{++}$$

To induce metamorphosis it is necessary to adjust the medium to an exactly defined equilibrium. Expressed in analogy to the law of mass action (or Gibbs-Donnan principle) we arrive at the equation:

$$\frac{[P_2Ca] \times [K^+]^2}{[PK]^2 \times [Ca^{++}]} = k;$$

Taking the sum of binding sites as constant and considering only the ions within the solution the term is simplified to:

$$\frac{[K^+]^2}{[Ca^{++}]} = k'; \quad or: \frac{[K^+]}{\sqrt{[Ca^{++}]}} = k'';$$

In practice, metamorphosis is initiated by applying solutions with a ratio $[K^+] : \sqrt{[Ca^{++}]} = 40$.

(2) A second, even more convenient procedure is to enrich seawater with Cs^+-ions at the expense of all other cations up to a final dose of about 40 mval Cs^+. Cesium leads to an enhanced operation of the membrane-bound Na^+-K^+-ATPase (May and Müller, 1975) as does the bacterial agent. The enhanced activity of the Na^+-K^+-pump, in turn, is a prerequisite, if not the cause, for the stimulating action of these agents.

Irrespective of the method employed, mechanisms comparable to excitable systems are triggered. We do not know whether the bacterial or ionic stimuli affect specialized sensory cells or 'ordinary' epithelial cells which are equipped with cilia and microvilli. Excitability, in any event, is not restricted to a limited receptive field. Reception of and response to inducing stimuli do not depend on the structural integrity of the larva. Thus, fragments of planulae are able to undergo metamorphosis (Fig. 3). By means of isolation experiments it can be shown, however, that excitability decreases from the anterior to the posterior pole. Anterior fragments can readily be stimulated to metamorphose, but not posterior fragments. The percentage of successful transformations remains constant when posterior material is removed but it decreases the more the planula is deprived of anterior tissue (Fig. 3).

In line with this gradient of excitability the anterior pole assumes the role of a pacemaker firing facilitating signals to the posterior region, which lower the threshold of excitability. The transmission of the signal is a slow process. It is followed by

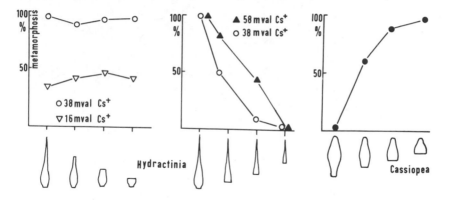

Fig. 3. Development of fragmentated planulae of <u>Hydractinia</u> and
of fragmentated swimming buds of <u>Cassiopea</u>.

a wave of contraction. The passage of excitation can be arrested
by transverse cuts or restrictions. With doses of stimulating
agents below the threshold of the posterior part the planula
persists, behind the barrier, in the larval state.

In <u>Cassiopea</u>, such experimental interference brings about
exactly the opposite effects. Removal of anterior tissue enables
the posterior part of the swimming bud to resume its arrested
development. The posterior fragment will differentiate into a free
swimming scyphistoma without being presented with a particular
releasing stimulus. It appears from these findings that in
<u>Cassiopea</u> an inhibitory centre is located at the anterior pole
blocking further development. In contrast to <u>Hydractinia</u>, the
more anterior material is removed, the more the animals will develop into tentaculated polyps. The difference in the underlying
physiological mechanism is also manifested by the failure of
<u>Cassiopea</u> buds to respond to ionic stimuli with transformation.
<u>Bowerbankia</u>, on the other hand, does respond. All artificial
manipulations applicable to <u>Hydractinia</u> can also be employed in
<u>Bowerbankia</u> but do not work in <u>Cassiopea</u>. These findings should
warn us that unifying concepts must be handled with care. Every
species has found its own way to meet ecological demands.

REFERENCES

Crisp, D.J. 1974. Factors influencing settlement of marine
    invertebrate larvae. In: Chemoreception in marine organ-
    isms (ed. by P.T. Grant and A.M. Mackie), Academic Press,
    London, New York.

Baier, R.E.   1970.   Surface properties influencing biological
    adhesion.   In: Adhesion in biological systems (ed. by R.S.
    Manly), Academic Press, New York, London.

Eiben, R.   in press.   Einfluß von Benetzungsspannung und Ionen auf
    die Substratbesiedlung und das Einsetzen der Metamorphose
    bei Bryozoenlarven (Bowerbankia gracilis).   Marine Biology,
    in press.

May, G. and W.A. Müller   1975.   Aktivitäten von Enzymen des
    Kohlenhydrat-Stoffwechsels und der Na^+-K^+-ATPase im Zuge
    der Embryonalentwicklung und Metamorphose von Hydractinia
    echinata (Hydrozoa).   Wilhelm Roux' Archiv 177: 235-254.

Müller, W.A.   1973.   Metamorphose-Induktion bei Planulalarven. I.
    Der bakterielle Induktor.   Wilhelm Roux' Archiv 173: 107-121.

Müller, W.A. and G. Buchal   1973.   Metamorphose-Induktion bei
    Planulalarven. II.   Induktion durch monovalente Kationen.
    Wilhelm Roux' Archiv 173: 122-135.

Müller, W.A., A. Mitze, J.P. Wickhorst, H.M. Meier-Menge, in press.
    Polar morphogenesis in early hydroid development: Action of
    cesium, neurotransmitters and of a head-activating morphogen
    on pattern formation.   Wilhelm Roux' Archiv.

Weiss, L.   1970.   A biophysical consideration of cell contact
    phenomena.   In: Adhesion in biological systems (ed. by R.S.
    Manly), Academic Press, New York, London.

Wieker, F.   1975.   Bildung und Metamorphose der Schwimmknospen von
    Cassiopea xamachana.   Diplomarbeit, Naturwissenschaftliche
    Fakultät, Technische Universität, Braunschweig.

SOME MICROENVIRONMENTAL INFLUENCES ON ATTACHMENT BEHAVIOR OF THE

PLANULA OF CYANEA CAPILLATA (CNIDARIA: SCYPHOZOA)

Robert H. Brewer

Department of Biology
Trinity College
Hartford, Connecticut 06106

The selective value of prudent settlement by larval stages is important for the persistence of sedentary species. The factors influencing larval behavior of marine invertebrates are reviewed by Crisp (1974). In many species reversal of taxes occurs prior to attachment on a suitable substrate; thus, the larva may be initially geonegative and/or photopositive, but as development proceeds changes occur in its response to the environmental factors so that the larva attaches upside-down in a shaded location (Crisp, 1974).

The planula of Cyanea capillata attaches upside down, consistent with the position usually occupied by the sedentary scyphistoma in the field, and this orientation appears to be the consequence of planular behavior (Brewer, 1976). The stimulus effecting this orientation may be the relative concentration of dissolved gases to which the planula is exposed. Brewer (1976) observes that planulae are more active, and attach rightside up, under conditions of presumed low oxygen and high carbon dioxide content; i.e., both in cultures containing large populations of protozoa, and in those with decomposing tissue of medusae. Perez (1920-21) notes activity of planulae in collecting buckets containing medusae. Chemical influence on larval attachment is shown for several species of Hydrozoa (e.g., by Nishihira, 1968; Müller, 1969) and is frequent among larvae of higher metazoans (e.g., Meadows and Campbell, 1972; Crisp, 1974).

The purpose of this paper is to more clearly establish that planulae are affected by populations of microorganisms, and to determine the extent to which planular behavior is influenced by different concentrations of oxygen or carbon dioxide.

Planulae are cultured with different densities of protozoa, and

the orientation and rate of attachment is determined. The orientation and activity of planulae exposed to different concentrations of
oxygen or carbon dioxide are measured. Since the planula must engage
in active movement to attach upside down on a substrate, the relationship between the level of planular activity and attachment
is assessed to provide another criterion for evaluating the effects
of the dissolved gases on behavior.

MATERIALS AND METHODS

Material and General Experimental Conditions

Planulae were obtained from gravid females of C. capillata (7
to 15 cm diameter) collected from Smith's Cove, Niantic, Connecticut,
during May, 1975. The seawater (22 0/00) used in the experiments
was obtained at the time the medusae were collected. The experiments
were set up with equal-aged planulae which developed from blastulae
in the laboratory, and were run at room temperature (18-20°C); the
illumination was fluorescent and constant. Stender dishes (50 mm
diameter x 25 mm), each with 20 planulae in 10 ml seawater, were used
in all experiments.

Protozoan Density and Planular Attachment

Eighteen dishes were set up. Different protozoan densities occurred in response to variation in culture (filtered seawater, unfiltered seawater, and seawater with small portions of tissue from a
medusa) and were not predetermined. The number of planulae attaching, their orientation, and the number of protozoa (estimated from
hemocytometer counts) was tallied for each container at 72 hours.

Planular Activity and Attachment Rate

The measure of activity recorded was the number of planulae
swimming at least 1 mm above the dish bottom counted during a thirty
second interval; this count was made within 1-2 hours of setting up
86 cultures for observation. The number of planulae attaching was
enumerated for each dish after the elapse of five hours and compared
with the level of activity observed during the initial observation.

The Effect of Oxygen Concentration

Planulae were exposed to four concentrations of oxygen: 0%;
7%; 14%; and 21%. The balance of the gas (to make 100%), except at
21% (air), was nitrogen. The sources of the mixtures were commercial

cylinders. These were connected to culture dishes with tubing of decreasing diameter. Surgical tubing, 0.05 inches in diameter, passed through a hole of that size drilled in a plastic lid that fitted tightly on the stender dish; an additional hole in the lid allowed excess gas to escape. The gas was bubbled into the seawater containing the planulae for thirty minutes, and thereafter the tube was withdrawn so that the gas concentration remained constant in the space above the water, but did not agitate the surface. A treatment control was maintained for 21% oxygen in which no bubbling nor administration of gas under pressure was employed. There were eight replicates for each treatment and the control. The numbers of planulae swimming above the bottom during a thirty second interval were counted on eight occasions beginning with the first hour after setting up the experiment, and continued at equal intervals for a period of forty hours; the sum of these eight observations are the data for each replicate. In addition, the orientation of attached planulae was recorded for each treatment after settlement (93 hours).

The Effect of Carbon Dioxide Concentration

Planulae were exposed to two concentrations of carbon dioxide: 0.03% (air); and 5% (95% air). The gas was introduced in the manner described in the previous section, and a treatment control was maintained for the cultures exposed to air. The number of planulae swimming above the bottom in each of eight replicates was counted during seven 15 second observation periods beginning at the eighth hour and spaced equally over the following twenty hours. The sum of these observations for each replicate are the data used to assess the effects of carbon dioxide. The pH was monitored, but the seawater was not buffered.

RESULTS

Protozoan Density and Planular Attachment

Table 1 shows the proportion of planulae attaching at different protozoan densities at 72 hours. There is a positive correlation (R = + 0.93; 0.81, 0.97) between density of protozoa and number of planulae attaching; this is significant ($P<10^{-6}$).

More planulae attach rightside up in the six cultures with the highest protozoan density (34.0 $\pm$ 21.04% rightside up) than in those six with the lowest density (18.6 $\pm$ 20.89%), but the difference is nonsignificant (0.10<P<0.25).

Table 1

The Effect of Protozoan Density on Proportion of
Planulae Attaching at 72 Hours

Protozoan density (log No./ ml)	Proportion planulae attaching (Arcsine $\sqrt{P}$)	Protozoan density (log No./ ml)	Proportion planulae attaching (Arcsine $\sqrt{P}$)
2.4440	0.00	3.6100	36.27
2.5682	0.00	3.8887	12.92
2.7024	12.92	4.0969	33.21
2.7451	0.00	4.4138	36.27
2.7451	0.00	4.3979	67.72
2.9335	12.92	5.2109	56.79
3.0453	12.92	5.2109	63.43
3.1804	18.43	5.2730	56.79
3.2674	30.00	5.3979	77.08

Planular Activity and Attachment Rate

Table 2 shows the relationship between activity and attachment
rate; the data reported are paired moving averages. The regression
($\hat{Y}$ = log 0.1583 + 1.21544X) is significant (P<0.001); the deviation

Table 2

The Relationship Between Activity of Planulae
and Number Attached at Five Hours

No. of cultures in interval	Mean Number of Planulae at Interval Midpoint	
	Active, hour 0 (log)	Attached, hour 5 (log)
29	0.1761	0.4253
20	0.3979	0.5320
15	0.5441	0.8323
12	0.6531	0.9314
3	0.7404	1.1412
2	0.8129	1.2553
4	0.8751	1.2913
1	1.0792	1.5315

from regression is nonsignificant (0.50<P<0.75). Thus, activity of planulae may be used to assess the likelihood of subsequent attachment even if the latter is not observed.

The Effect of Oxygen Concentration

The percent concentration of oxygen and, in parentheses, the mean number ($\pm$ s.d.) of planulae active are: 0% (1.1 $\pm$ 1.73); 7% (2.0 $\pm$ 1.85); 14% (4.0 $\pm$ 3.46); 21% (3.5 $\pm$ 2.00); and 21% [control] (2.4 $\pm$ 2.07). These differences are nonsignificant (P$\approx$0.10).

The percent planulae (in parentheses) attaching rightside up (on dish bottom) at the different concentrations of oxygen are: 0% (66 $\pm$ 57.4); 7% (41 $\pm$ 35.7); 14% (25 $\pm$ 23.0); 21% [control and experimental combined] (14 $\pm$ 17.9). The regression (Y = 53.0401 − 1.87179X) describing this relationship is significant (0.01<P<0.025). (Note: the high mortality at 0% oxygen precluded its inclusion in the regression calculation.)

It appears that oxygen concentration does not influence activity of planulae, but that it does influence orientation of planulae at time of settlement.

The Effect of Carbon Dioxide Concentration

The percent concentration of carbon dioxide and, in parentheses, the mean number ($\pm$ s.d.) of active planulae are: 0.03% [control] (0.04 $\pm$ 0.48; 0.03% (0.32 $\pm$ 0.285); and 5% (1.6 $\pm$ 0.45). The difference between the control and the experimental cultures (0.03% carbon dioxide) is nonsignificant (0.50<P<0.75); the difference between these cultures at ambient conditions and those of high carbon dioxide concentration is significant (P<0.001). The introduction of carbon dioxide reduced the pH 1.415 units (from 7.710 to 6.295).

High levels of carbon dioxide (or low pH) stimulate planulae to swim off the bottom of their container. It was not possible to observe orientation at time of attachment in these cultures because of early mortality.

Discussion

The planula of C. capillata responds in the laboratory to concentrations of dissolved oxygen or carbon dioxide as predicted by Brewer (1976), and the correlation between activity and attachment rate, with the data on protozoan density, suggests that orientation of the scyphistoma upside down in the field is brought about by the effect of the microorganismal environment on the planula. In nature, the conditions of high carbon dioxide concentration, the product of the activity of microorganisms, may occur in areas of calm water at the interface between the layer of silt and the water column (e.g., Fenchel, 1969; Odum, 1971). Similarly, pH declines in this layer; thus, the lack of control of pH in the experiment reported, although raising the obvious question of proximate causation of planular activity, does not in itself invalidate this extrapolation of laboratory results to field situations.

The mortality of planulae at high concentrations of carbon dioxide, which prevents observation of their orientation at time of attachment, is possibly the result of uniform distribution of toxic concentrations not expected to occur either under exposed conditions in the laboratory or in the field. Levels of carbon dioxide intermediate to those used should be tried to determine if increased activity under these conditions leads to a high proportion of planulae attaching upside down. A basic assumption for such speculation is that planula activity necessarily leads to attachment upside down; this has intuitive appeal, but still lacks rigorous demonstration. However, exposure to different levels of oxygen does not affect activity of planulae, but does influence orientation at attachment in a different way.

Attachment rightside up seems to be due to low levels of oxygen. Even though this conclusion is based upon a regression calculated for only three concentrations, its validity is supported by the position occupied by two additional data for orientation at 0% oxygen determined in unreplicated pilot experiments on two separate occasions: these fall within the 0.95 confidence interval calculated for the regression, extrapolated backward from 7% oxygen to include 0% oxygen. Presumably, when oxygen is withheld planulae can no longer maintain their activity, sink, and attach rightside up where they land.

The specific stimulus for attachment is not examined in this paper, but the importance of bacteria during this phase in the Hydrozoa (e.g., Müller, 1969, 1973) and possibly some Anthozoa (Ian Johnston, Department of Biology, University of California, Los Angeles, personal communication) suggests that the positive correlation between protozoan density and attachment rate may be spurious. Most of the protozoa counted were ciliates, and some were observed ingesting bacteria. It is possible that the population size of protozoa reflects the density of bacteria, and that it is the latter which facilitates the attachment of the planulae of C. capillata. It is also possible that specific populations of bacteria are differentially affected by the concentrations of dissolved gases used, above, effecting the results observed.

Most work on settlement behavior of marine larvae deals with the description of tactic responses and the analysis of attachment phenomena once the larva contacts a suitable substrate (e.g., see Crisp, 1974). Few seek to determine what stimulus is responsible for changes in tactic response in the period prior to surface contact by the larva. However, in many complex larvae of the higher Metazoa such a question may be inappropriate, for the shifts in taxes may be associated with their ontogeny and thus are endogenous changes taking place against a background of relatively constant stimuli (e.g., direction of light and gravitational force). The simpler larvae of the lower Metazoa, lacking the sensory apparatus of their more advanced relatives, may be caused to encounter potential attachment sites by response to exogenous factors through an avoidance reaction similar to that observed in the Protozoa (e.g., Fraenkel and Gunn, 1961; Jennings, 1962) which the planula superficially resembles.

I am grateful for funds received from Trinity College in partial support of this work.

Summary

The planulae of C. capillata are exposed in the laboratory to different concentrations of dissolved oxygen, carbon dioxide, and to protozoan cultures of varying density to determine if their response to these factors is consistent with the observed distribution of their scyphistoma stage in the field; i.e., usually suspended upside down.

Planulae must swim above the substrate to effect the inverted orientation of the scyphistoma, so both their level of activity as well as their position of attachment are used to assess their response to different treatments where possible.

Planular activity is positively correlated with protozoan density, and is greater at high concentrations of carbon dioxide than at ambient concentrations; the level of activity of planulae is not affected by different concentrations of oxygen.

More planulae attach upside down at high than at low protozoan densities, but the difference is not statistically significant; the sample size is small, and the variability, high. Orientation of attachment at different concentrations of carbon dioxide could not be determined with the conditions used. The proportion of planulae attaching upside down declines with increasing concentrations of oxygen.

These data tentatively support the hypothesis that the observed

orientation of scyphistomae is the consequence of planula response
to levels of carbon dioxide, though the precise details of the relationships between planulae, dissolved gases, and microorganisms
require further analysis.

The possible effect that bacterial populations have on the results and the significance of a probable similarity in behavioral
repetoire in planulae and protozoa are briefly discussed.

Literature Cited

Brewer, R.H., 1976. Larval settling behavior in Cyanea capillata
    (Cnidaria: Scyphozoa). Biol. Bull., 150: 183-199.
Crisp, D.J., 1974. Factors affecting the settlement of marine in-
    vertebrate larvae. Pages 177-265 in P.T. Grant and A.M.
    Mackie, Eds., Chemoreception in marine organisms. Academic
    Press, New York.
Fenchel, T., 1969. The ecology of the marine microbenthos. Part
    IV. Ophelia, 6: 1-182.
Fraenkel, G.S., and D.L. Gunn. 1961. The orientation of animals.
    Dover Publications, New York, 376 pp.
Jennings, H.S., 1962. Behavior of the lower organisms. Indiana
    University Press, Bloomington, 366 pp.
Meadows, P.J., and J.I. Campbell, 1972. Habitat selection by aqua-
    tic invertebrates. Advan. Mar. Biol., 10: 271-382.
Müller, W.A., 1969. Auslösung der Metamorphose durch Bakterien bei
    den Larven von Hydractinia echinata. Zool. Jahrb. Anat. Ont.,
    86: 84-95.
Müller, W.A., 1973. Induction of metamorphosis by bacteria and ions
    in the planulae of Hydractinia echinata; an approach to the
    mode of action. Pages 195-208 in T. Tokioka and S. Nishimura,
    Eds., The proceedings of the second international symposium on
    Cnidaria. Publ. Seto Mar. Biol. Lab., 20.
Nishihira, M., 1968. Brief experiments on the effect of algal ex-
    tracts in promoting settlement of the larvae of Coryne uchidai
    Stechow (Hydrozoa). Bull. Mar. Biol. Sta. Asamushi, 13: 91-
    101.
Odum, E.P., 1971. Fundamentals of ecology, 3rd Ed. W.B. Saunders
    Company, Philadelphia, 574 pp.
Perez, C., 1920-21. Un élevage de scyphistomes de Cyanea capillata.
    Bull. Biol. Fr. Belg., 54: 168-178.

A NEW TYPE OF LARVAL DEVELOPMENT IN THE ACTINIARIA: GIANT LARVAE.
MORPHOLOGICAL AND ECOLOGICAL ASPECTS OF LARVAL DEVELOPMENT
IN ACTINOSTOLA SPETSBERGENSIS

Karin Riemann-Zürneck

Biologische Anstalt Helgoland (Taxonomische
Arbeitsgruppe). Address: Institut für Meeres-
forschung, 285 Bremerhaven, West-Germany

Actinostola spetsbergensis, a medium-sized Arctic sea anemone
with circumpolar distribution, was first described by Carlgren
(1893a). Soon after (1893b) Carlgren dissected a female and found
some young stages in the gastral cavity of this animal which
looked rather uncommon. Although these stages have subsequently
been found several times (Carlgren, 1901, 1902, 1913, 1921), they
have never since been studied.

Material and Methods

I acknowledge the help of those who have provided material:
K.W. Petersen (Universitetets Zoologiske Museum, Kopenhagen),
R. Oleröd (Naturhistoriska Riksmuseet, Stockholm), J.W. Wacasey
(Arctic Biological Station, Sainte-Anne de Bellevue, Canada),
F.M. Bayer (Smithsonian Institution) and S.D. Grebelny (Zoological
Institute of the Academy of Sciences, Leningrad).

About eighty animals were dissected in order to ascertain the
sex and the presence of developmental stages. Eleven females were
selected for investigations of the larval development (Table 1).

Results

During the course of taxonomic studies I observed these developmental stages of A. spetsbergensis within the gastral cavity of
a female. At first glance, I could not detect any similarity with
either young anemones or with actiniarian larvae as known from
several descriptions. The stages in question are characterized by

355

their almost spherical shape, the transparency of their bodies even
in the fixed condition and their enormous size (up to one centimeter).
Moreover, it is remarkable that stages from one female show different
degrees of development as to size, the number of septa and the formation of tentacles (Fig.1).

Table 1
Material of Actinostola spetsbergensis

Number	Locality	Museum
3	60°42'N 7°04'W, 650m. Leg. Behrmann "Anton Dohrn", 12.6.1972	Bremerhaven
2	64°58'N 11°12'W, 500m, 25.8.1902	Kopenhagen
1	75°49'N 24°25'W, 80m, 21.6.1898	Stockholm Nr.613
1	69°32'N 177°41'W, 20m, 7.-8.9.1878	Stockholm Nr.19
2	Frobisher Bay, Baffin Island, from blade of Laminaria floating at surface, 27.7.1967	Bremerhaven
2	67°50'N 166°55'W, 50m, August 1959	Smithsonian Nr. 52 105, 52 131

If we search for comparable stages among the Actiniaria, we
should review in brief our knowledge about development and developmental stages in these Coelenterata. At present we know the reproduction cycles of about 20 species of Actiniaria (see Nyholm, 1943;
Riemann-Zürneck, 1969). All except Bolocera tuediae live in littoral
and sublittoral habitats. The results of these investigations reveal
that all these species go through a certain developmental stage, the
well-known Edwardsia-stage. This particular stage occurs in two
larval forms (Carlgren, 1906, p.85-87):

1. The small larva (100 to 250 µm) with an aboral tuft of cilia
besides its common ciliation. This larval form occurs in many families
but is particularly characteristic of many of the acontiate anemones
(e.g. Sagartia, Metridium). The tufted larva is a typical plankton
larva with a short pelagic life.

2. The rather big, barrel-shaped larva (500 up to 1500 µm)
without an aboral tuft of cilia (the larva of Halcampa is without
any cilia). This larva usually develops from eggs rich in yolk and
is found in several families. During a short pelagic life the larva
remains in the near-bottom water layers. This larva may also be
found in actinians with internal brood-caring.

If we return now to the larvae found in A. spetsbergensis, we
can recognize that these larvae are much larger than the tufted and
the barrel-shaped larvae. The "giant larvae" may exceed them in size

by ten to a hundred times. In addition, contrary to the known fact
that only one larval form occurs in each species, A. spetsbergensis
shows different larval stages which apparently represent a developmental series.

Thus, there is no congruence with the larval forms hitherto
known. However, the A. spetsbergensis larvae closely resemble certain
enigmatic larvae of Actiniaria, found since Van Beneden (1898), in
different regions of the open ocean (Table 2). In total there are
about 300 larvae of this type recorded and - as I assume from experience - there may be some additional hundreds in the stores of
museums. Table 2 shows that these larvae were mostly collected from
considerable depths.

Table 2
Records of spherical pelagic actiniarian larvae

Author		Locality	Depth(m)	Number of larvae
Beneden	1898	Tropical Eastern Atlantic, Subarctic Northern Atlantic	0 - 600	15
Senna	1907	Gulf of Mexico, Indopacific, Indian Ocean, South Pacific off Chile	0 - 800	10
Bamford	1912	Indian Ocean	150 - 500	6
Gravier	1920	North Atlantic Ocean	0 - 4800	12
Carlgren	1924	North Atlantic Ocean, Indian Ocean, Gulf of Guinea, off South Africa, Subantarctic	200 - 2500	50
Dantan	1927	Mediterranean off Algeria	surface	6
Carlgren	1934	North Atlantic Ocean	150 - 3000	174
Widersten	1968	Mediterranean, Adriatic	0 - 100	4

The pelagic giant larvae are strikingly similar to the A. spetsbergensis giant larvae: the common features are their size, their
general shape and the presence of different developmental stages.
Both the origin and the fate of the pelagic giant larvae is totally
unknown up to now, and so far no adult anemone is known which can
be linked with these larvae. Anatomic investigations (mainly by
Carlgren, 1924, 1934) have demonstrated, however, that pelagic giant
larvae occur not just in one species, but even in different families
of the Actiniaria.

Thus, Actinostola spetsbergensis is the first actinian known as

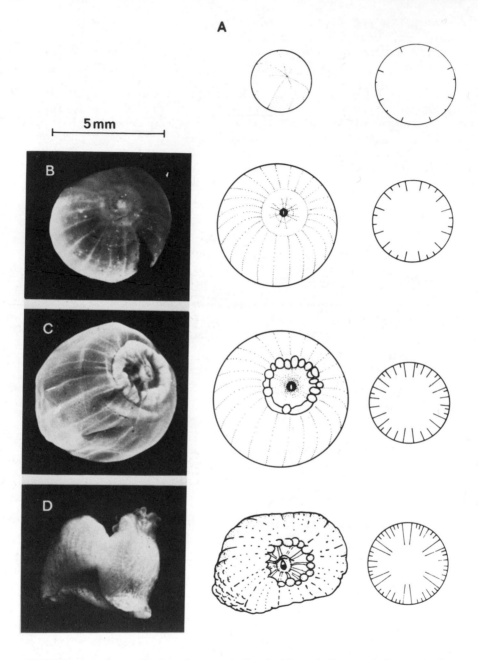

FIGURE 1. Larval development in <u>Actinostola spetsbergensis</u>.
A, smallest larval stage observed (8 plus 4 septa). B, 24 septa
stage showing cyclic bulge around the mouth opening. C, stage
with more than 24 septa during tentacle formation. D, first
anemone stage.

a producer of such spherical giant larvae. The Arctic species shelters
its offspring, and it was thus possible to follow the development from
the egg to the young anemone stage. There is only one observation
lacking: the release of the larvae or the young from the maternal
animal.

Besides my studies on the larval development I investigated
reproduction, oogenesis and early development of A. spetsbergensis
including some ecological aspects. These results will be published
elsewhere. For the understanding of larval development it is important to know that the early development takes place in the upper part
of the maternal animal; there, all stages from the fertilized egg up
to the 24 septa stage are found inside a peculiar embryonal envelope
by which they adhere to the endoderm of the female. In the bottom of
the gastral cavity only stages from 24 septa onwards are found; these
stages are always without embryonal envelopes. From these facts it
may be assumed that it is the 24 septa stage which breaks the envelope and hatches into the gastral cavity.

The smallest larva observed measures 2.5 mm in diameter and
possesses eight plus four septa, which are still rudimentary (Fig.1A).
The development of the larva proceeds with enlargement and additional
formation of septa, arising in the regular hexamerous way but remaining at a rudimentary stage up to the 24 septa stage. Somewhere
between the stages with 12 and 24 septa the small mouth opening and
the short actinopharynx is formed.

In contrast to the stages inside the embryonal envelope which
show only histological differentiations in the actinopharynx, the
hatched stages on the bottom of the gastral cavity possess all the
histological differentiations present in adult animals: musculature,
mesenteric filaments, gland cells and nematocyst cells. At a certain
distance from the mouth opening a cyclic bulge is formed (Fig.1B)
from which develops a double row of tentacles comprising at least
24 tentacles (Fig.1C). Such a simultaneous formation of a great
number of tentacles has never been observed during sexual reproduction in Actiniaria.

The described differentiations represent the first steps of
metamorphosis from the spherical larva into the young anemone. The
next steps are changes of body shape: the flattening of the aboral
pole leads to the pedal disc, and body wall contractions in horizontal
and vertical directions result in a stage smaller than the last
larval stage, but with all external features of a small Actinostola
spetsbergensis (Fig. 1D).

Discussion

Comparing the A. spetsbergensis larvae with the often recorded
pelagic larvae, it should be emphasized that we only know three

pelagic stages, those with 8, 12 and 24 septa (exceptionally one with
48 septa). All these free-living pelagic larvae differ in some respects from the Actinostola larvae by possessing histological differentiations even in the smallest stages. In A. spetsbergensis such
differentiations occur only after the larvae have left the embryonal
envelope; this suggests that there may be a certain inhibiting
influence due to the elaborate brood care in A. spetsbergensis. This
is confirmed by similar observations in Halcampa duodecimcirrata,
where the early development takes place also in a non-cellular envelope (Nyholm, 1949, p.489).

    I stated earlier that the spherical giant larvae differ not only
by their enormous size from the other actiniarian larvae, but also,
and above all, by presenting a developmental series of different
larval stages, while in all other cases so far known there is only
a single larval stage. With the exception of Protanthea all these
well-known larvae represent the Edwardsia-stage, a stage which is
not only characterized by the special arrangement of septa but also
by certain physiological achievements connected with metamorphosis
and settlement. Such a stage – regarded as typical for the development of the Actiniaria up to now – exists neither in the Actinostola
larvae nor in the pelagic giant larvae. The larval stage with 8 septa
has to be regarded as a transitional stage without any particular
physiological specializations. It is more likely in the giant larvae
that the stage with 24 septa may be analogous to the Edwardsia-stage,
because in A. spetsbergensis it is the first stage showing features
connected with metamorphosis and in the pelagic giant larvae it is the
last stage collected in the open ocean.

    Development via several pelagic stages was hitherto unknown
among Actiniaria. However, looking at the closest relatives we find
an astonishingly similar development in the Ceriantharia, especially
in Cerianthus lloydii (see Nyholm, 1943). Besides corresponding
morphological features there is a developmental series of several
larval stages in Cerianthus as well.

    The ecological significance of larval types and actiniarian dis-
persal types have already been the object of a discussion by Nyholm
in 1949 (p.467-470, p.501) who proposed four different types of dispersal in the Actiniaria. This scheme of Nyholm needs corrections as
to the number of dispersal types: the Peachia type has to be dropped,
because it is a special, unusual case of dispersal, and the dispersal
via medusa is not even obligatory (p.500). The Halcampa type has to
be omitted as well, because there is no fundamental difference between
it and the Tealia-type. Lastly, it is necessary to slightly alter
Nyholms Tealia-type (Urticina is a synonym of Tealia), because the
Tealia larva remains in the near-bottom water layer during pelagic
life. Thus, we can distinguish the following dispersal types (Fig.
2a, 2b):

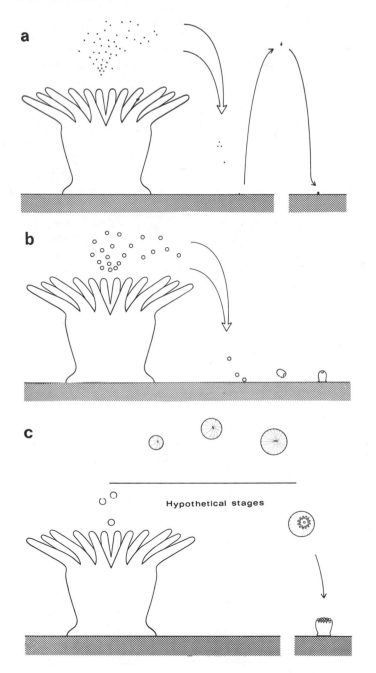

FIGURE 2. Actiniarian dispersal types. a, tufted larva type (Sagartia type of Nyholm); b, barrel-shaped larva type (Tealia type of Nyholm); c, giant larva type.

1. The tufted larva type of dispersal (Sagartia type of Nyholm).
Small eggs, emitted in large quantities, develop into small planktonic larvae with larval locomotion and sense organs. After a short
pelagic life the larvae approach the near-bottom layers and settle
down on suitable substrates. Ecological advantages of this type are
an effective dispersal together with some physiological ability to
respond to unsuitable substrates. Disadvantages include the relatively high loss of larvae during pelagic life and in the attempt to
settle (Thorson, 1951, p.287).

2. The barrel-shaped larva of dispersal (Tealia type of Nyholm).
Rather big eggs, rich in yolk, emitted in fewer numbers than the
former, becoming big, barrel-shaped larvae without special larval
organs. During a short pelagic life they remain in the near-bottom
layers (in turbulent currents they may also be found in plankton
samples, see Widersten, 1968). The advantages of the barrel-shaped
larvae are that they are usually not exposed to the dangers of pelagic life and that they remain in a suitable settling substrate near
the adult animals. However, the barrel-shaped larva has almost no
dispersal abilities.

To this scheme we can add now the third, novel type of dispersal
(Fig. 2c):

3. The giant larva type of dispersal. Very big eggs, rich in
yolk, emitted probably in small numbers and floating during early
development, giving rise to a line of successive lecithotrophic
larvae. The larval phase is probably rather long (up to some months)
and metamorphosis may begin after reaching a certain stage (probably the 24 septa stage). The distributional pattern of the pelagic
giant larvae, together with their absence in littoral and sublittoral
sea anemones, makes it evident that the adult actinians, which are
linked with the pelagic giant larvae are inhabitants of the deep-sea
bottoms. Lecithotrophic larvae with long pelagic life represent a
type of dispersal new for the Actiniaria, but rather common in the
other marine invertebrates (Thorson, 1951); in the Ceriantharia it
seems to be the principal type of dispersal.

In relating dispersal types to the habitats where the adult
animals are found, it is obvious that the first two types are possibilities for sublittoral and littoral actinians. In such hetero-
geneous and limited habitats the finding of patches for settling
has priority. Tufted larvae solve this problem by occuring in large
numbers in order to secure the settlement onto suitable substrates
by chance; slight corrections are possible because of the physiological abilities of the tufted larva. In the case of the barrel-
shaped larva there is an abandonment of true pelagic life and at the
same time of wide dispersal. Their near-bottom life (Mileikovski,
1971, p.196: demersal larva type) saves the larva from drifting away
from the favorable habitat. By contrast the new giant larva type of

dispersal achieves effective dispersal in extensive habitats with
stable ecological factors and constant bottom substrate. Such habitats are the deep-sea bottoms; the localities where the pelagic
giant larvae were collected confirm the assumption that they are the
larvae of deep-sea actinians. Thus, the giant larvae are another
example of lecithotrophic pelagic larvae in deep-sea invertebrates.
The existence of this larval type has earlier been denied for deep-
sea inhabitants (Mileikovski, 1971, p.197).

Summary

The giant larvae found in females of <u>Actinostola spetsbergensis</u>
resemble certain enigmatic larvae which have been found during the
last eighty years in different regions of the open ocean.

The larval development of <u>A. spetsbergensis</u> is described. In
contrast to all other actinarian larvae hitherto known, there exist
different, successive larval stages, and a simultaneous formation
of tentacles after the larvae reach the 24 septa stage. It is assumed
that the pelagic giant larvae collected in the open ocean represent
the same larval type with corresponding larval development.

The ecological significance of larval types and actiniarian
dispersal are discussed. A new scheme of dispersal types is given
including the "tufted" larva type, the"barrel-shaped" larva type
and the new giant larva type.

It is suggested that giant larvae are the dispersal stages of
deep-sea anemones, achieving wide dispersal in this almost stable
habitat.

References

BAMFORD, E., 1912. Pelagic Actiniarian larvae. <u>Trans</u>. <u>Linn</u>. <u>Soc</u>.
    <u>Lond</u>. Zool., 15: 395-406.
BENEDEN, E. van, 1898. Les Anthozoaires de la "Plankton-Expedition".
    <u>Erg</u>. <u>d</u>. <u>Plankton-Exp</u>., Kiel und Leipzig, 2: 1-222.
CARLGREN, O., 1893a. Studien über nordische Aktinien. <u>K</u>. <u>svenska</u>
    <u>VetenskAkad</u>. <u>Handl</u>., 25: 1-148.
CARLGREN, O., 1893b. Über das Vorkommen von Bruträumen bei Aktinien.
    <u>Övers</u>. <u>K</u>. <u>Vet.-Akad</u>. <u>Förh</u>., Stockholm, 50: 231-238.
CARLGREN, O., 1901. Die Brutpflege der Actiniarien. <u>Biol</u>. <u>Zbl</u>.,
    Leipzig, 21: 468-484.
CARLGREN, O., 1902. Die Actiniarien der Olga-Expedition. <u>Wiss</u>.
    <u>Meeresunters</u>., <u>Abt</u>. <u>Helgoland</u>, 5: 33-56.
CARLGREN, O., 1906. Die Aktinien-Larven. <u>Nord</u>. <u>Plankt</u>., Kiel,
    6: 65-89.
CARLGREN, O., 1913. Actiniaria. <u>Report Sec</u>. <u>Norwegian Exped</u>. "<u>Fram</u>"
    1898-1902, No.31: 1-6.
CARLGREN, O., 1921. Actiniaria Part I. <u>Dan</u>. <u>Ingolf-Exped</u>. 5: 1-241.

CARLGREN, O., 1924. Die Larven der Ceriantharien, Zoantharien und
        Aktiniarien, mit einem Anhang zu den Zoantharien. Wiss.
        Erg. "Valdivia", Jena, 19: 342-476.
CARLGREN, O., 1934. Ceriantharia, Zoantharia and Actiniaria. Report
        Sc. Results "Michael Sars" North Atlantic Deep-Sea Exped.
        V: 1-52.
DANTAN, J.-L., 1927. Remarques sur quelques larves pélagiques
        d'actinies. Assoc. Franc. Avanc. Sciences, 51 session,
        Paris, 249-250.
GRAVIER, C.J., 1920. Larves d'Actiniaires provenant des Campagnes
        scientifiques de Prince Albert I de Monaco. Rés. Camp. Sci.
        Monaco, 57: 1-25.
MILEIKOVSKI, S.A., 1971. Types of larval development in marine bottom
        invertebrates, their distribution and ecological signifi-
        cance: a re-evaluation. Marine Biology 10: 193-213
NYHOLM, K.G., 1943. Zur Entwicklung und Entwicklungsbiologie der
        Ceriantharien und Aktinien. Zool. Bidr.,Uppsala,22: 87-248.
NYHOLM, K.G., 1949. On the development and dispersal of athenaria
        actinia with special reference to Halcampa duodecimcirrata,
        M. Sars. Zool. Bidr. Uppsala, 27: 465-506.
RIEMANN-ZÜRNECK, K., 1969. Sagartia troglodytes (Anthozoa). Biologie
        und Morphologie einer schlickbewohnenden Aktinie. Veröff.
        Inst. Meeresforsch. Bremerh., 12: 169-230.
SENNA, A., 1907. Larve pelagichi di Attiniari. Pubbl. Ist. Studi sup.
        prat., Firenze, 1907: 79-198.
THORSON, G., 1952. Zur jetzigen Lage der marinen Bodentier-Ökologie.
        Verh. zool. Ges., Wilhelmshaven 1951: 276-327.
WIDERSTEN, B., 1968. On the morphology and development in some
        cnidarian larvae. Zool. Bidr., Uppsala, 37: 139-182.

STOLON VS. HYDRANTH DETERMINATION IN <u>PENNARIA TIARELLA</u> PLANULAE:

A ROLE FOR DNA SYNTHESIS

Georgia E. Lesh-Laurie

Department of Biology, Case Western Reserve University

Cleveland, Ohio 44106

Larvae of the gymnoblastic hydroid <u>Pennaria tiarella</u> develop
and metamorphose in a synchronous and predictable manner. They
progress from ovoid pre-planulae 12 hours after fertilization to
definitive, filiform planulae by 24 hours of development. Planulae
are free living for a defined period of time (approximately 5 days
at $21^{\circ}C$) and then metamorphose with the anterior regions of larvae
becoming stolons, the posterior regions forming hydranths. The
uniformity of this developmental pattern allows the utilization of
<u>Pennaria</u> as an experimental system to investigate the principal
morphogenetic determination occurring in hydroid larvae, that of
stolon versus hydranth.

Previous autoradiographic studies (Haynes <u>et al</u>., 1974)
revealed regional patterns of ^{3}H-thymidine incorporation during
larval development in <u>Pennaria</u>. Early planulae exhibited concentrations of precursor incorporation in the epidermis and posterior
portions of the larvae. In later planulae, however, the greatest
level of incorporation was reported in the gastrodermis and at the
anterior regions of the larvae. Because the autoradiographic
observations are consistent with a possible involvement of DNA
synthesis and/or mitotic activity in the regional commitment to
stolon and hydranth development that occurs in planula larvae,
attention was focused on the role of DNA synthesis in this morphogenetic determination.

Initially, the kinetics of ^{3}H-thymidine incorporation into
DNA in <u>P. tiarella</u> had to be examined to verify the efficacy of
this probe to analyze synthetic activity in this organism. Once
linearity of incorporation was established, morphological and biochemical events of larval development were compared in unmanipulated

365

planulae, half planulae and planulae exposed to an inhibitor of DNA
synthesis (hydroxyurea) or a mitotic blocking agent (colchicine).

MATERIALS AND METHODS

Sexually mature colonies of <u>Pennaria</u> were collected at Flatts
Islet, Bermuda. After collection all colonies and subsequent larvae were maintained at 21±1°C in Millipore-strep-sea water (MSSW).
MSSW is Bermuda Biological Station sea water that has been passed
through both a glass fiber filter and Millipore filter (0.45 μ pore
size) and to which streptomycin sulfate (10 μg/ml) and sulfadiazine
(10 μg/ml) have been added. To stimulate and synchronize spawning
8 inch fingerbowls containing 1 male and 5 female fronds were placed
in light tight containers. Cleaving embryos were harvested 5 hours
later, washed to remove excess sperm and placed in fresh MSSW.
Developing larvae were transferred to fresh MSSW daily.

Surgical Manipulation

To evaluate the capacity of half planulae to incorporate ^{3}H-
thymidine into DNA and to complete larval development, planulae
were sectioned transversely at 12 hour intervals from 12-96 hours
of development. The cut was made so that anterior and posterior
halves each contained similar amounts of tissue, as measured by
protein content. Anterior and posterior half planulae were maintained separately in MSSW which was changed daily.

Labeling Procedures

Incorporation of ^{3}H-thymidine (methyl-labeled, specific activity 6.7 Ci/mM, New England Nuclear) was selected as a marker for
DNA synthesis. Larvae were exposed to specific concentrations of
^{3}H-thymidine as a component of the culture solution for a defined
time interval. After removal from the radioactive solution, planulae were washed 3x over a 5 min period in non-radioactive thymidine
(400 μM) and fixed in fresh Carnoy's fixative.

DNA analysis was performed according to the method of Schneider
(1945). Fixed planulae were uniformly homogenized in 5% ice cold
trichloroacetic acid (TCA) and refrigerated at least 4 hours at 2°C.
The TCA insoluble material was washed twice with ice cold 5% TCA,
extracted once with 95% ethanol and once with 95% ethanol:ether
(3:1). The nucleic acids were hydrolyzed in 5% TCA for 20 min at
90°C. After cooling to 4°C the remaining macromolecular components
were removed by centrifugation and the supernatant divided into 2
aliquots, one for DNA determination and one for measurement of

radioactivity. Although the Schneider method does not separate DNA
from RNA, evidence that, in hydroids, the labeled material isolated
by this procedure is DNA has been presented by Lesh-Laurie et al.
(1976). Using DNase digestion, it was demonstrated that >90% of
the labeled material was enzyme sensitive.

Concentrations were determined using the Burton (1968) modification of the diphenylamine reaction. Standard curves were plotted
using calf thymus DNA (Type I sodium salt highly polymerized, Sigma)
and were linear to concentrations as low as 2.0 μg/ml. Liquid scintillation counting was executed in Bray's solution (Bray, 1960).
Although data from only a single experiment will be presented each
biochemical experiment was performed at least once during each of
2 summers. If the results from these experiments were equivocal,
additional experiments were executed with the data presented representing neither the high nor the low extreme.

Inhibition Studies

Clarification of any relationship existing between DNA synthesis and/or mitotic activity and stolon or hydranth determination in
P. tiarella planulae requires observation of the latter processes
in the absence of DNA synthesis and/or mitosis. Hydroxyurea (HU)
was selected as an inhibitory agent for DNA synthesis. Hydroxyurea
reportedly interferes with the reduction of ribonucleotides to deoxy-
ribonucleotides (Young and Hodas, 1964). This effect is temporary
and is overcome when the agent is removed. In addition HU has been
effective in inhibiting DNA synthesis in hydroids (Clarkson, 1969;
Lesh-Laurie et al., 1976).

Planulae were pulsed with 1 mg/ml HU for 96 hours at 12 hour
intervals from 12-96 hours of development. Following exposure to
the inhibitory compound larvae were washed several times with MSSW
and allowed to continue their development in MSSW. The 1 mg/ml concentration was chosen following preliminary biochemical investigations. (see Results) which revealed it to be a dosage capable of
severely restricting DNA synthesis, yet allowing 100% survival.

Colchicine was selected as a mitotic inhibitor. It binds re-
versibly to microtubular protein (Borisy and Taylor, 1967a, 1967b),
interfering with mitosis through its effect on spindle microtubules.
In addition, it reportedly has been employed successfully to induce
mitotic arrest in hydroid cells (Ham and Eakin, 1958; Corff and
Burnett, 1969). Larvae were pulsed with 0.2 mg/ml colchicine for
24 hours at 12 hour intervals from 12-96 hours of development.
After exposure to the alkaloid, planulae were washed several times
in MSSW and allowed to continue their development in MSSW. This
effective dosage was determined following preliminary experiments

which revealed it to be one capable of arresting mitosis, and ultimately restricting ^{3}H-thymidine incorporation into DNA.

RESULTS

Kinetics of ^{3}H-thymidine Incorporation

To establish the kinetics of ^{3}H-thymidine incorporation in
Pennaria planulae groups of 20 12 hour and 24 hour planulae were
incubated in 10 µCi/ml ^{3}H-thymidine for 10 min intervals from 10-90
min. Specific activities (cpm/µg DNA) were computed for each 10 min
interval and a representative experiment for each developmental
stage is plotted graphically in Figure 1. Regression lines were calculated and at both developmental stages the relationship between exposure time and specific activity was linear with the x and y inter-
cepts not significantly different from zero. Therefore, Pennaria
planulae incorporate ^{3}H-thymidine from the culture medium at a constant rate over a 90 min period.

To determine whether the rate of ^{3}H-thymidine incorporation
varied with the concentration of externally supplied precursor,
20 planulae were incubated 1 hour in varying concentrations of a
[3]H-thymidine solution of constant specific activity. The results
from one experiment are shown in Figure 2, with the calculated regression line. Although the rate of ^{3}H-thymidine incorporation was
a linear function of external precursor concentration over the range
of concentrations examined, the y intercept was not zero. This departure from the predicted result is probably correlated with the
mechanism of initial thymidine uptake by the organism; however, no
data are presently available that address this problem.

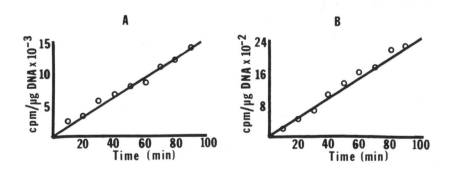

Figure 1. Kinetics of ^{3}H-thymidine incorporation into DNA.
(A). Incorporation into 12 hour planulae, (B). Incorporation
into 24 hour planulae.

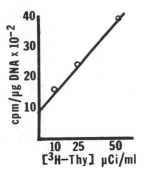

Figure 2. Influence of $[^3\text{H-thy}]$
on its incorporation into DNA in 24
hour planulae.

To confirm the autoradiographic data which suggested considerable and regional incorporation of ^{3}H-thymidine into DNA during larval development in <u>Pennaria</u>, the pattern of ^{3}H-thymidine incorporation was examined at 12 hour intervals from the formation of the
definitive planula (24 hours of development) until the initiation
of metamorphosis (96 hours of development). Embryos from a single
spawn were pulsed in groups of 50 with 50 µCi/ml ^{3}H-thymidine for
1 hour immediately preceeding each 12 hour interval of development.
After the exposure interval the larvae were washed in non-radioactive thymidine, fixed immediately in Carnoy's fixative and their DNA
subsequently isolated. The resulting pattern of incorporation is
shown in Figure 3, where it may be compared with data obtained from
half planulae.

Normal, unmanipulated larvae always exhibited a bimodal pattern
of ^{3}H-thymidine incorporation into DNA. Incorporation was highest
early in larval development (24 hours) and then dropped precipitous-
ly over the next 2 days. A second but smaller burst of incorporation
occurred late in larval life (84 hours) immediately preceding metamorphosis. The relative height of the latter elevation in incorporation varied from experiment to experiment but it was always < ½ the
incorporation level observed at 24 hours of development.

Organizational and Synthetic Capacities of Half Planulae

To qualitatively determine whether the regional pattern of ^{3}H-
thymidine incorporation observed autoradiographically was reflected
in the developmental potential of specific portions of the larva,
planulae were sectioned transversely at 12 hour intervals from 12-
84 hours of development in the manner described previously. From
the data (Table I) it was evident that the anterior halves of

planulae from any developmental stage possessed the ability to re-
organize entire planulae, capable of undergoing a normal, but temporally delayed, metamorphosis. Alternatively, posterior half planulae were incapable of reconstituting metamorphosing individuals and
all eventually succumbed. Both types of half planulae remained fully
motile throughout the experiment and evidenced no sloughing of cells.

If one examined the capacity of half planulae to incorporate
^{3}H-thymidine into DNA, a picture similar to the morphological results
emerged (Fig. 3). To facilitate both developmental synchrony and
accuracy in sectioning the number of animals required to perform the
biochemical experiment, a single developmental stage, the 24 hour
planula, was selected for this analysis. Three hundred 24 hour
planulae were sectioned transversely and allowed to heal. Fifty
anterior and 50 posterior half animals were then exposed to 50μCi/ml
^{3}H-thymidine for 1 hour, fixed in Carnoy's fixative and their DNA
extracted at each 12 hour interval from 36-96 hours of development.
Specific activity values (Fig. 3) revealed that the anterior halves
of 24 hour planulae maintained the capacity to synthesize DNA and
exhibited a pattern of synthesis virtually identical to unmanipulated
planulae within 24 hours of being excised (48 hours of development).
The posterior halves of 24 hour planulae however, continued to synthesize significant amounts of DNA for 12 hours following amputation,
but then their capacity to incorporate ^{3}H-thymidine into DNA declined
until barely perceptible amounts of incorporation were detected 3
days after isolation (96 hours of development).

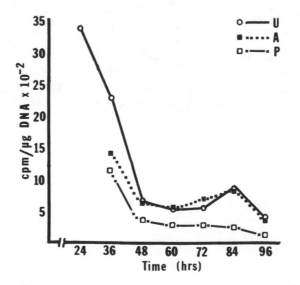

Figure 3. ^{3}H-thymidine uptake during larval development in unmanipulated planulae, anterior and posterior half planulae.

TABLE I

Survival of half planulae

Developmental stage	No. animals	No. animals completing metamorphosis within 2 weeks
Unmanipulated	20	20
12 hour anterior	22	20
12 hour posterior	25	0
24 hour anterior	25	22
24 hour posterior	25	0
36 hour anterior	10	6
36 hour posterior	10	1
48 hour anterior	10	3
48 hour posterior	10	0
60 hour anterior	10	7
60 hour posterior	10	0
72 hour anterior	10	7
72 hour posterior	10	0
84 hour anterior	7	5
84 hour posterior	7	0

^{3}H-thymidine Incorporation and Developmental Potential
in Planulae Exposed to HU or Colchicine

It is apparent therefore, from previous results, that despite the occurrence of regional stolon and hydranth determination and a regional pattern of ^{3}H-thymidine incorporation during larval development in P. tiarella, DNA synthesis is not sufficient to allow the expression of stolon or hydranth morphogenesis. If it were, posterior half planulae would predictably be able to form some definitive hydranth parts. The question of whether DNA synthetic activity is even necessary for these morphogenetic events can be approached by examining the potential for stolon and/or hydranth determination when DNA synthesis is severely restricted.

The ability of planulae exposed to 1 mg/ml HU to incorporate ^{3}H-thymidine into DNA fell quickly and markedly, so that by 4 hours of incubation in the inhibitor the level of incorporation into 12 hour planulae had dropped to <1/9 of that observed in unmanipulated larvae (Fig. 4). Longer incubation periods caused no significant further reduction in incorporation. The developmental potential of HU-treated planulae exhibited a clear and distinct limitation;

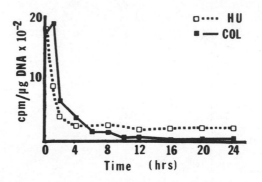

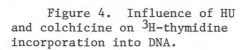

Figure 4.   Influence of HU
and colchicine on ^{3}H-thymidine
incorporation into DNA.

TABLE II

Developmental potential of planulae exposed to HU
n = 20

Developmental stage when placed in HU	Developmental potential expressed 10 days after removal from HU	
	Stolon and upright, no hydranth	Normal metamorphosis
12 hours	16	0
24 hours	20	0
36 hours	13	0
60 hours	10	0
72 hours	16	0
84 hours	11	0

hydranths failed to form (Table II). Groups of 20 planulae incubated 4 days in 1 mg/ml HU at developmental stages from 12-84 hours
of development revealed a complete absence of hydranth formation 10
days after removal from HU. These planulae attached normally to the
substratum, elaborated stolons and uprights, yet definitive hydranths
did not develop. On occasion abortive blisters appeared at the apex
of the upright, but neither tentacles nor hydranth morphology ever
became evident.

     Because numerous hydroid cell types reportedly possessed long
and variable G_2 periods (David and Campbell, 1972; Campbell and
David, 1974) it was possible that DNA synthesis and mitosis represented distinctly separable events in Pennaria planulae. Therefore,
stolon and hydranth determination were also investigated in the presence of the mitotic inhibitor, colchicine. Interestingly, colchicine
proved to be a more potent, although slower, inhibitor of ^{3}H-thymi-
dine incorporation into DNA than HU (Fig. 4). Twelve hour planulae
placed in 0.2 mg/ml colchicine maintained normal DNA specific activity levels for at least 1 hour; after that time the level of ^{3}H-thymi-
dine incorporation dropped to <1/20 the value observed in unmanipulated larvae.

     The morphological manifestation of the colchicine response was
also particularly significant. Several planulae pulsed with colchicine for 24 hours during periods of elevated ^{3}H-thymidine incorporation into DNA (i.e. early or late larvae) formed hydranths without
attaching to the substratum or exhibiting demonstrable evidence of
stolon growth (Table III).

TABLE III

Developmental potential of planulae exposed to colchicine
n = 15

Developmental stage when placed in colchicine	Developmental potential expressed 10 days after removal from colchicine	
	Normal metamor-phosis	Normal hydranth for-mation, no attachment or stolon development
12 hours	4	5
24 hours	2	2
36 hours	0	0
48 hours	0	1
60 hours	2	0
72 hours	0	1
84 hours	4	1
96 hours	1	2

DISCUSSION

It is evident therefore, that Pennaria planulae exhibit both
a regional stolon and hydranth determination and regional patterns
of DNA synthesis, as measured by ^{3}H-thymidine incorporation into
DNA. Although DNA synthesis does not appear to be a sufficient
process for morphogenetic expression, it is clearly a necessary one,
as evidenced by the modifications in the normal elaboration of developmental potential observed in animals in which DNA synthesis was
restricted. The morphogenetic significance of DNA synthesis in stolon versus hydranth determination in Pennaria larvae cannot be fairly
assessed at this time, however. Autoradiographic and cell population
studies corresponding to the biochemical experimentation are necessary to provide a cellular basis for the synthetic data. These
studies are in progress and will be presented in subsequent papers.

ACKNOWLEDGMENTS

The author acknowledges support from National Science Foundation Grant No. GM 34211 and a Brown-Hazen Grant from Research Corporation. This paper represents Contribution No. 663 from the Bermuda Biological Station for Research.

REFERENCES

Borisy, G.G., and E.N. Taylor, 1967a. The mechanism of action of
colchicine. Binding of colchicine-^{3}H to cellular protein.
J. Cell Biol., 34:525-533.

Borisy, G.G., and E.N. Taylor, 1967b. The mechanism of action of
colchicine. Colchicine binding to sea urchin eggs and the
mitotic apparatus. J. Cell Biol., 34:535-554.

Bray, G.A., 1960. A single efficient liquid scintillator for counting aqueous solutions in a liquid scintillation counter. Anal.
Biochem. 1:279-285.

Burton, K., 1968. Determination of DNA concentration with diphenylamine. Pages 163-166 in L. Grossman and K. Moldave, Eds.,
Methods in Enzymology, Volume 12. New York, Academic Press.

Campbell, R.D., and C.N. David, 1974. Cell cycle kinetics and development of Hydra attenuata. II. Interstitial cells. J. Cell
Sci., 16:349-358.

Clarkson, S.G., 1969. Nucleic acid and protein synthesis and pattern regulation in Hydra. II. Effect of inhibition of nucleic acid and protein synthesis on hypostome formation. J. Embryol. Exp. Morphol., 21:55-70.

Corff, S.C., and A.L. Burnett, 1969. Morphogenesis in hydra. I. Peduncle and basal disc formation at the distal end of regenerating hydra after exposure to colchicine. J. Embryol. Exp. Morphol., 21:417-443.

David, C.N., and R.D. Campbell, 1972. Cell cycle kinetics and development of Hydra attenuata. I. Epithelial cells. J. Cell Sci., 11:557-568.

Ham, R.G., and R.E. Eakin, 1958. Time sequence of certain physiological events during regeneration in hydra. J. Exp. Zool., 139: 35-54.

Haynes, J.F., R.G. Summers and M.I. Kessler, 1974. The application of autoradiography to developmental studies in the Hydrozoa. Amer. Zool., 14:783-795.

Lesh-Laurie, G.E., D.C. Brooks and E.R. Kaplan, 1976. Biosynthetic events of hydrid regeneration. I. The role of DNA synthesis during tentacle elaboration. Wilhelm Roux' Archiv., in press.

Schneider, W.C., 1945. Phosphorus compounds in animal tissues. I. Extraction and estimation of deoxypentose nucleic acid and of pentose nucleic acid. J. Biol. Chem., 161:293-303.

Young, C.W., and S. Hodas, 1964. Hydroxyurea: Inhibitory effect on DNA metabolism. Science, 146:1172-1174.

SEM Studies of Strobilating Aurelia

Dorothy B. Spangenberg and William Kuenning

University of Colorado
Department of Molecular, Cellular & Developmental Biology
Boulder,Colorado 80309

INTRODUCTION

Strobilation in Texas Aurelia provides a rapid, controlled developmental model for the study of morphogenetic changes in biological organisms. The Texas Aurelia polyps (scyphistomae) are induced to strobilate using thyroxine which initiates segmentation in
the organisms in 2-3 days. The segments metamorphose into ephyrae
in 3 days.

Basic morphological changes associated with strobilation of
Aurelia were described in a review (Spangenberg, 1968). The availability of the Scanning Electron Microscope (SEM) provided opportu-
nities to investigate morphological changes in strobilating Aurelia
in greater detail than is possible with the light microscope. The
use of stereoscopy with the SEM provides greater than life images
of the jellyfish.

METHODS

The Texas Aurelia (Spangenberg, 1965) are induced to strobilate using 10^{-5}M thyroxine (Spangenberg, 1971) at 30°C. The jellyfish are prepared for SEM studies at different stages of development.
SEM Preparation: The jellyfish are anesthetized with 1% $MnCl_2$.
The following steps are performed at 7°C using freshly prepared solutions. The specimens are fixed for 45 min. in 6% glutaraldehyde
in .2M s-collidine buffer with 8.6% sucrose and a trace of $CaCl_2$.
They are rinsed 2X with buffer, retained in buffer for 20 min., and
rinsed again. The specimens are placed in .2% OsO_4 in collidine
buffer for 5 min. and rinsed 2X again with buffer. 30% acetone is

377

added to the buffered specimens (1:1 ratio). Then 100% acetone is
added to bring the acetone concentration to 95% in 45 minutes. The
specimens are rinsed 2X in 100% acetone, soaked in 100% acetone for
10 minutes, and are critical point dried. The specimens are then
removed, mounted on nickel screen (500 mesh) with silver paint,
coated with about 600 Å gold palladium, and viewed with a Cambridge
SEM.

RESULTS

Jellyfish in various stages of strobilation (a stage showing at
least one new feature or marked growth of a feature) were chosen for
study. In this manner, the major phases of strobilation were arbitrarily divided into sequential stages which together exemplify the
entire process of strobilation:

Stage I (Fig. 1, 2A) is the normal non-strobilating polyp. Of
special interest is the fact that the tentacles appear identical and
arise from the body at equal planes forming a circlet around the oral
disc. The round contractile mouth crowns a cone-shaped hypostome.
Four peristomial pits mark the position of the internally located
gastral septa.

Stage II strobila (Fig. 2B, 3A) has a completed segment at the
distal end. A constriction divides the swollen portion of the lower
part of the perradial and interradial tentacles from the upper portion of the tentacles. The mouth may be round or quadrangular.

Stage IIIa strobila (Fig. 3B) shows that tentacle reduction has
begun in the perradial tentacles. Rhopalia formation is occurring
at the base of the perradial and interradial tentacles.

Stage IIIb strobila (Fig. 4A) has completely lost all of its
perradial tentacles but still has its interradial and adradial tentacles. The rhopalia are more elongate than in the earlier Stage
III and the large mouth is quadrangular. Bifurcation of the arms
has begun.

Stage IV strobila (Fig. 4B) has lost its perradial tentacles
and one interradial tentacle (arrow) is almost gone. Rhopalia are
longer than in the previous stage and lappet formation has begun.

Stage V strobila (Fig. 5A) has no tentacles and exemplifies
more advanced lappet growth than is seen previously. The mouth is
growing out to form a manubrium. An indentation of the arms on the
exumbrella side of the organism has begun along with lappet growth.

Stage VI strobila (Fig. 5B) shows prominent lappet growth and
increased indentation of the arms on the exumbrella side of the
ephyra. The mouth has the typical configuration of the ephrya manubrium.

Stage VII ephyra (Fig. 6A) is almost completed while still attached to the strobila. The arms and lappets have grown in length
and have thinned. The coronal and radial muscles are quite distinct.
The mouth is a fully developed manubrium and the rhopalia have assumed their normal position between the lappets. The peristomial
pits are located closer to the mouth than in the polyp.

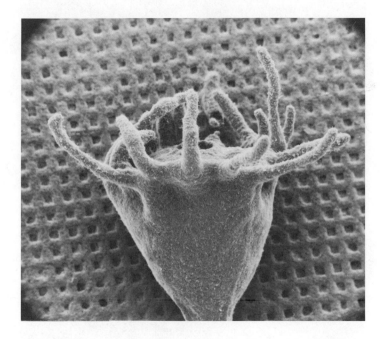

Fig. 1. Stage I. Polyp. Note tentacle morphology. 89x.

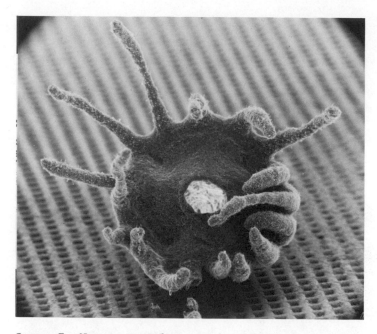

Fig. 2A. Stage I. Note tentacles, peristomial pits, mouth. 89x.

Fig. 2B. Stage II. Strobila. Note bulges of alternate tentacles. 162x.

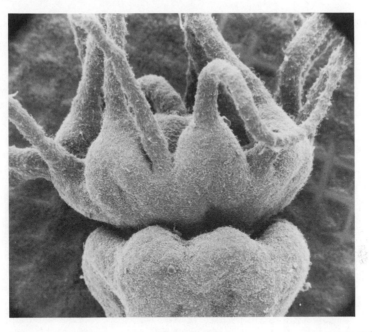

Fig. 3A. Stage II. Note tentacle swellings, segmentation. 161x.

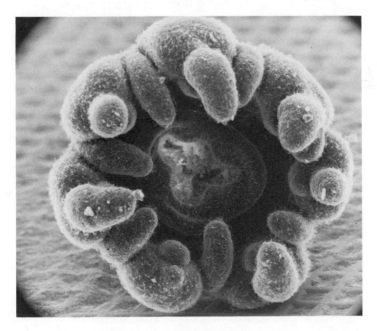

Fig. 3B. Stage IIIa. Strobila. Note rhopalia formation, mouth. 178x.

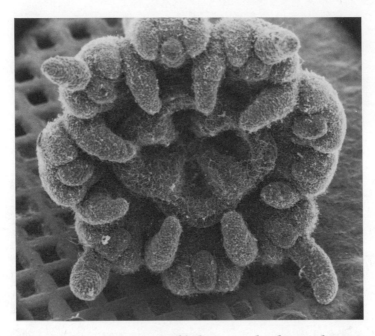

Fig. 4A. Stage IIIb. Note perradial tentacle loss, lappets. 178x.

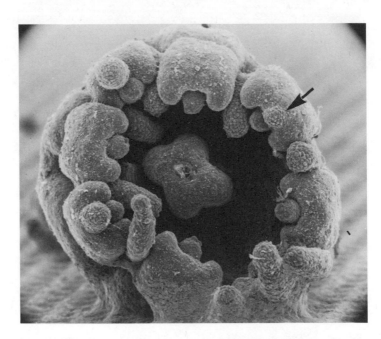

Fig. 4B. Stage IV. Strobila. Note interradial tentacle loss. 178x.

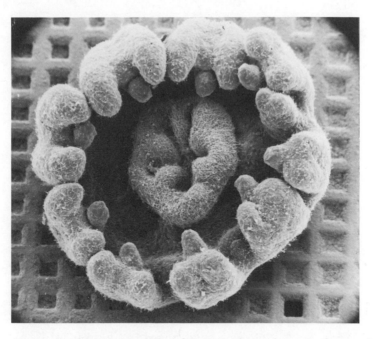

Fig. 5A. Stage V. Note absence of tentacles, lappet formation. 161x.

Fig. 5B. Stage VI. Strobila. Note lappet growth, mouth shape. 122x.

Fig. 6A. Stage VII. Ephyra. Note muscles, peristomial pits. 106x.

Fig. 6B. Stage VIII. Free ephyra. Note loss of peristomial pits. 53x.

Stage VIII (Fig. 6B) is a free ephyra which is very much like the Stage VII organisms although the arms and lappets appear somewhat longer. Of special interest is the loss of the peristomial pits which are seen in the earlier stages.

DISCUSSION

Few jellyfish have been viewed with the SEM thus far in spite of their suitability for SEM studies. In addition to the Aurelia work, Blanquet and Wetzel (1975) illustrated the surface morphology of Chrysaora with the SEM.

The metamorphic changes in the Aurelia are most dramatically illustrated by the SEM technic especially when viewed stereoscopically. Early in strobilation, a swelling occurs at the base of the perradial and interradial tentacles and a constriction appears above the swelling. Chuin (1930) suggested that the cells from the tentacles are resorbed into the swellings to give rise to lappets. This is clearly not the case in the Aurelia since the swellings appear prior to tentacle reduction. Rhopalia growth also occurs prior to tentacle loss in the Texas Aurelia as reported in Cassiopea by Bigelow (1898). Tentacles are lost in a sequential manner whereby those in the perradius disappear first, followed by the interradial and then the adradial tentacles. Sugiura (1963) described this sequence of tentacle loss for Mastigias papua. Not all tentacles in each radius disappear at the same rate, however. It is especially interesting that the peristomial pits are present in all developmental stages except the released ephyra. We believe that these pits are formed by the attachment of the columellae (modified gastral septa) to the oral disc. When the ephyra severs this connection to break free from the strobila, the pits disappear. Radial and coronal muscles likewise are especially distinct in ephyrae prepared with the SEM technic.

The authors thank Mr. Bob McGrew, Mr. George Wray and Mr. Richard Carter for their assistance with the SEM usage and stereoslide preparation. This research is funded by Grants #De03796-04 and #HD07844-03.

SUMMARY

SEM was applied to the demonstration of sequential changes in strobilating Aurelia following thyroxine treatment. Special emphasis is placed on the sequential phases of tentacle loss, the metamorphosis of the polyp mouth into the manubrium of the ephyra, and the growth stages of the arm, rhopalium and lappets.

The special technics described in this paper permit the viewing of whole intact organisms. The information obtained from these

studies provides a base line for future studies of organisms strobilating under various experimental conditions. The SEM method
may also be applied to organisms from various normal and abnormal
ecological settings and to identification of organisms, particularly
polyps, collected from nature.

BIBLIOGRAPHY

Bigelow, R.P. 1898. The Anatomy and Development of Cassiopea
xamachana. Mem. Boston Soc. Nat. Hist. 5:191-236.
Blanquet, R.S. and B. Wetzel. 1975. Surface ultrastructure of the
scyphopolyp, Chrysaora quinquecirrha. Biol. Bull. 148:181-192.
Chuin, T.T. 1930. Le cycle evolutif du scyphistome de Chrysaora.
Trav. Stn. Biol. Roscoff. 8:1-179.
Spangenberg, D.B. 1968. Recent studies of strobilation in jelly-
fish. Oceanogr. Mar. Biol. Ann. Rev. 6:231-247.
Spangenberg, D.B. 1971. Thyroxine induced metamorphosis in Aurelia.
J. exp. Zool. 178:183-194.
Spangenberg, D.B. and W. Kuenning. 1975. Technic for the study of
small soft-bodied Aurelia with Scanning EM. "33rd Ann. Proc. Elec-
tron Microsc. Soc. Amer." Las Vegas, Nevada. pp. 682-683.
Sugiura, Y. 1963. On the life-history of rhizostome medusae. I.
Mastigias papua L. Agassiz. Annot. Zool. Jap. 36:194-202.

Associations

THE QUANTITATIVE ANALYSIS OF THE HYDRA-HYDRAMOEBA HOST-PARASITE SYSTEM[1]

Alan E. Stiven

Department of Zoology, University of North Carolina

Chapel Hill, N. C. 27514

INTRODUCTION

Several species of the freshwater hydras have been reported infected with the protozoan amoeba parasite Hydramoeba hydroxena Entz. Entz (1912) and Reynolds and Looper (1928) found the greatest level of infection in late summer through December. The latter author suggested the hydramoeba posed a "serious threat" to the hydra population. Epidemic levels of the infection have also been reported by Welch and Loomis (1924) and Miller (1936) in Douglas Lake, Michigan, and by Bryden (1952) in his work on Hydra oligactis in Tennessee. All reported highest levels of the parasite in the autumn, considerable deterioration of the hydra populations by the amoeba, and even complete disappearance of one hydra species due to the infection (Bryden, 1952). Further field studies of this host-parasite system are generally lacking.

Laboratory research on hydra susceptibility and hydramoeba pathogenicity was almost equally deficient. The parasite was first described by Entz (1912). Early cytological observations were provided by Wermel (1925). It was described again by Hegner and Taliaferro (1924) near Baltimore, and in Virginia by Reynolds and Looper (1928). Its cosmopolitan distribution was confirmed by Ito (1949, 1950) with his report of its occurrence in Japan.

Hydramoeba hydroxena is a large, variable shaped amoeba containing a spherical nucleus of about 11-15 μ in diameter (see Beers,

---

[1] National Science Foundation support for this research is gratefully acknowledged.

1964, for an excellent description of the nucleus). It appears most infective in the proteus and verrucosa forms although its general infectivity has varied depending upon the author. Entz (1912) suggested that it had little virulence for hydra while Reynolds and Looper (1928) found it very infective depending on the condition of the culture medium.

Our own work began in the early 1960's in an attempt to understand the ecological and population processes that lead to the epizootic state of the hydramoeba infection observed in the field, and to shed some light on the problems of hydra species susceptibility, parasite pathogenicity, and the response of this system to environmental changes.

HYDRAMOEBA PATHOGENICITY AND HYDRA SUSCEPTIBILITY

In the hydra–hydramoeba infection process the sticky proteus-form amoeba usually first become attached to the hydra stalk or tentacles. The parasite population then increases rapidly until the tentacles disappear, and the remaining stalk is covered by amoeba. At this point, death (disintegration) usually follows within two days. (See Fig. 1 which depicts the infection sequence).

The pathogenicity of hydramoeba and the susceptibility or resistance of the hydra host are closely linked. Both can be measured in terms of (a) the rate of population increase of the amoeba on a single host, (b) the time of death of the host as a result of the infection, or (c) the percentage mortality of an infected population of hosts after a given time period. Table 1 summarizes some of these characteristics for several species of host. Clearly one of the largest of the hydras, H. littoralis, is practically immune to the attack of the hydramoebae, while H. pseudoligactis is quite susceptible. Our work has also shown that the symbiotic algae in Chlorohydra irridissima (= Hydra viridis) contribute to this species' level of resistance, with the albino form significantly more susceptible than normal green hydra (Stiven, 1965).

One should, of course, use caution in comparing results of hydra species susceptibility tests among different studies, since more than one "strain" of Hydramoeba hydroxena probably exists.

HYDRA–HYDRAMOEBA POPULATION PROCESSES

The hydra–hydramoeba system has many features that make it ideal as a laboratory experimental unit for the study of host-parasite population phenomena, particularly ecological processes involved in disease outbreaks. Both host and parasite can easily be

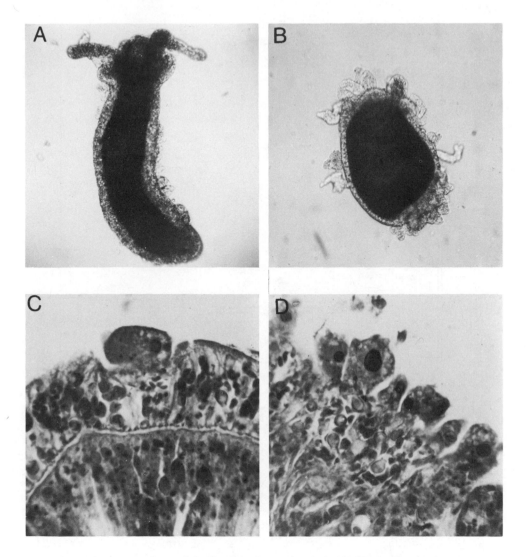

Fig. 1. Photographs of hydra infected with Hydramoeba hydroxena.
(A) As the amoeba population increases on the hydra, the tentacles
retract and are consumed; the amoebae eventually are distributed over
the hydra stalk. (B) During the latter stage of infection, the host
decreases in size and eventually disintegrates releasing many amoebae.
Sections of Hydramoeba and H. pseudoligactis. (C) Amoeba pseudopods
extend down through epidermis causing cellular disintegration. X288.
(D) Heavily infected hydra just prior to disintegration, cells in
disarray. X288. Adapted from Stiven (1973) with permission of
publisher.

Table 1. Comparative levels of resistance to _Hydramoeba hydroxena_ among the more common North American hydrozoan species. Initial infection of 1-2 hydramoeba/host for that species.[1]

Species	% Mortality (number Individuals)	Mean Day of Death ± 1S.E.	Mean Rate of Hydramoeba Increase ± 1S.E.	Species Size Dry Weight/ Bud (µg)	Comment
H. pseudoligactis	100 (10)	6.30 ± 0.153	0.957 ± 0.076	15	25°C, Stiven (1965)
H. viridis (Normal)	72 (18)	9.77 ± 0.902	0.702 ± 0.030	8	25°C, Stiven (1965)
H. viridis (Albino)	100 (18)	7.00 ± 0.385	0.778 ± 0.040	6	25°C, Stiven (1965)
H. littoralis	0 (10)	0	0.059 ± 0.025	22	25°C, Stiven (1965, 1971)
H. oligactis	92 (48)	6.41 ± 0.544	–	–	Reynolds and Looper (1928)
H. cauliculata	relatively resistant, no quantitative data				Rice (1960)
Craspedocusta sowerbii (medusa)	susceptible, dies within 6 days				Rice (1960)
C. sowerbii (polyp)	completely resistant				Rice (1960)

[1]From Stiven (1973) with permission of the publisher.

raised in large numbers (Stiven, 1962), the latent period of infection can be as little as a day or two, the state of host infection
is readily detected since the amoebae are easily visible, and
several host species with varying levels of susceptibility exist
which allow construction of host "communities" containing different
proportions of the various species.

Our approach to the quantitative analysis of this epizootic
system has involved the compartmentalization of the process into the
basic components leading to an epizootic. These include for the
host such components as resistance, population growth, density,
available space, and mortality. For the parasite we must consider
such facets as inoculation size, pathogenicity and infectivity, and
survivorship. The outcome of the interaction of host and parasite
can, of course, be greatly influenced by the environmental conditions of the aquatic medium, such as temperature, ion concentration,
etc.

The Infection Process

As discussed earlier (Fig. 1) a vulnerable hydra infected with
amoebae gradually deteriorates until death ensures. Acquisition of
the first hydramoeba by the host can be largely by random encounter
with free floating amoebae or by direct movement of amoebae onto the
surface of the hydra. The latter process is apparently nonrandom
with the host actually "attracting" amoebae (Stiven, 1964). The
growth form of the hydramoebae population on an individual hydra
depends upon the initial number of amoebae present, the nutritional
state of the host, and the species of host. Fig. 2 depicts a
comparison of typical hydramoebae population growth on individual
hosts of H. pseudoligactis and the more resistant green H. viridis
at 25°C. On the latter species the initial instantaneous growth
rate of the amoebae is slower, the length of survivorship of the
host longer, and the number of amoebae present at the death of the
host less (Stiven, 1964). At 20°C the process is slowed, the maximum
number of amoebae at the death of the host is reduced, and the probability of surviving the attack goes from 0 to 0.6 for green hydra
(Stiven, 1964, 1973).

All amoebae produced by an infected host are potentially capable of leaving the host and infecting new hosts. The actual number
that can become detached from the host increases as the infection
progresses. This number can go as high as 40 per day in the case of
H. pseudoligactis to 3-4 per day in H. viridis (Stiven, 1964). In
one experiment at 20°C infected H. pseudoligactis produced a mean
total of about 125 detached amoebae over 10 days; H. viridis under
corresponding conditions produced a mean total of 18 amoebae. This
time sequence of detached amoebae from an individual host represents
the potential number of new cases, assuming each detached amoeba

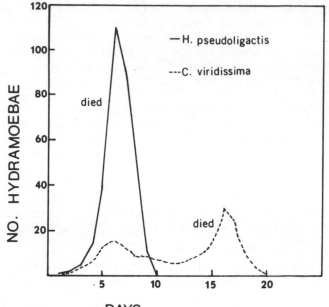

Fig. 2.   An example of the growth form of the hydramoeba population
on the susceptible H. pseudoligactis and the more resistant H. viridis
(25°C).   Initial infection with one amoeba.   Both hosts died (disin-
tegrated) releasing hydramoeba built up during the infection process.

produces a new infection.

     Now if an infected host is placed within a population of sus-
ceptible hosts, the number of new infections that result solely from
that first infected host represents a sequence that can be analyzed
quantitatively.   Such a sequence is analogous to the production of
offspring by a female individual, and has been explored by use of
the "generation law" model by Cole (1954).   If $2T^x$ represents 2 new
infections (offspring) produced during time interval x, then a
generation law (G) for an infected host will be, for example,
     $G = 1T^1 + 3T^2 + 2T^3 + 3T^4 + 3T^5 + 1T^6$
The expression 1/1-G generates the number of new infections through
time to be expected under a given set of conditions.   The above
generation law can also be written in terms of r, the instantaneous
infection rate, and a specific value of r computed (Stiven, 1973).
In the above example r = 0.97.   Over 85% of the value of r is
determined by the number of new cases occurring during the first two
time intervals.

Using the above approach, the effects of susceptible host population density and level of initial infection on the rate of infection were explored in populations of H. pseudoligactis and H.
viridis using Syracuse watch glasses as containers (Fig. 3).
Initial infection was varied by changing the number of attached
amoebae on a single infected inoculant host. Host species, host
density, and level of initial infection were all found to affect
significantly the instantaneous infection rate (Stiven, 1964).

If one assumes that each detached amoeba from an infected host
yields a new infection given sufficient susceptible hosts, one can
compute the maximum potential infection rate using the generation
law approach. For example, in the host, H. viridis, 20°C, and an
initial infection of 5 hydramoebae per host, the computed instantaneous infection rate was 1.5 (Stiven, 1967). This compares with
values of 017 to 1.1 in host densities of 0.5 and 1.0 individuals/
cm^2 respectively, and indicates that only a portion of the detached
amoebae actually produce new infections and/or that several combine

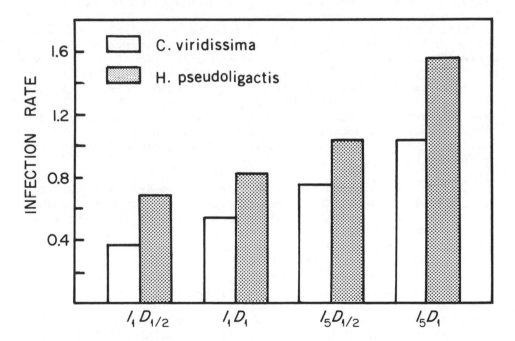

Fig. 3. Comparison between brown and green hydra of the instantaneous infection rate as a function of temperature, host population density, and level of initial infection. Infection rate
computed from the generation law approach. (From Stiven, 1964).

to produce a new infection.

Hydramoeba Epizootics

When an infected hydra is placed in a population of suscepti-
bles, and new infectives are allowed to accumulate, the accumulated
proportion of hosts infected through time assumes a sigmoid curve.
This infection curve is well characterized by the logistic function
(Stiven, 1964) and curves have been fitted by the Berkson (1953)
minimum logit chi-square procedure. This simple epizootic curve
has an upper asymptote of 1.0 (all host eventually become infected)
for all hydra species examined except the immune form H. littoralis.
A typical logistic function derived from an epizootic of hydramoeba
in a population of H. viridis yielded the following equation,

$$p = \frac{1.0}{1 + e^{-0.465(t - 10.42)}}$$

where p = the fraction of the host population infected at time t,
0.465 is the initial instantaneous infection rate, r, and 10.42 is
the time in days for half the population to become infected. When
values of r from the logistic fits were compared to values of r
derived from the generation law approach under identical experimental
conditions of host density, level of initial infection, and temperature, differences were generally not significant (see Fig. 4)
except in the high density, high initial infection ($I_5 D_2$) treatment.
This confirms the use of the instantaneous infection rate r as a
good parameter characterizing hydra-hydramoeba epizootics, and one
that is clearly responsive to different biological and environmental
treatments.

Utilizing the parameters of the logistic model, processes of
infection were also explored in host populations containing large
numbers (up to 668 hosts) in culture containers of different sizes
(Stiven, 1967). This permitted an assessment of the effects of
both host density and available substrate space on the infection
rate, and helped clarify the conditions that bring about the maximum possible instantaneous infection rate. In the largest containers
used (56 cm^2), doubling of host density produced an increase of 1.3
times in the infection rate in host densities ranging from 1.5 to
12 hosts/cm^2 of substrate. The maximum value of the infection rate
attained was around 1.0 and did not significantly differ from that
of 0.93 produced from a generation law experiment under identical
experimental conditions with the assumption that each detached
amoeba produced a new infection.

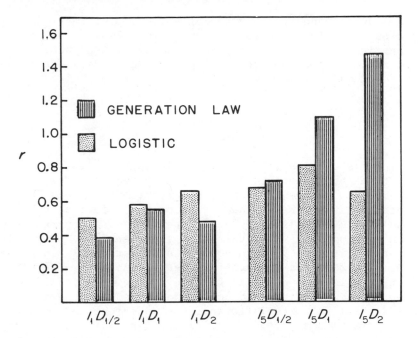

Fig. 4. Comparison of r, the instantaneous infection rate computed
from the generation law and the logistic function methods under
three levels of host density and two levels of initial infection;
20°C. (From Stiven, 1964).

Epizootics in Mixed Hydra Species Systems

The hydra-hydramoeba system has also been used to explore
properties of epizootics in mixed host species systems (Stiven,
1971, 1973) and to determine the contribution of each host species
to a measure of overall host community resistance. "Communities"
were constructed from four hydra species (including the albino
H. viridis) by varying the proportions of each. The instantaneous
infection rate was used as a measure of overall community susceptibility to the infection. An index of Community Resistance was also
derived by combining an independent measure of the resistance of
each species to the hydramoeba and its corresponding relative
abundance in the community.

Significant differences in infection rates were found among
the different communities with the proportions of most susceptible
species (H. pseudoligactis) and the most resistant species (H.
littoralis) accounting for most of the variation in infection rate.
The Community Resistance Index was also predictive of the behavior

of the community under the hydramoeba stress. Interactions among
the species did exist with some accumulation of hydramoeba on the
immune species. The use of such experimental mixed species systems
clearly provides insight into the complicated process of epizootics
in host communities with varying levels of susceptibility.

CONCLUSIONS

Hydramoeba hydroxena (Entz) is an obligate parasite on
several species of freshwater hydras, although at least one species,
Hydra littoralis, appears to be practically immune. This host-
parasite system has been investigated experimentally and the interactive processes leading to the spread of the infection through
host populations quantified. The growth form of the parasite population on susceptible individual hosts is essentially geometric,
and the epizootic wave through susceptible host populations has
been characterized by the sigmoid logistic function. The population
infection rate is also directly related to the infection life
history of the single infected host. Host population density,
level of initial parasite infection, host species, and such environmental variables as temperature significantly influence the
population infection rate. In addition, the variation in susceptibility to infection among several hydra species has permitted
the experimental analysis of the infection process in hydra
"communities" possessing differing levels of community resistance.

This research has clarified some of the biological and
ecological interactions between hydra host species and its protozoan parasite as well as helped explain the reported Hydramoeba
epizootics in natural populations of hydra.

REFERENCES

Beers, C. D. 1964. Cytochemical observations on the interphase
    nucleus of the parasite amoeba Hydramoeba hydroxena (Entz).
    J. Parasitol. 50: 630–633.

Berkson, J. 1953. A statistically precise and relatively simple
    method of estimating the bio-assay with quantal response,
    based on the logistic function. J. Amer. Stat. Assoc. 48:
    565–599.

Bryden, R. R. 1952. Ecology of Pelmatohydra oligactis in
    Kirkpatricks Lake, Tennessee. Ecol. Monogr. 22: 45–68.

Cole, L. C. 1954. The population consequences of life history
    phenomena. Quart. Rev. Biol. 29: 103–137.

Entz, Geza. 1912. Über eine neue amöbe auf Süsswasser polypen (Hydra oligactis Pall.). Archiv. f. Protist. 27: 19–47.

Hegner, R. W. and W. H. Taliaferro. 1924. Human Protozoology. Macmillan Co., New York.

Ito, T. 1949. On Hydramoeba hydroxena discovered in Japan. Science Reports of the Tohoku Univ. Ser. 4 (Biology) 18: 205–209.

Ito, T. 1950. Further notes on Hydramoeba hydroxena (Entz) from Japan. Mem. Ehime Univ. Sec. II (Science) 1: 27–36.

Miller, D. E. 1936. A limnological study of Pelmatohydra with special reference to their quantitative seasonal distribution. Trans. Amer. Micros. Soc. 55: 123–193.

Reynolds, B. D. and J. B. Looper. 1928. Infection experiments with Hydramoeba hydroxena nov. gen. J. Parasit. 15: 23–30.

Rice, N. E. 1960. Hydramoeba hydroxena (Entz), a parasite on the freshwater medusa, Craspedacusta sowerbii Lankester, and its pathogenicity for Hydra cauliculata Hyman. J. Protozool. 7: 151–156.

Stiven, A. E. 1962. Experimental studies on the epidemiology of the host-parasite system hydra and Hydramoeba hydroxena (Entz). I. The effect of the parasite on the individual host. Physiol. Zool. 35: 166–178.

Stiven, A. E. 1964. Experimental studies on the host parasite system hydra and Hydramoeba hydroxena (Entz). II. The components of a simple epidemic. Ecol. Monogr. 34: 119–142.

Stiven, A. E. 1965. The association of symbiotic algae with the resistance of Chlorohydra viridissima (Pallas) to Hydramoeba hydroxena (Entz). J. Inverteb. Pathol. 7: 356–367.

Stiven, A. E. 1967. The influence of host population space in experimental epizootics of Hydramoeba hydroxena (Entz). J. Inverteb. Pathol. 9: 536–545.

Stiven, A. E. 1968. The components of a threshold in experimental epizootics of Hydramoeba hydroxena in populations of Chlorohydra viridissima. J. Inverteb. Pathol. 11: 348–357.

Stiven, A. E. 1971. The spread of Hydramoeba infections in mixed hydra species systems. Oecologia 6: 118–132.

Stiven, A. E.   1973.   Hydra-hydramoeba: a model system for the study
     of epizootic processes.   Current Topics Comparat. Pathobiol.
     2: 146-212.

Welch, P. S. and H. A. Loomis.   1924.   A limnological study of Hydra
     oligactis in Douglas Lake, Michigan.   Trans. Amer. Micros. Soc.
     43 : 203-235.

Wermel, E.   1925.   Beiträge zur Cytologie der Amoeba hydroxena
     Entz.   Arch. Russes de Protistol.   4 : 95-120.

ASPECTS OF LIGHT IN THE BIOLOGY OF GREEN HYDRA

R. L. Pardy

Department of Developmental and Cell Biology

University of California, Irvine, California 92717

A wide variety of cnidarians maintain algal symbionts within their tissues. These symbionts are photosynthetically active when illuminated and translocate large percentages of their photo-synthetic products to the host (see Muscatine, 1974, for a comprehensive review). In addition to the light requirements for photosynthesis, the symbionts, like other plants, appear to have light requirements for cell division, cell development, and other processes. By altering light intensities, the wavelength of radiant energy and the duration of exposure (photoperiod), the experimenter can attempt to selectively manipulate some of the symbiont's biology and thereby dissect or characterize subsystems functioning within the intact association. Using this strategy I have examined the effects of light on the green hydra symbiosis and have found that symbiont multiplication and ultrastructure exhibit a light dependency, and that high-intensity light may cause expulsion of symbionts. In terms of the physiology of the association, the light-dependent photosynthetic reactions may alter the association's metabolic and respiratory processes. Finally, the light-stimulated and light-dependent processes of the symbiotic algae may be superimposed on the host's normal light-sensitive behavior or phototaxis. What follows is a brief description of the green hydra symbiosis followed by a review of some recent work on the role of light in various aspects of the association ending with a description of current experiments on green hydra phototaxis.

Green hydra maintain unicellular green algae in the bases of their digestive epitheliomuscular cells. These symbionts, called zoochlorellae, are known to be photosynthetically active and capable of translocating reduced carbon, probably as maltose

(Muscatine, 1965), to the animals' tissues. The net movement of
reduced carbon from zoochlorellae to host is of great importance
to the survival of the association. Green animals which are maintained in light demonstrate an ability to survive starvation that
is significantly greater than aposymbiotic (= symbiont-free) hydra
(Muscatine and Lenhoff, 1965). How hydra recognize potential
algal symbionts and establish a viable, hereditary endosymbiosis
has been described by Pardy and Muscatine (1973), and a growing
body of research on this aspect of the symbiosis has been reviewed
recently by Muscatine, Cook, Pardy and Pool (1975). Moreover,
Pool (in this volume) has defined some intriguing immunological
properties of the symbiotic algae.

     To say that light is important, even crucial, in the biology
of green hydra is to say nothing that has not already been demonstrated or suggested by experimentation. In terms of the normal
reproduction of the zoochlorellae, light appears to be a critical
factor. When a population of Hydra viridis is fed to repletion
with Artemia nauplii every 24 hours, the hydra grow, bud, and the
population doubles in size approximately every two days. Pardy
(1974) has shown that under these conditions the symbionts also
grow and double their population though they never appear to
overgrow the host. When grown in the absence of light, the
animals reproduce and their numbers increase normally but the
algae cease dividing. Under these dark conditions the hydra lose
their green color and become pale. Examination of the host's
digestive cells shows that the number of symbionts per cell is
reduced and that the rate of reduction is approximately the same
as the growth rate of the population. No evidence exists for the
expulsion of algae from the host or digestion of symbionts by
host cells. What is observed is a continued dilution of the algae
standing crop, which has ceased multiplying, by the repeated
divisions of the host's cells (Pardy, 1974). When the hosts are
returned to the light after several weeks in the dark, they
rapidly turn green as the remaining symbionts undergo division,
and replenish the normal standing crop which is approximately 20
symbionts per digestive cell (approximately 4.5×10^4 cells per
hydranth). Thus growth of hydra in the light is essential for
normal algal reproduction.

     If hydra are not fed, whether maintained in the light or dark,
the number of symbionts remains constant (Pardy, 1974). Hence the
presence of light alone, though important, is not the only factor
involved in symbiont growth. Apparently factors or conditions
associated with the host are also essential and these principles
are operative when the host is growing or undergoing sexual reproduction and may represent regulatory systems which are organized
to maintain a certain symbiont/host cell ratio. Further support
for the notion of host control of symbiotic algae comes from
grafting experiments (Pardy and Heacox, 1976) that suggest that

the head region of the hydra exerts an inhibiting influence on
algal multiplication.

Within individual zoochlorellae the usual configuration of
subcellular organelles, at least in symbionts from the English
strain of green hydra, appears to be light dependent (Pardy, 1976a).
Normally these symbionts evidence a prominent pyrenoid, vesicu-
lated polyphosphate bodies and a cup-shaped chloroplast. When
the hosts are maintained in the dark for several days, however,
the algae lose their pyrenoids, the polyphosphate bodies are
replaced by a massive, shield-shaped structure, and the chloroplast membranes become extremely compacted (Pardy, 1976a). Such
gross ultrastructural changes are not observed in symbionts from
the Florida strain of Hydra viridis grown under similar conditions.
As in the previous example, host processes may be involved as the
ultrastructure changes described for the English symbionts are not
manifested when the algae are transferred to Florida strain aposymbionts, nor are changes in Florida algae observed when resident
in English hosts (Pardy, 1976a). Thus the effect of a lack of
light on the disposition of symbiont organelles depends upon both
a sensitive algae strain and the residence of those symbionts in
a specific strain of hydra.

At the other extreme it has been shown (Pardy, 1976b) that
intense light ($\approx$ 600 watts M^{-2}) can cause the expulsion of algae
from the hosts. Action spectra have indicated that light of
550-650 nanometers (red region of the spectrum) is most effective.
There is some evidence that hydra tissue, especially in the
English strain, may act as a light screen. It takes more radiant
energy to bring about expulsion of English symbionts ($\approx$ 2000
watts M^{-2}) and this may be due to the larger (hence thicker)
digestive and epidermal cells of this strain (Pardy, 1976b).

Animals that have expelled their algae are referred to as
bleached hydra and it has been hypothesized (Pardy, 1976b) that
high intensity light kills or disrupts the symbionts and that the
host reacts to these moribund symbionts by expelling them. Following exposure to intense light, the algae look disrupted and appear
to have been dislodged from their normal basal location and
gathered into a large, apical, digestive cell vacuole (Pardy, 1976b).
A high percentage of bleached animals remain symbiont-free and
from them aposymbiotic clones have been developed.

To summarize the preceding paragraphs, light has been shown
to be important for optimal symbiont growth, for the normal maintenance of subcellular organelles and that intense light brings
about the expulsion of symbionts. These effects of light on the
hydra symbiosis have been manifested and observed mainly in the
behavior of the symbiotic algae though in each case the involvement of the host is suspected.

An important interface of light with the association is via
the photosynthetic activity of the algal endosymbionts. The two
major products of photosynthesis, oxygen and carbohydrate, are
available to enter the host's metabolic network and thereby
influence the animal's cellular physiology. Pardy and Dieckmann
(1975) have shown that oxygen evolved by the symbionts as a result
of the photosynthetic process may serve 50 to 100% of the association's aerobic respiratory needs. Furthermore, Pardy and White
(in preparation) demonstrate that the photosynthetic activities
of the symbionts may, depending upon the state of the host's
nutrition, cause the hydra to exhibit either a carbohydrate
metabolism with a respiratory quotient (R.Q.) of ≈ 1 or a mixed
metabolism with an R.Q. of ≈ 0.8. Aposymbiotic hydra or green
hydra that have been maintained in the dark exhibit a predominantly
fat metabolism, whether fed or starved. When the hosts are fed
and maintained in the light, experiments suggest that the major
fraction of photosynthetically fixed carbon is directed towards the
synthesis of new algal protoplasm as the host and symbiont population expand. The host's growth and metabolism is fueled mainly by
fat derived from exogenous food (Artemia nauplii). When the hosts
are starved, however, animal and symbiont growth ceases and it
appears that photosynthetic products formerly directed to the
growth of algae are shunted into the host's metabolic pathways.
As these products are mainly carbohydrate (Muscatine, 1965) the
R.Q. of the association is pushed up, approaching 1. Thus light,
through the medium of the photosynthetic apparatus of the symbionts,
may influence the metabolism of the host-zoochlorellae complex.

Since light is such an important factor in the symbiosis,
especially under the conditions where exogenous food is limiting,
it is not surprising that the association has a mechanism to
assure harvesting of this vital commodity. Hydra, including some
non-symbiotic species, exhibit a positive phototaxis. Trembley,
(1744) performed elegant experiments which unequivocally demonstrated green hydra's attraction to light. Subsequently a number
of workers over the years have studied the phototactic response
and their findings are reviewed in Kanaev (1955). Of curiosity was
the question as to whether the presence of algal symbionts had a
modulating effect on the phototaxis of green hydra.

To study the phototactic behavior of symbiotic hydra I devised
a chamber consisting of a 9 cm plastic petri dish that had been
painted black on the outside and wrapped with black tape to prevent the entry of light. In like manner the plastic lid was
covered except for a central circular window or port having a
diameter of 3.3 cm. To test for phototactic activity, hydra
(usually about 20, 5-day starved, English strain) were placed
randomly in the dishes and allowed to settle and attach. After
3 hours the lids were put in place and the dishes oriented 75 cm
under a pair of 40-watt fluorescent, cool-white lamps. At 24-hour

intervals the lids of the experimental dishes were removed and
replaced with a plastic grid made from a petri dish cover. The
grid was divided into eight equal sectors and four concentric
rings spaced 1.2 cm apart. The rings were numbered 1 to 4 with 1
being the outer ring and 4 being the central disc located directly
under the transparent port. The area of each ring was calculated
and found to be 4.5, 12.1, 28.8, 30.9 cm^2 , 1-4 respectively.

Table I shows the distribution of hydra and their density in
the rings at various times during the experiment. It is clear
that the animals moved progressively to the area under the window
with 77% of the animals being located in ring 4 at the fourth day.
The position of the window was of no importance, for in other
chambers where the port was located eccentrically no difference in
the response was detected.

To assess the possible role of algal symbionts in the
phototactic response, I repeated the above experiments using aposymbiotic hydra and green hydra exposed to 10^{-6} M DCMU [3--(3,4-
dichlorophenyl)-1,1-dimethylurea]. DCMU is a specific inhibitor
of photosynthesis which has been shown to block photosynthesis of
the symbionts and to be non-injurious to the host (Pardy and
Dieckmann, 1975). Green animals without inhibitor served as
controls. The behavior of green animals in chambers without lids
was also examined.

In Table II is shown the results of the experiments described
above. As can be seen, aposymbiotic hydra exhibit a positive
phototaxis that appears equal to that of algal-bearing animals.
Moreover DCMU had no apparent effect on the migration of green
hydra towards light -- a strong indication that the photosynthetic
process probably plays no role in the response. Hence, it appears
that these hydra have an intrinsic positive response to light and
that the presence of algal symbionts does not seem to alter the
host's behavior. At this time, however, final conclusions are not
warranted as neither the action spectrum, the effect of various
states of host starvation or the intensity of light have not been
examined in relation to the phototaxis. Any one of these factors
acting through one or both partners could potentially affect the
response.

In this review I have tried to indicate the role of light in
various plant and algal processes in the green hydra symbiosis.
With the exception of hydra's phototaxis the sensitivity of the
association to light appears to reside mainly in the symbiotic
algae. Light plays a direct role in the multiplication and
ultrastructure of the green symbionts though the effects are
apparently modulated by the host. Products of the photosynthetic
reaction of the symbionts which enter the host's metabolic network represent chemical links to light that may alter the hydra's

TABLE I. The average number of green hydra in each of the four rings of the experimental chamber at various times after placement of the animals in the chamber. Data is expressed as the average number of hydra ± s.d., n = four chambers, or average density (hydra cm^{-2}) in each ring.

| | Ring Number | | | |
Day	1	2	3	4
1	3.5 ± 1.0 (0.12)	7.3 ± 2.9 (0.36)	6.3 ± 3.6 (0.52)	1.3 ± 0.5 (0.31)
2	1.3 ± 1.5 (0.05)	6.0 ± 2.1 (0.30)	3.3 ± 1.5 (0.27)	8.0 ± 0.8 (2.01)
3	1.0 ± 1.1 (0.04)	2.2 ± 0.9 (0.11)	2.7 ± 1.5 (0.23)	14.2 ± 1.5 (3.58)
4	0	2.5 ± 0.5 (0.12)	2.0 ± 0.5 (0.12)	15.2 ± 2.6 (3.83)

TABLE II. The average number of hydra in each of the four rings of the experimental chamber after three days exposure to various conditions. Data is expressed as the average number of hydra ± s.d., n = four chambers (control, n = 2) or average density (hydra cm^{-2}) in each ring.

| | Ring Number | | | |
Condition	1	2	3	4
Aposymbiotic hydra	0	0.8 ± 0.5 (0.03)	0	14.0 ± 1.6 (3.5)
Green hydra, 10^{-6} M DCMU	1.3 ± 1.5 (0.39)	5.0 ± 1.5 (0.25)	0.03 ± 0.5 (0.002)	11.5 ± 3.5 (2.88)
Green hydra	1.0 ± 1.1 (0.04)	2.2 ± 0.9 (0.10)	2.7 ± 1.5 (0.23)	14.2 ± 1.5 (3.58)
Control-green hydra (open dish)	2.0 ± 0 (0.04)	3.5 ± 0.7 (0.16)	7.0 ± 1.4 (5.3)	3.5 ± 2.1 (0.88)

physiology. Finally, it is evident that the ability to control
or manipulate this important environmental factor offers the
experimenter a valuable tool with which to probe the algae/hydra
association.

REFERENCES

Kanaev, I.I. 1952. Hydra: Essays on the Biology of Fresh Water
    Polyps. Soviet Academy of Sciences, Moscow.

Muscatine, L. 1965. Symbiosis of hydra and algae. III. Extracellu-
    lar products of the algae. Comp. Biochem. Physiol. 16: 77–92.

Muscatine, L. 1974. Endosymbiosis of cnidarians and algae. Pages
    359–395 in: L. Muscatine and H.M. Lenhoff, Eds., Coelenterate
    Biology: Reviews and New Perspectives. Academic Press, New
    York.

Muscatine, L., and H.M. Lenhoff 1965. Symbiosis of hydra and algae.
    II. Effects of limited food and starvation on growth of symbio-
    tic and aposymbiotic hydra. Biol. Bull. 129: 316–328.

Muscatine, L., C.B. Cook, R.L. Pardy, and R.R. Pool 1975. Uptake,
    recognition and maintenance of symbiotic Chlorella by Hydra
    viridis. Symp. Soc. Exp. Biol. 29: 175–203.

Pardy, R.L. 1974. Some factors affecting the growth and distribu-
    tion of the algal endosymbionts of Hydra viridis. Biol. Bull.
    147: 105–118.

Pardy, R.L. 1976a. The morphology of green Hydra symbionts as
    influenced by host strain and host environment. J. Cell. Sci.
    20: 655–669.

Pardy, R.L. 1976b. The production of aposymbiotic hydra by the
    photodestruction of green hydra zoochlorellae. Biol. Bull.
    (in press).

Pardy, R.L., and L. Muscatine 1973. Recognition of symbiotic algae
    by Hydra viridis. A quantitative study of the uptake of living
    algae by aposymbiotic H. viridis. Biol. Bull. 145: 565–579.

Pardy, R.L., and C. Dieckmann 1975. Oxygen consumption in the
    symbiotic hydra, Hydra viridis. Exp. Zool. 194: 373–378.

Pardy, R.L. and A.E. Heacox 1976. Growth of algal symbionts in
    regenerating hydra. Nature 260: 809–810.

Trembley, A. 1744. Mémoires pour servir à l'histoire d'un genre
    de polypes d'eau douce à bras en forme de cornes. Leyden, J.
    and H. Verbeck.

SURVIVAL DURING STARVATION OF SYMBIOTIC, APOSYMBIOTIC, AND NON-SYMBIOTIC HYDRA

Marcia Ornsby Kelty and Clayton B. Cook

Department of Zoology
Ohio State University
Columbus, Ohio 43210

Many experimental studies on the benefit of endosymbiotic algae to their coelenterate hosts have utilized so-called "aposymbiotic" animals as controls. These individuals, although from a species which normally contains algal symbionts, have no algae. Algae-free individuals may be obtained through experimental treatment; aposymbiotic strains of Hydra viridis have been produced in culture by the use of 0.5 per cent glycerol (Whitney, 1907). Populations of aposymbionts may also be obtained through the culture of eggs of H. viridis which have not been infected with algae. Although these algae-free eggs are apparently produced in nature, no naturally occurring aposymbiotic H. viridis have been reported. Differential survival during starvation has been proposed as one factor which may favor symbiotic green hydra over aposymbiotic forms (Muscatine and Lenhoff, 1965).

Clearly, other species of hydra exist without the benefit of symbiotic algae. No comparison of these non-symbiotes with aposymbiotic H. viridis in terms of survival during periods of food deprivation has yet been made. In this paper we compare symbiotic, aposymbiotic, and non-symbiotic hydra with respect to their ability to survive during periods of starvation. Our data suggest that aposymbionts fare less well than non-symbiotic forms; this result suggests that in the course of evolution H. viridis may have lost certain essential metabolic pathways needed to withstand periods without food.

MATERIALS AND METHODS

Four strains of hydra were used in this study. Two strains were symbiotic strains of Hydra viridis (English and Florida)

while a third was an aposymbiotic strain derived from the Florida
hydra (Muscatine et al., 1975). The fourth hydra used was a non-
symbiotic strain, Hydra littoralis, obtained from the Carolina Biological Supply Company, Burlington, N. C. All hydra were kept at
20°C with a 12 hour light/12 hour dark cycle; illumination was produced by a General Electric Cool-White fluorescent lamp (2700 lumens
per m²).

    All hydra were fed to repletion for at least five days before
the beginning of the starvation experiments. Samples of ten "uniform" hydra (Lenhoff and Bovaird, 1961) each were placed in 100 x
15 mm. Petri dishes containing forty ml. of "M" solution (Muscatine,
1961). The number of hydranths was counted and the culture solutions were changed daily. The dishes were thoroughly cleaned every
three days to remove any accumulation of organic matter on the
bottoms of the dishes.

    Starving hydra were periodically tested for feeding ability.
Hydra were given generous amounts of newly hatched brine shrimp
nauplii for a period of one hour. At the end of that time, the
hydra were scored on ability to (1) capture prey and (2) ingest
prey. Capture of prey was defined as retention of prey on the tentacles at the time of observation.

                                RESULTS

    Both non-symbiotic brown hydra and symbiotic green hydra sur-
vive extended periods of starvation better than aposymbiotic hydra
(Figure 1). (Although statistical parameters are not given, the
standard errors for all points in Figure 1 are less than two hydranths). Our data for the survival of Florida green and aposymbiotic hydra confirm the results of Muscatine and Lenhoff (1965).
Both of these strains produce buds during the first 3 days of starvation, after which hydranths are lost. Aposymbiont populations
show a drastic decline, with no hydra surviving longer than 27 days;
in contrast, cultures of Florida green hydra contained virtually
the same number of hydranths after 4 weeks of starvation as started
the experiment. We have kept starving English green hydra alive
for at least 6 weeks under these conditions; this strain shows better than 130 per cent survivorship after this time.

    Brown hydra also survive starvation much more successfully than
aposymbiotic hydra; after 4 weeks, better than 50 per cent of the
original population remains, while more than 30 per cent survive
after 6 weeks. The data in Figure 1 show that brown hydra appear
to produce no buds when starved. This is not a general feature of
starvation in these hydra, since subsequent repetitions of this experiment indicate that additional buds may be produced. We ascribe
the low budding rate of the hydra in Figure 1 to be perhaps due to

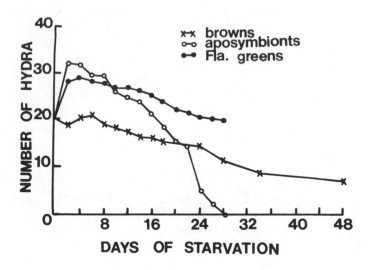

Figure 1. Survival of symbiotic, aposymbiotic, and non-symbiotic hydra during starvation. Values given are the number of hydranths counted each day; each point represents the mean of 5 replicate samples.

insufficient feeding during the meals just preceding starvation. If this is so, then better-fed hydra might show even greater survival during starvation than our results indicate.

In addition to survival during starvation, the ability to feed after starvation is also much greater for green and brown hydra than it is for aposymbionts (Table 1). Starvation affects both the ability to capture prey and to ingest food. All strains show an eventual decrease in feeding efficacy following starvation, but the aposymbionts are most drastically affected. Loss of feeding ability in these hydra is evident within 8 days. After sixteen days of starvation, less than 6 per cent of the surviving aposymbionts could ingest food; ability to capture or ingest prey is completely lost within 24 days. In contrast, 42 per cent of the surviving Florida green hydra could ingest prey following 2 weeks of starvation, and the larger English strain are able to feed after 6 weeks of starvation. Non-symbiotic hydra, like the symbiotic strains, can capture and ingest food for at least 6 weeks after starvation.

Table 1

Strain: Florida aposymbiont

Days of starvation	n	% capturing prey	% ingesting prey
5	10	90.0	90.0
8	11	36.4	36.4
9	8	75.0	75.0
11	10	70.0	00.0
12	11	81.8	18.2
16	18	22.2	05.6
24	10	00.0	00.0

Strain: H. littoralis

Days of starvation	n	% capturing prey	% ingesting prey
12	28	100.0	100.0
28	11	81.8	81.8
34	9	66.7	66.7
42	7	57.1	57.1

Table 1. Ability of aposymbiotic Florida hydra and H. littoralis
to feed after starvation. Hydra were scored on ability to capture and ingest prey after being presented Artemia nauplii for a
period of one hour.

In all strains the ability to ingest food is lost before the
hydra become ineffective in capturing prey. Brine shrimp nauplii
can be immobilized and held at the tentacles, indicating that at
least some nematocysts are functional in hydra which capture
shrimp but cannot ingest them. However, none of the behaviors
usually associated with the hydra feeding response (tentacle
writhing, column bending, mouth opening) are seen, and the hydra
make no attempt to bring prey to the mouth.

DISCUSSION

The ability to withstand periods of starvation is critical
for organisms which are faced with an unpredictable food supply.
The ability of symbiotic hydra to survive longer periods without
feeding than aposymbionts is most certainly due to augmentation of
food reserves with translocated algal products (Muscatine and
Lenhoff, 1965). Previous work (Roffman and Lenhoff, 1969) has
shown that the maltose translocated to green hydra by symbiotic
algae is rapidly converted into glycogen. This glycogen should
provide a continual energy reserve which allows the hydra to

sustain itself during starvation without resorting to the catabolism of proteins. Our unpublished data show that while starving
aposymbiotic hydra rapidly catabolize glycogen stores, starving
green hydra actually accumulate them (Kelty, in prep.).

Since aposymbiotic hydra have no such continually produced
carbohydrate reserves, they must turn to a greater degree to the
use of protein during starvation. Muscatine and Lenhoff (1965)
have shown that the catabolism of proteins in aposymbionts proceeds
at a rate over two times that seen in symbiotic hydra over a five
day period of starvation. This increased use of proteins by starving aposymbionts may explain their early loss of feeding ability.
These hydra become unable to ingest prey soon after starvation begins; they can capture prey, but display none of the behavior which
is characteristic of the hydra feeding response. It may be that
protein catabolism affects sensory, nervous, or effector elements
which are involved in this behavioral repertoire. The number of
glutathione receptors could be reduced; these receptors are evidently located on the cell surface, and provide the initial sensory clue for the feeding response (Lenhoff, 1968). The effects of
long-term starvation on the glutathione response have not been
tested. In addition, the loss of nematocysts, resulting either
from protein catabolism or general cell loss, could result in the
release of quantities of glutathione from the prey which are insufficient to elicit feeding behavior. Behavior associated with the
feeding response may also be limited by the breakdown of myofibrils
in epithelial cells, or by the loss of nerve cells. These changes
in cell populations could be examined by comparing cell macerations
of starving hydra from symbiotic, aposymbiotic, and non-symbiotic
strains. Whatever the cause, early loss by aposymbionts of the
ability to ingest prey further limits the survival of these hydra,
for it ensures the death of the individual well before evidence of
such is seen.

Although the non-symbiotic species, H. littoralis, also lacks
algal symbionts, these hydra survive and feed longer than do aposymbionts. We suggest that this may be due to the amount and
timing of protein catabolism. Our unpublished data show that while
glycogen reserves per unit biomass and rate of glycogen utilization
are approximately equal in non-symbiotic and aposymbiotic hydra,
protein is not substantially catabolized by non-symbiotic hydra
during the first week of starvation. Aposymbionts, however, begin
catabolism of proteins on the first day of starvation (Kelty, in
prep.). This ability to delay protein utilization in starving non-
symbionts may be an important factor in their ability to survive
and feed during periods of starvation. Aposymbionts, however, must
turn quickly to the use of proteins as an energy source. This catabolism of protein leads to the loss of necessary functions, such
as the feeding response, serves to hasten death in these organisms,
and may thus be the key to their apparent absence in nature.

SUMMARY

Algae-free green hydra (aposymbionts) show much lower survival during starvation than either symbiotic green hydra or non-
symbiotic hydra. Aposymbionts also lose the ability to feed much
earlier than do the other hydra studied. It is proposed that these
differences may be due to more rapid catabolism of proteins in aposymbionts. The consequences of such differences are discussed in
light of their possible role in limiting the occurrence of natural
populations of aposymbiotic hydra.

LITERATURE CITED

Lenhoff, H. M., 1968. Chemical perspectives on the feeding re-
    sponse, digestion and nutrition of selected coelenterates.
    Pages 158-221 in B. Scheer and M. Florkin, Eds., Chemical
    Zoology, Vol. 2. Academic Press, New York.
Lenhoff, H. M., and J. Bovaird, 1961. A quantitative chemical
    approach to problems of nematocyst distribution and replace-
    ment in Hydra. Develop. Biol., 3: 227-240.
Muscatine, L., 1961. Symbiosis in marine and fresh water coelen-
    terates. Pages 255-268 in H. M. Lenhoff and W. F. Loomis,
    Eds., The Biology of Hydra. University of Miami Press.
Muscatine, L., C. B. Cook, R. L. Pardy and R. R. Pool, 1975. Up-
    take, recognition and maintenance of symbiotic Chlorella by
    Hydra viridis. Symp. Soc. Exp. Biol., 29: 175-203.
Muscatine, L., and H. M. Lenhoff, 1965. Symbiosis of hydra and
    algae. II. Effects of limited food and starvation on growth
    of symbiotic and aposymbiotic hydra. Biol. Bull., 129: 316-
    328.
Roffman, B., and H. M. Lenhoff, 1969. Formation of polysaccharides
    by Hydra from substrates produced by their endosymbiotic algae.
    Nature 221: 381-382.
Whitney, D. D., 1907. Artificial removal of the green bodies of
    Hydra viridis. Biol. Bull., 13: 291-299.

SULFATE UTILIZATION IN GREEN HYDRA

Clayton B. Cook

Department of Zoology
Ohio State University
Columbus, Ohio 43210

Recent work on the symbiosis between cnidarians and endosymbiotic algae has demonstrated that algae release soluble photosynthate to a variety of coelenterate hosts (Muscatine, 1974). The flow of organic carbon from algae to animal has reasonably been investigated as a major nutritional benefit of algal symbiosis. Lewis and Smith (1971) have drawn attention to the possibility that other nutrients could be transferred between algae and host. In this paper, I wish to suggest that sulfate may be important in the metabolism of coelenterates which harbor algal symbionts, and I present preliminary evidence which shows that the possession of symbiotic algae enhances the ability of green hydra to metabolize sulfate.

It is generally thought that the reduction of inorganic sulfate to sulfhydryl is an ability which is found widely among bacteria, algae and higher plants but one which is lacking in animal cells (Roy and Trudinger, 1970). One or another of the sulfur amino acids (cysteine, homocysteine and methionine) appears to be a dietary essential for animals (e. g., Prosser, 1973, p. 118). It appears likely that symbiotic algae are capable of sulfate reduction; free-living Chlorella can reduce sulfate to a variety of sulfur amino acids (Schiff, 1959, 1964), and zooxanthellae can be cultured in media which contain sulfate but not sulfur amino acids (McLaughlin and Zahl 1959). However, direct evidence for sulfate reduction in either zoochlorellae or zooxanthellae is lacking.

If symbiotic algae are capable of sulfate reduction and coelenterate hosts are not, the translocation of sulfate reduction products from algae to animal tissue would present a situation which is analogous to the translocation of reduced carbon. The experiments which follow show (1) that symbiotic hydra have a greater affinity for sulfate than do non-symbiotic hydra, and (2) that sulfate is used by green hydra for nematocyst synthesis and it is not used for this purpose by hydra which lack algae. Nematocyst capsules contain sulfur-amino acids (Blanquet and Lenhoff, 1966), and it is suggested that the algae may contribute sulfate reduction products for nematocyst synthesis.

MATERIALS AND METHODS

Hydra used in this study include two strains of Hydra viridis containing symbiotic algae (English and Florida),an aposymbiotic strain of the Florida hydra and a strain of a non-symbiotic hydra, H. littoralis. The sources of the H. viridis strains are given in Muscatine et al. (1975), while the nonsymbiotic strain was purchased from the Carolina Biological Supply Co. (Burlington, N.C.). All experimental and stock cultures were maintained in M-solution (Muscatine, 1961) at 20° under a 12 hour light/12 hour dark photoperiod. Stock cultures were fed Artemia larvae daily for at least a week prior to use.

$^{35}SO_4$ was used to study the uptake and incorporation of sulfate. Hydra used in all experiments had one small bud and had not fed for 24 hours. $Na_2^{35}SO_4$ (New England Nuclear) was added to culture solution to give a concentration of 10 $\mu c \cdot ml^{-1}$ (1.6 μg $Na_2^{35}SO_4 \cdot ml^{-1}$). In all experiments hydra were incubated in this solution for 24 hours, then were thoroughly rinsed and transferred to sulfate-free culture solution for a 24 hour chase period. Radioactivity in hydra was measured either by drying intact hydra or pieces of hydra on filter paper (Pardy and Muscatine, 1973) or by counting aliquots of hydra homogenates; these samples included symbiotic algae. Filter paper samples were counted in a toluene-based scintillation cocktail; a dioxane-based cocktail (8g "Omni-fluor", 100 g naphthalene per liter) was used for aqueous samples. All samples were corrected for quenching with a channels ratio curve derived from a quenched series of ^{35}S standards. Protein content of radioactive homogenates was determined by the Lowry procedure (Lowry et al., 1951).

Autoradiography was used to study the incorporation of ^{35}S into the nematocysts of regenerating hydra. Tentacles and hypostomes were removed from hydra prior to radiosulfate incubation. After the 24 hour chase period cells from the digestive zone were macerated on slides (David 1973) and dried overnight. The cells were post-fixed with 3:1 ethanol/acetic acid (Merchant et al., 1964) and dried prior to coating with Kodak NTB-2 emulsion. The slides were exposed for 5 days at 4°; following development, they

were stained with 0.25 per cent Toluidine Blue and mounted.
Counts were made of grains located over 30 isolated stenoteles
from each hydra; background levels were determined for each slide
by counting the grains within 30 tissue-free areas outlined by a
tracing of a stenotele using a microscope drawing tube. The mean
value of these counts was subtracted from each stenotele count.

RESULTS

Uptake of radiosulfate by hydra. Both symbiotic and aposymbiotic hydra have higher levels of sulfate uptake and incorporation than non-symbiotic hydra (Table 1). Aposymbiotic hydra take
up twice as much sulfate per hydra as do non-symbiotic hydra, even
though non-symbiotic hydra are much larger. This comparison is
more striking when the data are expressed on a per unit protein
basis; the specific activity of uptake by aposymbionts is at least
12 times greater than that of the non-symbiotic strain. The distribution of ^{35}S between algae and animal tissue in symbiotic
hydra can be estimated from the amount of label in symbiotic and
aposymbiotic hydra. This comparison suggests that perhaps 50 per
cent of the radioactivity in green hydra is associated with the
algae. One puzzling aspect of the data in Table 1 is the difference in sulfate uptake between the two strains of green hydra;
English hydra appear to be less efficient at sulfate uptake (per
unit protein) than Florida hydra.

Table 1

Strain	dpm per hydra*	dpm per μg protein*
Florida green	35978 ± 4518 (3)	3707 ± 435 (4)
English green	47103 ± 10850 (5)	2254 ± 460 (4)
Florida aposymbiont	14924 ± 3141 (5)	1984 ± 289 (4)
H. littoralis	7755 ± 1633 (5)	121 ± 36 (5)

* Including algae and attached buds

Uptake of radiosulfate by symbiotic and non-symbiotic hydra. Each
strain was incubated in $Na_2{}^{35}SO_4$ (10 μc·ml^{-1}) for 24 hours, followed by a 24 hour chase in sulfate-free medium. Results are expressed as $\overline{X}$ ± S. D.; sample size in parentheses.

Counts of radioactivity in pieces of hydra from various body
regions indicate that most of the label is found in the tentacles,
hypostome and budding regions of all strains (Table 2). The pre-

Table 2

Strain	N	Tentacles + Hypostome	Digestive	Budding	Peduncle
Florida green	3	29.6 ± 9.0	21.7 ± 6.0	33.5 ± 4.2	15.1 ± 4.5
English green	5	23.1 ± 5.5	27.2 ± 5.6	32.4 ± 6.9	17.2 ± 3.6
Florida aposymbiont	5	31.2 ± 5.7	23.4 ± 8.4	29.7 ± 7.7	15.8 ± 8.7
H. littoralis	5	33.9 ± 2.6	24.6 ± 4.2	26.1 ± 3.6	15.5 ± 1.6

Incorporation of radiosulfate into body regions of symbiotic and non-symbiotic hydra; incubation procedures as in Table 3. Data expressed as the percentage of total activity in one hydra incorporated into each region. ($\overline{X} \pm$ S. D.)

dominant cell types in these regions are nematocytes and epithelial cells; these cell types should therefore contain most of the
label. Although the peduncle contains many mucous cells, this region shows the lowest incorporation of ^{35}S. Since acid mucopolysaccharides contain sulfated sugar residues, sulfate incorporation
should be seen in mucous cells in all strains. This result also
suggests that there is little contamination of samples by radioactive micro-organisms which attach to the pedal disc.

Incorporation of ^{35}S into nematocysts. Preliminary results
of grain counts from autoradiographs show that green hydra incorporate more label into stenoteles than either aposymbionts or non-
symbiotic hydra (Table 3). If half of the radioactivity in green
hydra is assumed to be located in algal cells, then the total ^{35}S
in hydra tissue is roughly similar for symbiotic and aposymbiotic
hydra. Since the nematocysts of green hydra are at least four
times as radioactive as those from aposymbionts, it appears that
either the algae are supplying some labeled component for the synthesis of green hydra nematocysts, or the algae are promoting the
incorporation of sulfate into nematocysts. The sulfur amino acids
cysteine and methionine are obvious candidates for the first possibility. It is not known whether sulfate is present in nematocyst
capsules; however, histochemical and chromatographic evidence suggests that mucopolysaccharides are associated with capsules (Blanquet and Lenhoff 1966). Such moities may contain sulfate residues.
How algae would augment any sulfate incorporation into nematocysts
is not clear.

Table 3

Strain	N	Grains over stenoteles*
Florida green	3	10.9 ± 4.4
Florida aposymbiont	5	2.4 ± 2.3
H. littoralis	3	1.3 ± 0.7

* $\overline{X}$ ± S. D.

Incorporation of radiosulfate into stenoteles of symbiotic and non-
symbiotic hydra. Counts were determined from radioautographs of
cell maceration preparations of hydra incubated as in the previous
tables and corrected for background as explained in the text. Each
individual value is the mean of 30 nematocysts from the budding
and digestive zones of a single hydra.

DISCUSSION

The metabolism of sulfate by coelenterates has not received much attention, but those studies which are relevant suggest that coelenterates without algal symbionts do not incorporate sulfate into tissues. Sulfate residues have not been found in analyses of Metridium body wall, even though most collagenous tissues usually contain acidic polysaccharides (Katzman and Jeanloz, 1970). Studies of the sulfate exchange between hydromedusae and seawater suggest that little sulfate is bound to tissue, or at least to mesoglea (Mackay, 1969). These findings are consistent with data in this paper which show that non-symbiotic hydra have a relatively low affinity for sulfate and do not incorporate appreciable amounts of sulfate into nematocysts. Symbiotic and aposymbiotic hydra on the other hand have a much higher affinity for sulfate; green hydra use either sulfate or sulfate reduction products in nematocyst synthesis. There is no evidence that sulfur amino acids resulting from algal sulfate reduction are incorporated into hydra protein; this must be demonstrated by the comparison of amino acids in protein hydrolysates from symbiotic and aposymbiotic hydra which have been incubated with radiosulfate.

The high affinity of symbiotic and aposymbiotic hydra for sulfate can be viewed as an adaptation of the hydra for algal symbiosis. Zoochlorellae in green hydra are located in membrane-bound vacuoles in digestive cell cytoplasm (Wood, 1961; Muscatine et al., 1975); any materials which reach the algae from the external environment must first pass through or be taken up by the hydra cell membrane. If increased sulfate uptake by H. viridis is of benefit to the animal either by the direct utilization of algal sulfate reduction products or by the general stimulation of algal metabolism, then the evolution of an efficient sulfate uptake mechanism by the hydra would be reasonable. In this context, it is interesting to note that although aposymbiotic hydra retain this high sulfate affinity, they apparently do not use the sulfate which they accumulate for nematocyst synthesis. This would suggest that the algae are providing sulfur amino acids, or at least sulfate reduction products, for nematocyst synthesis in green hydra.

Acknowledgements. I am grateful to Dr. S. B. Cook for assistance. This work was supported by grants from the Graduate School and the College of Biological Sciences, Ohio State University.

SUMMARY

Studies of sulfate incorporation by symbiotic and aposymbiotic green hydra show that both strains have a high affinity for sulfate. When both strains are incubated with radiosulfate, hydra with algae incorporate more label into nematocysts than aposymbiotic hydra. Non-symbiotic hydra show limited uptake of sulfate relative to the symbiotic and aposymbiotic strains; the affinity of the symbiotic and aposymbiotic strains for sulfate is considered as a possible adaptation for sulfate utilization by the algae. The use of algal sulfate reduction products by hydra tissue is discussed.

LITERATURE CITED

Blanquet, R., and H. M. Lenhoff, 1966. A disulfide-linked collagenous protein of nematocyst capsules. Science, 154: 152-153.

David, C. N., 1973. Quantitative method for maceration of hydra tissue. Wilhelm Roux' Arch. Entwicklungsmech. Organismen, 171: 259-268.

Katzman, R. L., and R. W. Jeanloz, 1970. Are acidic polysaccharides involved in collagen fibril formation or stabilization? Biochim. Biophys. Acta, 220: 516-521.

Lewis, D. H., and D. C. Smith, 1971. The autotrophic nutrition of symbiotic marine coelenterates with special reference to hermatypic corals. I. Movement of photosynthetic products between the symbionts. Proc. Roy. Soc. London Ser. B., 178: 111-129.

Lowry, O., N. Rosebrough, A. Farr and R. Randall, 1951. Protein measurement with the Folin phenol reagenat. J. Biol. Chem., 193: 265-275.

Mackay, W. C., 1969. Sulphate regulation in jellyfish. Comp. Biochem. Physiol., 30: 481-488.

McLaughlin, J. J. A., and P. A. Zahl, 1959. Axenic zooxanthellae from various invertebrate hosts. Annals N. Y. Acad. Sci., 77: 55-73.

Merchant, D. J., R. H. Kahn and W. H. Murphy, 1964. Handbook of Cell and Organ Culture. Burgess Publ. Co., Minneapolis, 263 pp.

Muscatine, L., 1961. Symbiosis in marine and fresh water coelenterates. Pages 255-268 in H. M. Lenhoff and W. F. Loomis, Eds., The Biology of Hydra. University of Miami Press.

Muscatine, L., 1974. Endosymbiosis of cnidarians and algae. Pages 359-395 in L. Muscatine and H. M. Lenhoff, Eds., Coelenterate Biology: Reviews and New Perspectives. Academic Press, New York.

Muscatine, L., C. B. Cook, R. L. Pardy and R. R. Pool, 1975. Uptake, recognition and maintenance of symbiotic Chlorella by Hydra viridis. Symp. Soc. Exp. Biol., 29: 175-203.

Pardy, R. L. and L. Muscatine, 1973. Recognition of symbiotic
    algae by Hydra viridis. A quantitative study of the uptake
    of living algae by aposymbiotic H. viridis. Biol. Bull., 147:
    105-118.
Prosser, C. L., 1973. Comparative Animal Physiology. W. B. Saun-
    ders, Philadelphia, 966 pp.
Roy, A. B., and P. A. Trudinger, 1970. The Biochemistry of Inor-
    ganic Compounds of Sulphur. Cambridge University Press,
    400 pp.
Schiff, J. A., 1959. Studies on sulfate utilization by Chlorella
    pyrenoidosa using sulfate S-35; the occurrence of S-adenosyl
    methionine. Plant Physiol., 34: 73-80.
Schiff, J. A., 1964. Studies on sulfate utilization by algae.
    II. Further identification of reduced compounds formed from
    sulfate by Chlorella. Plant Physiol., 39: 176-179.
Wood, R. L., 1961. The fine structure of intercellular and meso-
    gleal attachments of epithelial cells in hydra. Pages 51-68
    in H. M. Lenhoff and W. F. Loomis, Eds., The Biology of Hydra.
    University of Miami Press.

SPECIFICITY OF SYMBIOSES BETWEEN MARINE CNIDARIANS AND ZOOXANTHELLAE

D. A. Schoenberg and R. K. Trench

Department of Biology, Yale University

New Haven, Ct. 06520, USA

The dinoflagellate zooxanthella, Symbiodinium (=Gymnodinium) microadriaticum Freudenthal, is harbored by more than 80 species of marine cnidarians, tridacnids and protists (Taylor, 1974). It is the only described zooxanthella species found in benthic symbiotic cnidarians. Its large host distribution is exceptional among algal symbionts of invertebrates (Taylor, 1974; Hollande and Carré, 1974; Karakashian, 1975). This wide phylogenetic range for a single species of zooxanthella suggests that the morphological and ultrastructural criteria used for its identification (Freudenthal, 1962; Kevin, et al., 1969; Taylor, 1969) might not distinguish between varieties of S. microadriaticum with narrower host or ecological distributions. This paper describes a series of studies using other methods to characterize genetic variation and specificity among isolates of S. microadriaticum.

Previous reports that varieties of S. microadriaticum exist have been based on studies using the intact symbiosis (Lang, 1970; Lewis and Smith, 1971; Trench, 1971a; Dustan, 1975), freshly isolated algae (Trench, 1969; 1971b; 1971c), or possibly contaminated cultures (Kinzie, 1974). Differences in the behavior of S. microadriaticum under these conditions cannot be solely attributed to differences in the alga's genetic makeup. To avoid such artifacts, experiments were performed on zooxanthellae cultured axenically for one to two years in a defined medium (McLaughlin and Zahl, 1966) under uniform conditions of light and temperature (Schoenberg, in prep.). S. microadriaticum from 17 of the 70 host species collected grew under such conditions. The variation described below may thus represent a small fraction of the variability of S. microadriaticum.

GENETIC VARIATION IN S. MICROADRIATICUM

Electrophoresis of proteins has often been used to compare gene expression among closely related organisms (reviewed in Markert, 1975). In this analysis, algae isolated from a single host individual or colony are referred to as a population; algae which exhibit identical enzyme patterns are referred to as a strain.

Single cell clones and multiple cell cultures from the same population produced identical isozyme patterns on starch gels and total soluble protein patterns on acrylamide gels. These observations suggest 1) each population is homogeneous in respect to these proteins, and 2) a single cell may give rise to a clone containing all isozymes found within a population.

The isozyme patterns of 40 populations of S. microadriaticum were compared. Zooxanthellae from these 40 populations were placed into 12 strains, each with a unique combination of the isozyme patterns for three enzymes: malate dehydrogenase (MDH), glucose phosphate isomerase (GPI), and α-esterase (EST) (Fig. 1). Only two of 17 other enzymes were detected (Schoenberg, in prep.). β-esterase gave the same pattern as α-esterase, although the slower bands had greater β-esterase activity. Tetrazolium oxidase, where present, produced

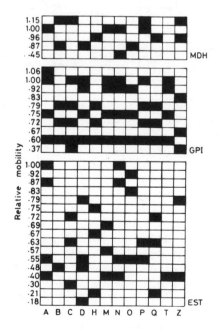

Fig. 1. Schematic diagram of the MDH, GPI and EST patterns of 12 strains of S. microadriaticum.

a single band with a different mobility in each strain. Soluble
protein patterns on disc acrylamide gels were also strain-specific.

    A value for similarity between each pair of strains was derived
by dividing the number of isozyme bands (Fig. 1) in common by the
total number of bands in both strains. Clustering by the single-
linkage method (Sneath and Sokal, 1973) produced the dendrogram in
Fig. 2. Strains A and N (72% similarity) and strains Q and H are
the most closely related. Strain Z (11% similarity to strain D) is
the least related to the others. The patterns of the GPI triplet at
.72-.75-.79 mobility (Fig. 1) suggests a division of these strains
into three groups. Strain Z is one group. The other two groups
separate at 28% similarity, which is the next fork on the diagram.

    The strains of S. microadriaticum also vary in cell size (Fig.
3). Strains in the same 28% similarity group as strain A have dimensions below 9.5 μm x 10.0 μm. Strains in the other 28% similarity
group and strain Z are larger. The standard deviations between the
groups separated at 28% similarity hardly overlap.

                  SPECIFICITY OF S. MICROADRIATICUM

    Both Weiss (1953) and Dubos and Kessler (1963) have emphasized
the subtle, relative nature of biological specificity. The latter
described specificity in symbiotic associations as being concerned
with the "overall pattern of adjustment between the two components,
and is an expression of the complementariness of all their dynamically interacting attributes." In evaluating specificity, not only
the isolated alga but also the host invertebrate must be considered.
We examined specificity by determining both the distribution of the

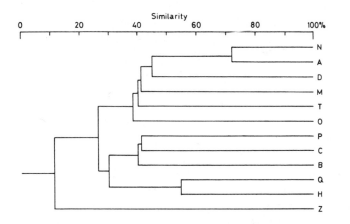

Fig. 2. Dendrogram of 12 strains of S. microadriaticum by single-
linkage clustering of a similarity matrix (see text).

strains in natural hosts and the compatibility of one test host with
a variety of strains.

Table 1 lists the hosts and localities from which these strains
were obtained. There is no one-to-one correspondence between strain
and host species. Five strains were isolated from more than one
host species. Different individuals of three species each harbored
two distinct strains belonging to the larger 28% similarity group.
There may be a geographic factor involved with strain distribution
in Bartholomea annulata. Our data indicate that the symbionts of
eight Atlantic cnidarians are more similar to that of a tridacnid
from the Great Barrier Reef than to that of Zoanthus sociatus from
Jamaica. This last strain uniquely releases leucine after isolation
(Trench, 1974).

Strain distribution does not follow host taxonomy. Each of the
three samples from the gorgonian genus Pseudopterogorgia contained
a different strain. Bartholomea and Heteractis, hosts of strain H,
are in the same family as Aiptasia, a host of strain A. Factors
which are likely to determine host taxonomy do not appear to contribute to compatibility.

The host distribution in Table 1 is only a sample of some of
the hosts in which these strains may be found in nature. These
strains may, no doubt, be found in other species we have not examined. However, the distribution of the strains within the limited
sample is non-random, and suggests that strains have specific host

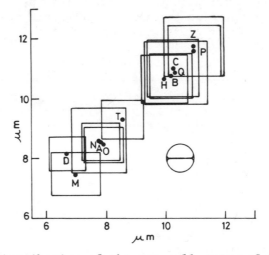

Fig. 3. Size distribution of the two-cell stage of 12 strains of
S. microadriaticum (means and standard deviations). x-axis: length
parallel to cleavage plane. y-axis: length perpendicular to cleavage plane.

Table 1

Sources of Symbiodinium microadriaticum strains

Strain	Host species	Location	No. of hosts
A	Aiptasia tagetes	Berm/Fla	4/1
	Gorgonia ventalina	Berm/Jam	1/1
	Pseudopterogorgia bipinnata	Jamaica	1
D	Condylactis gigantea	Florida	1
M	Meandrina meandrites	Jamaica	1
N	Mussa angulosa	Jamaica	1
	Pseudopterogorgia americana	Bermuda	1
O	Oculina diffusa	Bermuda	4
T	Tridacna gigas	Gr. Barrier	1
B	Bartholomea annulata	Bermuda	2
C	Cassiopeia xamachana	Berm/Fla	3/3
	Protopalythoa grandis	Bermuda	1
	Pseudopterogorgia acerosa	Florida	1
H	Bartholomea annulata	Barbados	3
	Heteractis lucida	Jamaica	1
	Rhodactis sancti-thomae	Barbados	1
P	Palythoa mammilosa	Bermuda	1
Q	Palythoa mammilosa	Berm/Jam	2/1
	Protopalythoa grandis	Berm/Jam	4/1
Z	Zoanthus sociatus	Jamaica	1

distributions.

We have also looked at the ability of these strains to form an association with aposymbiotic individuals of the small Bermuda sea anemone, Aiptasia tagetes (Duch. and Mich.), a natural host of strain A. Such experiments could demonstrate whether the potential range of a strain is limited to those hosts in which it is found in nature. The test host's original zooxanthellae were removed by six months of either dark or DCMU treatment (Vandermuelen, Davis and Muscatine, 1972). After determining that the anemones were free of zooxanthellae, they were fed a 10 µl pellet of cultured zooxanthellae in Artemia extract. The subsequent course of infection was followed by cutting off tentacles and counting the number of zooxanthellae/mm every four days for eight weeks. Counting was also discontinued once cell counts in all anemones infected with algae from a particular host species were greater than 10,000 cells/mm.

Fig. 4 shows the median algal cell densities in Aiptasia tentacles of four groups of strains in the first eight weeks after infection. The density of strain A algae (from all three hosts) rapidly

increased and stabilized at 15,000 cells/mm after day 16. Strain N
algae, most similar in isozyme pattern to strain A, showed a slower
increase in density, stabilizing at 5,000 cells/mm by day 32. The
median of the other small strains which infected (M, O and T) stabilized at about 300 cells/mm by day 12 before increasing to 700
cells/mm after day 40. The median of the large strains which infected (B, C, H and Q) increased gradually to a density of 1,500 cells/
mm by day 56. Strains D (n=4), P (n=4) and Z (n=6) did not infect
Aiptasia. Thus, Aiptasia may establish associations with strains
other than strain A, but these associations differ in one aspect of
"the pattern of adjustment between the two components,..." i.e.,
cell proliferation. These associations may differ functionally
from the association between Aiptasia and strain A.

     Zooxanthellae from successfully infected anemones were isolated
and again placed in axenic culture. These cultures, when electro-
phoresed and when used for infections a second time, provided the
identical isozyme and infection patterns as did the original cultures. No other strain behaved as did strain A. Cell size was also
unchanged inside Aiptasia tentacles during infection and after re-
isolation. Thus, passage through a host modifies neither the

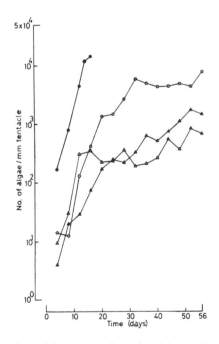

Fig. 4. Median algal cell densities in Aiptasia tentacles to 56
days after infection with cultured S. microadriaticum. ●—● Strain
A (n=16); o—o strain N (n=4); ᐃ–ᐃ strains M, O and T (n=7); ▲—▲
strains B, C, H and Q (n=30).

Table 2

Genetic and infective variation between strains of
Symbiodinium microadriaticum

Electrophoretic group	Size group	Strain	Infectivity
1	small	A	high
		N	intermediate
		M, O, T	low
		D	none
2	large	B, C, H, Q	low
		P	none
3	large	Z	none

characteristics upon which we have established the existence of
strains of S. microadriaticum nor their host specificity.

Table 2 compares trends in isozyme patterns, size and infectivity in Aiptasia of the 12 strains under study. Electrophoretic
differences between groups 1 and 2 correlate with size differences.
The ability to infect Aiptasia rapidly to a high algal density,
however, is a strain characteristic.

OTHER FACTORS IN SPECIFICITY: MODES OF NATURAL INFECTIONS

The specificity demonstrated above does not completely explain
the narrow host range of these strains. Other aspects of these associations may also influence their distribution. In some cnidarians, zooxanthellae are present in the planula during release from
the parent (reviewed in Schoenberg, in prep.). Direct passage of
zooxanthellae between host generations would guarantee the transmission of an established strain. There have been just as many
reports of cnidarians and tridacnids which release reproductive
stages that lack algae. These larvae are infected by presumably
motile S. microadriaticum in the water column. Healthy zooxanthellae which are periodically egested from cnidarians (Taylor, 1969;
Reimer, 1971; Trench, 1974; Steele, 1975) may provide a local concentration of suitable zooxanthellae. Larvae infected with other
strains may be selected against or retarded in development (cf.
Sugiura, 1964; Ludwig, 1969). No multiple infections have been
observed. One of the infecting strains may be better adapted to
this host and might eventually outnumber and exclude the other
strains (Provasoli, Yamasu and Manton, 1968).

Alternatively, the recognition process in larvae may be more subtle than our assay performed on adults would demonstrate. Kinzie's (1974) observation that any S. microadriaticum culture with motile cells would infect Pseudopterogorgia bipinnata larvae casts doubt on this hypothesis. Unfortunately, Kinzie's experimental hosts were lost after one week and he did not quantitatively compare successful infections.

The mechanism and the signals of zooxanthella-cnidarian host recognition, whether carried out by the host, the alga, or both, are unknown. Variation in the soluble proteins discussed above may be related to differences in enzyme function between strains, or may reflect differences in cell surface macromolecular structure (Pool, 1976) and thus may play some role in recognition. Our demonstration of genetic variation and differences in specificity between strains of Symbiodinium microadriaticum may provide a useful tool toward a better understanding of the relationship between recognition and specificity in algal-invertebrate symbiosis.

ACKNOWLEDGMENTS

Contribution No. 667 from the Bermuda Biological Station. We thank W. H. Brakel and J. B. Schoenberg for suggestions on the statistical treatment of our data; L. Provasoli for advice on culture procedures; the Bermuda Biological Station, Discovery Bay Marine Laboratory and Bellairs Research Institute for the use of their facilities. This research was supported by USPHS Training Grant HD-00032 (to D.A.S.), a Sigma Xi Grant-in-Aid (to D.A.S.), and by NSF grants DES 74-14672 (to R.K.T.), GB-34211 (to R.K.T.) and NIH grant 5 ROI GM 20177-03 (to R.K.T.).

LITERATURE CITED

Dustan, P., 1975. "Genecological differentiation in the reef-building coral Montastrea annularis." Ph. D. dissertation. SUNY at Stony Brook. 300 pp.
Dubos, R., and A. W. Kessler, 1963. Integrative and disintegrative factors in symbiotic associations. Symp. Soc. Gen. Microbiol., 13: 1-11.
Freudenthal, H. D., 1962. Symbiodinium gen. nov. and Symbiodinium microadriaticum sp. mov., a zooxanthella: taxonomy, life cycle and morphology. J. Protozool., 9: 45-52.
Hollande, A., and D. Carré, 1974. Les xanthelles des radiolaires sphaerocollides, des acanthaires et de Velella velella: Infrastructure -cytochimie -taxonomie. Protistolog., 10: 573-601.
Karakashian, M. W., 1975. Symbiosis in Paramecium bursaria. Symp. Soc. Exptl. Biol., 29: 145-173.
Kevin, M., Hall, W. T., McLaughlin, J. J. A., and P. A. Zahl, 1969.

Symbiodinium microadriaticum Freudenthal, a revised taxonomic description, ultrastructure. J. Phycol., 5: 341-350.

Kinzie, R. A., III., 1974. Experimental infection of aposymbiotic gorgonian polyps with zooxanthellae. J. exptl. mar. Biol. Ecol., 15: 335-345.

Lang, J., 1970. "Inter-specific aggression within the scleractinian reef corals." Ph. D. dissertation. Yale University. 80 pp.

Lewis, D. H. and D. C. Smith, 1971. The autotrophic nutrition of symbiotic marine coelenterates with special reference to hermatypic corals. I. Movement of photosynthetic products between the symbionts. Proc. R. Soc. London, B178: 111-128.

Ludwig, L.-D., 1969. Die Zooxanthellen bei Cassiopeia andromeda Eschscholz 1829 (Polyp-Stadium) und ihre Bedeutung für die Strobilation. Zool. Jb. (Anat.), 86: 238-277.

Markert, C., Ed., Isozymes. IV. Genetics and evolution. Academic Press, New York. 965 pp.

McLaughlin, J. J. A., and P. A. Zahl, 1966. Endozoic algae. Pages 257-297 in S. M. Henry, Ed., Symbiosis. I. Academic Press, New York.

Pool, R. R., jr., 1976. "Symbiosis of Chlorella with Chlorohydra viridissima." Ph. D. dissertation. UCLA. 122pp.

Provasoli, L., Yamasu, T., and I. Manton, 1968. Experiments on the resynthesis of symbiosis in Convoluta roscoffensis with different flagellate cultures. J. mar. Biol. Ass. U.K., 48: 465-479.

Reimer, A. A., 1971. Observations on the relationships between several species of tropical zoanthids (Zoanthidea, Coelenterata) and their zooxanthellae. J. exptl. mar. Biol. Ecol., 7: 207-217.

Sneath, P. H. A., and R. R. Sokal, 1973. Numerical taxonomy. W. H. Freeman & Co., San Francisco.

Steele, R. D., 1975. Stages in the life history of a symbiotic zooxanthella in pellets extruded by its host Aiptasia tagetes (Duch. and Mich.) (Coelenterata, Anthozoa). Biol. Bull.,149: 590-600.

Sugiura, Y., 1964. On the life-history of rhizostome medusae. II. Indispensability of zooxanthellae for strobilation in Mastigias papua. Embryologica, 8: 223-233.

Taylor, D. L., 1969a. Identity of zooxanthellae isolated from some Pacific Tridacnidae. J. Phycol., 5: 336-340.

_____, 1969b. On the regulation and maintenance of algal numbers in zooxanthellae-coelenterate symbiosis, with a note on the nutritional relationship in Anemonia sulcata. J. mar. Biol. Ass. U.K., 49: 1057-1065.

_____, 1974. Symbiotic marine algae: taxonomy and biological fitness. Pages 245-262 in W. B. Vernberg, Ed., Symbiosis in the Sea, Univ. South Carolina Press, Columbia.

Trench, R. K., 1971a. The physiology and biochemistry of zooxanthellae symbiotic with marine coelenterates. I. The assimilation of photosynthetic products of zooxanthellae by two marine coelenterates. Proc. R. Soc. London, B177: 251-264.

_____, 1971b. _____. II. Liberation of fixed ^{14}C by

zooxanthellae _in vitro_. Proc. R. Soc. London, B177: 237-250.
_____, 1971c. _____. III. The effect of homogenates of
  host tissues on the excretion of photosynthetic products _in vitro_
  by zooxanthellae from two marine coelenterates. Proc. R. Soc.
  London, B177: 251-264.
_____, 1969. Zooxanthellae: a problem in taxonomy. Ass. Island
  Marine Laboratories of the Caribbean. Kingston, Jamaica (Abstract).
_____, 1974. Nutritional potentials in _Zoanthus sociatus_ (Coelen-
  terata, Anthozoa). Helgolander wiss. Meeresunters., 26: 174-216.
Vandermuelen, J. H., Davis, N. D., and L. Muscatine, 1972. The effects of inhibitors of photosynthesis on zooxanthellae in corals
  and other marine invertebrates. Mar. Biol., 16: 185-191.
Weiss, P., 1953. Specificity in growth control. Pages 195-206 in E.
  G. Butler, Ed., _Biological specificity and growth_. Princeton,
  Univ. Press, Princeton, N.J.

Macromolecular Mimicry: Substances Released by Sea Anemones
and Their Role in the Protection of Anemone Fishes

Dietrich Schlichter

Zoological Institute

University of Cologne, Germany FRG

A. Introduction

To provide a background for the investigations dealt with in this
paper, it is necessary to summarize some fundamental experiments
on the protection of anemone fishes.

*a) The effect of isolation of acclimated anemone fishes from
     anemones*

Anemone fishes isolated from their anemones are harmed by their
former anemones if they are brought together with them again.
However, after an acclimation process the fishes can again live
in the midst of the stinging tentacles. The results of such
isolation experiments are the same whether or not the anemones
were associated with other anemone fishes in the meantime.
Furthermore, the species, the size and the mode of locomotion of
the fishes associating with the anemones have no effect on the
outcome of the experiment. Thus, it is clear that the fishes
lose their protection during the period of isolation, while the
capacity of the anemones to discharge their cnidocytes seems not
to be altered fundamentally (the term cnidocyte is used deliberately in contrast to nematocyst, see Section B, p. 3).

b) The specificity of the acquired protection

The specificity of protection which anemone fishes possess after
their acclimation depends on the species of anemone to which
they were acclimated; i.e., an individual anemone fish acclimated successively to several anemone species acquires a different specificity of protection from each species, being either
effective against only one species or effective against several

species simultaneously. The anemones themselves thus partici-
pate in the acquisition of the protection of the fishes.

c) *Chemical interactions between anemones and anemone fishes*

Acclimated or isolated fishes, shut in perforated plastic
boxes, were placed either in contact with anemones or at a distance of 5-10 cm from them. When placed in contact with
anemones, acclimated fishes did not lose their protection and
unacclimated fishes became protected or needed only a very
short time for acclimation. Acclimated anemone fishes placed
at a distance lost their protection and isolated anemones
could not become acclimated. Thus, in order to maintain the
protection, intimate contact with anemones is absolutely
necessary.

d) *Transfer experiments with* ^3H*-labelled amino acids*

Dissolved ^{3}H-amino acids were accumulated and incorporated by
anemones into proteins and glycoproteins. When fish were
acclimated to those anemones, ^{3}H-labelled substances were
detected on the surface of the fishes.

e) *"Unnatural" associations*

In the aquarium it is possible to form "unnatural" associations
between anemones and anemone fishes. Anemone fishes from the
Indian Ocean accept even anemones from the Mediterranean, though
these actinians never live in symbiosis with fishes in their
biotope. This constitutes evidence that in the course of evolution, no special "symbiotic anemones" arose. We know other
fishes, labrids (*Thalassoma amblycephalus*), not closely related
to the "true anemone fishes", which can also live unharmed in the
midst of anemones. The ability to acquire protection is therefore not unique, suggesting that the underlying principles of
acquired protection may be common and perhaps not sophisticated.

f) *Intra- and interspecific recognition*

If tentacles of anemones of different species or even those of
different clones come in contact, they show more or less strong
recognition (aggression) reactions. If tentacles of the same
individual touch one other, they do not react in this manner.
This is trivial, but it demonstrates again that, in accordance
with the organization of coelenterates, chemical signals are the
most relevant signals for these animals. The discrimination
between self and non-self is realized on the basis of chemical
information. This discrimination can be accomplished most
effectively on the surface of the anemone, for cnidarians do not
possess a central coordinating system which could give appropriate
commands.

B. Possible principles underlying the protection of anemone fishes

The findings and considerations listed above, together with others not mentioned here, show that to understand the protection of anemone fishes one must study the chemical alterations which take place on the fishes during the process of acclimation. In a more general sense we have to answer the question: how are the cnido-cytes switched on and off? how is the discharge controlled or influenced? The answers must recognize the cnidocytes are integral parts of the organism, and that earlier investigators have warned that the reactions of nematocysts and cnidocytes should not be confused (quoted in Mariscal, 1974). Several authors (e.g., Westfall, 1965) have shown that around the flagel-lum of the cnidocyte are microvilli which surely play an impor-tant role in chemosensitivity.

Starting from the fact that the alterations which take place on the integument of the fishes during the acclimation process are of a chemical nature, there are two quite different processes which could lead to the protection of the fishes: 1) the fishes themselves might synthesize protecting substances during the process of acclimation; 2) chemical interactions might take place between fishes and anemones. Most of the investigators studying the symbiosis of fishes and anemones have favored the first possibility above to explain their results (e.g. Davenport and Norris, 1958; Eibl-Eibesfeldt, 1960; Mariscal, 1965, 1971). However, the findings summarized in points A. *b-f* above indicate that the anemones are essentially involved in the acquisition of the protection of the fishes and that any one of a number of possible mechanisms may be used to give this protection (Schlichter, 1967, 1968, 1972, 1975). One possible mechanism is that the anemones induce the secretion of protecting substances in the fish integument (*induction*). A second possibility is that substances from the anemones together with those from the fishes react to form a protecting product (*formation*), e.g. by "neutral-izing" the capacity of the fish mucus to stimulate nematocysts. The reaction product remains either on the surface or diffuses from the surface into the sea water and thus no longer stimu-lates the nematocytes. Another possibility is that during the acclimation process the fishes adsorb anemone secretions onto their integument (*impregnation*) and acquire in this way the same (or the decisive) chemical signals that the anemones have. In effect, the anemone fish become chemically disguised as anemones. These possibilities were explored in the work presented in this paper.

C. Results and discussion

The following investigations were undertaken to look for char-acteristic differences between isolated and acclimated anemone

fishes on a chemical level, bearing in mind that differences may
be causally connected with the protection but not necessarily so.
To provide convincing results, investigations of the receptor/
effector unit, i.e. the cnidocytes, must follow and such tests
should be done under conditions which simulate the natural
conditions completely.

a) *Substances released by anemones*

The fact that cnidarians release special secretions is well
known (e.g., Lane, 1968; Martin, 1968). Specific functions of
the secretions have been demonstrated (e.g., Howe & Sheikh,
1975). A good starting point for comparative biochemical studies
was the fact that there are anemone species on which acclimation
is possible ("symbiotic anemones") and others on which it is not
("nonsymbiotic anemones"). Thus, one might expect to be able to
define certain categories of substances involved in the acclimation process.

The different species of anemones were placed in small volumes
of water (50-400 ml, depending on the size) and were occasionally gently touched with glass rods. The water from the cultures was sterilized by membrane filtration, and lyophilized,
its protein and amino acid content determined.

By disc-gelelectrophoresis and isoelectric focusing, proteins,
glyco-lipoproteins and mucopolysaccharides were separated.
After acid hydrolysis the amino acid composition was determined.
Material was analyzed from several "symbiotic" anemones and
one "nonsymbiotic" species. Of special interest was *Condylactis
aurantiaca*, a "symbiotic" species from the Mediterranean, never
associated in the biotope with fishes (see Section A. *e*).

Proteins. The protein pattern of "symbiotic" species is more
complex (this is also true of *C. aurantiaca*) than that of "non-
symbiotic" ones (Fig. 1). "Symbiotic" anemones show, in contrast
to "nonsymbiotic" ones, the presence of free acid mucopolysaccharides. Glyco- and lipoproteins are present in both anemone
types. In the future these two classes of substances will be
studied intensively.

Amino acids. Obvious differences were observed amongst the
anemones in the relative amounts of 3 amino acids. "Symbiotic"
anemones have a higher content of ser and gly (!). "Nonsymbiotic"
species contain much more (glu + glyn) (Table 1, left side).
The analyses demonstrate that there are, in fact, differences in
the composition of substances released by "symbiotic" and "non-
symbiotic" anemones. It is necessary now to show whether these
differences are connected in any way with the protection of
anemone fishes.

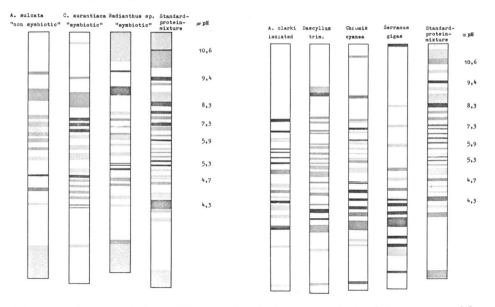

Fig. 1. The protein-pattern of substances released by "nonsymbiotic" and "symbiotic" anemone-species (isoelectric focusing).

Fig. 2. The protein-pattern of the mucus of different fish-species (isoelectric focusing).

Table 1

The amino acid composition of substances released by "nonsymbiotic" and "symbiotic" anemone-species (average of 3 determinations; % of the total amount)

Amino acid or compound	A. sulcata "nonsymbiotic"	C. aurantiaca "symbiotic"	Radianthus sp "symbiotic"
Asp + Asn	9,9	11,0	10,7
Thr	6,4	7,2	6,6
Ser	9,7	12,5	12,5
Glu + Gln	26,2	14,2	9,8
Pro	5,8	4,4	6,2
Gly	11,9	14,9	16,5
Ala	6,6	7,7	7,0
Half Cys	4,3	5,3	9,2
Val	5,0	5,5	5,3
Ile	4,4	3,4	4,6
Leu	5,2	4,5	5,1
Tyr	0,5	2,1	+
Phe	2,6	2,0	2,5
Lys	+	?	+
His	+	1,3	+
Arg	4,1	3,0	5,2

The amino acid composition of mucous of acclimated and isolated Amphiprion clarki (average of 3 determinations; % of the total amount)

Amino acid or compound	A. clarki isolated	A. clarki acclimated
Asp + Asn	10,1	10,3
Thr	6,4	6,1
Ser	7,8	9,4
Glu + Gln	14,9	15,6
Pro	5,5	4,8
Gly	11,8	14,1
Ala	7,7	7,9
Val	5,3	5,6
Ile	4,6	4,1
Leu	7,8	7,1
Tyr	2,7	3,0
Phe	3,0	3,0
His	2,4	2,6
Arg	4,1	3,5
Others, not identified	5,9	2,9

b) The composition of mucus from various fishes

These experiments have not yet been done as intensively as those described above on the substances released by anemones. Mucus can be easily obtained from the fishes (removed with surgical wool) but in small and variable quantities.

Mucus from non-anemone fishes and from isolated anemone fishes

On a histological/histochemical level, Bremer (1972) investigated the composition of the integument of 50 species of marine and freshwater teleosts. The different types of secretory cells, as well as undifferentiated epithelium cells, produce: proteins, glycoproteins, neutral and acid mucopolysaccharides and polysaccharides with sialic acids and with sulfate groups. I recently began histochemical studies of the epithelia of different species of fishes and anemones.

The material taken from the surface of the fish was analyzed for protein and the amino acid composition.

Proteins. The protein patterns (isoelectric focusing) of isolated *Amphiprion clarki*, isolated *Dascyllus trimaculatus* (the juvenile fishes are associated with anemones), and *Chromis cyanea* and *Serranus gigas* (neither of which ever live in symbiosis) were compared. The deviations between potentially symbiotic fishes and *C. cyanea* are not as large as the deviations between the symbionts and *S. gigas* (Fig. 2).

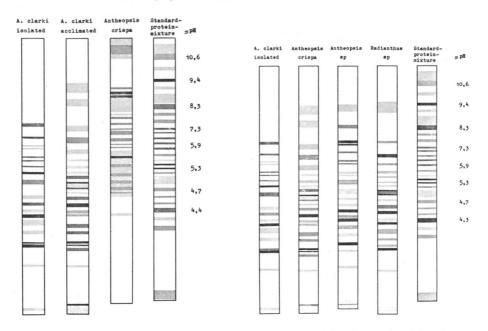

Fig. 3. The protein-pattern of mucus of *Amphiprion clarki* after isolation acclimated to *Antheopsis crispa* (isoelectric focusing).

Fig. 4. The protein-pattern of mucus of *Amphiprion clarki* acclimated to different species of anemones (isoelectric focusing).

Mucus from isolated and acclimated anemone fishes

Proteins. A comparison of the protein patterns of acclimated
and isolated *Amphiprion clarki* clearly demonstrates differences,
especially in the basic region. Acclimated fishes possess 3-4
more protein bands. It is notable that when we look at the
protein pattern of the anemone species to which the fish were
acclimated, we observe those same basic protein bands (Fig. 3).

Amino acids. On the level of the amino acid composition, there
are corresponding differences. The content of ser and gly is
elevated in acclimated fishes (Table 1, right side). The differences in the protein fractions and amino acid composition between
isolated and acclimated anemone fishes can be considered in two
ways. These differences could be either important evidence for
the view that anemone substances are involved in the acquisition
of the protection of anemone fishes, for ser and gly are the
amino acids in which "symbiotic" anemones differ from "nonsymbiotic".
The differences may originate in the fact that during the acclimation, the anemone fishes may produce substances with a higher
content of ser and gly.

The alterations which take place during the process of acclimation are not as fundamental as perhaps expected. This is probably
due to the fact that the protecting substance (or substances) is
not a pure protein, but a glyco- or lipoprotein containing only
a small fraction of polypeptide. If mucopolysaccharides are
involved, the amino acid content is even smaller. Glyco- and
lipoproteins, as well as mucopolysaccharides, are well known as
substances responsible for cell adhesion and cell recognition
(e.g., Allen & Minnikin, 1975). Further analysis should resolve
this dilemma.

*Mucus from anemone fishes acclimated to different species of
anemones*

The analysis of mucus of anemone fishes of the same species acclimated to different species of anemones should show conclusively
whether protection arises by induced syntheses in the fish integument or by acceptance of anemone substances. The very first
investigations delivered some evidence for the specificity of the
protection determined by the anemone species. According to the
species of the host anemone, the mucus of *Amphiprion clarki* is
altered specifically (Fig. 4). The different specificities of
the protection of anemone fishes (see A. *b*, p. 1) seem therefore
to be a real phenomenon of different quality and not of quantity,
a hypothesis which could also have explained the data. The
possibility that one species of anemone fish can synthesize
protecting substances of different specificity seems rather
unlikely.

Comparative studies on the composition of mucus of different species of anemone fishes, isolated and acclimated to different species of anemones, are planned, as well as additional analyses of the mucus of close and distant relatives of the anemone fishes. Two important and conclusive results can be expected: 1) further evidence for the participation of anemone substances in the protection; 2) whether the anemone fishes possess special structural adaptations for symbiosis, in addition to ethological adaptations.

E. Concluding remarks

Assuming the results presented above are true, we can formulate the following scheme or model for the control of the discharge of cnidocytes, at least in their role in the protection of anemone fishes. The model may have the advantage that it handles the problem immediately on the nematocytes. Substances with inhibitory qualities are produced by the anemones for their own purposes, e.g., protection against self-stinging. These inhibitory substances cover the whole body. If these substances are present, the cnidocytes do not discharge. Sensitization of the cnidocytes takes place as soon as the anemones come into contact with "stimulating substances". This happens if anemones are touched by food objects or by anemone fishes which have been isolated from anemones. The surface of these fishes does not contain the protecting substances from the anemones.

After acclimation the fishes possess the decisive inhibitory chemical signals on their surfaces and can then deceive the receptive structures of the anemones. Structures which could act as chemosensitive organelles are the microvilli around the flagellum of the cnidocyte. These apical structures of the cnidocytes could, after stimulation, e.g., by food, lower the threshold for mechanical stimulation received by the flagellum. The sensitization could be accomplished by altering the permeability of membranes for certain ions, and according to the "osmotic hypothesis" or the "contraction hypothesis" of discharge, the changed conditions would lead to the explosion.

The principle of macromolecular mimicry likewise forms the basis of a symbiosis fundamentally different from the one mentioned here, but we can draw conclusions by analogy. Smither and coworkers (1972) are studying the parasitic symbiosis of *Schistosoma*, the causative organism of schistosomiasis. They were able to show that *Schistosoma* incorporate in their integument antigens from the host organism. By this means the parasites disguise themselves chemically and are able to avoid recognition by the host antibodies.

These investigations were supported by the Deutsche Forschungsgemeinschaft. The assistance of Dr. D.M. Ross in preparing the manuscript is gratefully acknowledged.

D.   References

Davenport, D. and K.S. Norris, 1958. Observations on the symbiosis
    of the sea anemone Stoichactis and the pomacentrid fish Amphi-
    prion percula. Biol. Bull., 115: 397-410.

Eibl-Eibesfeldt, I., 1960. Beobachtungen und Versuche an Anemonen-
    fischen (Amphiprion) der Malediven und Nicobaren. Z. Tierpsychol.
    17: 1-10.

Howe, N.R. and Y.M. Sheikh, 1975. A sea anemone alarm pheromone.
    Science 189: 386-388.

Lane, C.E., 1968. Chemical aspects of ecology: Pharmacology and
    Toxicology. In: Chemical Zoology. Vol. II, pp 263-284. Ed.
    by M. Florkin and B.T. Scheer. New York: Academic Press.

Mariscal, R.N., 1965. Observations on acclimation behavior in the
    symbiosis of anemone fish and sea anemones. Am. Zool. :694.

Mariscal, R.N. 1971. Experimental studies on the protection of
    anemone fishes from sea anemones. In: The biology of symbiosis,
    Ed. by T.C. Cheng, University Park Press, Baltimore, 283-315.

Mariscal, R.N., 1974. Nematocysts. In: Coelenterate Biology pp
    129-166. Ed. by L. Muscatine and H.M. Lenhoff. New York:
    Academic Press.

Martin, E.J., 1968. Specific antigens released into sea water by
    contracting anemones (Coelenterata). Comp. Biochem. Physiol.
    25: 169-176.

Schlichter, D., 1967. Zur Klärung der Anemonen-Fisch-Symbiose.
    Naturwissenschaften, 21: 569-576.

Schlichter, D., 1968a. Das Zusammenleben von Riffanenmonen und
    Anemonenfischen. Z. Tierpsychol., 25: 933-954.

Schlichter, D. 1972. Chemische Tarnung. Die stoffliche Grundlage
    der Anpassung von Anemonenfischen an Riffanemonen. Mar. Biol.
    12: 137-150.

Schlichter, D., 1975. Produktion oder Übernahme von Schutzstoffen
    als Ursache des Nesselschutzes von Anemonen-fischen. J. exp.
    mar. Biol. Ecol., 20:49-61.

Smithers, S.R., 1972. Recent advances in the immunology of
    schistosomiasis. Br. Med. Bull., 28: 49-54.

Westfall, J.A., 1965. Nematocysts of the sea anemone Metridium.
    Amer. Zool., 5: 377-393.

APOSEMATIC COLORATION AND MUTUALISM IN SPONGE-DWELLING TROPICAL ZOANTHIDS

David A. West

Smithsonian Tropical Research Institute

Box 2072, Balboa, Canal Zone

Consolidated substrate appears to be at a premium for sessile coral reef inhabitants. Studies of tropical reef faunas suggest that organisms are competing for space in a number of ways (e.g., Lang, 1971; Connell, 1976; Jackson and Buss, 1975; Sebens, 1976). Many sessile organisms, however, appear to have lost out in the more open regions of reefs to superior competitors such as the scleractinian corals (Jackson, et al, 1971). Still others have evolved mechanisms which enable them to utilize other organisms as substrate for attachment. Such associations are often complex in nature, involving adaptations by coevolving symbionts to facilitate mutualistic interaction. Found on the coral reefs of Puerto Rico is one genus of zoanthideans, Parazoanthus, which has exploited such an interaction as a solution to substrate scarcity.

Parazoanthus is a worldwide tropical and temperate genus of primarily epizoic zoanthids. In addition to the four recognized species from Puerto Rico, P. tunicans, P. swiftii, P. parasiticus, and P. catenularis, one new species has been found and is being described (West, in preparation). With the exception of P. tunicans, which occurs only on the arborescent hydroid Plumularia sp., all Puerto Rican species are epizoic on sponges.

During two years of study, the curious observation was made that sponge-dwelling Parazoanthus almost always contrast markedly in color with the sponges on which they dwell. In addition, comparison of a list of host sponges with an inventory of West Indian sponges preyed upon by reef fish (Randall and Hartman, 1968) revealed that the majority of these host sponges are among those favored by sponge feeding fish (Table 1). These two observations ask the

TABLE 1

Occurrence of Host Sponges in West Indian Reef Fish Stomachs[1]

Sponge	Rank[2]	Occurrences[3]	Dominance[4]
Callyspongia vaginalis	1	27	15
Xestospongia sp.	4	11	1
Gelliodes ramosa	5	9	4
Agelas sp.	6	8	4
Spheciospongia sp.	12	5	2
Callyspongia armigera	15	5	1
Iotrochota birotulata	16	4	1

[1] Data are from Table 2, Randall and Hartman (1968)
[2] Sponges are ranked according to the incidence at which they are found in fish stomachs from a list of the 22 sponges most frequently eaten by fish.
[3] Number of occurrences in fish stomachs; these data are from 176 individuals of those species having over 30% sponge material in their stomachs.
[4] Numbers are the instances in which the sponge was the most abundant one in the stomach.

questions: 1) of what significance is this presumed advertisement of the zoanthids' presence; and 2) how do they avoid being eaten along with their host sponges which are under such predation pressure? One possible explanation is that the coloration is aposematic, serving to prevent predation. This hypothesis requires that the zoanthids be either unpalatable or toxic to sponge predators, firstly, to permit the evolution of the association, and thereafter, at a level sufficient to maintain it. In this paper, I report the results of experiments designed to test this hypothesis. The first series of experiments examined two species of zoanthids, Parazoanthus swiftii and Parazoanthus n. sp., for toxic compounds. The second group of experiments attempted to: 1) determine the degree of predation on the zoanthid, P. swiftii, and its host, Iotrochota birotulata, by the rock beauty, Holacanthus tricolor; and 2) test if predator avoidance is a response to a visual stimulus.

MATERIALS AND METHODS

Zoanthids and sponges were collected from the inner fringing reefs and outer deep reefs at the edge of the insular shelf off La Parguera, Puerto Rico, in water depths ranging from 10 to 50 meters. Specimens were maintained in various sizes of aquaria with running sea water in the Department of Marine Sciences, University of Puerto

Rico, laboratory at La Parguera.

The toxicity of extracts from Parazoanthus swiftii and P. n. sp. was assayed by intraperitoneal injection into white mullet, Mugil curema. These zoanthid species were chosen because of their abundance and of their relatively large size. Sponges bearing the two species were immediately frozen after collection to facilitate separation of zoanthid from sponge. Zoanthids were then separated from their hosts and extracted following the initial procedures of Moore and Scheuer (1971) with Hawaiian Palythoa. The collected and cleaned sample of each species was placed in 70% ethanol (4ml/g wet weight) and ground in a Waring blender for 15 minutes. The materials were extracted for 12 hours at 4° C. Extracts were next centrifuged (27,000 X g) for one hour at 4° C. The supernatant was decanted and the sediment washed once with 70% ethanol. The extracts (supernatant) and washings were combined and evaporated to low volume under reduced pressure (50° C) and freeze-dried.

The white mullet, Mugil curema, was chosen as the assay organism for several reasons: 1) it is a teleost as are potential predators of zoanthids; 2) it does not prey upon sponges nor inhabit the same environment as the zoanthids, and thus can be assumed to lack specific detoxification mechanisms; and 3) it is both abundant and easily collected. Mullet were collected at Playa Papayo, La Parguera, by seine, and maintained in holding cages in the sea until used.

Injection experiments were four hours in duration. Dosages and volumes of innocula were adjusted to fish weight. Each dose was delivered in a constant proportion to fish weight, in distilled water, at 0.05 ml/g fish weight. Injection controls consisted of distilled water injected in the same proportion as the experimental innocula (0.05 ml/g fish weight). Assay mullet ranged from 7 to 35 grams at each dosage. The osmolality of the highest dosages was determined with a Fiske osmometer to examine extracts for an abundance of salts which could cause osmotic stress during the experimental periods. Both mortality and death times were recorded.

The degree of predation on a host sponge and its zoanthid symbiont was examined by feeding experiments using P. swiftii and its predominant host, Iotrochota birotulata (Figure 1a), and the sponge feeder, Holacanthus tricolor. The high predation by H. tricolor on this sponge was the primary criterion in the selection of experimental animals. Randall and Hartman (1968) found that of over 20 species of sponge which comprise 97.1% of the rock beauty's diet, I. birotulata was the most commonly found sponge in its stomach contents (15.6% by volume). H. tricolor is a common deep reef pomacanthid in Puerto Rico, most often found below 10 meters depth. Experimental fish were captured in traps and kept in holding cages until their survival was assured.

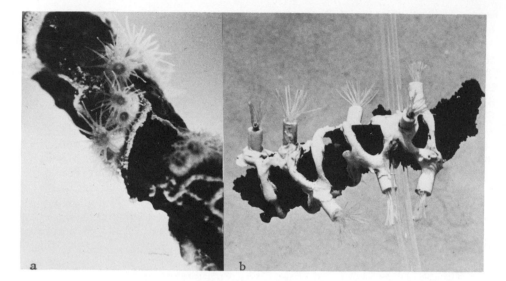

Figure 1. <u>Iotrochota</u> <u>birotulata</u> with its epizoite, <u>Parazoanthus</u> <u>swiftii</u> (a), and suspended on a glass rod with artificial zoanthid (b).

Feeding experiments were conducted in running sea water in out-door cement tanks, 10 x 4 x 1.5 feet. Each tank was partitioned in-to five equal divisions, 2 x 4 feet, one division for each fish. Predators were acclimated to tank conditions for two days without food before being used in the experiments.

Two sets of laboratory experiments were carried out in which in-dividual predators were presented prey of different types in week long trials. In the first set, fish were given one of three types of prey: 1) bare sponge only (single choice); 2) sponge with zoan-thid (single choice); or 3) both bare sponge and sponge with zoanthid (mixed choice). These experiments tested whether or not the predator would eat prey sponge with zoanthid. The second set of experiments was identical to the first, with the exceptions that plastic, fabri-cated zoanthid colonies were used in place of <u>P. swiftii</u> colonies and the bare sponge trials were not repeated. This set was designed to determine if predator avoidance of zoanthid-sponge prey is a visual response.

Sponges were cut into uniform-sized pieces, weighed, and sus-pended on glass rods in the experimental tanks. From separate con-trol material, a dry to wet weight ratio was calculated for bare sponge and sponge with zoanthids by drying the material for 24 hours at 100°C. This ratio was used to determine the dry weight of the

material at the start of the experiments. After one week, the sponges were removed, dried, and weighed. Additional samples of sponges
were suspended in a tank without a predator for one week to determine the change of weight due to experimental procedure. These samples served as controls. Within each experimental prey type, weight
changes of prey exposed to fish were statistically compared to the
changes in weight of the corresponding controls with the t-test.

The artificial zoanthids were made of chromate tinted Caulk-Tex
(Travaco Industries), yellow wire insulation and yellow polypropylene
rope fibers. The plastic zoanthids were formed on a cylindrical rod,
removed, and soaked in running sea water for one week before use.
Weighed pieces of sponges were then placed inside the "zoanthid"
clone and presented to predators. The ends of the sponges were allowed to protrude from the clones in order to present unobstructed
sponge for predation (Figure 1b). After one week, the sponges were
removed, separated from the plastic zoanthids, dried, and weighed.

The last set of experiments was conducted in the field. Pieces
of sponge with and without zoanthid symbionts were weighed and suspended on glass rods. The samples were then attached to stakes dri-
ven into the reef at the edge of the insular shelf, 10 kilometers
south of La Parguera, and left exposed to predators for one week. At
the end of the week, the samples were returned to the laboratory,
dried, and weighed. The weight changes of the two field groups (bare
sponge and sponge with zoanthid) were then statistically compared.

RESULTS

Intraperitoneal injection of _Mugil_ _curema_ revealed that both species of zoanthids contained toxic substances. After the initial dose
showed the presence of toxic substances, additional dosages were chosen to establish a dosage-mortality curve and a 50% lethal dose. Dosages, mortality (%), and mean death times for the two zoanthid extracts are shown in Table 2.

Lethal and sub-lethal symptoms of injected mullet, for both extract series, were loss of balance, increased breathing rate and apparent loss of use of the acoustico-lateralis system. Some of the
surviving fish in each of the four dosages below 100% mortality displayed the same predeath symptoms of the fish that died. Those fish
which did not regain their normal orientation and behavior within 60
minutes after injection eventually died.

The results of the laboratory and field predation experiments
are summarized in Tables 3-4. All mean weight changes are negative
(loss) except for one control (Table 3) which experienced a mean gain
in weight. However, some individual test and control sponges from all

TABLE 2

Dosages, Mortality, and Mean Death Times of
Innoculations of Zoanthid Extracts into Mugil curema

Dosage mg/g	Mortality (%)	N	Mean Death Time minutes (±SE)	
		P. swiftii		
.50	100	9	23.5	(3.9)
.43	78	9	37.0	(11.9)
.33	22	9	28.0	(6.0)
.30	11	9	32.0	
.25	0	9	----	
		P. n. sp.		
.65	100	5	6.8	(0.8)
.50	56	9	15.6	(6.9)
.35	22	9	16.5	(7.5)
.25	11	9	21.0	
.10	0	9	----	

TABLE 3

Mean Weight Loss (%) in Single and
Mixed Choice Feeding Experiments (±SE)

Prey Type	Weight Loss		
	Prey	Control	
Single choice:			
1) sponge without zoanthid	25.4 (3.5) (n=5)	12.8 (2.6) (n=6)	p<.05
2) sponge with zoanthid	1.3 (2.4) (n=5)	5.5 (6.0) (n=6)	N.S.
Mixed choice:			
3) sponge without zoanthid	28.5 (1.3) (n=6)	13.2 (3.2) (n=6)	p<.05
sponge with zoanthid	5.4 (3.6) (n=6)	5.7 (2.2) (n=6)	N.S.
Single choice:			
4) sponge with plastic zoanthid	11.7 (4.0) (n=5)	5.5 (3.1) (n=6)	N.S.
Mixed choice:			
5) sponge without plastic zoanthid	18.0 (2.7) (n=5)	−8.8 (5.3) (n=6)	p<.05
sponge with plastic zoanthid	0.1 (4.3) (n=5)	5.5 (3.1) (n=6)	N.S.

TABLE 4

Mean Weight Loss (%)
in Field Experiments

1)	sponge without zoanthid	14.9 (1.8) (n=5)	
2)	sponge with zoanthid	9.6 (1.7) (n=5)	p < .06

categories except that of test prey without zoanthids increased in weight during the week period. In the latter category, all sponges suffered a loss in weight.

Examination of mean weight losses in the laboratory experiments with live zoanthids (Table 3) indicates that the test bare sponges in both single and mixed choice experiments lost more weight than their respective controls. A similar loss was not found for the zoanthid-bearing sponges. The mixed choice experiments allowed a direct comparison of the two prey types. Here, a significant difference was found between the two prey types, but not between the control groups.

The feeding experiments utilizing artificial zoanthids showed similar results (Table 3). In the single choice experiments (presentation of sponge with plastic zoanthid) no significant difference in weight loss was found between controls and prey sponges. Within the mixed choice experiments, sponges with plastic zoanthids lost more than controls, whereas the experimental bare sponges lost significantly more than their controls.

The field experiments suggest that zoanthid-bearing sponges are not preyed upon on the reef. These results (Table 4) show that bare sponge again lost more weight than sponge with zoanthids. Weight loss here is compared between the two prey types as it was not possible to run controls simultaneously.

DISCUSSION

The bioassay of zoanthid extracts on the white mullet, Mugil curema, demonstrated the toxicity of the two common Puerto Rican sponge-dwelling zoanthids, Parazoanthus swiftii and P. n. sp. The symptoms displayed by the injected fish were essentially identical. The relationship between oral and intrperitoneal dosages of toxins has been reported (see Milstein, 1971). Various assay organisms have been found 15 to 37 times more tolerant orally than intraperitoneally.

Consequently, from the data presented in this study, it is conceivable that a fish, if it did feed upon sponges bearing zoanthids

used in the bioassays, could ingest the equivalent of the sublethal or
lethal intraperitoneal dosages. If a similar intolerance is assumed
for sponge predators such as Holacanthus tricolor, even sublethal
amounts could be deleterious to the predator if it developed the symptoms displayed by the mullet at sublethal dosages. Fish suffering
from such symptoms could easily be eaten by larger predators. Assuming that potential predators possess no adaptations for detoxification
of such substances, and that an effective amount of toxin could be
ingested orally, predators would be expected to learn to avoid these
zoanthids if sensory reinforced cues were present.

     After establishment of toxicity, feeding experiments were design-
ed to investigate the extent of predation by a sponge-feeding fish on
a zoanthid and its host sponge in the laboratory and field, and to
determine if discrimination between sponge with and without zoanthids
is accomplished visually. The results of these experiments support
the hypothesis that contrasting coloration is aposematic. Predators
presented with only bare sponge fed on the available food, but those
presented with only sponge bearing zoanthids endured the experimental
period without eating. Those with a choice ate only bare sponge.

     Another possible explanation for predator avoidance of zoanthids
is that fish are responding to the zoanthids' nematocysts. In order
to eliminate that possibility as well as a chemotactic one, the feeding experiments with artificial zoanthids were carried out. The results were similar to those of the experiments using live zoanthids,
i.e., H. tricolor did not prey upon sponge which bore imitation zoanthids. As there was no chemical or nematocyst stimuli present, the
lack of predation suggests that predators are responding visually.

     Finally, the field experiments suggest that sponge with zoanthids
is under less predation pressure in nature than is sponge alone. It
is not known what sponge predators other than fish may exist nor what
role the coloration might have in interaction with such predators.
However, the site on which these experiments took place was inhabited
by the densest populations of pomacanthids seen by this author in the
La Parguera area. Personal observations of the relative abundances
of I. birotulata with and without P. swiftii showed that virtually no
I. birotulata occurred on these deep reefs without its zoanthid symbiont. However, on the inner shallow reefs where H. tricolor and
other sponge predators are uncommon, Iotrochota was often found without P. swiftii in less than three meters of water. In water this
shallow, adult pomacanthids were rare and juveniles there feed little
on sponges (see Randall, 1967; Feddern, 1968).

     Although virtually all zoanthid-sponge associations in Puerto
Rico were aposematic, an exception existed frequently enough to warrant examination. The maroon zoanthid, Parazoanthus n. sp., was
often found on a similarly colored, undescribed sponge. Preliminary
bioassay by intraperitoneal injection of extracts into mullet,

indicated that this sponge contains toxic compounds. This is further
substantiated by the testing of sponge for antibiotic activity
against 82 species of marine bacteria (see this author's data, Burkholder, 1972). Intense inhibition of all bacteria species was shown,
indicating the presence of a potent biological agent. These two
pieces of information suggest that this sponge may contain compounds
toxic enough to afford protection from predation. Consequently, contrasting zoanthid color would be of no additional value to either
symbiont.

The other species of sponge-dwelling zoanthids in Puerto Rico,
appear to take advantage of this aposematic system either directly,
or possibly by mimicking the contrasting coloration. The best evidence for this is the observation that all of these species do contrast with their host sponges. Parazoanthus parasiticus is found on
virtually all individuals of its host species below three meters with
the exception of the extremely thin-layered, encrusting, boring
sponges of the genus Cliona. P. catenularis, a less common species,
is found on two species of Xestospongia, X. muta and X. sp. The
latter sponge also occurs virtually always with the zoanthid symbiont.
An undescribed species of the typically deep water genus, Epizoanthus,
is also found on Xestospongia sp., often on the same individual
sponge with P. catenularis. The fact that most of the host sponge
species occur exclusively with zoanthids suggests that the protective
value of the association is high in all cases.

Studies of symbiotic associations usually attempt to elucidate
benefits and costs to each member. This paper has reported experimental data which allows some comment on the evolution and maintenance of the zoanthid-sponge associations in Puerto Rico. As mentioned before, consolidated substrate is at a premium on coral reefs.
The obvious benefit derived from the association by the zoanthid is
substrate for attachment. This source of substrate does not seem to
have been extensively exploited by sessile organisms in the West Indies. No other epizoites were observed living on the zoanthid host
sponges. Another possible benefit to the zoanthids may be at a cost
to the host. Zoanthids living on sponges with large exhalant oscula
are always concentrated on the exterior surfaces where inhalant openings predominate. This ensures that oxygenated, unfiltered water
passes by the zoanthids before entering the sponges. The possibility
is thus raised that zoanthid prey availability is enhanced by these
sponge-generated currents. Although it is not known if overlap in
food requirements exists between the two symbionts, there is the possibility that availability of food is reduced for the sponges by the
zoanthids' presence.

Although the association may exist at some cost to the host
sponges, such cost appears to be far outweighed by the benefits demonstrated in this study. Predation on zoanthid-bearing sponges does
not seem to occur. As only this most obvious aspect of the

associations was examined, it is not known what other interactions
exist between the symbionts. What role might these have played in
the development of these associations? When more studies are completed, answers may be found for this question and the more obvious
one: why have the zoanthids chosen for hosts the most edible
sponges in the first place?

REFERENCES

Burkholder, P. R., 1973. The ecology of marine antibiotics and coral
    reefs. Pages 117-182 in O. A. Jones and R. Endean, Eds., Biol-
    ogy and Geology of Coral Reefs, 2. Academic Press, New York.
Connell, J. H., 1976. Mechanisms determining diversity of reef-
    building corals. In G. O. Mackie, ed., Coelenterate Ecology
    and Behavior. Plenum Publishing Co., New York.
Feddern, H. A., 1968. Systematics and ecology of western Atlantic
    angelfishes, family Chaetodontidae, with an analysis of hybrid-
    ization in Holacanthus. Ph.D. dissertation, University of
    Miami, Coral Cables. 211 pp.
Jackson, J. B. C., and L. Buss, 1975. Allelopathy and spatial com-
    petition among coral reef invertebrates. Proc. Nat. Acad. Sci.,
    USA, 72: 5160-5163.
Jackson, J. B. C., Goreau, T. F., and W. D. Hartman, 1971. Recent
    brachiopod-coralline sponge communities and their paleoecolog-
    ical significance. Science 173: 623-625.
Lang, J., 1971. Interspecific aggression by scleractinian corals.
    1. The rediscovery of Scolymia cubensis (Milne-Edwards and
    Haime). Bull. Marine Science 21: 952-959.
Milstein, C. B., 1971. Puffer toxin of Sphoeroides testudineus
    (Tetraodontinae) in predator-prey interaction. M.S. thesis,
    University of Puerto Rico, Mayaguez. 85 pp.
Moore, R. E., and P. J. Scheuer, 1971. Palytoxin: a new marine
    toxin from a coelenterate. Science 172: 495-497.
Randall, J. E., 1967. Food habits of reef fishes of the West Indies.
    Stud. Trop. Oceanogr. (5): 665-847.
Randall, J. E. and W. D. Hartman, 1968. Sponge-feeding fishes of
    the West Indies. Marine Biol. 1: 216-225.
Sebens, K. P., 1976. The ecology of Caribbean sea anemones: util-
    ization of space on a coral reef. In G. O. Mackie, ed.,
    Coelenterate Ecology and Behavior, Plenum Publishing Co., New
    York.

SKELETAL MODIFICATION BY A POLYCHAETE ANNELID IN SOME SCLERACTINIAN CORALS[1]

R. H. Randall and L. G. Eldredge

University of Guam Marine Laboratory

Agana, Guam

Background and Introduction

The presence in some faviid corals of deep intercalicular grooves and subphaceloid corallites united by a system of tubercules and epitheca-like tubes (groove-and-tube structure) has been the subject of considerable controversy and speculation in reef-coral taxonomy since the feature was first described by Milne Edwards and Haime (1848) in their diagnosis of the genus *Phymastrea*. During the past ten years, the present authors have been studying and collecting reef corals at Guam and at other Micronesian islands and have never observed features which could be related to such a unique mode of corallite junction. The first encounter with this structural feature came while the senior author was a member of a joint Sino-American team conducting a biological survey of the coral reefs of southern Taiwan. During this survey the coral wall junction was hastily noted while cleaning and packing coral specimens, but it was not until later when the coral collection was being studied in detail, that the significance and possible association of the coral and some other organism was realized. This has been a problem noted by several others who had studied similar structures (Quelch, 1886; Rosen, 1968). We thus found ourselves in the same situation as previous workers studying this structural feature, without preserved coralla with intact polyps and without the opportunity to observe *in situ* specimens known to possess the structure. Our field data and specimen notes indicate that this common and conspicuous coral, here placed with *Favia valenciennesii* (Milne Edwards and Haime, 1848), was collected and observed in a wide range of biotopes along the Southern Taiwan coast (Jones et al., 1972).

---

[1]Contribution No. 85, University of Guam Marine Laboratory.

During 1974 and 1975 the senior author again had the opportunity to collect corals from different localities in Taiwan. During the course of this work a number of Favia valenciennesii colonies were freshly collected and preserved and many in situ specimens observed. Examination of the preserved specimens revealed the presence of numerous polychaete worms believed to belong to the Family Syllidae, genus Toposyllis, within the tube structure. Subsequent examinations of the coral collection from Taiwan have revealed the same phenomenon in the related faviid genera Favites and Cyphastrea and in the unrelated mussid genus Symphyllia.

The first mention of a coral with deep groove-and-tube structure was made by Milne Edwards and Haime (1948) in describing the type of Phymastrea valenciennesii. Here the feature was given generic consideration because of its departure from the normal mode of corallite junction. Duncan (1883) took the same view as Milne Edwards and Haime on the significance of the corallite junction and added a new species Phymastrea irregularis. Quelch (1886) erected a new species Phymastrea aspera for his specimen showing a groove-and-tube structure and was the first to suggest the feature might be those of worm tubes. In his revision of the "Astraeidae," Matthai (1914) believed that the groove-and-tube structure was simply variation within the species with no generic significance so he combined a number of species including the Phymastrea species of Milne Edwards and Haime and Quelch, under the name Favia bertholleti (Valenciennes). Vaughan (1918) agreed with Matthai in regard to the significance of the groove-and-tube structure, but regrouped his species of Favia and pointed out that Valenciennes' name F. bertholleti was invalid, being only a manuscript name, and suggested the use of the next available name, Favia valenciennesii (Milne Edwards and Haime, 1848). In addition to those species previously grouped together, Matthai (1924) added Duncan's Phymastrea irregularis to the synonymy of F. valenciennesii. In formal descriptions F. valenciennesii has been used by nearly all subsequent authors since Vaughan's (1918) regrouping of Favia. Exceptions to this usage are Yabe, Sugiyama, and Eguchi (1936) who retained the generic name Phymastrea, but gave no reason for doing so, Umbgrove (1939) used Phymastrea as a subgeneric rank for species of Favia possessing groove-and-tube structures. Rosen (1968) revised the species within the genus Phymastrea and redistributed part of them questionably to Plesiastrea and the remainder among several species of Favia.

In addition to the significance of the groove-and-tube structure, a major point of confusion exists on the mode of colony increase. In the original generic description of Phymastrea given by Milne Edwards and Haime (1848), the method of colony increase was reported as "extracalicular and subapical." The same authors' generic diagnosis of Phymastrea published in 1857 is similar to their 1848 description except they then reported it as "calicular and submarginal." Duncan (1883) noted the change in the method of colony increase reported in

the 1857 diagnosis and, based upon drawings and descriptions of the
species published in the first diagnosis of Phymastrea and from
examination of his own Phymastrea irregularis specimen, emended the
genus by stating the method of colony increase as that of "extracalicular gemmation."

Rosen (1968) remarked that species corresponding to Milne
Edwards and Haime's genus Phymastrea were similar to species of
Favia but because of the frequent presence of extratentacular budding
he questionably placed them in the genus Plesiastrea. Although superficially the many Taiwan specimens examined appeared to increase by
extratentacular budding, close examination failed to reveal that this
was actually the case except in rare occasions. Division usually
takes place in the small marginal angles of the polygonal-shaped
calices, although subequal division is also present, and a new wall
is rapidly established while the newly divided calice is quite small.
A number of newly developing calices were sectioned and it was found
that they all retained direct communication with the parent polyp
below the newly formed wall. Duncan (1883) observed this direct
communication between newly developed calices and the parent polyp,
remarking that it gave the appearance of fission, but explained it
as the consequence of a thinning of the wall when the buds arise near
the margin of the calice. A number of polyps were also observed in
the distomodaeal condition, especially near the colony margin. We
therefore agree with Milne Edwards and Haime's (1857) diagnosis that
the mode of colony increase is primarily by marginal intratentacular
budding and retain the specimens referrable to Phymastrea in the
genus Favia.

From the Taiwan collection of corals, the groove-and-tube
structure has been observed in four genera and nine species as listed
below:

In addition, groove-and-tube structures have been examined in
two specimens of Favia favus from a collection of corals from Heron
Island, Australia, presented to the authors by Dr. Masashi Yamaguchi.

> Favia valenciennesii (Milne Edwards and Haime, 1848)
> Favia favus (Forskaal, 1775)
> Favia speciosa (Dana, 1846)
> Favia sp. 1
> Cyphastrea gardineri Matthai, 1914
> Cyphastrea microphthalma
> Favites pentagona (Esper, 1794)
> Favites sp. 1
> Symphyllia nobilis (Dana, 1846)

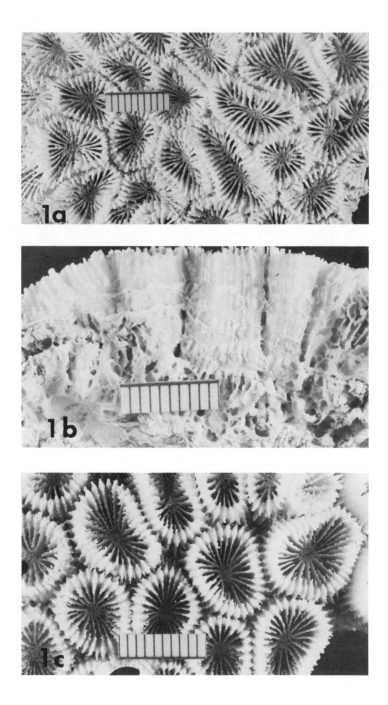

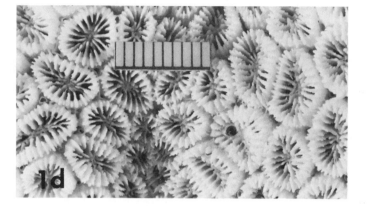

Figure 1a, _Favia favus_ with narrow intercalicular grooves and some-
    what inconspicuous tubes, note the subequal fission; 1b, broken
    vertical section of _Favia favus_ showing the subphaceloid
    corallites; 1c, _Favia_ sp. 1 showing very deep grooves and
    numerous but inconspicuous tubes; 1d, _Favia valenciennesii_
    showing deep grooves and tubes.

An interesting feature of most faviid colonies possessing the groove-
and-tube structure is that all or nearly all of the corallites within
a colony are bounded by the tubes, whereas in Cyphastrea this feature
surrounds clusters of corallites many times. With the exception of
Symphyllia nobilis and the two Favites species, the colonies possessing the tubes could be recognized in the field by the presence of
narrow, deep intercolicular grooves.

     Detailed descriptions of coral species with associated groove-
and-tube structure have been given by Milne Edwards and Haime (1848,
1857), Duncan (1883), Quelch (1886), and Rosen (1968). Rosen's
descriptions are especially comprehensive and are augmented by
plates and text figures showing the feature in a number of different
species and a developmental sequence of the structure. Details of
the groove-and-tube structure in our specimens are essentially the
same as those described by Rosen (1968). In general the structure
is characterized by a vertical and horizontal system of intercommuni-
cating epithecal-like tubes situated between adjacent corallites.
The upper ends of the tubes form openings which range from small
circular to elongate trough-like apertures. Although the groove is
prominent in Favia, it may be lacking or poorly developed in Favites
and Symphyllia. Figures 1-2 show the general details of these
features for several genera.

Observations of the Polyps

     The living polypoid layer covering the skeleton was found to be
interrupted at the locations of tube openings in F. valenciennesii;
a possibility first suggested by Rosen (1968). The interruptions
are confined to the coenosarc lying between the corallites and, thus,
would be analogous to edge zone found at the peripheral margin of the
colony. This suggests that the tube material is epithecal in origin,
and the internal growth lines represent intervals of successive
deposition similar to the lines observed in basal epitheca deposited
on the skeletal tissue at the margin of the colony. The presence of
the worm in the upper part of the tubes apparently maintains the
openings in the coenosarc which induces the coral to secrete the tube
material.

     Rosen (1968) felt that at least some of the corals possessing
the groove-and-tube structure were adversely affected in their overall
growth, pointing out that in Crossland's Leptastrea, in which he found
groove-and-tube structures, the septal cycles were fewer and the
general character less spinulose. In our field observations there was
no evidence to indicate adverse growth, as affected specimens were
observed within many sizes and appeared to be able to grow normally
and compete effectively for space. In fact some of the largest
faviids observed had the groove-and-tube feature.

Observations on the Worms

The worms examined from $\underline{F}$. $\underline{valenciennesii}$ are members of an undescribed species of the genus $\underline{Toposyllis}$ of the family Syllidae. Specimens are long, to as much as 11 mm, and have a very small diameter, less than 0.5 mm. Their tubes are normally no more than 1 mm in diameter. The family is characterized by having a distinct prostomium, four eyes and three antennae, palps of variable shape and size, and an eversible proboscis with various arrangements of teeth. The body segments have uniramous parapodia, and in the present specimens the setae are all compound (Imajima, 1966a).

The polychaete family contains many species from throughout the Indo-West Pacific; eighty-six species are known from Japanese waters (Imajima, 1966a). Polychaetes are known to have a destructive effect on corals and coral reefs (Hartman, 1954) mainly because of their ability to penetrate the substrate or because of their affinity to live in crevices or other narrow passages. A number of polychaetes live in close association with other invertebrates, mainly echinoderms or sipunculans (Gibbs, 1971; Grassé, 1959). Nowhere has the association between any polychaete and any scleractinian coral been noted. Systematic literature of Palau (Takahasi, 1941), the Marshall Islands (Hartman, 1954; Reish, 1968), the Solomon Islands (Gibbs, 1971), and Japan (Imajima, 1966a, 1966b, 1966c) has been examined and this association has not been recorded, although a number of species were collected from unidentified coral substrate. These papers were based on field-collected specimens where the collections were made extremely carefully in order to note the substrate. In some, particular care was taken to break open corals to remove the polychaetes found as infauna.

For members of the Syllidae, direct and indirect methods of reproduction are known (Imajima, 1966a). The indirect method is carried out by asexual stolons which result from budding of the posterior part of the adult. These are filled with gametes, and change morphologically, becoming detached from the original. This means of budding may account for the high relative abundance of individuals, even in a small coral colony. The stimulus to bud may be brought about by the coral's growth.

The feeding habits of syllids are virtually unknown. No zooxanthellae cells were observed following a squash and smear preparation of an entire worm. Perhaps the worm is a mucus feeder. Although numerous corals were observed in the field, no worms were seen freely moving about on the surface of the polyps.

Figure 2a, <u>Favia speciosa</u> showing shallow but wide grooves and
numerous small tubes; 2b, detail of 2a; 2c, <u>Favites pentagona</u>
showing very shallow grooves and numerous elongate tubes; 2d,
detail of <u>Favites</u> sp. 1 showing elongate tubes with no grooves.

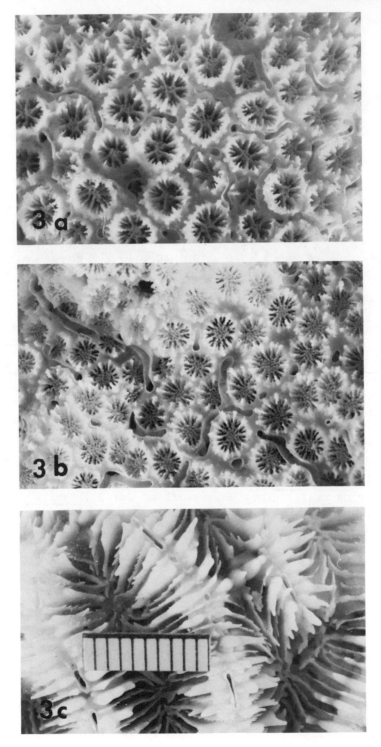

Figure 3a, <u>Cyphastrea gardineri</u> showing normal protuberant coral-
    lites with elongate tubes, corallite diameter about 2 mm;
    3b, <u>Cyphastrea microphthalma</u> showing normal protuberant coral-
    lites and elongate tubes encircling clusters of corallites,
    corallite diameter about 1.3 mm; 3c, <u>Symphyllia nobilis</u> showing
    elongate tubes with no grooves; 3d, anterior end of polychaete
    annelid (<u>Toposyllis</u> sp.) collected from <u>Favia valenciennesii</u>,
    worm diameter less than 0.5 mm.

Summary

A definite association between a polychaete annelid and <u>Favia</u>
<u>valenciennesii</u> was found. This association results in a tube-like
structure between the corallites. In some species a deep groove is
found, whereas in others the groove may be lacking or poorly
developed. This feature was also found in three other genera of
corals. The historical background of this feature was reviewed and
the questionable mode of colony increase was found to be actually
intratentacular. Judging from the appearance of living colonies
in the field and skeletal morphology, this association does not
appear to be detrimental.

Acknowledgements

The authors wish to express their gratitude to the National
Science Foundation U. S.-R.O.C. Cooperative Science Program Grant
No. GF-42381 under whose auspices the Taiwan field work was carried
out, to Ying-Min-Cheng, Department of Geology, National Taiwan University for his assisting in the collection of the specimens, and to
Dr. Minoru Imajima, National Science Museum, Tokyo for his verify-
ing the generic determination of the worm and for his stating that
it is an undescribed species.

Literature Cited

Duncan, P. M., 1883. On the madreporarian genus <u>Phymastrea</u> of
    Milne Edwards and Jules Haime, with a description of a new
    species. Proc. Zool. Soc. Lond., 1883:406-412.

Gibbs, P. E., 1971. The polychaete fauna of the Solomon Islands.
    Bull. Brit. Mus. (Nat. Hist.), 21(5):99-211.

Grassé, P., 1959. Annelides, myzostomides, sipunculiens, echiuriens,
    priapuliens, endoproctes, phoronidiens. Traité de Zoologie,
    5(1):1-1053.

Hartman, O., 1954. Marine annelids from the northern Marshall
    Islands. Geol. Surv. Prof. Pap., 2600:619-644.

Imajima, M., 1966a. The Syllidae (polychaetous annelids) from
    Japan. I. Exogoninae. Publ. Seto Mar. Biol. Lab., 8(5):385-404.

Imajima, M., 1966b. The Syllidae (polychaetous annelids) from Japan
    (IV). Syllinae (1). Publ. Seto Mar. Biol. Lab., 9(3):219-252.

Imajima, M., 1966c. The Syllidae (polychaetous annelids) from
    Japan (V). Syllinae (2). Publ. Seto Mar. Biol. Lab., 9(4):253-
    294.

Jones, R. S., R. H. Randall, Ying-Min-Cheng, H. T. Kami, and Shin-Min-Mak, 1972. A marine biological survey of southern Taiwan with emphasis on corals and fishes. Institute of Oceanography, National Taiwan University, Special Pub. No. 1, 93 p.

Matthai, G., 1914. A revision of the recent colonial Astraeidae possessing distinct corallites. Trans. Linn. Soc. Lond., Zool., 17(2):1-140.

Matthai, G., 1924. Report on the madreporarian corals in the Indian Museum. Mem. Indian Mus., 8:1-59.

Milne Edwards, H., and J. Haime, 1848. Recherches sur les polypiers. Mém. 2: Monographie des turbinolides. Ann. Sci. Nat. (Paris), 9(3):211-344.

Milne Edwards, H., and J. Haime, 1857. Histoire naturelle des coralliaires. Paris.

Quelch, J. J., 1886. Report on the reef corals. Repts. Sci. Res. H.M.S. "Challenger" 1873-76, 16(3):1-203.

Rosen, B. R., 1968. An account of a pathologic structure in the Faviidae (Anthozoa): A revision of Favia valenciennesii (Edwards and Haime) and its allies. Bull. Brit. Mus. (Nat. Hist.), Zool., 16(8):325-352.

Takahasi, K., 1941. Polychaeta of the Palao Islands of the South Sea Islands. Palao Trop. Biol. Sta. Stud., 2(2):157-217.

Umbgrove, J. H. F., 1939. Madreporaria from the Bay of Batavia. Zool. Meded., 22:1-64.

Vaughan, T. W., 1918. Some shoal-water corals from Murray Island (Australia), Cocos-Keeling Islands and Fanning Island. Pap. Dept. Mar. Biol. Carnegie Inst. Wash., 9(213):51-234.

Yabe, H., T. Sugiyama, and M. Eguchi, 1936. Recent reef-building corals from Japan and the South Sea Islands under the Japanese Mandate. Tohoku Imp. Univ., Sci. Repts. (Sendai), ser. 2, spec. vol., 1:1-66.

THE ASSOCIATION OF HYDRACTINIID HYDROIDS AND HERMIT CRABS, WITH

NEW OBSERVATIONS FROM NORTH FLORIDA

Claudia E. Mills [1]

Dept. of Biological Science, Florida State University

Tallahassee, Florida 32306 U.S.A.

The hydrozoan family Hydractiniidae, composed of the genera
Hydractinia, Podocoryne, and Stylactis, has a cosmopolitan distribution. Hydractiniids are often found in association with other marine
animals; perhaps the most common association involves hydractiniid
polyps colonizing shells occupied by hermit crabs. The hydroid-
hermit crab association is not obligatory for either animal; both the
polyps and the hermit crabs are also found freeliving or on other
available substrates.

Hydractiniid polyp colonies are highly polymorphic. Colonies
may produce nutritive gastrozooids, sexual gonozooids, and dactylozooids, the so-called "defensive" polyps that include both spiralzooids and tentaculozooids. Based on their morphology and location
overhanging the edge of the shell aperture, spiralzooids, which are
found only in hydractiniid colonies associated with hermit crabs,
have generally been considered to be defensive zooids. Observation
of polyp colonies and the resident hermit crabs does not support this
presumption, as will be discussed later, and the function of spiralzooids remains unclear.

Podocoryne selena Mills, 1976 and Hydractinia echinata (Fleming,
1828) occur syntopically in the northern Gulf of Mexico, most commonly, but not exclusively, on hermit crab shells. The exact nature of
the association, i.e. whether commensal or mutualistic, has been
unclear. I have made field collections to investigate this association, primarily with respect to the hydroid life-cycles. Only the
two above-mentioned species of hydroids commonly are found on gastropod shells on the Gulf of Mexico coast. At least five species of

[1] Present address: 7044 50th Ave. N.E., Seattle, Washington 98115

467

hermit crabs occupy empty shells in north Florida. I have only included the three large hermit crabs in my study: Clibanarius vittatus, Pagurus longicarpus, and Pagurus pollicaris. The two similar, smaller hermit crabs, Pagurus annulipes and Pagurus bonairensis are common in some of the study areas and undoubtedly utilize some of the same shells as young individuals of the other three species, but the extent of their participation in the association has not been studied.

METHODS

The following locations on the Gulf coast of north Florida (approximately 30° N. Lat., 85° W. Long.) were sampled for colonies of the hydrozoans Hydractinia echinata and Podocoryne selena on hermit crab shells and other substrates: in Franklin County at Alligator Point, Alligator Harbor, Baymouth Bar, St. Teresa, and Turkey Point; and in Gulf County in St. Joseph Bay. Animals were collected on sandy bottoms either by wading or snorkeling in 0 to 3 meters of water.

The animals were kept in the laboratory in 20-gallon glass aquaria containing natural sea water and were fed pieces of fish or Artemia nauplii. The water was maintained at room temperature, approximately 20° C, and was filtered with both a sub-gravel filter and a Dyna-Flow outside filter.

RESULTS

It appears that Podocoryne selena and Hydractinia echinata utilize the same species of shells in Florida. It should be noted, however, that in no instances have both species of hydroid been found on the same shell, although a few shells appeared to have two colonies of Hydractinia, separated by a distinct boundary. The specific identification of the gastropod shells on which hydroids have been collected and the hermit crabs that were occupying the shells at the time of collection are shown in Table I. In addition, Podocoryne selena has been found on the legs and carapace of the horseshoe crab Limulus polyphemus and on the dactyls and chelae of one specimen of the hermit crab Pagurus pollicaris.

McLean (1975) found significant hydroid associations with three species of snail shells collected in the same part of Florida as was sampled in my study; more than 40% of the Polinices duplicatus, Murex pomum and Cantharus cancellarius shells inhabited by Pagurus pollicaris in his study areas had hydroids living on them. My collections indicate that hydroids also occur very commonly on Murex florifer dilectus and Busycon contrarium shells.

Sea water temperatures for the north Florida coastal area vary

between about 5° C and 35° C (Shier, 1965), with the lowest temperatures occurring in December and January and the highs between May and
September. Including the records of Defenbaugh and Hopkins, 1973,
from Galveston Bay, Texas, Hydractinia has been collected during every
month of the year in the Gulf of Mexico, and Podocoryne has been
collected in every month except September.

Hermit crab and hydroid collections in north Florida are summarized in Table II. Seasonal changes in the distribution of hydroids on
hermit crab shells from January through May, 1975, are summarized in
Table III. It may be noted that: 1) More hermit crabs were collected in May than in the winter months. This does not reflect the time
spent in collecting, which was relatively constant, but represents an
increasing number of near-shore crabs. 2) The percentage incidence
of hydroids on hermit crab shells increased considerably as the spring
progressed. As would be expected, more shells were also found with
new hydroid colonies in the later months. 3) Relative frequencies of
Podocoryne and Hydractinia on gastropod shells remain fairly constant
during the study period from January through May, 1975. 4) Cliban-
arius-occupied shells have no hydroids.

Gastropod Shell Substrate	Podocoryne selena	Hydractinia echinata	Juvenile Hydroid Colony
Busycon contrarium	P	P	P
Busycon spiratum		P	P
Calotrophon ostrearum		L	
Cantharus cancellarius	P,L live snail	P,L	L
Fasciolaria lilium hunteria	P	P	
Littorina irrorata		L	
Melongena corona		P	
Murex florifer dilectus	P,L	P,L	
Murex pomum	P	P	
Nassarius vibex		L	
Pleuroploca gigantea	P		
Polinices duplicatus	P,L	P,L	P,L
Terebra dislocata		P	
Turbo castanea		live snail	P

Table I. Distribution of Podocoryne and Hydractinia on hermit crab
shells and some live shells in the north Florida Gulf of Mexico, by
species of substrate-shell and of hermit crab. Letters refer to
hermit crabs occupying shells at the time of collection. P = Pagurus
pollicaris, L = Pagurus longicarpus. Juvenile colonies of hydroids
were too young to differentiate species.

	Pagurus pollicaris	Pagurus longicarpus	Clibanarius vittatus
Total number of hermit crabs collected	152	137	130
Percent of hermit crabs collected with hydractiniid colonies on their shells	71%	30%	0
Total number of hermit crabs observed with hydractiniid colonies	112	39	0
Percent of shells with Podocoryne selena	36%	20%	0
Percent of shells with Hydractinia echinata	49%	77%	0
Percent of shells with hydroid colonies too young to determine species between P. and H.	15%	3%	0

Table II. Summary of hermit crab and hydroid collections on the north Florida Gulf of Mexico from January through May, 1975.

DISCUSSION

The few hermit crab shells supporting hydroid colonies in December, January, and February, had healthy colonies of either Podocoryne or Hydractinia that covered the entire shell surfaces and were producing many gonophores. Although there is no record of Podocoryne being collected in the Gulf of Mexico in September, or of colonies of either species having gonophores in November, it is probable that a small number of colonies of both species overwinters in an actively feeding, fully reproductive condition. In the spring, as the water temperature increases, the planular settlement rate probably increases; perhaps planulae cannot settle during the winter, thus accounting for my observing only mature colonies during these months. In the late spring (Table III), many hermit crab shells were collected with young hydroid colonies which in many cases had not developed gonophores, spines, or spiralzooids, the characteristics by which they can be generically distinguished, although the polyps could be identified as

	Summary of all collections	Collections from January through March 1975	Collections from May 1975
Total number of hermit crabs collected	365	147	145
Percent of total hermit crabs with hydractiniid colonies on their shells	35%	6%	61%
Total number of hermit crabs observed with hydractiniid colonies	151	9	88
Percent of shells with Podocoryne selena	·32%	33%	27%
Percent of shells with Hydractinia echinata	56%	67%	63%
Percent of shells with hydroid colonies too young to determine species between P. and H.	12%	0	10%

Table III. Seasonal changes in the distribution of hydroids on gastropod shells occupied by the hermit crabs Pagurus pollicaris, Pagurus longicarpus, and Clibanarius vittatus, from January through May, 1975, on the north Florida Gulf of Mexico.

members of the Hydractiniidae. On Polinices and similar naticid shells, new colonies were almost invariably found growing in the pit of the umbilicus, which was often also occupied by the commensal polynoid scale-worm Leptonotus sublevis. On shells like Busycon, which possess long anterior (siphonal) canals, new colonies were found in the canal. These two sites of planular settlement corres-pond, in the order discussed, to the two most preferred sites for larval settlement of epizoic fauna, especially spirorbid worms and ectoprocts, on hermit crab shells as observed by James Carlton (personal communication) in California.

As has been shown in Table I, the shell habitats for Podocoryne selena and Hydractinia echinata do not seem to be mutually exclusive.

Crowell (1945) found that at Woods Hole, P. carnea and H. echinata
both could settle on several different species of snail shells, but
that in nearly all cases, hydroid colonies located on Nassarius
trivittatus shells were P. carnea and hydroids on Littorina littorea
shells were H. echinata. I found no such clearly defined difference
in Florida. Teitelbaum (1966) conducted settling studies with H.
echinata at Woods Hole and found that planulae preferred to settle in
the pits and grooves of the sculptured Nassarius obsoletus shells
rather than on the smoother Polinices duplicatus or L. littorea
shells. Yet Crowell reports that at Woods Hole H. echinata is commonly found on L. littorea and in Florida it is commonly located on
smooth P. duplicatus shells. Rees (1967) suggests that there is some
evidence that P. carnea planulae prefer rugged surfaces, but I found
that in Florida, P. selena seems to be about equally common on smooth
and sculptured shells. Müller (1973) has shown that for H. echinata,
planular metamorphosis is induced by the presence of certain marine
bacteria. Clearly, more work needs to be done in order to elucidate
settling behavior and habitat selection in the Hydractiniidae.

My data indicating that no Clibanarius shells have hydroids
agrees with that of McLean (1975). Wright (1973) found about 0.1
percent of the Clibanarius shells in coastal Texas covered with
hydroids or remnants of hydroid colonies. McLean and Wright agree
that Clibanarius shells provide inappropriate substrates for hydroids
since these crabs, unlike their fellow pagurids, often spend hours
out of the water at low tide in full sun, exposing the hydroids to
excessive desiccation and temperature fluctuations. This exposure
is probably sufficient to prohibit growth of settled planulae.

Using the same species of hydroids and hermit crabs that are
found in Florida, Wright (1973) in Texas found that the two species
of Pagurus never seemed to be stung by the hydroids on shells, but
that Clibanarius was apparently stung and was unlikely to enter a
hydroid-covered shell. He also found both species of Pagurus to
prefer hydroid-covered shells when given a choice. In shell-changing
experiments (Mills, 1976b), I have found that some individuals of all
three species of hermit crabs appear to be stung by the hydroids,
although only Clibanarius will not enter the shells. There was evidence that all species of hermit crabs tested preferred clean shells.
Wright and I have both observed that pagurid crabs will eat the hydroids on occasion, but in my laboratory this occurred only when the
crabs had not been fed for several days. Wright proposes that an
intermittent hydroid diet may somehow provide the hermit crabs with
either an immunity or an inhibitor for the hydroid nematocysts.

Jensen (1970), using Pagurus bernhardus and H. echinata in
Europe in "two-choice-tests", found that most tested hermit crabs
showed a preference for hydroid-covered shells. Jensen interpreted
this behavior as evidence for recognition by the hermit crabs of some
chemical cue diffusing from the hydroids. Grant et al. (1973, 1974)

found that in Maine, where the hermit crabs Pagurus acadianus and
P. pubescens are both present in the shallow subtidal, that P. acadianus prefers shells with Hydractinia on them, that it will take a
smaller than adequate shell if it is covered with hydroids, and that
the presence of hydroids protected hermit crabs from predation by
rock crabs. Pagurus pubescens does not occupy shells with hydroids,
so presence of hydroids on shells may conveniently partition the
shell-resource in this situation. Grant found no hydroids on live
snails in Maine.

It has been assumed that spiralzooids serve some defensive purpose that may be advantageous to either the hydroid colony or the
hermit crab. By shaking or squeezing a hermit crab shell, the spiralzooids can be made to lash out over the aperture edge in synchronization several times. Under less-contrived circumstances, however,
spiralzooids are less impressive. In an extensive series of experiments, Schijfsma (1935), trying to determine their function, submitted
spiralzooids to a variety of stimuli. The spiralzooids only reacted
to movements of the shell or to some sudden retreats of the crab; no
tactile stimulation of spiralzooids caused this behavior. As a
hermit crab moves in and out of its shell, it brushes the spiralzooids
nearly constantly and no lashing occurs. I never observed spontaneous lashing out of spiralzooids in laboratory situations including
shell-changing experiments involving three species of hermit crabs
(P. pollicaris, P. longicarpus, and C. vittatus) and hydroid-covered
shells. Wright (1973) also does not report any special defensive
behavior by H. echinata or P. carnea (probably P. selena), nor does
Max Braverman (personal communication) for P. carnea.

Spiralzooids are a unique feature of some hydractiniid colonies
associated with hermit crabs. Of over 70 living species of hydractiniids, at least 15 colonize hermit crab shells. Less than half of
the hermit crab-associated species produce spiralzooids, and among
these, spiralzooids are not always present on a hermit crab's shell.
Spiralzooids are not found in hydroid colonies on live snail shells
or other substrates. Located at the rim of the shell aperture,
spiralzooids may encircle the entire aperture or be present only in a
portion of this area, in a band one to several polyps deep. Spiralzooids are long slender polyps that are normally coiled. Most are
2 to 5 mm long when extended. The morphology varies somewhat by
species, but nematocysts are concentrated primarily at the tip of the
polyp and may also extend down the column, especially on the side
that faces into the shell aperture when uncoiled.

Schijfsma (1935) noted that the great variation in development
of spiralzooids within a species indicates that they may not be a
special adaptation of the polyp colony for the hermit crab. Braverman
(1960), however, showed that upon removal of the hermit crab, Podocoryne carnea spiralzooids degenerated within a few days and were

transformed to either feeding or sexual polyps. When crabs were
returned to the empty shells, spiralzooids reformed at the aperture
edge within 5 days. Burnett, Sindelar, and Diehl (1967) confirmed
Braverman's observations and through another set of experiments
concluded that spiralzooids were induced and maintained by a combination of extrinsic factors and that some diffusible material produced
by the crab may concentrate near the shell aperture and be a contributing factor to spiralzooid formation. Braverman (1974) later
showed by plugging up shell apertures with wax and allowing shell-
less hermit crabs to examine the shells for several days, that
spiralzooids formed all over the shells, not just at the apertures,
in response to the constant turning and mechanical handling by the
crabs trying to enter the shells. He concluded that spiralzooids
form normally at shell apertures as a result of the movements of the
crabs in and out of the shells.

Spiralzooids are apparently derived from sexual zooids whereas
tentaculozooids are modified feeding zooids whose appearance resembles a long contractile tentacle with scattered nematocysts (Müller,
1964). Tentaculozooids are found both in polyp colonies on hermit
crab shells and in other hydractiniid colonies, but again are not
apparently produced by all hydractiniid species. Tentaculozooids
seem to occur most commonly at colony edges. Braverman (personal
communication) also has observed tentaculozooids between two colonies
on the same shell and adjacent to limpet shells on snail shells.
Although tentaculozooids seem to be an edge feature, they are also
found scattered about polyp colonies either singly or in small patches
(Burnett et al., 1967).

Christensen (1967) made a field study of the feeding biology of
H. echinata collected on Pagurus bernhardus shells in Denmark. On the
basis of contents of gastrovascular cavities, he concluded that Hydractinia is almost exclusively a benthic feeder, eating primarily
nematodes, ophiuroid arm joints, and harpacticoids. In the laboratory,
however, Hydractinia can exist on a variety of diets. It has sometimes been thought that one advantage for hydroids living on hermit
crab shells is the availability of food lost by the hermit crab as it
eats. Christensen's studies have shown that the crab's contribution
in this manner is minimal, although the hydroids are known to ingest
an undetermined amount of hermit crab eggs and young larvae. The
hermit crabs, however, do not seem to eat any of the newly released
hydromedusae, even if these are drawn into the crabs' feeding currents.

The benthic mode of feeding well explains the distribution of
polyps on hermit crab shells, where feeding polyps are oriented on
the bottom of the shells so that they are dragged over the benthos as
the crab moves. On shells of the live snail Cantharus, which moves
just under the surface of the sand, feeding polyps are evenly distributed over the entire shell surface. Hydroid colonies on Limulus are
found either on the legs or on the carapace, locations which would

also enable benthic feeding. On rocky coasts, <u>Hydractinia</u> <u>echinata</u> is also commonly found on rocks, <u>Mytilus</u> shells, wood, <u>Laminaria</u>, and barnacles (Schijfsma, 1935). Similarly, <u>P</u>. <u>carnea</u> in Britain is found on rocks as well as on hermit crab shells. In these locations, the hydroids are presumably planktivorous. This is a more common mode of feeding in the Hydrozoa, living in areas where good water circulation carries sufficient food to the polyps.

Predation on hydroid colonies on hermit crab shells in the field is probably minimal. In New Hampshire, the nudibranch <u>Cuthona</u> commonly feeds on <u>Hydractinia</u> colonies on hermit crab shells. The nudibranch feeds on polyps for a few weeks, then leaves the colony to lay eggs. It apparently rarely grazes a colony so severely that this relationship is fatal to the hydroid, and the grazed tissues are rapidly regenerated (Brian Rivest, personal communication). In Florida, the incidence of nudibranchs on hydractiniid colonies is very low. The snails <u>Cantharus</u> <u>cancellarius</u> and <u>Calliostoma</u> <u>jujubinum</u> were observed eating hydroid colonies in the lab, but it is doubtful that many similar contacts would occur in nature. The extent of predation in the field by hermit crabs eating polyps off shells as reported earlier in this paper is unknown.

SUMMARY

Although many studies have been undertaken to investigate various aspects of the symbiosis between hydractiniid hydroids and hermit crabs, a holistic understanding of the association fails to emerge. The association is not obligatory; many hermit crabs occupy clean shells, and the hydroids will settle on other hard substrates when these are available. By living on hermit crab shells, however, the otherwise planktivorous hydroids may adopt a benthic mode of feeding. The hydroids also avoid being silted over as long as their substrate-shell remains in the hermit crab shell pool. Some species of hermit crabs apparently prefer hydroid-covered shells to clean ones, but other species of crabs will not enter hydroid-covered shells. In some areas, the presence of hydroids may partition the shell resource, although the advantages of possessing hydroid-covered shells are not entirely clear. The effects of epizooic hydroids for hermit crabs in intra- and interspecific competition and in avoiding predation are not well understood.

LITERATURE CITED

Braverman, M., 1960. Differentiation and commensalism in <u>Podocoryne</u> <u>carnea</u>. <u>Amer</u>. <u>Midland</u> <u>Natur</u>., 63: 223-225.
Braverman, M., 1974. The cellular basis of morphogenesis and morphostasis in hydroids. <u>Oceanogr</u>. <u>Mar</u>. <u>Biol</u>. <u>Annu</u>. <u>Rev</u>., 12: 129-221.

Burnett, A. L., W. Sindelar and N. Diehl, 1967. An examination of polymorphism in the hydroid Hydractinia echinata. J. Mar. Biol. Ass. U. K., 47: 645-658.

Christensen, H. E., 1967. Ecology of Hydractinia echinata (Fleming) (Hydroidea, Athecata). I. Feeding biology. Ophelia, 4: 245-275.

Crowell, S., 1945. A comparison of shells utilized by Hydractinia and Podocoryne. Ecology, 26: 207.

Defenbaugh, R. E., and S. H. Hopkins, 1973. The occurrence and distribution of the hydroids of the Galveston Bay, Texas area. Center for Marine Resources, Texas A. and M. University, 202 pp.

Grant, W. C. Jr., and P. J. Pontier, 1973. Fitness in the hermit crab Pagurus acadianus with reference to Hydractinia echinata. Bull. Mt. Desert Is. Biol. Lab., 13: 50-53.

Grant, W. C. Jr., and K. M. Ulmer, 1974. Shell selection and aggressive behavior in two sympatric species of hermit crabs. Biol. Bull., 146: 32-43.

Jensen, K., 1970. The interaction between Pagurus bernhardus (L.) and Hydractinia echinata (Fleming). Ophelia, 8: 135-144.

McLean, R. B., 1975. A description of a marine benthic faunal habitat web: a behavioral study. Ph.D. Dissertation, Florida State University, Tallahassee, Florida, 176 pp.

Mills, C. E., 1976a. Podocoryne selena, a new species of hydroid from the Gulf of Mexico, and a comparison with Hydractinia echinata. Biol. Bull., in press.

Mills, C. E., 1976b. Studies on the behavior and life histories of several species of hydroids and hydromedusae. M. S. Thesis, Florida State University, Tallahassee, Florida, 111 pp.

Müller, W. A., 1964. Experimentelle Untersuchungen über Stockentwicklung, Polypendifferenzierung und Sexualchimaeren bei Hydractinia echinata. Wilhelm Roux' Arch. Entwickl.-mech. Org., 155: 181-268.

Müller, W. A., 1973., Induction of metamorphosis by bacteria and ions in the planulae of Hydractinia echinata; an approach to the mode of action. Publ. Seto Mar. Biol. Lab., 20: 195-207.

Rees, W. J., 1967. Symbiotic associations of Cnidaria with Mollusca. Proc. Malac. Soc. London, 37: 213-231.

Schijfsma, K., 1935. Observations on Hydractinia echinata (Flem.) and Eupagurus bernhardus (L.). Arch. Néerlandaises Zool., 1: 261-314.

Shier, C. F., 1965. A taxonomic and ecological study of shallow water hydroids of the northeastern Gulf of Mexico. M. S. Thesis, Florida State University, Tallahassee, Florida, 128 pp.

Teitelbaum, M., 1966. Behavior and settling mechanism of planulae of Hydractinia echinata. Biol. Bull., 131: 410-411. (Abstract)

Wright, H. O., 1973. Effect of commensal hydroids on hermit crab competition in the littoral zone of Texas. Nature, 241: 139-140.

Functional Morphology

LOCOMOTION IN SEA ANEMONES: THE PEDAL DISK

Elaine A. Robson

Zoology Department

University of Reading, England

Many sessile actinians have a pedal disk which is concerned alternately with adhesion and with locomotion (Stephenson, 1928; Batham and Pantin, 1951).   This paper is about the effector mechanisms involved.

Among anemones there exist at least four modes of locomotion: burrowing (not considered here), creeping on the pedal disk (e.g. Metridium (Pantin, 1952); and see Edmunds, Potts, Swinfen and Waters, 1976), walking with use of the tentacles (e.g. Gonactinia,Protanthea (Robson, 1971); Calliactis uses the whole crown to transfer to a new shell (Ross and Sutton, 1961), and swimming (in unusual anemones, i.e., Stomphia, Gonactinia and Boloceroides (Ross, 1974).   These patterns of coordinated movements have been analysed in a few species but their physiological basis, that determines the order in which different muscles contract, is only in part understood.   It is to be hoped that electrophysiological studies of anemone behaviour will provide further elucidation (McFarlane, 1969; 1975).

At the same time, two features essential to locomotion cannot be accounted for by muscular activity alone: release of the anemone from the substrate and its reattachment which involve the pedal ectoderm and this has no muscle fibres (Hertwig and Hertwig, 1879). Non-muscular ectodermal effectors, then, must act in conjunction with the system of endodermal muscles.   Their role is discussed here, chiefly with reference to Metridium.

MATERIAL AND METHODS

The anemones used were Metridium senile (L.) unless stated otherwise, and Anemonia sulcata (Pennant), Actinia equina (L.), all from Plymouth.   Observations on Stomphia coccinea (Müller) were made at Friday Harbor, Washington and Helsingør, Denmark, and on Gonactinia

prolifera (Sars) and Protanthea simplex (Carlgren) at Kristineberg,
Sweden.

Methods employed include observation under a dissecting microscope, where possible with photographic recording; ordinary histological methods for sections and maceration, together with standard
light microscopy; and routine methods for electron microscopy,
using a 5% glutaraldehyde fixative in 0.2M cacodylate buffer at
pH 7.3 with 10mM $CaCl_2$. Principles are as in Grimstone and Skaer
(1972).

RESULTS

Adhesion

Anemones were observed in a glass dish using a dissecting microscope. Specimens removed from rocks or weed were allowed to relax
before testing the adhesive properties of the foot with a fine glass
rod. Small Metridium were also allowed to reattach to pieces of
thin polythene, through which the surface of the pedal disk could
be observed.

Adhesiveness of the pedal disk is easily demonstrated. In the
normal course of events a detached anemone will fasten itself to the
nearest substrate and when it is about to do so the pedal surface
is remarkably adhesive. At this stage the pedal disk is usually
relaxed and expanded, and a newly-drawn glass probe adheres to it
so firmly that the pedal tissue can be pulled out into a narrow
tongue (Fig.1). This property is not shown by the column.

The tenacious ectodermal adhesion illustrated in Fig.1 cannot
be attributed to nematocysts of the pedal disk. The adhesive
material is not mucus and it does not show metachromasia with
toluidine blue. That the secretion of mucus is a different process
is shown by placing carbon particles on the pedal disk. The
particles are soon lifted away from the surface by its protective
layer of mucus, i.e., whether or not carbon particles stimulate the
production of mucus they demonstrate its presence. The particles
do not move laterally, and so do not provide evidence for ciliary
action.

When fully relaxed the pedal disk of a detached anemone may be
inflated to such an extent that it becomes transparent, allowing
the mesenteries to be seen. From time to time vertical mesenteric
muscles contract, forming a radial pattern of shallow grooves (Fig.2).
Hydrostatic pressure usually causes the pedal disk to bulge outwards
between the grooves and the rim may appear scalloped. A more pronounced muscular wave may then spread across the foot, and when the
pedal disk contracts in this way the surface becomes opaque and
corrugated. Local, non-propagated contractions may follow mechanical
stimuli such as contact with a glass rod: either a groove forms as
the retractor or parietobasilar muscles of a mesentery contract, or
else surface corrugations indicate the action of basilar or circular
muscles. Especially after strong or repeated stimulation such

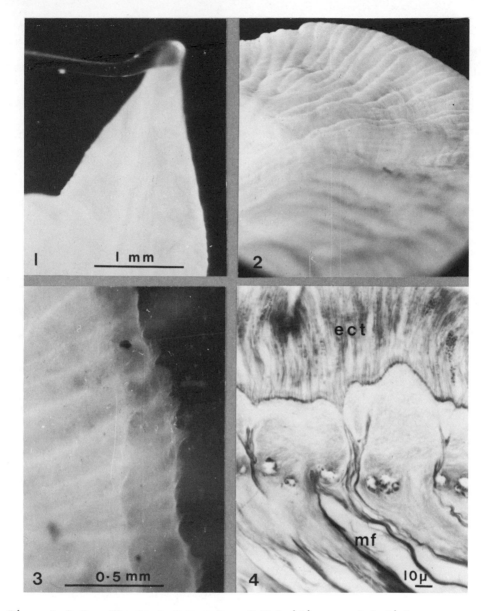

Figs. 1,2,3. Flash photographs of _Metridium_, pedal disk.
Fig.1.   Adhesion to glass rod.
Fig.2.   Pedal surface with grooves and dimples. Size 3cm approx.
Fig.3.   On polythene from below: adhesive peaks located between
         mesenteries.
Fig.4.   _Stomphia_: radial section of pedal disk showing muscle
         fibres from mesentery (mf) in contact with ectodermal
         supporting cells (ect).

contractions may spread to larger areas.

Optimal conditions for detecting adhesiveness seem to occur when the pedal muscles are relaxed, and a few observations on Metridium indeed suggest that adhesiveness may be lost as a wave of contraction passes over the pedal disk. The impression, which could be wrong, is that adhesiveness is a local, possibly all-or-none property which is affected in some way by the contraction of pedal disk muscles. A similar impression was gained from a specimen of Anemonia, when a fine glass rod adhering to the expanded surface of the pedal disk appeared to inhibit adhesion within a radius of perhaps 1 cm. This continued for some seconds after the probe was removed, although adhesiveness returned to a given area very quickly. The passage of muscular waves over the foot also seemed to inhibit adhesiveness.

Whether or not it is subject to inhibition, adhesiveness seems to behave as an all-or-none state of the pedal disk in several other anemones. After swimming Stomphia, for example, rests expanded and inert, but when muscular activity resumes and the anemone is about to reattach itself the pedal disk is so adhesive that the anemone can be lifted about by the tip of a glass rod touching the foot (Robson, 1961). In Stomphia contact between the tentacles and Modiolus shells promotes adhesiveness (Ellis, Ross and Sutton, 1969). In Gonactinia, the pedal disk is similarly adhesive after swimming or stepping and will fasten to a strand of weed or a glass fibre (Robson, 1971). Cnidae are probably not involved: there were none at the spot on a coverslip where a specimen had remained for two days. In Calliactis parasitica, the pedal disk becomes adhesive before the anemone settles on a new Buccinum shell. Ross and Sutton (1961) suggested that a condition of 'stickiness' was produced when the unattached pedal disk first touched the shell; and the behaviour of several other species supports this idea (Ross, 1974).

Such observations suggest that adhesiveness is due to an active state of the pedal ectoderm which can be excited and possibly inhibited or masked by suitable stimuli. The phenomenon is clear but difficult to measure. It is tentatively regarded as an all-or-none state produced by secretion of an unknown material. Once evident, adhesiveness persists in the presence of Mg^{++} ions (present observations; Ellis, Ross and Sutton (1969)).

Detachment

It is suggested that release of the pedal disk from the substrate is a different active process, during which cells of the ectoderm contract.

By means of a glass rod adhering to the pedal surface a narrow peak of tissue can be pulled out (Fig.1). A limit of extension is reached when the apical angle approaches 30° and the ectoderm is then released, leaving no visible trace on the glass. The peak subsides completely during the next 10 seconds or so. In Metridium attached to polythene or to finely-branched algae, comparable

peaks of ectoderm are seen (Fig.3), and during locomotion they are
released from the substrate as the anemone moves forward.   Peaks
are up to 300μ long, and represent a two-or-three fold extension of
the epithelium.   Under a binocular microscope they show refringent
striations, which represent epithelial cells.

Contractility of the ectoderm is suggested but not proved by
observing the release of adhesive peaks on stretching.   When extended to the limit they let go suddenly and within a few seconds
can hardly be distinguished from the surrounding epithelium.   The
impression is gained repeatedly that the epithelial peaks are
contractile and that they detach from the substrate when their
constituent elements shorten.   The response may spread within a
local area as a number of peaks may detach at about the same time.

The release of epithelial peaks is prevented by adding an
equal volume of isotonic $MgCl_2$ ($7\frac{1}{2}\%$) to the seawater.   Within 3-5
minutes peaks adhering to strands of Cladophora, for example,
instead of detaching now rupture when they are pulled.   When fresh
seawater is replaced the detachment response returns after a similarly short time.   The rapid action of Mg^{++} ions suggests that the
ectoderm is affected directly.   Detachment of the pedal disk is
prevented by Mg^{++} ions in other anemones, for example, Epiactis
prolifera (own observation) and Stomphia (Ellis, Ross and Sutton,
1969).

Species such as Anemonia, Actinia and Tealia are also attached
to natural substrates by ectodermal peaks.   If the anemones are
prised off carefully their pedal disks appear to be fastened by
numerous adhesive peaks all of a similar kind: they are of about
the same dimensions as in Metridium and detach, apparently by contracting, when stretched.   The pedal disk may thus adhere at a
large number of small points and Parker's impression (1917) that
"points of adhesion represented minute well localised organs of
attachment" seems correct.   It must be borne in mind also that in
stationary specimens the pedal disk secretes a cuticular material
of unknown composition (Batham and Pantin, 1951).

Structure of the Ectoderm

The structure of the pedal ectoderm is specialized in relation
to adhesion and detachment (see Hertwig and Hertwig, 1879;   Batham
and Pantin, 1951).   In Metridium it lacks muscle fibres and possibly
nerve cells.   Sense cells occur and there are scattered nematocysts.
Most numerous are the so-called supporting cells (e.g., Stephenson,
1928), whose structure might possibly confer both adhesiveness and
contractility.   There are granular gland cells whose secretion might
also be adhesive, and mucus cells.

The pedal ectoderm appears to be carpeted by microvilli, which
must be the peripheral agents of attachment and release.   Most of
the microvilli belong to supporting cells.   In Metridium they are
relatively short (2-3μ x 0.15-0.25μ: Figs.5,6,9), whereas in
Protanthea and Gonactinia they are very long (up to 5μ: Figs.7,10)
and cilia, rare in Metridium, are invariably present (Carlgren, 1893).

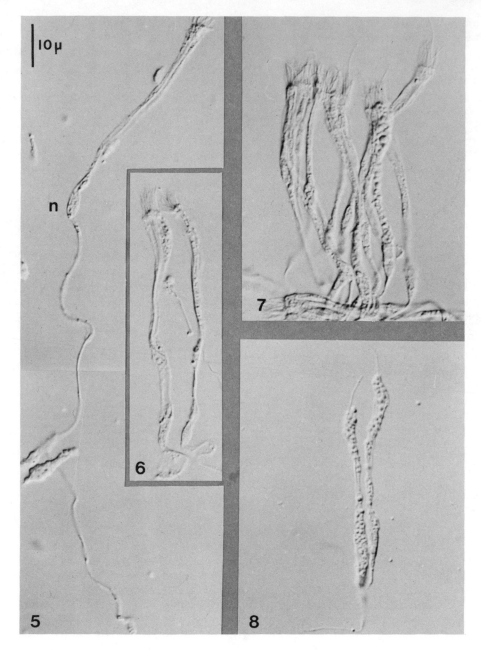

Cells of pedal ectoderm (Hertwig method – Nomarski interference contrast).

Figs. 5,6.　<u>Metridium</u>: supporting cells with microvilli, axial fibre, nucleus (n).

Figs. 7,8.　<u>Protanthea</u>: supporting cells (note cilia and microvilli) and granular gland cells.

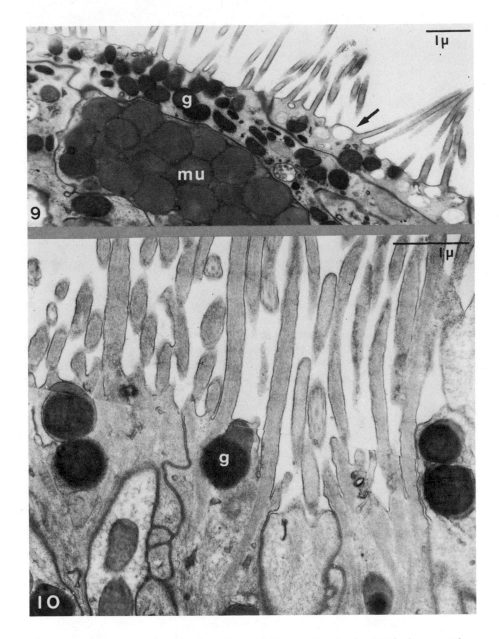

Detached pedal disks: supporting cells and microvilli in section.
Fig.9. Metridium: dark granules (g), empty vesicles (arrow), and
        a mucus cell (mu).
Fig.10. Protanthea: bundles of filaments from cores of microvilli
        converge towards the main axis. Note granules (g).

Fig.12 shows the pedal epithelium of a small <u>Metridium</u> attached
to <u>Ulva</u>: the microvilli are packed into a layer which adapts even
to bacteria on the substrate.

In <u>Metridium</u> each supporting cell has a well-marked axial
fibre approximately 0.3-0.5µ wide. It runs from the core of apical
microvilli to stout rootlets which anchor the cell to the mesoglea
(Figs.5,6,11,12). This fibre must provide mechanical support but
it might also account for ectodermal contractility. The possibly
contractile nature of such fibres has been discussed before, since
they take up the same histological stains as muscle (Batham and
Pantin, 1951; Robson, 1961); and it now seems that their sinuous
outlines, a feature not supporting the suggestion, could perhaps
be due to the use of anaesthetic before fixation. With the electron
microscope a fibre is seen as a bundle of microfilaments under 7nm
in diameter. In a variety of other tissues microfilaments of
similar dimensions consist of actin and form part of a contractile
system, for example, in intestinal microvilli (Mooseker and Tilney,
1975). An analagous situation is reported for the mucous cells of
a sea pen (Crawford and Chia, 1974).

The supporting cells may also secrete adhesive material as their
distal cytoplasm contains fine granules which take up eosin or acid
fuchsin. In the electron microscope they are usually oval (about
0.75 x 0.5µ) and appear dark or with a dark core (Figs.9,10,11).
They appear to release non-hygroscopic, somewhat filamentous contents
leaving empty vesicles as evidence of the process (Fig.9, arrow).
Granular gland cells (Fig.8) are another possible source of sticky
material. Their granules also stain with eosin and acid fuchsin but
they are larger and form long lines. In the electron microscope
they are oval (1.2 x 0.6µ), striated, and aligned between microtubules
(Figs.12,13). Cutress and Ross (1969) attribute pedal adhesion to
these cells in <u>Calliactis</u> <u>tricolor</u>.

The Ectoderm and Locomotion

In creeping locomotion the front edge of the pedal disk moves
forward in a series of convex wave-fronts, and as each muscular
wave sweeps forward from the hind edge there is local detachment,
displacement, and reattachment (Parker, 1917; Pantin, 1952;
Edmunds et al., 1976). During locomotion the activity of ectodermal
elements is linked to that of endodermal muscles. How are ectoderm
and endoderm coordinated? If ectodermal contraction accounts for
release of the pedal disk does it precede, accompany or follow the
muscular wave? And how is it excited?

In <u>Metridium</u> on a polythene surface peaks occur between the mesenteries rather than along their lines of insertion (Fig.3). This
means that in itself the mechanical pull of mesenteric muscles would
tend to cause their release, as a serial process in the case of a
muscular wave. Along the junctures of mesenteries with pedal disk,
moreover, some of the septal muscle fibres are definitely connected
to the pedal ectoderm (Fig.4). Sections of <u>Metridium</u> (preparations

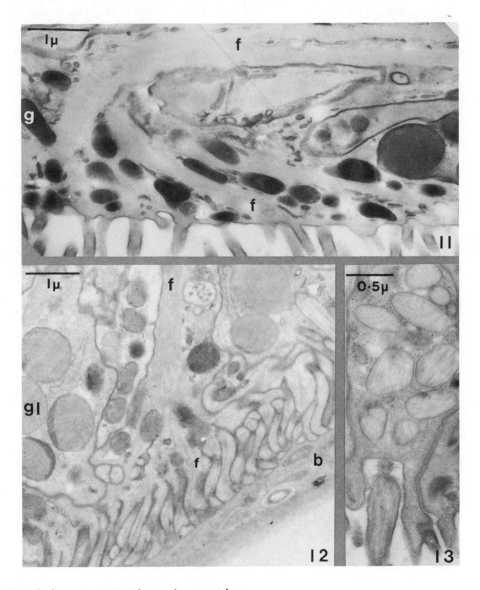

Metridium: pedal disks in section.
Fig.11. Detached. Shows bundles of filaments (f) converging from
        the cores of microvilli, and dark granules (g).
Fig.12. On Ulva. Note packing of microvilli, filaments (f), a
        granular gland cell (g) and bacteria (b).
Fig.13. Part of a granular gland cell, through cilium.

of E.J. Batham) and Stomphia (Robson,1961) show that both retractor
and parietobasilar muscle fibres are in contact with the bases of
supporting cells.   Similar connexions occur in Calliactis (Robson,
1965) and possibly in Gonactinia and Protanthea.   The tapered
bundles of exocoelic and endocoelic muscle fibres which reach the
ectoderm correspond to rows of dimples at the pedal surface when
contraction begins (Fig.2).   Assuming that the histological contacts represent physiological connexions it is possible that they
could transmit excitation or inhibition.

From observations on swimming anemones, however, detachment is
a propagated ectodermal reaction which would precede rather than
follow muscular activity.   Cinefilm records of Stomphia detaching
from glass first show a wave of opacity rapidly crossing the pedal
disk, which could imply a wave of ectodermal contraction.   Detachment behaves as a propagated response in that it may be total and
quick, or else slow, or incomplete.   It can occur before any action
of muscles is seen (Robson,1961).   In Gonactinia, release of the
pedal disk is rapid and also seems to involve no muscles.   That
detachment spreads from the edge can be seen in Protanthea,which
is larger.   In these ánemones, then, muscular action is not evident
until after detachment.   Further observations are needed to
ascertain whether this is so in creeping anemones such as Metridium.

DISCUSSION

The effector mechanisms discussed here need further investigation,
for until the properties of ectodermal fibres and of the adhesive
matrix are better known their speculative role is uncertain.   Evidence for a secretion facilitating detachment should also be sought
(Ross,1974).   If ectodermal contractility is demonstrated microfilaments may prove to be of wider significance among Anthozoa
(Crawford and Chia, 1974).

Such effectors are excited or influenced by distant stimuli.
Thus in Calliactis the pedal disk is released on stimulation of the
slow ectodermal conduction system SS1, and further SS1 pulses prevent
attachment.   The SS1 appears to be epithelial, and it may influence
endodermal systems as well (McFarlane,1969; 1975).   Present observations would favour a similar interpretation for detachment in other
species, i.e., release would also be caused by SS1 pulses propagated
over the pedal disk.   The control of adhesiveness is less certain,
if only because several factors could be involved.   If secretory
cells resemble cnidoblasts, for example, distant stimuli may affect
their threshold without actually causing discharge.   There are at
least three pathways by which distant stimuli might influence the
pedal ectoderm: via the epithelium itself, via contacts with endodermal muscle fibres, and via neurites, i.e., the processes of sense
cells or nerve cells, including the possibility that these might come
from the endoderm (Robson,1965).   It is safe to assume that the
endodermal nervous system plays a major role in locomotion, besides
which it interacts,at least indirectly,with other conduction systems

such as the SS1 (Pantin,1952; McFarlane,1975). To carry the
matter further a strategic combination of electrophysiological
experiments with electron microscopy seems called for.

    It is a pleasure to acknowledge help from many sources.   I
am indebted especially to the late Dr. E.J. Batham for the use of
her time-lapse film of Metridium (taken with Dr. P.M.B. Walker)
and of her histological sections, and to Dr. I.D.McFarlane, Dr.
G.A.B. Shelton and Mr. P. Anderson for the priviledge of dis-
cussing their unpublished work.   I would like to thank the Director
and Staff of the Marine Laboratory, Plymouth, for facilities and
material and especially Dr. A.J. Southward for advice and the loan
of photographic equipment; all who helped me during visits to the
Friday Harbor Laboratories, the Marinbiologisk Labatorium, Helsingør,
and the Zoological Station Kristineberg; and the Research Board of
Reading University for financial support.

SUMMARY

    Ectodermal effectors are responsible for detachment and adhesion
of the pedal disk.   Their action is coordinated with that of endodermal muscles, with which they form histological connexions.
    It is possible to attribute release of the pedal disk to ecto-
dermal supporting cells, which may be contractile.   Microvilli
are the peripheral agents of this process.   Adhesive material could
be secreted either by supporting cells or by granular gland cells.
Further evidence is required about the nature of ectodermal effector
mechanisms and about their control, especially in relation to muscular
activity and the properties of conduction systems.

REFERENCES

BATHAM,E.J. and PANTIN,C.F.A., 1951. The organization of the muscular
        system of Metridium senile. Quart.J. micr.Sci., 92: 27-54.
CARLGREN,O., 1893. Studien über Nordische Actinien I.
        Kongl. Sv. Vet.Akad. Handl. Stockholm, 25: 23-36.
CRAWFORD,R.J. and CHIA,F.-S., 1974. Fine structure of the mucous
        cell in the sea pen, Ptilosarcus guerneyi, with special
        emphasis on the possible role of microfilaments in the control
        of mucus release. Can.J.Zool., 52: 1427-1432.
CUTRESS,C.E. and ROSS,D.M., 1969. The sea anemone Calliactis
        tricolor and its association with the hermit crab Dardanus
        venosus. J.Zool., 158: 225-241.
EDMUNDS,M., POTTS,G.W., SWINFEN,R.C. and WATERS,V.L., 1976.
        Defensive behaviour of sea anemones in response to predation
        by the opisthobranch mollusc Aeolidia papillosa (L.).
        J.mar.biol.Ass.U.K., 56: 65-83.
ELLIS,V.L., ROSS,D.M. and SUTTON,L., 1969. The pedal disc of the
        swimming sea anemone Stomphia coccinea during detachment,
        swimming, and resettlement. Can.J.Zool. 47: 333-342.

GRIMSTONE,A.V. and SKAER,R.J., 1972. _A guide book to microscopical methods_. University Press, Cambridge.

HERTWIG,O. and HERTWIG,R. 1879. Studien zür _Blättertheorie_. Heft 1. _Die Aktinien_. G.Fischer, Jena.

MOOSEKER,S. and TILNEY,L.G., 1975. Organization of an actin filament-membrane complex. _J.Cell.Biol._, 67: 725-743.

McFARLANE,I.D., 1969. Coordination of pedal-disk detachment in the sea anemone _Calliactis parasitica_. _J.exp. Biol._,51: 387-396

McFARLANE,I.D., 1975. Control of mouth opening and pharynx protrusion during feeding in the sea anemone _Calliactis parasitica_. _J.exp.Biol._ 63: 615-626.

PANTIN,C.F.A., 1952. The elementary nervous system. _Proc.Roy.Soc.B_, 140: 147-168.

PARKER,G.H., 1917. Pedal locomotion in actinians. _J.exp.Zool._, 22: 111-124.

ROBSON, E.A., 1961. Some observations on the swimming behaviour of the anemone _Stomphia coccinea_. _J.exp.Biol._, 38: 343-363.

ROBSON, E.A., 1965. Some aspects of the structure of the nervous system in the anemone _Calliactis_. _Am.Zool._, 5: 403-410.

ROBSON, E.A., 1971. The behaviour and neuromuscular system of _Gonactinia prolifera_, a swimming sea-anemone. _J.exp. Biol._, 5: 611-640.

ROSS, D.M., 1974. Behaviour patterns in associations and inter-actions with other animals. Pages 281-312 in L.Muscatine and H.M. Lenhoff, Eds., _Coelenterate Biology_. Academic Press, New York.

ROSS, D.M. and SUTTON,L. 1961. The response of the sea anemone _Calliactis parasitica_ to shells of the hermit crab _Pagurus bernhardus_. _Proc.Roy.Soc.B_, 155: 266-281.

STEPHENSON,T.A., 1928. _The British sea anemones. Vol.1._ Ray Society, London.

REFLECTIONS ON TRANSPARENCY

GARTH CHAPMAN

QUEEN ELIZABETH COLLEGE

CAMPDEN HILL ROAD, LONDON   W8   7AH

INTRODUCTION

The transparency of pelagic organisms has been remarked upon
by many naturalists, most recently by Hamner et al., (1975) but nowhere is it possible to find much more than casual comment and speculation about the physical means by which this is achieved, the visibility of organisms to their predators or their prey and the part
which transparency may play in the lives of such organisms. Is
transparency the primitive state of organisms and opacity the more
highly evolved condition? I think not, because the number of organisms which are strikingly transparent is small compared with the non-
transparent ones and those parts of the body of vertebrates which
are transparent, the cornea and the lens, are constructed in a specific way which has been shown to provide a basis for a physical
explanation.

PHYSICAL BASIS OF TRANSPARENCY

Since, in 1967, I knew of no measurements of the amount of light
transmitted by pelagic animals I attempted to measure the transmittance of medusae in as natural a state as possible using a Hitachi-
Perkin Elmer spectrophotometer covering the wavelengths 200-1000 nm.
The results are shown graphically in Figs. 1 and 2 and are in general
agreement with those of Greze (1963 and 1964) which I have recently
found. He obtained a 'transparency coefficient', $T = \emptyset n/\emptyset p$ where
$\emptyset n$ is the transmitted light value and $\emptyset p$ that of the incident light
as recorded by a light meter used in conjunction with a photomicro-
graphic apparatus. His results were made on an ocean cruise and are
evidently approximate ones. Indeed caution is needed in the interpretation of any transmittance measurements under conditions in which

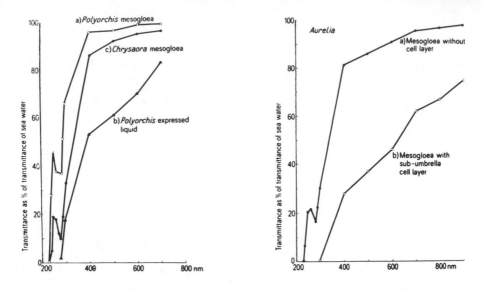

Fig. 1.   Transmittance of mesogloea of <u>Polyorchis</u>, (a) expressed
liquid, (b) and mesogloea of <u>Chrysaora</u>, (c) cut to fit cuvette of
spectrophotometer
Fig. 2.   Transmittance of mesogloea of <u>Aurelia</u> without, (a) and
with, (b) cell layer placed at right angles to the light path (Both
from Chapman, G., Experientia 32, 123 (1976) with permission)

light scattering also occurs (Ross & Birley, 1973). Greze's values
for T range from 0.91 for <u>Rhopalonema</u>, through 0.87, 0.85 and 0.74
for the siphonophores <u>Lensia</u>, <u>Diphyes</u> and <u>Chelophyes</u> to 0.48 for an
unidentified ctenophore. Thus it seems that, while some organisms
have a transmittance approaching that of the cornea (i.e. in the 90-
99% range) most do not, even although they are regarded as 'transparent organisms'.

     When light falls on any organism it can be transmitted, scat-
tered, reflected or absorbed: what the organism actually looks like
depends upon the fate which befalls the light or rather on which fate
predominates. If all the materials of which the organism is made had
the same refractive index there would be no optical <u>inhomogeneities</u>
at which light scattering could occur and the organism would be transparent. However, animals and cells are characterized by possessing
a great many inhomogeneities at all size levels. If these discontinuities are very few the amount of light scattered is small but if
they are numerous, and especially if they are comparable in size with
the wavelength of light, the amount of scattering is great. The vertebrate cornea which contains about 20% of its wet weight of collagen
of refractive index 1.55 held in a matrix of refractive index 1.335
has many inhomogeneities but nevertheless has a transmittance of up
to 99%. Maurice (1957) proposed an explanation of this which required

a critical and regular spacing of the fibrils such that any light
scattered by a fibril interferes with that scattered by others:
only the direct ray passes on. Cox et al., (1970) showed that such
precision in the fibrillar spacing was not necessary and that a degree of randomness of the fibrils is permissible, at any rate for
fibrils below a certain size. In a recent paper (Chapman, 1976) I
tried to examine whether the features of structure and composition
which give rise to the transparency of the cornea are general among
transparent organisms. While they may contribute to an explanation
of the transparency of highly structured animals such as Sagitta and
Tomopteris, such a degree of structural complexity is probably absent
from the pelagic coelenterates which depend for their high transmittance more on their having a very small amount of structural protein and thus possessing few homogeneities which can scatter light.
Moreover the cells, being spread out in thin layers on the bulky
mesogloeal jelly, offer only a small impediment to light transmission,
probably no more than that of the six or so cell layers of the cornea.
However it will be obvious that more measurements and precise observations are needed before these suggestions can be substantiated.

                    VISIBILITY OF OBJECTS UNDER WATER

        Are ctenophores and medusae 'visible' to the receptor systems
of their predators and prey under the illumination and at the orientation in which they live? Lacking direct observations to cite, it
is nevertheless worth attempting to answer the question although
there is much that is speculative in what follows.

        In general the visibility of an object depends upon its contrast
with the background and hence is dependent on both the nature of the
object and of the background as well as on the optical properties of
the medium and the direction and intensity of the illumination. Hence
any factor diminishing contrast can diminish visibility. Contrast
can be of brightness or colour but for the present only brightness
will be considered. Denton (1970) has analysed exhaustively the
principles on which the silvery reflecting scales of fish are arranged so that the light reflected from them matches the direct light
entering the eye of an observer and hence renders them non-contrasting
with their environment and has shown that these conditions can be
made to apply to all directions except from directly underneath:
however neither he nor Duntley (1963) has much to say on the visibility of transparent objects as opposed to opaque ones. By defin-
ition a transparent object is one which does not reflect or scatter
light but, since few organisms are completely transparent both the
transmission of light through them and the reflection of incident
light must be considered. In the case of some marine animals we have
already seen that their transparency probably does not result from
the arrangement of fibrils that enables the cornea to scatter only
1% of incident light, but rather from the very small protein content

leading to the presence of few or small structural fibres and other
light scattering inhomogeneities. Any animal which has a light
transmittance of less than 100% will appear darker than the background when viewed in the direction from which the light is coming
provided that the eye is able to perceive the incremental difference
at the intensity to which it is adapted.

However in the natural environment of the sea or a lake, illumination is not unidirectional owing to a number of factors. Since
the radiance at any point may be due to direct sunlight, skylight
including light reflected from or scattered by clouds and light scattered in the water by solvent and solute molecules and by particles.
While the first two decrease logarithmically with depth (Jerlov,
1966), the third is zero at the surface and at a maximum at a certain depth dependent upon attenuation and upon reflection from the
bottom in shallow water (Duntley, 1963). The variation in radiance
in any vertical plane can be represented by a vector or polar diagram.
It is markedly asymmetrical since illumination is, of course, chiefly
from above the free surface of the water and it is further skewed by
any zenith angle of the sun of more than 0°, i.e. when the sun is not
directly overhead. The polar diagram will also vary according to the
angle that the plane in which it is drawn subtends to the azimuthal
angle of the direct ray. Accordingly it seems likely that visibility
from different directions may vary also.

The contrast of an object with its background can now be considered in terms of the polar diagram illustrated in Fig. 3 and in
terms of the transmittance, t and the reflectance, r. Let the radiance (which can be taken to be approximately the same as light intensity) along the direct path be D and the radiance of the upwelling
light reaching the object along the indirect path be I. Light reaching the eye direct will be D, equal to the background radiance (ig-
noring attenuation between the object and the eye). Light reaching
the eye from the object will be Dt + Ir. The contrast is the ratio
of the radiance from object and background and can be written as the
ratio, $(Dt + Ir)/D$. However in sense organs it is the ratio of the
increment, or radiance difference, to the absolute radiance which
determines perception, so the effective contrast can be written as,
$(D - (Dt + Ir))/D$. The more transparent the object the higher will be
the value of Dt and the lower the value of Ir. However the relative
values of D and I will vary with the position on the polar diagram
at which the observer is situated but however high Dt becomes it
will always be less than D and, in a natural situation, it is unlikely that I and Ir will ever be greater than Dt when the observer
is in position 1. Consequently the object will appear dark against
a lighter background.

However, variations of the position of the object and observer
with respect to the polar diagram will affect the visibility and the
polar diagram itself will be affected by factors including the depth

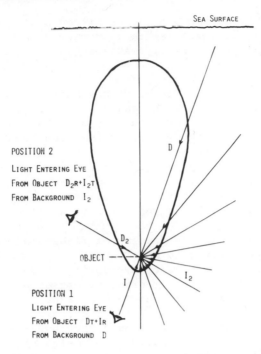

Fig. 3. Polar diagram of radiance at an object in the sea and of light reaching the eye when the object is seen from different angles.

and turbidity. Consider now position 2 (Fig. 3) in which no ray from the surface enters the eye direct. The background radiance, now the upwelling light, say I_2 and the direct (downwelling) radiance, say D_2 give rise to light received at the eye which can be written $D_2r + I_2t$. In this case since D_2 is again greater than I_2 it is possible for $D_2r + I_2t$ to be greater than I_2 (the background radiance). In this case the object would appear bright against a darker background. It would seem possible, if the observer's eye were at a certain angle, for $D_2r + I_2t$ to be equal to I_2 in which case the object, although not perfectly transparent would be "invisible". The conclusion would appear to be that a degree of transparency renders the contrast of an object less at all angles of viewing and may reduce it to zero at certain angles.

At present it seems impossible to check these suggestions owing to the lack of any measurements of the transmittance and reflectance of pelagic organisms but it should be possible to investigate their visibility in the laboratory (at least to the human eye) by the use of suitable apparatus operating at the known radiance levels of the sea. The direct observation of transparent organisms during 800 h. of offshore SCUBA diving has been described by Hamner et al., (1975) and Hamner (1974) but the only previous report of such underwater

observations is that of the pioneering attempts of Bainbridge (1952).
Perhaps the most significant remark in Hamner's papers is that "any
condition that impairs visibility makes it difficult to see gelatinous transparent animals".

So far the visibility of objects has been considered from the
geometrical point of view, rather than with respect to the perceiving
apparatus. Information is very scarce about the dioptrics and sensitivities of the eyes of invertebrates and even about those of fishes
owing to the fact that each of the two methods available for their
investigation, the neurophysiological and the behavioural, present
their own difficulties of interpretation. Nevertheless an assessment of some of the limits to the visibility of transparent organisms
can be made using what information there is available.

The role of light in the feeding of fishes is summarized by
Blaxter (1970) especially from his own work on young and larval
herrings. The laboratory studies are mainly on small fishes which
are unlikely to prey on large transparent organisms but might well
be the prey of such and might therefore be expected to benefit from
avoiding them. The data on fishes is, therefore, likely to be about
the best available and of the best-developed underwater visual systems. As the threshold illumination for vision Blaxter (1970) gives
as a typical value 10^{-5} to 10^{-6} lux (or metre candles) corresponding
to an irradiance of 10^{-5} to 10^{-6} μ W cm^{-2} (cf. threshold for man of
10^{-8} μ W cm^{-2}) probably bettered by some deep sea fishes but, as
Blaxter says, different methods of estimation (ERG or behaviour) give
very different values. Also the increment of distinguishable intensity is known to vary too and also to be widely different for different fishes e.g. Anguilla japonica 540, Kareius bicolorata 2, Gadus
morhua 1, Mallotus villosus (capelin) <0.01, all expressed as $\Delta I/I\%$.
A very low figure will facilitate perception of small differences
such as those likely to be due to transparency. Figures for the
human eye vary between 50 and 2% according to the illumination
(Starling 1962). The fact that some fishes can do better than this
could assist them in avoiding predatory transparent organisms or
parts of organisms. Visual acuity also comes into the perception
of objects and figures quoted by Blaxter varied from 5'-7' for species of tunny and from 50' to 7' for the pure cone area of the retinas
of 1-30 cm long herrings. An angle of 6' implies discrimination of
objects of 0.17 mm at 10 cm which is perhaps near to the feeding
distance. In other words it is possible that a young herring could
perceive the tentacles of a siphonophore at a distance away which
would enable the fish to avoid bumping into them.

Apart from those of cephalopods, the non-compound eyes of invertebrates are inferior to those of vertebrates in most respects.
Although it seems that their sensitivity to light has not been
measured, visual acuity can be estimated from published data. Thus
the retinal cup of the polychaete Alciopa, which has larger and more

elaborate eyes than the cubomedusae, measures 400 μm in diameter
(Demoll, 1909) and is composed of 10^4 retinal cells (Pumphrey, 1961).
If these are 5 μm in width and the lens lies at the centre of the
aperture then the angle subtended at the lens centre by two visual
cells separated by one unstimulated cell would be $1^o26'$. At a distance of 10 cm discrimination between points 2.5 mm apart should be
possible, a performance greatly inferior to the fish eyes just considered.

The foregoing arguments have, it seems to me, tended to show
that if transparency tends to make animals invisible, the light receptor systems of some other animals living in the same environment
are efficient enough to counteract the advantages gained, at least
to some extent.

ECOLOGICAL IMPLICATIONS

Just as it seems to be difficult to find a convincing role for
bioluminescence in many phyla, not least in coelenterates, (Morin,
1974) so it is difficult to find a convincing role for the transparency of planktonic members but the belief remains that it is useful in some way since it, also, is shared to a greater or a lesser
degree by pelagic forms from many phyla including fishes (Meyer-
Rochow, 1974). Hamner et al (1975) considered that the gelatinous
zooplankton which they studied and which included hydromedusae,
siphonophores, scyphomedusae, ctenophores, heteropods, pteropods,
thaliaceans and larvaceans constituted an ecological group on the basis of predator-protection strategies and Greze, (1963 and 1964) also
considered that, among the transparent forms which he investigated
there was a predominance of prey. Especially in those zones of the
sea in which illumination is only a little above the threshold for
visibility he reckoned that an organism which transmitted 50% of the
incident light had a "sphere of visibility" possessing a volume of
one eighth that of the sphere within which an opaque, reflecting
animal could be seen. As Hamner says, the organism "must avoid predators in an environment unusually devoid of hiding places". However one could argue that being transparent, like other forms of
crypsis, gives an advantage to predators or prey alike although not
all predators of transparent animals prey by sight, e.g. ctenophores
(Swanberg, 1974).

It seems difficult to obtain from the literature a reliable
picture of the part which gelatinous, transparent animals play in
any food web. Their irregular occurrence in time and space and
their lack of hard, indigestible remains may partly account for this.
Also, as Lebour (1922 and 1923) commented, one is never sure that
what they appear to have been feeding on when caught in nets, was
not taken while they were crushed together in the net. She attempted
to see what the carnivores of the plankton would eat by offering them

a variety of prey in the laboratory and she found that, indeed,
hydro- and scyphomedusae and ctenophores captured a variety of organisms including young fishes. Her studies still remain some of the
most informative available on the food and feeding of planktonic
organisms but field studies like those of Hamner et al. (1975) should
provide direct observations.

The crypsis of animals which results from the matching of their
colour and pattern with that of the background is effective only
when they are stationary and is immediately upset by movement. This
type of concealment is useless for floating forms since an animal
can only remain stationary by being neutrally buoyant (which places
it at the mercy of water movements) or by swimming. Moreover, in an
environment "unusually devoid of hiding places" the matching of the
background must be of the uniform type which, in opaque animals, is
provided by reflective fish scales (Denton, 1970). The advantage
of a transparent body is that it provides crypsis against any background and at the same time permits movement and locomotion. This
would seem to be the chief advantage of transparency but since this
characteristic is frequently linked with a bulky gelatinous construction it is, perhaps, difficult to ascribe certain of the advantages
to any one of the associated and interrelated characteristics. As
a result of the bulky construction the cell layers can be spread out
instead of being folded as they frequently are in, for example, actinians. The thin cell layers cause little light scattering and help
to render the body transparent. The bulky construction means that
large size can be attained with a minimum of organic material and
that growth of the individual and of the population can be rapid
(Chapman, 1974). Madin (1974) notes that the transparent salp
Thalia democratica also has very high rates of individual and population growth. Finally flotation is facilitated by the provision
of a large volume in which a small diminution in concentration of a
heavy ion (e.g. SO_4^{2-}) can counteract the higher density of the very
small quantity of protein in the organism (Denton and Shaw, 1961).

However not the last word has been said about the usefulness of
transparency since some transparent animals have evolved ways in
which their invisibility can be modified. The siphonophore Hippo-
podius hippopus responds to touch by becoming milky white in 1 or 2
seconds and only becomes transparent again after 15-30 min (Mackie
and Mackie, 1967). The same animal, as are many other hydrozoans,
is also luminescent. It is difficult to reconcile these attributes
with transparency crypsis as defensive strategy!

REFERENCES

BAINBRIDGE,R., 1952. Underwater observations on the swimming of
        marine zooplankton. J. Mar. Biol. Assoc. U.K., 31:107-112.
BLAXTER,J.H.S., 1970. Light: Animals: Fishes. pp. 213-285 in Marine
        Ecology Vol. 1, Pt. 1. Ed. O.Kinne. Wiley-Interscience, London.

CHAPMAN,G., 1974. The Skeletal System. pp. 93-128 in Coelenterate
    Biology, Reviews & New Perspectives. Ed. L.Muscatine & H.M.
    Lenhoff, Academic Press. New York & London.
CHAPMAN,G., 1976. Transparency in Organisms. Experientia, 32:123-125.
COX,J.L., R.A.FARRELL, R.W.HART AND M.E.LANGHAM, 1970. The transpar-
    ency of the mammalian cornea. J. Physiol., 210:601-616.
DEMOLL,R., 1909. Die Augen von Alciopa cantrainii. Zool. Jahrb. Abt.
    Anat. Ontog. Tiere, 27:651-686.
DENTON,E.J., 1970. On the organization of reflecting surfaces in some
    marine animals. Phil. Trans. R. Soc. B., 258:285-313.
DENTON,E.J. AND T.I.SHAW, 1961. The buoyancy of gelatinous marine
    animals. J. Physiol. (London), 161:14p-15p.
DUNTLEY,S.Q., 1963. Light in the sea. J. Opt. Soc. Am., 53:214-233.
GREZE,V.N., 1963. The determination of transparency among planktonic
    organisms and its protective significance. Dokl. Akad. Nauk.
    S.S.S.R., 151:435-438.
GREZE,V.N., 1964. The transparency of planktonic organisms in the equ-
    atorial part of the Atlantic Ocean. Okeanologiya, 4:125-127.
HAMNER,W.M., 1974. Blue-water plankton. Natl. Geog. Mag., 146:530-545.
HAMNER,W.M., L.P.MADIN, A.L.ALLDREDGE, R.W.GILMER, P.P.HAMNER, 1975.
    Underwater observations of gelatinous zooplankton:Sampling prob-
    lems,feeding biology and behaviour. Limnol.Oceanog.,20:907-917.
JERLOV,N.G., 1966. Aspects of light measurement in the sea. pp. 91-98
    in Light as an Ecological Factor, Ed. R.Bainbridge et al, Black-
    wall, Oxford.
LEBOUR,M.V., 1922. The food of plankton organisms. J. mar. Biol. Ass.
    U.K., 12:644-77.
LEBOUR,M.V., 1923. The food of plankton organisms 2. J. mar. Biol.
    Ass. U.K., 13:70-92.
MACKIE,G.O. AND G.V.MACKIE, 1967. Mesogloeal ultrastructure and re-
    versible opacity in a transparent siphonophore. Vie. Milieu.,
    Serie A: 18:47-71.
MADIN,L.P., 1974. Field observations on the feeding behaviour of salps
    (Tunicata: Thaliacea). Mar. Biol. (Berl.), 25:143-148.
MAURICE,D.M., 1957. The structure and transparency of the cornea.
    J. Physiol., 136:263-286.
MEYER-ROCHOW,V.B., 1974. Leptocephali and other transparent fish
    larvae from the south-eastern Atlantic Ocean. Zool. Anz. 192
    (3-4):240-251.
MORIN,J., 1974. Coelenterate Bioluminescence. pp. 397-438 in Coelen-
    terate Biology, Reviews & New Perspectives. Ed. L.Muscatine &
    H.M.Lenhoff. Academic Press. New York & London.
PUMPHREY,R.J., 1961. Concerning vision. pp.193-208 in The Cell and the
    Organism. Ed. J.A.Ramsay & V.B.Wigglesworth. Cambridge U.P. Lond.
ROSS,G. & BIRLEY,A.W., 1973. Optical properties of polymeric materials
    and their measurement. J. Phys. D: Appl. Phys. 6:795-808.
STARLING,E.H. AND LOVATT EVANS,C., 1962. Principles of human physio-
    logy. 13th Ed. Ed. by H.Davson & M.G.Eggleton. Churchill, London.
SWANBERG,N., 1974. The feeding behaviour of Beroë ovata. Mar. Biol.
    (Berl.), 24:69-74.

A QUALITATIVE AND QUANTITATIVE INVENTORY OF NERVOUS CELLS IN HYDRA ATTENUATA PALL

Pierre Tardent and Christian Weber

Zoological Institute, University of Zurich

Künstlergasse 16, 8006 Zurich / Switzerland

INTRODUCTION

The nervous system of Hydra (Athecatae, Hydridae) is, as has been shown by Hadzì (1909), an elementary network of loosely interconnected neurons and other cells believed to be of sensory nature (Burnett and Diehl,1964; Davis, Burnett and Haynes, 1968; Haynes, Burnett and Davis, 1968; Lentz, 1966; Lentz and Barrnett, 1965 et al.). In spite of the absence of a morphologically definable center of coordination it endows the polyp with a surprisingly rich behavioural pattern (Haug, 1933; Haase-Eichler, 1931; Passano and McCullough, 1962, 1963, 1964; Rushforth, 1971, 1973; Tardent and Frei, 1969; et al.). This behaviour is motivated by 2 fundamentally different components: One is the "spontaneous contraction-extension activity" which is controlled by 2 pace-maker centers, one of which is situated in the hypostome, the other in the stalk of the polyp (McCullough, 1962, 1965; Passano,1962; Passano and McCullough, 1963, 1964, 1965). The other component consists of dynamic reactions to external stimuli such as light, chemicals or mechanical stimuli (Haug, 1933; Feldman and Lenhoff, 1960; Passano and McCullough, 1962; Rushforth, Burnett and Maynard, 1963; Singer, 1963; Tardent and Frei, 1969; Frei, 1973 et al.).

After having discovered how susceptible Hydra is to various photic stimuli (Tardent and Frei, 1969; Frei, 1973; Tardent, Frei and Borner, 1976) we were anxious to

know more about the physical basis of light perception
in these animals. In view of the fact that the functional
identification of cells by electrophysiological methods
is not yet possible, we hoped that a careful qualitative
and quantitative inventory of neurons and supposedly
sensory elements might be useful in this respect. This
paper describes 11 different cell-types, their number and
axial distribution as they were found and counted in
dissociated polyps (David, 1973).

MATERIAL AND METHODS

The polyps of <u>Hydra</u> <u>attenuata</u> Pall. (Tardent, 1966)
were kept in Loomis solution (Loomis and Lenhoff, 1956)
at 18 ± 2°C in complete darkness. They were fed zooplankton
of the Lake of Zurich, or <u>Artemia</u>, every 2 days. Four
hours after feeding the culture solution was replaced.
The manipulations were carried out under a weak red light.
In order to avoid a disturbing development of buds, the
specimens chosen for the experiments were starved for 8
days before they were dissociated.

For the examination of nervous elements and for their
morphological description entire polyps were dissociated
according to the method described by David (1973), which
was slightly modified. For the quantitative investigations
25 selected Hydras, previously immobilized with MS 222
and Novocaine, were cut into 8 axial fragments (Fig. 3)
each of which was dissociated individually, so as to prevent any loss of cells. The total number of ectodermal
epithelio-muscular cells (EMC), and of the 11 types of
nervous cells (NC) classified morphologically (Fig. 1)
were counted in each preparation. The cells were examined
and counted with light-, phase-contrast- and interference-
phase contrast microscopes at a magnification of 400x
with the aid of an ocular grid.

RESULTS

a. The Qualitative Inventory of Nervous Cells

Taking purely morphological critera such as size,
shape, number of cytoplasmic processes, presence or absence
of cilia, etc., it was possible to distinguish between the
following 11 types of cells shown in Figure 1.

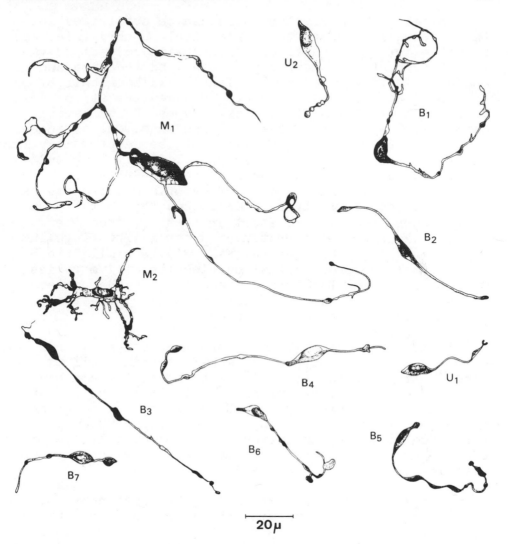

Fig. 1: Drawings of the 11 types (see lettering) of the nervous and neurosensory cells found in <u>Hydra attenuata</u> Pall. (The cells are drawn in scale).

<u>Multipolars</u>: Cells featuring more than 2 conspicuous cytoplasmic processes were classified as multipolar neurons, amongst which there is a large type (M_1) with branched neurites reaching a length of 200 μm (Fig. 2a). The small multipolars (M_2) have short processes the length of which does not exceed 50 μm.

Bipolars: The bipolar cells can be subdivided into
the following 7 types: the large bipolars (B_1) (Fig. 2e)
resemble the large multipolars (M_1) with respect to the
length (200 μm) of their 2 processes and the size and
shape of their perikarya. The length of the richly branched
processes of the small bipolars (B_2) reaches at the most
60 μm. The remaining bipolar cell-types (B_{3-7}) have in
common the asymmetry of their 2 processes. One of these
processes resembling the neurites of the large multipolars
(M_1) and bipolars (B_1), is always relatively long and
thin, and features a varying number of nodules and
thickenings (Fig. 2d). The other cytoplasmic process has
a specific shape. In type B_3 (Fig. 2d) this 5 to 15 μm
long process resembles a short drum-stick, from the
terminal knob of which protrudes a threadlike flagellum
5 to 10 μm long. The B_4-type (Fig. 2f) is similar to B_3.
The neck of the process that carries the cilium varies
as to its length. The terminal nodule from the center of
which emerges the cilium is larger (diam. 7 μm) than in
B_3, and has a flattened or concave tip.

B5 has a drumstick-like, short process (Fig. 2b). Unlike the previous 2 types it lacks a terminal cilium. This
also is true for the remaining 2 bipolars (B_6 and B_7). In
B_6 one process is particularly short. It tapers off and
lacks a terminal knob or cup-shaped ending. Type B_7 has
a distinct terminal swelling of the process which is
separated from the perikaryon by a thin cytoplasmic bridge
of 5-10 μm length, which in some cases shows one or two
swellings.

Unipolars: There are 2 kinds of unipolar neurons. The
perikaryon of U_1 is 5 μm wide and 15 μm long. The nucleus
takes up about half of its volume. The only cytoplasmic
process attains 40-80 μm. Type U_2 (Fig. 2c) is small and
reaches a total length of 30-50 μm; the cell body is
10-20 μm long. The short process features a varying number
of swellings.

b. The Number and Axial Distribution of Nervous Cells

The average numbers of cell-types, and their relative
frequencies, obtained from cell counts carried out on 25
specimens, are given in Table 1. This shows that the ratio
of ectodermal epithelio-muscular cells to the total number
of nerve cells is roughly 1:1. The average number of nerve

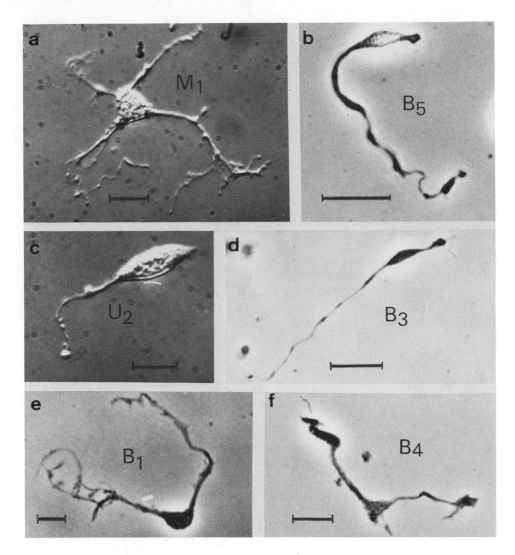

Fig. 2: Six types (see lettering) of isolated nervous and
        sensory cells of <u>Hydra</u> <u>attenuata</u> Pall. as
        appearing in the interference (a,c) and the normal
        (b, d-f) phase contrast (cfr. Fig. 1). Each bar
        corresponds to 10 μm (c = cilium).

cells per animal is 1266.2 ± 170.2. Amongst these the
most frequent types are the small ($M_2 = 39.05\%$), the
large ($M_1 = 16.26\%$) multipolars and the small bipolars
($B_2 = 13.92\%$).

Table 1: Absolute and relative numbers of epitheliomuscular cells (EMC), nerve cells (NC) and
nerve-cell types as counted in 25 different
specimens of Hydra attenuata Pall.

Cell-type	Average Numbers per animal	Relative Numbers	
EMC	1068.7 ± 163.8		
NC (pooled)	1266.2 ± 170.2	100%	100%
Multipolar NC			
M_1	205.8 ± 38.7	16.26%	
M_2	494.3 ± 61.5	39.05%	55.31%
Bipolar NC			
B_1	88.4 ± 14.5	6.97%	
B_2	176.4 ± 23.1	13.92%	
B_3	117.2 ± 17.8	9.25%	
B_4	96.9 ± 19.7	7.65%	
B_5	8.7 ± 2.2	0.71%	
B_6	16.7 ± 2.9	1.33%	
B_7	31.1 ± 6.8	2.44%	42.27%
Unipolar NC			
U_1	14.4 ± 3.3	1.17%	
U_2	16.2 ± 3.6	1.25%	2.42%

The axial distribution of the pooled nerve cells (Fig. 3)
as counted in 25 specimens gives a U-shaped curve reaching
the highest values in the distal- and proximal-most
regions, in which nearly 50% of all nervous elements are
located. The corresponding curves for the various types
of nerve cells (Fig. 4) show that - with the exception
of M_1 - the most numerous types, i.e. the small multipolars (M_2) and the small bipolars (B_2) are distributed
similarly. The other less numerous types exhibit totally
different distribution patterns, as shown in Figure 4.
It can be therefore said that the types classified above
differ not only with respect to their size and shape but

also with regard to their relative numbers and distribution patterns.

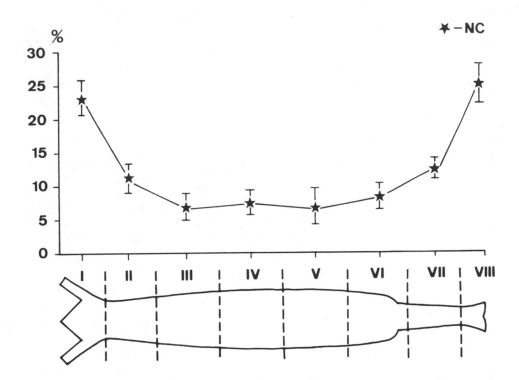

Fig. 3: Axial distribution of the pooled nervous elements as recorded in 25 different specimens. The average values given for each level correspond to the % of the total amount of nervous and sensory cells (100%) recorded in an animal (cfr. Table 1).

DISCUSSION

The maceration-method (David, 1963) used in this paper for establishing a "taxonomy" of nerve cells in Hydra based on morphological criteria, seems for the moment being to be the most satisfactory, although it does not absolutely exclude artefacts due to chemical (fixation) and mechanical damage to the isolated cells. It has the further disadvantage that it does not permit more than guesses as to the possible functions of the cells. Con-

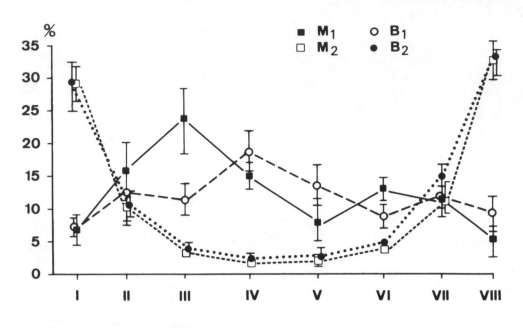

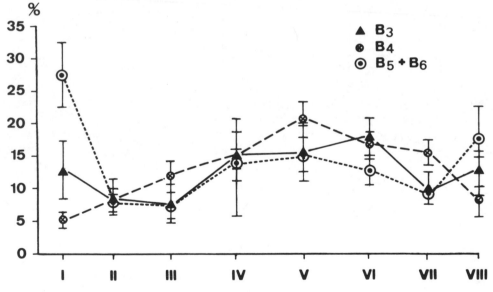

Fig. 4: Relative values of the axial distribution of
        different types of nervous and neurosensory cells
        as counted in 25 specimens (cfr. Fig. 3).

clusive evidence in this respect could only come from
appropriate electrophysiological studies, which, for
technical reasons, still seem to be out of reach.

Amongst the 11 types of nerve cells and supposed sensory cells there are 4 kinds, the large and small multipolars (M_1, M_2) and bipolars (B_1, B_2), which seem to be
ganglionic elements of the primitive nerve net. The method
does not allow to decide whether these also fulfill a
neurosecretory function or whether this function is limited to a particular type or to a stage of functional differentiation of one or the other of the types (cfr.
Lentz and Barrnett, 1965; Davis, 1969, 1973c).

We believe that sensory elements have to be looked for
amongst the asymmetrical bipolars of the described types
B_3-B_7. Particular attention must be drawn to the types B_3
and B_4 from one end of which protrudes a small cilium.
The existence of neurosensory cells provided with cilia
has been reported by Lentz (1966) and Davis (1973a). Our
electron-microscopical studies (unpublished) of these
cells indicate that the base of the cilium is anchored
deep in the cytoplasm of the perikaryon, and that the
cilium loses its 9 + 2-structure in the distal portion
protruding from the cell. The base of the cilium is surrounded by a large vesicle into which protrude 2 concentric rows of microvilli (cfr. Davis, 1973a). These cells
are found in the intercellular spaces of the ectoderm,
their longitudinal axis running parallel and not perpendicular to the mesoglea. The presence of cilia and internal microvilli made us suspect that these cells could be
rudimentary photoreceptors (cfr. Eakin, 1962, 1965). Their
axial distribution (Fig. 4) does, however, not coincide
with the axial pattern of regional photosensitivity in
Hydra attenuata (Zürcher, 1974). The other bipolar cells
($B_5,6,7$), which lack externally visible cilia, could well
be intermediate stages of differentiation leading to the
final stage represented by the types B_3 and B_4.

The same can be said about the unipolar cells U_1 and
U_2 which may not be functional as such, but which could
be on their way to becoming any kind of multipolar or bipolar cell. It may well be that the number of different
"types" of nervous and neurosensory cells, as they have
been described here on a merely morphological base may
be reduced as soon as the differentiation pathways, taking
their issue from pluripotent interstitial cells (Schneider,

1890; McConnell,1932; Burnett and Diehl, 1964; Davis,
1973a, b, c; Gierer,1974 et al.) is known. A study aiming
at the reconstruction of a "genealogical tree" of nervous
elements in <u>Hydra</u> is in progress.

The cell-counts do not match in all respects with
those done by Bode et al. (1973). The fact that the total
numbers of ectodermal epithelio-muscular cells and nerve
cells are much lower than those given by Bode et al.
(1973) may be due to the relatively long starvation period
imposed on our experimental animals, previous to maceration. However, the U-shaped distribution of nerve cells
confirms their findings. The present investigation has
shown that not all types of ganglionic and neurosensory
types are distributed in accordance with this pattern
(Fig. 4). It will therefore be of interest to investigate
the functional significance of these type-specific distribution patterns as well as their formation and maintenance.

This paper was supported by the Swiss National Science
Foundation (Grant No. 3.711.72). The authors are indepted
to Prof. J. Deak for reading the manuscript.

REFERENCES

Bode, H., S. Berking, C.N. David, A. Gierer, H. Schaller
        and E. Trenker, 1973. Quantitative analysis of
        cell types during growth and morphogenesis in
        <u>Hydra</u>. <u>Wilhelm Roux'Archiv</u>, 171: 269-285.
Burnett, A.L. and N. Diehl, 1964. The nervous system of
        <u>Hydra</u>. I. Types, distribution and origin of
        nerve elements. <u>J. Exp. Zool</u>., 157: 217-226.
David, N.C., 1973. A quantitative method for maceration
        of Hydra tissue. <u>Wilhelm Roux'Archiv</u>, 171: 259-
        268.
Davis, L.E., 1969. Differentiation of neurosensory cells
        in <u>Hydra</u>. <u>J. Cell Science</u>, 5: 699-726.
Davis, L.E., 1973a. Ultrastructure of Neurosensory Cell
        Development. Pages 271-298 in A.L.Burnett (Ed.),
        <u>Biology of Hydra</u>. Academic Press, New York and
        London.
Davis, L.E., 1973b. Ultrastructure of Ganglionic Cell
        Development. Pages 299-318 in A.L.Burnett (Ed.),
        <u>Biology of Hydra</u>. Academic Press, New York and
        London.

Davis, L.E., 1973c. Structure of Neurosecretory Cells with
     Special Reference to the Nature of the Secretory
     Product. Pages 319-342 in A.L.Burnett (Ed.),
     Biology of Hydra. Academic Press, New York and
     London.
Davis, L.E., A.L. Burnett and J.F. Haynes, 1968. Histolo-
     gical and ultrastructural study of the muscular
     and nervous systems in Hydra. II. The nervous
     system. J. Exp. Zool., 167: 295-332.
Eakin, R.M., 1962. Lines of evolution of photoreceptors.
     J. Gen. Physiol., 46: 357A - 367A.
Eakin, R.M., 1965. Evolution of photoreceptors. Cold
     Spring Harbor Symposia on Quantitative Biology,
     30: 363-370.
Feldman, M. and H.M.Lenhoff, 1960. Phototaxis in Hydra
     littoralis: Rate studies and localisation of
     the "photoreceptor". Anat. Rec., 137: 354-355.
Frei, E., 1973. Untersuchungen über die allgemeine und
     die spektrale Photosensibilität von Hydra atte-
     nuata Pall. Ph.D.thesis, University of Zurich,
     94 pp.
Gierer, A., 1974. Hydra as a Model for the Development of
     Biological Form. Scientific Amer. 231: 44-54.
Haase-Eichler, R.H., 1931. Beiträge zur Reizphysiologie
     von Hydra. Zool. Jb., Abt. Zool., 50: 265-312.
Hadzì, J., 1909. Ueber das Nervensystem von Hydra. Arb.
     Zool. Inst. Wien, 17: 225-268.
Haug, G., 1933. Die Lichtreaktionen der Hydren (Chloro-
     hydra viridissima und Pelmatohydra oligactis).
     Z. vergl. Physiol., 19: 246-303.
Haynes, J.F., A.L. Burnett and L.E. Davis, 1968. Histolo-
     gical and ultrastructural study of the muscular
     and nervous system in Hydra. II. J. Exp. Zool.,
     167: 295-302.
Lentz, T.L., 1966. The cell biology of Hydra. North Holland
     Publ. Co. Amsterdam.
Lentz, T.L. and R.J. Barrnett, 1965. Fine structure of the
     nervous system of Hydra. Amer. Zool., 5: 341-356.
Loomis, W.F. and H.M.Lenhoff, 1956. Growth and sexual
     differentiation of Hydra in mass culture. J. Exp.
     Zool., 132: 555-573.
McConnell, C., 1932. The development of ectodermal nerve
     net in the buds of Hydra. Quart. J. micr. Sci.,
     75: 495-509.
McCullough, C.B., 1962. Modification of Hydra pacemaker
     activity by light. Amer. Zool.,2: 436.

McCullough, C.B., 1965. Pacemaker interaction in Hydra.
        Amer. Zool., 5: 499-504.
Passano, L.M., 1962. Neurophysiological study of the co-
        ordinating systems and pacemakers of Hydra.
        Amer. Zool., 2: 435-436.
Passano, L.M. and C.B. McCullough, 1962. The light
        response and the rhythmic potentials of Hydra.
        Proc. Nat. Acad. Sci. (Wash.), 48: 1376-1382.
Passano, L.M. and C.B. McCullough, 1963. Pacemaker hier-
        archies controlling the behaviour of Hydra.
        Nature, 199: 1174-1175.
Passano, L.M. and C.B. McCullough, 1964. Coordinating
        systems and behaviour in Hydra. I. Pacemaker
        system of the periodic contractions. J. Exp.
        Biol., 41: 643-664.
Passano, L.M. and C.B. McCullough, 1965. Coordinating
        systems and behaviour in Hydra. II. The rhythmic
        potential system. J. Exp. Biol., 42: 205-231.
Rushforth, N.B., 1971. Behavioral and electrophysiological
        studies of Hydra. I. Analysis of contraction
        Pulse Patterns. Biol. Bull., 140: 255-273.
Rushforth, N.B., 1973. Behavior. Pages 3-41 in A.L. Bur-
        nett (Ed.), Biology of Hydra, Academic Press,
        New York and London.
Rushforth, N.B., A.L. Burnett and R. Maynard, 1963. Be-
        havior in Hydra: The contraction responses of
        Hydra pirardi to mechanical and light stimuli.
        Science, 139: 760-761.
Schneider, K.C., 1890. Histologie von Hydra fusca mit
        besonderer Berücksichtigung des Nervensystems
        der Hydropolypen. Arch. mikr. Anat., 35: 321-
        379.
Singer, R.H., N.B. Rushforth and A.L. Burnett, 1963.
        The photodynamic action of light on Hydra. J.
        Exp. Zool., 154: 169-174.
Tardent, P. and E. Frei, 1969. Reaction patterns of dark-
        and light-adapted Hydra to light stimuli. Expe-
        rientia, 25: 265-267.
Tardent, P., E. Frei and M. Borner, 1976. The reactions
        of Hydra attenuata Pall. to various photic
        stimuli. In this book.
Zürcher, F., 1974. Regionale Lichtempfindlichkeit von
        Hydra attenuata Pall. und Koordination des Kon-
        traktionsverhaltens. Thesis, Zoological Insti-
        tute, University of Zurich.

EVIDENCE FOR NEURAL CONTROL OF MUSCLES IN CTENOPHORES

Mari-Luz HERNANDEZ-NICAISE

Laboratoire d'Histologie, Université Claude Bernard

69621,. Villeurbanne, France

"Animals are essentially predatory behavior machines. They possess sense organs to receive information, and predictor machinery both to process it and to direct an appropriate apparatus of moving parts so as to catch food ... To meet these requirements organisms have standard materials at their disposal and their functional structures must conform to limited engineering possibilities."(Pantin, 1965)

Ctenophores are indeed efficient predators and feed upon various planktonic organisms. Beroid and cydippid species are propelled by the metachronal waves which run down the eight rows of ciliary plates. The contact of the prey with the key-zone of the epidermis (lips of Beroe, tentacles of Pleurobrachia) triggers the feeding behavior, setting at work various muscular systems. Lobates and cestids swim not only with their locomotory cilia but by undulations of, respectively, the oral lobes or the whole body, i.e. by muscular contractions.

Anyone who has watched the incredible swiftness and accuracy of Beroe and Pleurobrachia while catching a prey, is irresistibly led to postulate that some neuromuscular mechanism is involved. Again for any observer the elegant swim of lobates and cestids and its quick adaptation to any disturbance require some underlying neuromuscular coordination.

My purpose is to recapitulate the actual evidence for such neuromuscular control. Some of these results have been published in part previously, in publications dealing chiefly with the nervous system rather than the muscles (Hernandez-Nicaise, 1968, 1973a, b, c). Therefore I will only summarize them briefly.

MATERIALS AND METHODS

The material studied was collected over a period of several years, and subsequently observed or processed for electron microscopy, at the Station Zoologique, Villefranche-sur-Mer (06 France). The tissues were usually fixed in 3% glutaraldehyde in cacodylate-buffered saline solution, rinsed in the buffered saline and post-fixed in 1% OsO_4-cacodylate. Isosmotic conditions were retained throughout the different batches. The method of Karnovsky (1964) was used to demonstrate the presence of cholinesterases at neuromuscular junctions. The adaptation to electron microscopy of the chromaffin reaction (Wood, 1966, 1967) was used to detect the presence of aminergic neurons.

RESULTS

Ultrastructure of Muscles

The electron microscope survey clearly confirms that all muscles in ctenophores are true muscle cells without any epithelial component. Muscle cells can be observed in two locations: either underlying the epithelia, or running across the mesoglea where their branched extremities anchor upon the epithelial cells. There is a wide range of sizes and shapes from the giant intramesogleal cylindrical muscle cells of beroids and cestids (several mm long and up to 50 μm in diameter) to the minute ribbon-shaped subepithelial fibres of cydippids (less than 1 μm thick).

Fundamentally the organization of the contractile apparatus appears to be the same throughout the phylum. All muscles are smooth muscles and have thick and thin myofilaments, the diameter of which are respectively 15.5 nm and 6 nm. The thin filaments surround the thick ones in a more or less orderly fashion. In Beroe ovata I have found a ratio of 1 thick to 5-9 thin. There are neither dense bodies nor attachment plates and the thin myofilaments probably attach to the cell membrane.

A longitudinal smooth sarcotubular system is present, and is highly developed in the giant fibres. The cell membrane does not bear "caveolae intracellulares" (Gabella, 1973) as in vertebrates. However some small tubular cisternae may open into the mesoglea or the intercellular space. Mitochondria are more or less abundant and voluminous, depending upon the rate of activity of the muscle. The giant mesogleal muscle cells of Beroe ovata are multinucleated.

Intermuscular close contacts occur between mesogleal fibres of Beroe ovata and display the morphological features of gap junctions.

Neuromuscular Junctions

Neuromuscular junctions have been identified in eight species belonging to four sub-orders: Beroe ovata and B. forskali (Beroids), Pleurobrachia rhodopis, Hormiphora plumosa, Callianira bialata and Lampetia pancerina (Cydippids), Eucharis multicornis (Lobates) and Cestum veneris (Cestids).

I have described in a previous paper (Hernandez-Nicaise, 1973c) the ultrastructure of the synapses of ctenophores. There I pointed out that neuromuscular junctions do exist and that in such cases the neurite contains one or several "presynaptic triads" (synaptic vesicles – endoplasmic reticulum – mitochondrion) typical of the phylum. The total length of synaptic contact thus varies in the sections observed from 0.5 to 7 μm. These features are illustrated by Fig. 1.

These contacts are "close neuromuscular junctions" as defined by Burnstock (1970). The width of the synaptic cleft varies from 12 to 17 nm, and there is no basal lamina intercalated between the nerve and muscle. However the junctional cleft is filled by a dense material organized in the form of a thin dense lamina, equidistant from both plasma membranes.

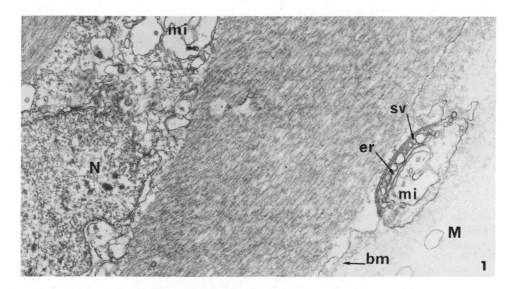

Figure 1. Neuromuscular junction on a mesogleal muscle cell in Beroe ovata. bm: basal lamina, er: endoplasmic reticulum, M: mesoglea, mi: mitochondrion, mu: muscle, N: nucleus, ne: neurite, sv: synaptic vesicle. X 18,000.

The size, shape and content of the synaptic vesicles are very variable from one species to another and even in a given species. In some cases <u>omega</u> profiles have been observed, each suggesting the opening of a vesicle into the synaptic cleft.

The post-synaptic muscle membrane appears definitely thickened by a dense intracellular coat. Post-synaptic sub-surface cisternae may occur, but are not the rule, and there is no permanent differentiation of the post-junctional area.

The ratio of neuromuscular junctions to muscle fibres is obviously related to the level of activity of the muscles. For instance the thin sub-epithelial fibres of <u>Pleurobrachia</u> act mainly by antagonizing the mesoglea so that the body shape remains unchanged; these muscles receive very few synapses. On the contrary in the same animal the tentacular muscles are liberally innervated. In <u>Beroe ovata</u> the different muscular systems appear richly innervated. No quantitative measurements have been made but an individual innervation of most cells of the longitudinal muscular fields is likely to occur.

Ultrastructural Evidence for Multiple Innervation of Muscle Fibres in <u>Beroe ovata</u>

There are four main muscular fields in <u>Beroe ovata</u>: 1) external longitudinal mesogleal (ELM) fibres, 2) pharyngeal circular mesogleal (PCM) fibres, 3) pharyngeal longitudinal subepithelial (PLE) fibres and 4) radial fibres.

Neuromuscular junctions have been observed upon muscle cells belonging to the four muscular systems.

As suggested in a previous paper (Hernandez-Nicaise, 1973) a morphometric ultrastructural analysis has shown that the external and pharyngeal plexuses of <u>Beroe ovata</u> consist of several types of nerve fibres. Each type is characterized by the size, shape and content of its intracytoplasmic (non-synaptic) and synaptic vesicles. The question now is: which neurones innervate which muscular field ?

Fortunately the PLE muscle cells run side by side with the neurites of the pharyngeal nerve net. In this particular case it has been possible to identify the nervous profiles which are part of the neuromuscular junctions.

The three morphologically distinct neuronal types (A, B, C) of the pharyngeal nerve net innervate the muscle cells of the PLE system as illustrated by Figs. 2, 3 and 4. Generally in a given section one profile can be observed. In a few cases I have been able to observe B and C neurites making a synaptic contact on the same muscle cell. Such morphological features strongly suggest a multiple innervation

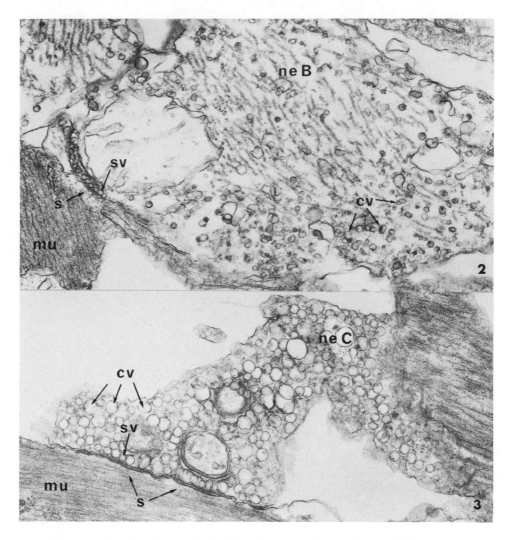

Figures 2, 3 and 4. Multiple innervation of the PLE muscles of Beroe ovata.

Figure 2. Type B neurite (ne B) making a synaptic contact (s) on a PLE muscle cell (mu). B neurites contain small (625 nm) agranular synaptic vesicles (sv), and round or oval cytoplasmic vesicles (cv), 75 nm in greatest diameter, enclosing an eccentric dense granule. X 20,000.

Figure 3. Type C neurite (ne C) making a synaptic contact on a PLE muscle cell. These neurites contain large granular cytoplasmic vesicles (130 nm) and smaller synaptic vesicles (110 nm) of same appearance. Same legend as in Fig. 1. X 20,000.

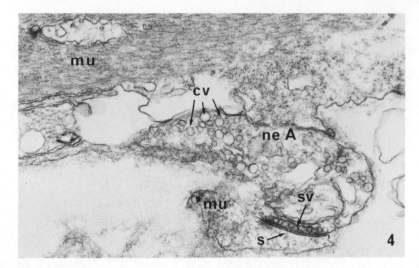

Figure 4. Type A neurite (ne A) making a synaptic contact on a
PLE muscle cell. Type A neurites contain medium-sized (85 nm), round,
agranular cytoplasmic vesicles. The synaptic vesicles are small
(70 nm) and agranular. Same legend as in Fig. 2. X 20,000.

of the PLE muscular system if not of each of its muscle cells.

The innervation of the intramesogleal muscles has not been
determined with certainty. At the contacts so far observed the section of the neurite generally contains very few if any cytoplasmic
vesicles (see Fig. 1) and thus cannot be unhesitatingly assigned to
any neuronal type. There is an obvious variation in the size of synaptic vesicles from different neuromuscular junctions. A situation
similar to that found at PLE muscles is likely to occur at least in
the ELM fibers.

Cytochemical Evidence for Dual Innervation
of Muscles in Beroe ovata

I have already assumed in a previous publication (Hernandez-
Nicaise, 1973c) that the synapses of ctenophores are chemical synapses. I have now tried to obtain cytochemical evidence for the presence of postulated neurotransmitters such as acetylcholine (ACh)
and monoamines (MA).

The presence of ACh and/or MA is reported in the nervous system
of a wide range of invertebrates (see the general review of Gerschen-

feld, 1973, and for cnidarians: Lentz, 1968). It is generally admitted
that cholinergic transmission can be inferred from the presence of
acetylcholinesterase (AChE) at a synaptic junction (Koelle, 1963).
Karnovsky's method (1964) for AChE has thus been applied to the
tissues of Beroe ovata.

A positive reaction occurs only in a portion of the synapses observed in both the external and the pharyngeal nerve nets. The precipitate was obtained at a variety of interneuronal and neuroeffector
synapses including neuromuscular junctions, so that the existence of
a cholinergic system may be postulated in both the external and the
pharyngeal plexuses. At these neuromuscular junctions the heavy to
moderate precipitate is localized on the pre- and post-synaptic membranes, and in some instances a moderate precipitate underlines the
non-junctional neuronal membrane (Fig. 5). However the fragile tissues
of this ctenophore are badly damaged by histochemical processing; a
reliable correlation of these cytochemical findings with the ultrastructural classification of neurones is not possible.

I have tried to localize MA in the plexuses of Beroe ovata using
a modification of the fluorescence method (Eranko and Raisanen, 1966).
This method does not give very satisfactory results with my material.
In some instances the external and pharyngeal plexuses display a
yellow-green fluorescence which is indicative of the presence of MA.

I have obtained more accurate data from the chromaffine reaction
adapted to electron microscopy (Wood, 1966, 1967). In the external
nerve net a catecholamine is probably stored in the large granulated
vesicles (130 to 160 nm) of one neuronal type. In the pharyngeal
nerve net the type B neurones show a positive reaction in their dense
cored vesicles after reaction for total MA content, and a lighter
but distinctive reaction after treatment for indolamines (Fig. 6).
There is a MA in these vesicles and it may be serotonine.

Therefore, there is now cytochemical evidence for a dual innervation, with cholinergic (type A neurones ?) and (indol)aminergic
(type B neurones) innervating the PLE muscular system. A similar dual
innervation is likely for the mesogleal muscular systems but at
present the evidence is incomplete.

CONCLUSIONS

The ultrastructural observations presented in this paper let me
conclude that ctenophore muscles are innervated and that there can
be a multiple innervation of a given muscular field. The cytochemical
data obtained from a study of Beroe ovata suggest that all muscular
fields possess at least a cholinergic innervation. In the particular
case of the PLE system there is apparently a dual innervation, cholinergic and aminergic.

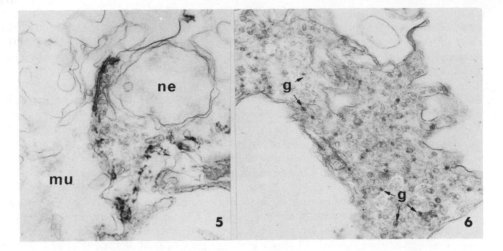

Figure 5. Pharyngeal nerve net of Beroe ovata. AChE activity at a synaptic contact on a PLE muscle cell, following incubation in Karnovsky's medium. The ferrocyanide precipitate is localized on both junctional membranes and on the neuronal membrane. mu: muscle, ne: nerve. X 20,000.

Figure 6. Pharyngeal nerve net of Beroe ovata. Type B neurite following formaldehyde-glutaraldehyde-dichromate treatment (followed by osmium tetroxide). Indolamine granules (g) are present in the cytoplasmic vesicles. X 20,000.

Behavioral observations have brought some evidence about the neural control of muscles in Beroe ovata. Mechanical as well as electrical stimuli applied to the external and the pharyngeal epithelia elicit a variety of muscular responses depending on the localization and intensity of stimulation.A preliminary electrophysiological study carried out in collaboration with G. O. Mackie (unpublished results) suggests that these responses are indeed mediated by the nerve net. Therefore I can deduce from these experiments that the ELM system is innervated by the external nerve net which also controls the ciliated locomotory cells, and that the PCM, PLE and radial systems are innervated by the pharyngeal nerve net.

In all cases I have observed that in a given muscular field the contraction spreads slowly, thus suggesting that the excitation is conducted in the muscle field. In the case of the PCM system muscle activity may be myogenic. Peristaltic waves may run down or up the pharynx for long periods even in the absence of food. The intermuscular junctions could be the cytological pathway of such propagation.

Stimulation experiments show that muscles are organized in distinct fields, the activities of which can be integrated in a complex pattern like feeding behavior. I believe that certain muscular fields correspond to particular sensory areas, assuming that the external stimulus is transmitted by the nervous system rather than by the muscles themselves. In other words I suppose the existence of an integrative pathway: sensory receptor - (inter)neuron(s) - muscle.

So far only the last link of this chain has been considered. What do we know about the processing of information by the nervous system of ctenophores? The behavioral studies of Horridge (1965) showed that some sensory receptors of Eucharis integument were mechanoreceptors. I have found such receptors by electron microscopy in the eight species listed in first section. In each case these receptors give and receive synapses (Hernandez-Nicaise, 1974), as is known in most animals (Vinnikov, 1970) and particularly in cnidarians (Peteya, 1973, Westfall, 1973). These reciprocal contacts provide the structural basis for feed-back between sensory receptors and one (or several?) components of the nerve net.

In the nervous system of the various species that I examined I noted the presence of polarized, reciprocal and symmetrical interneural synapses (Hernandez-Nicaise, 1968, 1973c). Considering that the net is made of several types of neurones, at least in Beroe ovata and Pleurobrachia rhodopis, these three types of junctions permit many theoretical circuits: divergent and convergent chains, loop connexions known in more evolved invertebrates (Kandel and Wachtel, 1968) with an additional plasticity attributable to the symmetrical synapses. The aggregates actually observed in these preparations show the existence of complex connexions. Further data supporting this conclusion are expected to come from the study of serial sections.

This histological survey thus shows that the ctenophores possess all the necessary components for an integrating circuitry. However, the morphological approach cannot by itself account for behavioral sequences. Electrophysiology and pharmacology are now needed to gain a better understanding of the nervous system of ctenophores.

REFERENCES

BURNSTOCK, G., 1970. Structure of smooth muscle and its innervation. In : Smooth muscle (E. BULBRING, A.F. BRADING, A.W. JONES and T. TOMITA Eds.), Edward Arnold, London, pp.1/69.

ERÄNKÖ, O. and RÄISÄNEN, L., 1966. Demonstration of catecholamines in adrenergic nerve fibers by fixation in aqueous formaldehyde solution and fluorescence microscopy. J. Histochem. Cytochem. 14: 690-691.

GABELLA, G., 1973. Fine structure of smooth muscle. I. Cellular structures and electrophysiological behaviour. Phil. Trans.

R. Soc., London, B, 265: 7-16.

GERSCHENFELD, H. M., 1973. Chemical transmission in Invertebrate central nervous systems and neuromuscular junctions. Physiol. Rev., 53: 1-119.

HERNANDEZ-NICAISE, M. L., 1968. Distribution et ultrastructure des synapses symétriques dans le système nerveux des Cténaires. C. R. Acad. Sci., Paris, 267: 1731-1734.

HERNANDEZ-NICAISE, M. L., 1973a. Le système nerveux des Cténaires. I. Structure et ultrastructure des réseaux épithéliaux. Z. Zellforsch. mikrosk. Anat., 137: 223-250.

HERNANDEZ-NICAISE, M. L., 1973b. Le système nerveux des Cténaires. II. Les éléments nerveux intra-mésogléens chez les Béroïdés et les Cydippidés. Z. Zellforsch. mikrosk. Anat., 143: 117-133.

HERNANDEZ-NICAISE, M. L., 1973c. The nervous system of Ctenophora. III. Ultrastructure of synapses. J. Neurocytol., 2: 249-263.

HERNANDEZ-NICAISE, M. L., 1974. Ultrastructural evidence for a sensory-motor neuron in Ctenophora. Tissue and Cell, 6: 43-47.

HORRIDGE, G. A., 1965. Non-motile sensory cilia and neuromuscular junctions in a ctenophore independant effector organ. Proc. R. Soc., B, 162: 333-350.

KANDEL, E. and WACHTEL, H., 1968. The functional organization of neural aggragates in Aplysia. In: Physiological and biochemical aspects of nervous integration (F.D. CARLSON Ed.), Prentice Hall, Englewood Cliffs, New Jersey, pp. 17-65.

KARNOVSKY, M. J., 1964. The localization of cholinesterase activity in rat cardiac muscle by electron microscopy. J. Cell Biol. 23: 217-232.

KOELLE, G. B., 1963. Cytological distributions and physiological functions of cholinesterases. In: Cholinesterases and anti-cholinesterase agents (G.B. KOELLE Ed.), Handb. exp. Pharmakol., Springer Verlag, Berlin, suppl. 15, pp. 189-298.

LENTZ, T. L., 1968. Primitive nervous systems. Yale Univ. Press, New Haven, 148 p.

PANTIN, C. F. A., 1965. The capabilities of the Coelenterate behaviour machine. Amer. Zool., 5: 581-589.

PETEYA, D. J., 1973. A light and electron microscope study of the nervous system of Cerianthopsis americanus (Cnidaria, Ceriantharia). Z. Zellforsch. mikrosk. Anat., 144: 301-317.

VINNIKOV, Y. A., 1969. The ultrastructural and cytochemical bases of the mechanism of function of the sense organ receptors. In: The structure and function of nervous tissue (G. H. BOURNE Ed.), Academic Press, New York, pp. 265-392.

WESTFALL; J. A., 1973. Ultrastructural evidence for a granule-containing sensory-motor-interneuron in Hydra littoralis. J. Ultrastruct. Res., 42: 268-282.

WOOD, J. G., 1966. Electron microscopic localization of amines in central nervous tissue. Nature (London), 209: 1131-1133.

WOOD, J. G., 1967. Cytochemical localization of 5-hydroxytryptamine (5-HT) in the central nervous system. Anat. Rec., 157: 343.

THE ORGANIZATION OF THE NEURO-MUSCULAR SYSTEM

OF THE DIADUMENE ACONTIUM

Tuneyo Wada

Department of Biology, Ochanomizu University, Tokyo
Tateyama Marine Laboratory, Tateyama, Chiba Prefecture
Japan

How is the neuro-muscular system organized in the acontium? This question arose in the course of a functional analysis (Wada, 1972, 1973) which showed decremental conduction of semi-rhythmic contractions from the proximal toward the distal end. This was interpreted in terms of interneural facilitation, and it was assumed that there were numerous junctions in the nervous pathway at which delay occurred. In this paper, an attempt is made to clarify the nervous and muscular organization of the acontium, which is a singularly simple, linear preparation.

MATERIAL AND METHODS

Acontial filaments were obtained from specimens of Diadumene luciae gathered from the shore of Tateyama Bay, near the Tateyama Marine Laboratory. Fresh filaments were stained which reduced methylene blue by Unna's method (Pantin, 1948). To observe the nerve cells satisfactorily, it is necessary to prevent the quick fading due to lack of oxygen when a coverslip is applied. Filaments were first immersed in the stain for 3 min. and then transferred to a hanging drop under a coverslip in 1 drop of sea water plus 1 drop of 1/3 M $MgCl_2$. The coverslip was placed on a depression slide for observation under the microscope at 600X.

To prevent shrinkage, filaments to be fixed for electron microscopy were stretched on a small rectangular framework made of filter paper and placed in the fixative (cold 6.25% glutaraldehyde in sea water) for one hour, followed by postfixation for one hour in cold 1% Os_4. After dehydration in graded alcohols pieces were embedded in Epon. Sections were cut with a diamond knife on a Porter-Blum MT 1 ultramicrotome and mounted on colodion-covered

grids. The unstained sections were examined with a Hitachi HU-IIA
electron microscope. Sensory and musculo-epithelial cells were
isolated from filaments by a 3-hour immersion in Goodrich's solution (1942) and stained with 1% eosin for light microscope study.

RESULTS

Light Microscope Studies

The sensory cells were well stained after about 5 mins. immersion. These cells were found only in the dorso-lateral regions
of the filaments. Spindle shaped small cells (Fig. 1) and elongate
comma-shaped ones (Fig. 2) were found. Each cell has a proximal
neurite which is sometimes smooth, sometimes beaded, and measures
between 50μ and 100μ in length. Diameters of these processes, even
at the widest part of the neurite beads, are less than 0.5μ. The
neurites run parallel to the mesogloeal core along the underside of
the arrayed nematocysts, although their somas lie perpendicular to
the core. Bipolar cells are also observed in the dorsum (Fig. 3).
These cells form several parallel rows (Fig. 4). No muscle cells
are found in the dorsum. Contacts between the neurites and the
bipolar cells are seen rather frequently. The fine structure of
these junctions will be described in connection with the electron
micrographs. Slender neurites having nodes, or varicosities, are
also often seen. Their somas are about 10μ in length and 5μ in
diameter. A few isosceles triangle-shaped sensory cells having two
basal neurites are also seen in the dorsum (Fig. 5). The neurites
of these cells could be followed for a length of about 100μ. These
sensory cells are very easily isolated by use of Goodrich's solution
(Fig. 6). A few bipolar cells were also isolated but their neurites
were usually entangled with the musculo-epithelial cells. In the
ventrum, in contrast to the dorsum, there are no sensory cells. The
bipolar nerve cells run in parallel with the pair of longitudinal
muscle bands. Their cell bodies are 50 to 300μ apart. These cells
which are similar to those on the dorsum, form nerve-chains, as shown
in Fig. 8. The cell bodies are 10μ in length and about 5μ in
diameter. The neurites are too slender to follow for their whole
length, so the length could not be estimated, but they can be followed
for 50μ to 100μ.

Low Power Electron Microscopy

As in the photomicrographs of cross-sectioned filaments of
Diadumene (Yanagita & Wada, 1959) and Metridium (Wada, 1972), a
single-layered sheet of longitudinal muscle is revealed (Fig. 9).
The muscle fibers develop in the basal processes of musculo-epithelial
cells which are located only in the ventrum. These cells do not form
a syncytium, although a few desmosomal junctions are seen between
the muscle fibers. The bipolar nerve cells consist of a soma

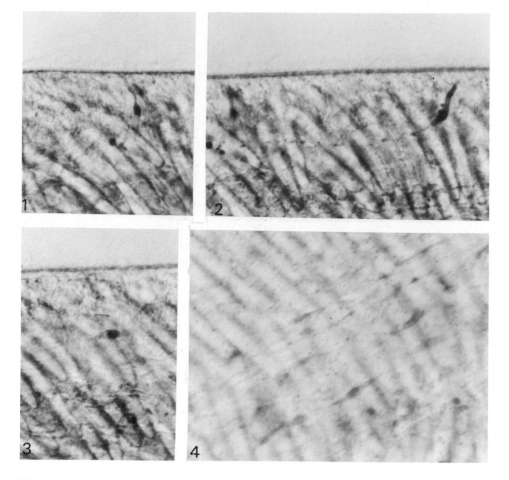

Figure 1. Sensory cell lying perpendicular to the long axis of the
      acontium.
Figure 2. Elongate, comma-shaped sensory cell with a beaded neurite.
Figure 3. Bipolar nerve cell oriented parallel to the long axis of
      acontium.
Figure 4. Bipolar nerve cells seen from the dorsal side of the
      filament, running in parallel with the long axis of the acontium.

All these cells, which are stained vitally with reduced methylene
blue, are seen in the dorsum of the acontium. X810.

containing large corded vesicles (500 to 800mμ in diameter) and
neurites (0.1 to 0.5μ in diameter). In longitudinally sectioned
filaments, the nerve cells run in parallel with the muscle fibers
(Fig. 10). In the dorsum, the somas of bipolar nerve cells are

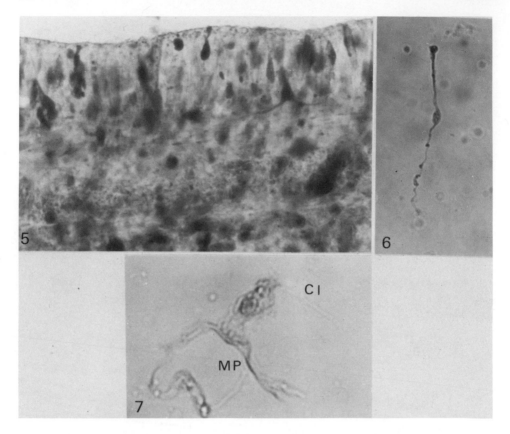

Figure 5. Isosceles triangle-shaped sensory cell having two
    neurites, vitally stained as described for preceding figures
    and in same region. X 810.
Figures 6-7. Cells isolated from acontium by treatment with
    Goodrich's solution. Figure 6. Isolated sensory cell having
    beaded neurite. X 810. Figure 7. Isolated musculo-epithelial
    cell. (CI) apical cilium; (MP) muscular process (muscle
    fiber). X 2020.

located on the underside of the nematocyst array. In this region
(Fig. 11), no musculo-epithelial cells are found. Several neurites
of sensory cells and/or bipolar nerve cells appear to form loose
bundles, oriented parallel to the longitudinal axis of the filament
(Fig. 12). On both sides of the mesogloeal axis, a network is found
which was not seen by light microscopy. It is suggested that the
interneural junctions in these nerve chains may belong to the
soma-dendritic synapse type, which is well known as a one-way
synapse. Such junctions would connect the postsynaptic soma with

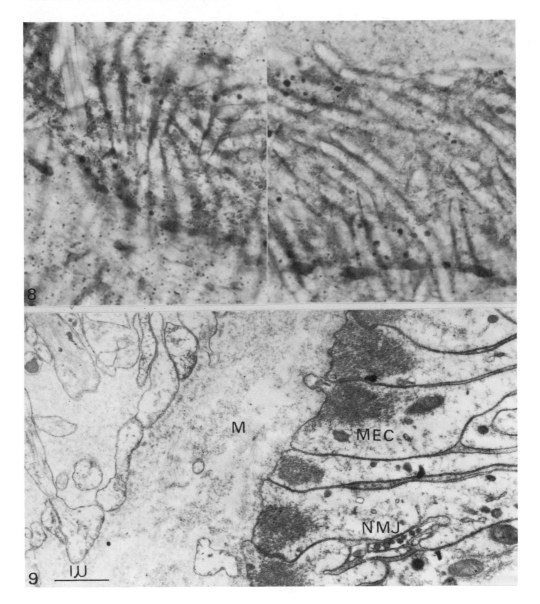

Figure 8. Ventral nerve-chain of acontium suggested as pathway
   of interneural-facilitated conduction. Vital staining as in
   Figures 1-5. X810.
Figure 9. Electron micrograph of transverse section of acontium
   showing single-layered musculo-epithelial cells in ventrum.
   (MEC) musculo-epithelial cell; (NMJ) neuro-muscular junction;
   (M) mesogloea.

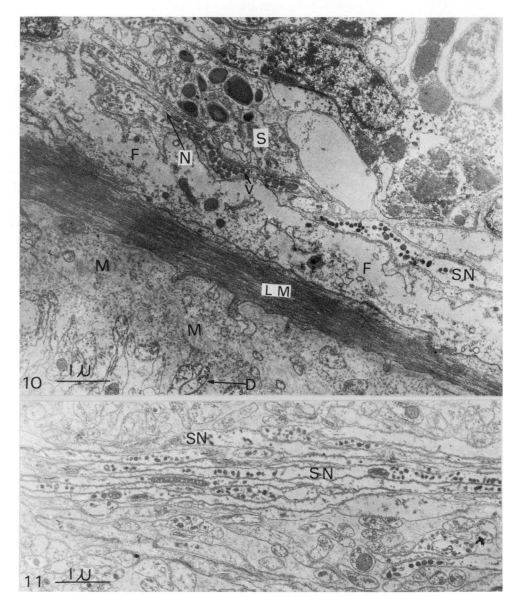

Figure 10. Electron micrograph of a bipolar nerve cell running
     parallel to a muscle fiber in the ventrum. (V) suggested
     synaptic vesicle; (S) soma of bipolar nerve cell; (N) neurite
     of another bipolar nerve cell; (SN) neurite of sensory cell;
     (LM) longitudinal muscle fiber; (F) adjacent foot of musculo-
     epithelial cell; (M) mesogloea; (D) dorsal tissue of acontium.
Figure 11. Longitudinal section of acontium showing a loose bundle
     of sensory neurites in the dorsum. (SN) sensory neurite.

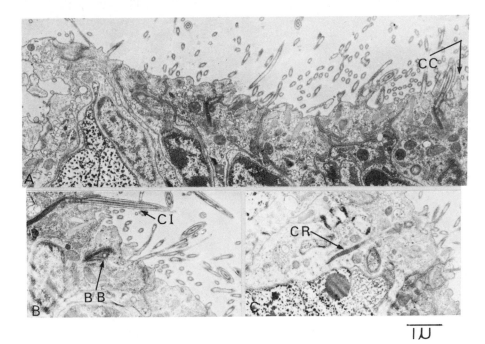

Figure 12 (a,b,c). Transverse sections of an acontium showing the
     apical part of a sensory cell. (CI) longitudinal section of
     cilium; (CR) ciliary rootlet; (CC) ciliary collar; (BB) basal
     body of cilium.

the presynaptic neurite of the adjacent bipolar nerve cell. The
presynaptic neurites contain many vesicles about 100mμ in diameter
(Fig. 10). The same type of junction is also found in the dorsum,
between the neurites of the sensory cells and the somas of the
bipolar nerve cells. In the ventrum, vesicles having a smooth
outline and dense inclusions are found at neuro-muscular junctions
(Fig. 9) and interneural junctions. Sensory cells having an
apical cilium are located in the dorso-lateral part of the filament.
Each cilium is surrounded by a cytoplasmic collar and appears to have
the usual structure of cilia. The very long ciliary rootlet has
a periodicity of about 700Å. Microtubules are also seen.

                              DISCUSSION

     It is well known that, while vital staining with methylene blue
can on occasion give a superb picture of individual nerve cells,
the technique as applied to cnidarians encounters many difficulties;
this is true of Diadumene, where staining of the nervous and sensory
cells in the acontium proved capricious and where the cells, even
when stained, tended to fade quickly. Since fading is due in part

to oxygen depletion, a hanging drop method was employed in the
present study. The success achieved by this approach may be due
both to prevention of rapid oxygen depletion and to flattening
of the acontium by surface tension in the hanging drop. This
flattening, which had the effect of reducing the thickness of
the acontium to less than 20μ, made the nervous and sensory
elements more readily visible, and easier to bring into focus.
The addition of magnesium ions to prevent spontaneous movement
in no way reduced the affinity of the tissue for the stain.
Silver staining of acontia was attempted, but is difficult, and
acontia appear to be too slender to be satisfactorily treated by
this method.

Synapses, neurosecretory material and sensory cilia are
characteristic features of nerve cells (Jha & Mackie, 1967).
The sensory cells of the acontium were recognized by the presence
of a sensory hair arising from an apical invagination. The musculo-
epithelial cells are also ciliated, but their cilia differ from
those of the sensory cells in lacking a cytoplasmic collar and
banded rootlets. The beading observed in some neurites appears
both in fresh and fixed material and is therefore not artefacutal,
although the significance of the beading is uncertain.

This study draws attention to several novel features of the
neuromuscular organization of the acontium, such as the arrangement of the bipolar neurons in longitudinal chains on the dorsal
and ventral sides. These chains can be tentatively regarded as
the longitudinal conduction pathway responsible for spread of
contractions along the acontium; since spread is decremental,
interneural facilition between the bipolar nerve elements must
be considered as a possibility. Certain other points require
clarification. Thus the sensory neurites are probably too short
to innervate the ventral muscle directly, which suggests that
there must be an intermediate link in the local reflex pathway.
This pathway, like the longitudinal conduction pathway, must
involve magnesium-sensitive synaptic junctions, since the
coiling response to local photic, chemical and tactile stimulation
is blocked by addition of magnesium ions. These points can best
be clarified by further high-resolution electron microscopy.

ACKNOWLEDGEMENTS

The writer wishes to express her sincere thanks to Professor
J.C. Dan of Ochanomizu University and to Professor G.O. Mackie
of the University of Victoria for reading the manuscript, also to
Mrs. S.O. Hashimoto for help with electronmicrography.

SUMMARY

Correlative light- and electron-microscopic observations were carried out on the acontia of Diadumene. Methylene blue-stained acontia were examined by a hanging-drop technique, which gave improved visualization of the nerve elements, and minimized fading of the stained elements due to oxygen depletion.

Sensory cells were found to be restricted to the dorso-lateral regions, and are absent from the ventral side. Bipolar neurons were found in chain-like configurations both on the dorsal side and on the ventral side, where chains are associated with each of the two ventral muscle bands. Under the electron microscope, components of these chains are seen running between the feet of the epithelio-muscular cells and structures tentatively identified as neuromuscular junctions have been observed in these regions. Musculo-epithelial tissue is found only on the ventral side of the acontium.

LITERATURE CITED

Goodrich, E.S. 1942. A new method of dissociating cells. Quart. J. Micr. Sci., 85: 245-258.

Jha, R.K. and Mackie, G.O. 1967. The recognition, distribution and ultrastructure of hydrozoan nerve elements. J. Morph. 123: 43-62.

Pantin, C.F.A. 1948. Notes on microscopical technique for zoologists. Cambridge University Press, Cambridge.

Wada, T. 1972. Contractile activities of the acontial filaments of Metridium senile var. fimbriatum VERRILL. Jour. Fac. Sci. Hokkaido Univ. Ser. VI, Zool. 18(3): 387-399.

Wada, T. 1973. Muscular activities of the acontium of sea anemone. Publication of the Seto Marine Biological Station. Univ. of Kyoto 20: 598-613.

Yanagita, T.M. and Wada, T. 1959. Physiological mechanism of nematocysts responses in sea-anemone. VI. Cytologia. 24: 81-97.

ULTRASTRUCTURE AND ATTACHMENT OF THE

BASAL DISK OF HALICLYSTUS

C.L. Singla

Department of Biology, University of Victoria

Victoria, B.C., Canada

The basal disk is the most proximal part of Haliclystus, which attaches to the substrate. Clark (1878) reported the basal disk as having 'collectocysts' (structures similar to suckers) but Migot (1922a) rejected that idea, and he suggested that secretory material was responsible for the attachment. The ultrastructure of the basal disk of Hydra was studied by Philpott et al. (1966), Davis (1973) and Davis and Bursztajn (1973) and that of the scyphistoma of Aurelia by Chapman (1968, 1969), but there is only a brief account of the fine structure of the basal disk of Haliclystus (Westfall, 1968). The present study of the basal disk of Haliclystus adds some new information which might help us to understand its attachment.

MATERIAL AND METHODS

The basal disk of Haliclystus stejnegeri attached to eelgrass (Zostera) was fixed in 2.5% glutaraldehyde in Millonig's phosphate buffer for one hour, rinsed and postosmicated for one hour. The tissue was dehydrated through graded alcohols and propylene oxide was embedded in epon (812) according to Luft (1961). The sections were stained with uranyl acetate and lead citrate (Venable and Coggeshall, 1965).

OBSERVATIONS

The basal disk is covered with a single layer of 70μ tall epithelial cells which in turn is covered by a 60-100μ thick layer of extracellular secretion. This secretion is almost homogeneous near the epithelium but shows fibrillar structure near the attachment site (Fig. 1). The fibrillar part contains 50Å thick fibres of variable length interspersed with dense granules of 200-500Å thickness. In this layer are embedded diatoms, algae and bacteria. The epithelium is differentiated into the following four categories in addition to having a rich nervous supply:- adhesive secreting cells, supporting cells, mucous secreting cells and cnidoblasts. The apical parts of these cells bear cilia, microvilli and fingerlike processes. These cells are joined with septate desmosomes.

The Adhesive Secreting Cells:- These cells are 6-10µ in thickness and are present only in the adhesive region of <u>Haliclystus</u>. The nucleus is usually located near the proximal region. The cytoplasm in the proximal region has an actively secreting rough E.R., several mitochondria and a number of Golgi elements. The rest of the cytoplasm is filled with dense cored elongated rods (Westfall's "dense-core vesicles") 2-3µ long and 0.5-1µ in thickness and a few microtubules. Fig. 2 shows large rods attached to a Golgi cisterna. The dense core is surrounded with a fibrillar matrix. The secretory material probably passes out through the finger-like processes at the apex of the cell, and is assumed to serve an adhesive role.

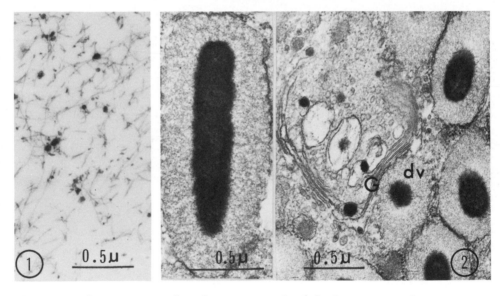

Fig. 1.   Fibrous material interspersed with dense granules.

Fig. 2.   Section of an adhesive secreting cell. dv, dense cored rod, G, Golgi. Insert shows a dense cored rod.

Supporting Cells:- These cells are narrow in the middle and fan out both at the proximal and distal ends. The maximum width is at the proximal end where these cells are attached to the mesoglea with half desmosomes. The nucleus, with its large nucleolus lies in the

centre.   The cytoplasm has many dense secretory vesicles of 0.3-
0.4µ diameter which are typically present in the apical region.   In
addition, these cells contain bundles of microtubules and microfilaments which run parallel along the longitudinal axis of these
cells (Fig. 3).   In the proximal region these microtubules and microfilaments are attached to the basal mesoglea (Fig. 4).

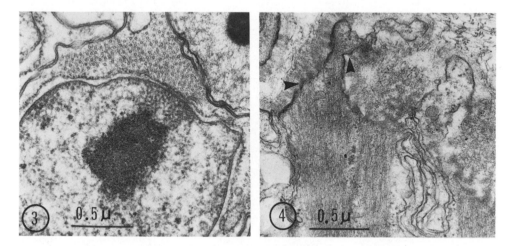

Fig. 3.   Cross section of a supporting cell showing microtubules
          and microfilaments.

Fig. 4.   L.S. of a supporting cell showing attachment of microtubules
          and microfilaments to mesoglea and half desmosomes (pointer).

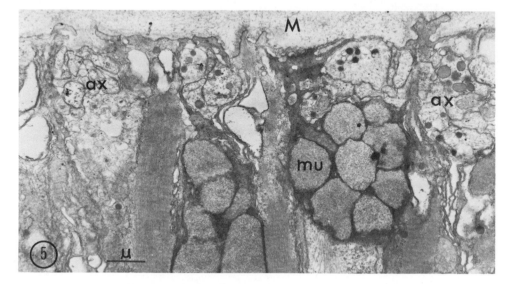

Fig. 5.   Shows location of axons.   ax, axon bundle; M, mesoglea;
          mu, mucous vesicles.

Mucous Secreting Cells:- These cells are as large as the adhesive secreting cells. The nucleus is present at or near the proximal part of the cell. The cytoplasm just above the nucleus contains rough E.R., Golgi elements and a number of mitochondria, whereas the rest of the cytoplasm is full of 1-3µ large mucous vesicles (Fig. 5) which seems to be produced in the cisternae of the Golgi apparatus. The vesicles, like the adhesive rods, show a fine fibrillar content but lack the central core and are almost round. The mucous also appears to pass out through finger-like processes.

Cnidoblasts:- The basal disk epithelium has a few cnidoblasts, most of these are young and lie near the proximal region close to the mesoglea. These cells contain a large nucleus, a nematocyst capsule, a number of mitochondria and several dense secretory vesicles. The fully grown cnidoblasts lie near the surface and bear a cnidocil.

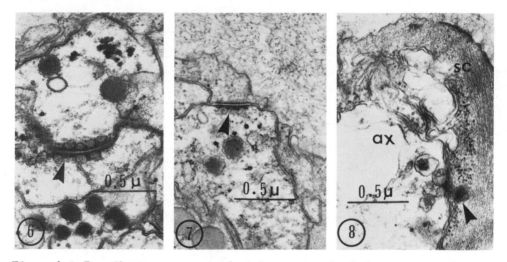

Figs. 6 & 7. Shows symmetrical and asymmetrical interneuronal synapses respectively (pointer).

Fig. 8. Axon lying adjacent to a supporting cell. ax, axon; sc, supporting cell and vesicles (pointer).

Nervous System:- The basal disk is richly supplied with axons of 0.25-1µ diameter and lie near the basal region of the epithelium very close to the mesoglea (Fig. 5). Most of these axons are present in bundles adjacent to the supporting cells and run parallel to the mesoglea. The axoplasm contains mitochondria, glycogen particles and three categories of vesicles: a) synaptic vesicles (500-700Å in diameter) with amorphous material in their lumen. These vesicles are present at both symmetrical (Fig. 6) and asymmetrical (Fig. 7) interneuronal synapses. The former type are very common. The mem-

branes in the synaptic region are thicker and denser than elsewhere.
b) dense cored vesicles, 1500-2000Å in diameter, with electron dense
material.   These vesicles resemble neurosecretory vesicles.   At some
places these vesicles are attached to the axonal membrane and similar
vesicles lie inside the adjacent supporting cell (Fig. 8).   Unlike
the synaptic membranes, the axonal membranes in this region do not
show any specialization.   c) large vesicles of 2500Å diameter with
membrane-enclosed amorphous material are quite common.

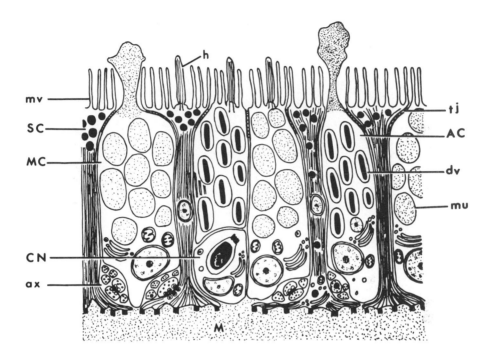

Fig. 9.   Diagram showing the structure of the basal disk. AC,
          adhesive secreting cell; ax, axon bundle; CN, cnidoblast;
          dr, dense cored rods; h, cilium; M, mesoglea; MC, mucous
          secreting cell; mu, mucous vesicle; mv, microvillus; sc,
          supporting cell; tj, tight junction.

DISCUSSION

     The structure of the basal disk is summarized in Fig. 9.   The
basal disk of Haliclystus stejnegeri has cnidoblasts and three types
of secretory cells i.e., mucous cells, presumed adhesive producing
cells and supporting cells.   The cnidoblasts, unlike those of the sea
anemone Stomphia (Ellis et al., 1969) and Calliactis (Mariscal, 1972)
are so few and so far apart that they could hardly play any significant role in the attachment of the basal disk.   The adhesive producing

cells contain columns of elongated dense cored rods similar to those
of Haliclystus auricula (Westfall, 1968). These unique bodies are
only present in the adhesive epithelium of the basal disk and anchors,
which strongly suggests that their contents help in attachment to the
substratum. The fine structure of these rods, except for the central
core, is similar to that of the mucous vesicles of Haliclystus and
type 2 secretory vesicles of Hydra (Davis, 1973). Furthermore, the
contents of both adhesive rods and mucous vesicles pass out through
similar finger-like processes, which agrees with Westfall's observations (1968). Davis (1973) suggested that some of the secretory
vesicles of the basal disk of Hydra provide adhesive material as they
contain acid mucopolysaccharides (Philpott et al., 1966). Ellis
et al. (1969) recognized the involvement of secretory material in
the attachment of the basal disk of Stomphia to the gastropod shell.
The present studies show that the basal disk of Haliclystus is
attached to the substratum with fibres of about 50Å diameter, which
are many times thicker than the filamentous contents of the dense
cored rods. Secondly, their presence close to the substratum suggests that the adhesive material after its release polymerizes to
form fibres. Similar but well oriented lamellar fibres have been
described in the cuticle covering the basal disk of the scyphistoma
of Aurelia (Chapman, 1968).

    The supporting cells of Haliclystus auricula were reported to
contain microfilaments (Migot, 1922b and Westfall, 1968) or microtubules (Chapman, 1974), but in Haliclystus stejnegeri both microtubules and microfilaments are present. Having contractile elements
and secretory vesicles these cells resemble the glandulomuscular cells
of Hydra. These microtubules and microfilaments being anchored at
the proximal end on the mesoglea could generate a force during contraction and probably help the epithelium to adjust itself according
to the contour of the substratum. These cells also show some similarity with the desmocytes of Aurelia in their outline, and in having
bundles of microfilaments. Thus the supporting cells of Haliclystus
may represent an intermediate stage between the glandulomuscular
cells of Hydra and the desmocytes of Aurelia.

    The basal disk of Haliclystus has large nerve bundles located
near the basal region of the epithelium. The axons form interneural
symmetrical and asymmetrical synapses, observed for the first time in
the basal disk of a cnidarian. These synapses may be responsible for
spreading of excitation during the attachment or detachment of the
basal disk. The axons lying immediately adjacent to the supporting
cells contain dense cored vesicles attached to the axonal membrane.
The same kind of vesicles have been observed in the adjacent part of
the supporting cell. A similar situation has been described between
the neurosecretory cells and epitheliomuscular cells of Hydra (Davis
and Bursztajn, 1973). As the supporting cells contain contractile
elements the release of the neurosecretory material could influence
the contraction of these cell elements.

ACKNOWLEDGEMENT

I am indebted to Dr. G.O. Mackie for supporting this investigation from his National Research Council of Canada Grant #A 1427.

SUMMARY

The basal disk of Haliclystus stejnegeri is attached to eelgrass by a layer of fibrous material.

The epithelium of the basal disk has adhesive secreting cells, supporting cells, mucous secreting cells and cnidoblasts. These cells bear cilia, microvilli and finger-like processes. The supporting cells contain microtubules and microfilaments, possibly representing a contractile system which allows the base to adjust to the contour of the substratum.

The secretory material appears to pass out through finger-like processes.

A number of nerve bundles lie adjacent to the base of the supporting cells and run parallel to mesoglea. These axons show interneuronal symmetrical and asymmetrical synapses.

BIBLIOGRAPHY

Chapman, D.M. 1968. Structure, histochemistry and formation of the podocyst and cuticle of Aurelia aurita. J. mar. biol. Ass. U.K. 48: 187-208.

Chapman, D.M. 1969. The nature of cnidarian desmocytes. Tissue and Cell. 1(4): 619-632.

Chapman, D.M. 1974. Cnidarian Histology. In "Coelenterate Biology" (L. Muscatine and H.M. Lenhoff, eds.) pp. 2-36. Academic Press, New York.

Clark, H.J. 1878. Lucernariae and their allies. A memoir on the anatomy and physiology of Haliclystus auricula. Smithsonian Contributions of Knowledge. XXIII, pp. 242.

Davis, L.E. 1973. Histological and ultrastructural studies of the basal disk of Hydra. I. The glandulomuscular cell. Z. Zellforsch. 139: 1-27.

Davis, L.E. and S. Bursztajn. 1973. Histological and ultrastructural studies of the basal disk of Hydra. II. Nerve cells and other epithelial cells. Z. Zellforsch. 139: 29-45.

Ellis, V.L., D.M. Ross and L. Sutton. 1969. The pedal disk of the swimming sea anemone _Stomphia coccinea_ during detachment, swimming, and resettlement. Can. J. Zool. 47: 333-342.

Luft, J. H. 1961. Improvement in epoxy resin embedding methods. J. Biophys. Biochem. Cytol. 9: 409-414.

Mariscal, R.N. 1972. The nature of the adhesion to shells of the symbiotic sea anemone _Calliactis tricolor_ (Leseur). J. exp. mar. Biol. Ecol. 8: 217-224.

Migot, A. 1922a. Sur la mode de fixation des Lucernaires à leur support. Compt. rend. Soc. Biol. (Paris) 86: 827-829.

Migot, A. 1922b. A propos de la fixation des Lucernaires. Compt. rend. Soc. Biol. (Paris) 87: 151-153.

Philpott, D.E., A.B. Chaet and Burnett, A.L. 1966. A study of the secretory granules of the basal disk of _Hydra_. J. Ultrastruc. Res. 14: 74-84.

Venable, H. John and R. Coggeshall. 1965. A simplified lead citrate stain for use in E.M. J. Cell. Biol. 25: 407-408.

Westfall, J.A. 1968. Electron microscopy of the adhesive organs of stalked jellyfish _Haliclystus_. Am. Zool. 8: 803-804. (Abst.).

THE ULTRASTRUCTURE OF THE MUSCLE SYSTEM

OF Stomphia coccinea

H. M. Amerongen and D. J. Peteya

Department of Zoology, University of Alberta

Edmonton, Alberta, Canada

In the forty years since Pantin's classic studies, there have been considerable changes in our concepts of the neuromuscular organization of sea anemones. In large part this is the result of behavioral studies illustrating the elaborate shell-climbing behavior of Calliactis (Ross and Sutton, 1961a,b) and swimming in Stomphia (Yentsch and Pierce, 1955; Sund, 1958). A great deal of attention has since been devoted to the structure (Robson, 1961, 1963; Peteya, 1976) and physiology (Ross, 1957; Hoyle, 1960; McFarlane, 1969a,b; Lawn, 1976; Lawn and McFarlane, 1976) of the nervous and neuroidal systems controlling these patterns; however, our knowledge of the muscular organization of anemones is still quite limited. We have undertaken an ultrastructural study of the muscle system of Stomphia coccinea in a search for anatomical correlations to the general physiology of actinians and to the swimming behavior of Stomphia. This first paper is a report of the occurrence of consistent differences in the structure of muscle fibers from the different endodermal muscle fields of Stomphia.

MATERIAL AND METHODS

Stomphia coccinea was dredged in the San Juan Channel of Puget Sound at depths of about 300 feet and maintained in artificial sea water (Instant Ocean) at 12° C. The study was based on 10 individuals with oral-disc diameters of 1-4 cm. Prior to fixation the anemones were anaesthetized in equal parts of 6.7% $MgCl_2 \cdot 6H_2O$ and sea water for 5 to 6 hours, beginning during the refractory period which follows swimming (Sund, 1958), thus eliminating the necessity for gradual addition of anaesthetic (Robson, 1963). Full expansion of the anemones was achieved by inflation, through the pharynx, using a hypodermic syringe during anaesthetization and subsequently,

541

during fixation. They were fixed for 2 hours in 2.5% glutaraldehyde
in phosphate buffer at pH 7.6, post-fixed in 2% OsO_4 in the same
buffer for 1 hour, then dehydrated in an ethanol series and embedded
in Epon. Sections were mounted on uncoated grids, stained in uranyl
acetate and lead citrate, and examined with a Phillips EM 201 at
80 kvs. Measurements were made on the printed micrographs.

RESULTS

The general organization of the muscular system of Stomphia
has been described previously (Sund, 1958). We would like to add
to this the observation of a mesogleal component in the retractors
and the column circulars. These will be called mesogleal muscles
in accordance with Carlgren's terminology (Carlgren, 1949). On the
other hand, muscle fields formed of the bases of epitheliomuscular
cells will be referred to as epithelial muscles rather than following
the detailed subdivisions of Carlgren. Earlier reports of anemone
muscle structure support the general belief that anemones have only
one type of muscle fiber (Batham and Pantin, 1954; Grimstone et al.,
1958). In contrast to these, the muscle fibers of Stomphia show
considerable variation in structure. There appear to be two different "types" of fiber, characterized by the size and distribution of
their myofilaments. In type A fibers (Pl. 1a) the thick "myosin"
filaments are approximately 150 Å in diameter, and the thin "actin"
filaments are approximately 70Å. The filaments of both size classes
show a consistent diameter and their distribution is uniform
throughout the contractile area. In general, the thin filaments
show no association with the thick filaments. In type B fibers
(Pl. 1b), the 150 Å filament is replaced by a large filament which
ranges, in cross-sectioned fibers, from 150 to 700 Å in diameter.
In longitudinal sections many of these large filaments resemble
paramyosin filaments in that they are spindle shaped and are banded
with a periodicity of 60 Å . In the type B fibers the 70 Å filaments
tend to be grouped less randomly than in A fibers and often show a
close association with the paramyosin-like filaments; the large
filaments are irregularly clustered, leaving spaces in the contractile area which contain only the thin filaments.

With the exception of the myofilament patterns, the general
organization of the fibers is the same in both types (Pl. 2). Note
that in the case of epithelial muscles, a cytoplasmic process from
each muscle fiber passes up toward the nerve plexus and that a
peduncle connects each fiber with the epithelial portion of the cell.
In the case of the mesogleal muscles, this intimate association with
the nervous system is absent. There is no clear correlation, however, between the fiber type present and the degree of association
with the nervous system.

The distribution of the two fiber types is puzzling. In all
animals examined the retractor contains only type A fibers and the
parietals and parietobasilars only type B. On the other hand, some
individual variation is seen in the oral disc and pedal disc circulars in that they may be all A or all B, depending on the animal.

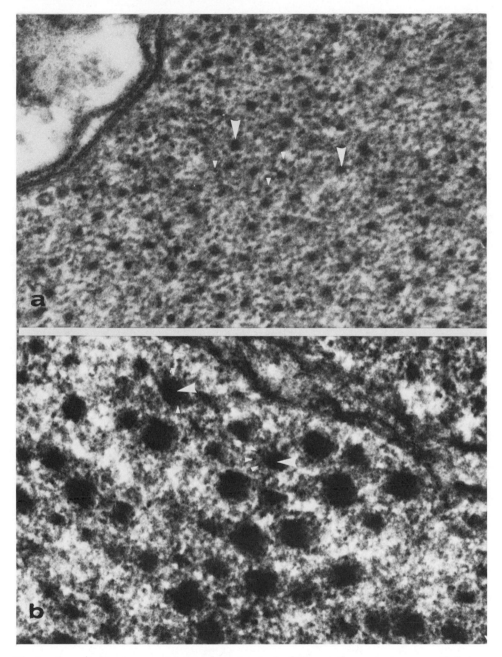

Plate 1. a) Cross-section of a type A fiber; large arrows: thick
(150 Å) filaments; small arrows: thin (70 Å) filaments. b) Cross-
section of a type B fiber; note the association of thin filaments
with thick filaments; large arrows: thick (150-700 Å) filaments;
small arrows: thin (70 Å) filaments. X 122,000.

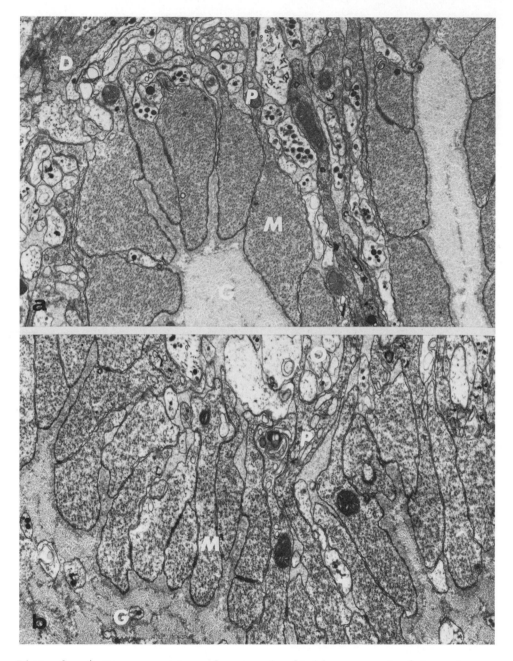

Plate 2. a) Cross-section of a muscle field composed of type A
fibers; b) Cross-section of a muscle field composed of type B
fibers. M, muscle fiber; P, peduncle; G, mesoglea; D, pedal disc
endoderm; O, oral disc endoderm. X 36,000.

In the column circulars, the epithelial fibers are type B and the mesogleal fibers are type A. It should be noted here that these two types are found under identical conditions of anaesthesia and fixation, therefore minimizing the possibility that their occurrence is a fixation artifact. The situation in the ectodermal muscles of the oral disc, tentacles, and pharynx is as yet undetermined.

DISCUSSION

There are two possible explanations for the occurrence of the two different fiber types observed here. First, they may represent two distinct morphological kinds of muscle. If this is the case, one would expect a behavioral correlate for the difference, the most likely being the existence of the typically fast and slow muscles reported in physiological studies (Pantin, 1935; Hall and Pantin, 1937; Batham and Pantin, 1954; Ross, 1957; Robson and Josephson, 1969). However, gross observations indicate that in muscle fields capable of fast or slow contractions, such as sphincter and retractor, the entire field is involved in each contraction (Batham and Pantin, 1954; Pantin, 1965), suggesting that there is no separation into fast and slow areas. In addition, no detectable differences in fiber structure have been found in any anemone muscle field with the exception of the column circulars of Stomphia. Finally, the variable occurrence of type A or type B fibers in the circulars of the oral and pedal disc of Stomphia strongly suggests that the two observed fiber types do not represent two distinct kinds of muscle.

The two types of fiber must then represent different functional states of one kind of muscle fiber. Similar variations associated with differences in contraction state have been seen in Chrysaora (Perkins et al., 1971) but not in Metridium (Grimstone et al., 1958). If this is the case in Stomphia, as is indicated by the variable composition of the oral disc and pedal disc circulars, it means that the inputs to the muscles show varying degrees of sensitivity to $MgCl_2$ anaesthetization and thus suggests the existence of a double innervation. Preliminary evidence indicates that there are two kinds of neuromuscular junction in the retractor of Stomphia (Peteya and Amerongen, unpublished). Further, it has been suggested that myoidal or neuroidal conducting pathways are present in anemones (Chapman et al., 1962; McFarlane, 1969a). Clearly, further study is necessary on the basic anatomy of the muscle system of actinians. A comparative study is in progress to determine the interspecific variation of the muscle ultrastructure. In addition, we are studying the effects of contraction states on the fine structure of the muscles of Stomphia.

SUMMARY

Two different types of muscle fiber found in the endoderm of Stomphia coccinea are described. The myofilaments of type A are 70 Å and 150 Å in diameter. In type B fibers, the 150 Å filament is replaced by a larger paramyosin-like filament with a diameter

varying from 150 to 700 Å. The retractors and mesogleal circulars
of the column are made up of type A fibers. The parietals and
parietobasilars are made up of type B. The oral disc and pedal
disc circulars may be all type A or all type B, depending on the
animal. The functional significance of the differences in muscle
fiber ultrastructure is discussed.

     We are grateful to Dr. D.M. Ross for his helpful comments on
the manuscript. This research was supported by a National Research
Council of Canada grant to Dr. D.M. Ross.

REFERENCES

BATHAM, E.J., AND C.F.A. PANTIN, 1954. Slow contraction and its
     relation to spontaneous activity in the sea-anemone Metrid-
     ium senile (L.). J. Exp. Biol., 31: 84-103.
CARLGREN, O., 1949. A survey of the Ptychodactiaria, Corallimorph-
     aria and Actiniaria. K.V.A. Handl., 1: 1-121.
CHAPMAN, D.M., C.F.A. PANTIN AND E.A. ROBSON, 1962. Muscle in
     Coelenterates. Rev. Canad. Biol., 21: nos. 3 et 4: 267-
     278.
GRIMSTONE, A.V., R.B. HORNE, C.F.A. PANTIN AND E.A. ROBSON, 1958.
     The fine structure of the mesenteries of the sea-anemone
     Metridium senile. Quart. J. Micr. Sci., 99: 523-540.
HALL, D.M. AND C.F.A. PANTIN, 1937. The nerve net of the
     Actinozoa. V. Temperature and facilitation in Metridium
     senile. J. Exp. Biol., 14: 71-78.
HOYLE, G., 1960. Neuromuscular activity in the swimming sea
     anemone Stomphia coccinea (Müller). J. Exp. Biol., 37:
     671-688.
LAWN, I.D., 1976. A slow-conduction system triggers swimming in
     the sea anemone Stomphia coccinea. Science, in press.
LAWN, I.D. AND I.D. MCFARLANE, 1976. Control of shell settling in
     the sea anemone Stomphia coccinea. J. Exp. Biol., in press.
MCFARLANE, I.D., 1969a. Two slow conduction systems in the sea
     anemone Calliactis parasitica. J. Exp. Biol., 51: 377-385.
MCFARLANE, I.D., 1969b. Coordination of pedal-disc detachment in
     the sea anemone Calliactis parasitica. J. Exp. Biol., 51:
     387-396.
PANTIN, C.F.A., 1935. The nerve net of the Actinozoa. I. Facil-
     itation. J. Exp. Biol., 12: 119-138.
PANTIN, C.F.A., 1965. Capabilities of the Coelenterate behaviour
     machine. Am. Zool., 5: 581-589.
PERKINS, F.O., R.W. RAMSEY AND S.F. STREET, 1971. The ultrastruc-
     ture of fishing tentacle muscle in the jellyfish Chrysaora
     quinquecirrha: a comparison of contracted and relaxed states.
     J. Ultrastruct. Res., 35: 431-450.
PETEYA, D.J., 1976. An anatomical study of the nervous system and
     some associated tissues of the sea anemone Stomphia coccinea
     (Müller). The University of Alberta, Ph.D. thesis.

ROBSON, E.A., 1961. The swimming response and its pacemaker
        system in the anemone Stomphia coccinea. J. Exp. Biol., 38:
        685-694.
ROBSON, E.A., 1963. The nerve-net of a swimming anemone, Stomphia
        coccinea. Quart. J. Micr. Sci., 104: 535-549.
ROBSON, E.A. AND R.K. JOSEPHSON, 1969. Neuromuscular properties of
        mesenteries from the sea-anemone Metridium. J. Exp. Biol.,
        50: 151-168.
ROSS, D.M., 1957. Quick and slow contractions in the isolated
        sphincter of the sea anemone Calliactis parasitica. J. Exp.
        Biol., 34: 11-28.
ROSS, D.M. AND L. SUTTON, 1961a. The response of the sea anemone
        Calliactis parasitica to shells of the hermit crab Pagurus
        bernhardus. Proc. Roy. Soc. B, 155: 266-281.
ROSS, D.M. AND L. SUTTON, 1961b. The association between the
        hermit crab Dardanus arrosor (Herbst) and the sea anemone
        Calliactis parasitica (Couch). Proc. Roy. Soc. B, 155:
        282-291.
SUND, P.N., 1958. A study of the muscular anatomy and swimming
        behavior of the sea anemone Stomphia coccinea. Quart. J.
        Micr. Sci., 99: 401-420.
YENTSCH, C.S. AND D.C. PIERCE, 1955. A 'swimming' anemone from
        Puget Sound. Science, 122: 1231-1233.

INCREASE IN NEMATOCYST AND SPIROCYST DISCHARGE IN A SEA ANEMONE

IN RESPONSE TO MECHANICAL STIMULATION

Edwin J. Conklin and Richard N. Mariscal

Dept. of Biological Science, Florida State University

Tallahassee, Florida U.S.A. 32306

Nematocysts have classically been considered to be independent effectors, that is, the nematocyst, nematocyte, and the associated sensory receptor(s) operate without the control or intervention of the nervous system (e.g., Parker, 1916, 1919; Parker and Van Alstyne, 1932; Pantin, 1942a, b; Pantin and Pantin, 1943; Ewer, 1947; Jones, 1947; Burnett et al., 1960; Lentz, 1966; Picken and Skaer, 1966).

Recent behavioral studies, however, have indicated that the animal itself might be able to control nematocyst discharge in some way (e.g., Davenport et al., 1961; Ross and Sutton, 1964; Ellis et al., 1969). These studies indicated that the discharge of tentacle nematocysts might depend on the substrate to which the pedal disk was attached.

In addition, studies on cnidarian feeding behavior suggest that prey capture and nematocyst discharge are reduced in animals fed to repletion (e.g., Burnett et al., 1960; Bouchet, 1961; Sandberg et al., 1971; Sandberg, 1972; Mariscal, 1973; Smith et al., 1974).

The apparent control over nematocyst and/or spirocyst discharge in previous studies has been shown to be inhibitory in nature, rather than stimulatory. During the collection of various anemones from hard substrates, however, we have frequently noted an apparent increase in the "stickiness" of the tentacles, and in some cases an increased stinging capability during contact with the hands and arms. Whether this was an "active" process in which the animal defensively increased the number or type of nematocysts discharged due to rough handling during collection, or whether it was a "passive" situation in which the contraction of the anemone's

549

tentacles simply increased the number of nematocysts and spirocysts
discharged per unit area, remained unknown.

One of the most striking responses in this regard was observed
in the sea anemone Anemonia sargassensis. This anemone was collected from hard substrates in areas of relatively high wave shock.
When detached either from its natural substrate in the field or from
the wall of an aquarium, its tentacles almost immediately because
extremely "sticky," so much so that it was frequently difficult to
dislodge the anemone from the hand or fingers of the observer without damage to the animal. This suggested to us that Anemonia, which
reaches several centimeters in oral disk diameter, might be a
suitable experimental animal in which to investigate the possibility
that the apparent increase in stickiness was due to tentacle nematocyst and/or spirocyst discharge and that it might be under the
control of the animal, perhaps due to some sort of feedback from
the pedal disk to the tentacles.

Anemonia also was attractive for such a study because it had
the "typical" anemone nematocyst complement (cnidom) consisting of
spirocysts, basitrichous isorhizas and microbasic p-mastigophores
(Carlgren, 1945). In addition, it was a relatively simple system
in that the tentacles only contained basitrichs and spirocysts,
thus simplifying the quantification procedures.

Based on the above observations, the present study was undertaken to determine (1) if tentacle nematocyst and/or spirocyst discharge could be measured in the intact animal, (2) if it was possible to detect any change in tentacle spirocyst and/or nematocyst
discharge following mechanical stimulation, and (3) if any differences could be detected in the relative proportions of spirocysts
and nematocysts discharged in response to different types of
mechanical stimulation.

METHODS AND MATERIALS

Anemonia sargassensis Hargitt was collected from rock jet-
ties in the north Florida Gulf of Mexico. The anemones were maintained in the laboratory on a diet of freshly thawed frozen shrimp
chunks and live Artemia nauplii, but starved for three days prior
to an experiment. Two to three well settled anemones on glass
plates were used during a single trial, with one acting as an
undisturbed control.

Nematocyst and spirocyst discharge was assayed after the
method of Sandberg et al. (1971) and Mariscal (1973). Glass
coverslips coated with a particle-free extract of Artemia nauplii
were presented to the tentacles of both experimental and control

anemones before and after each trial.

Coverslips were stained in an aqueous solution of either 0.1%
toluidine blue or 1% acid fuchsin for about 30 seconds. The toluidine blue stained only the basitrichous isorhizas while the acid
fuchsin stained only the spirocysts.

In order to quantify with more accuracy the larger numbers of
cnidae discharged to the coverslips, a new optical density method
was developed, employing a sensitive microphotometer. Light transmitted through an area of tentacle contact on the coverslip (called
a trace) in a field of view at 400X was measured with a Hacker
cadmium sulfide photometer. A standard curve was prepared by making
direct counts of the number of cnidae discharged to the coverslips
and comparing these with the transmitted light readings given by
the photometer.

Direct counts and photometer readings were plotted for basitrichs (Fig. 1), and for spirocysts (Fig. 2) and a best-fit line
calculated. Photometer readings could then be used to count indirectly the number of basitrichs or spirocysts at a particular magnification. This method was extremely useful for counting large
numbers of dense cnidae, and proved reliable when re-checked using
direct counts.

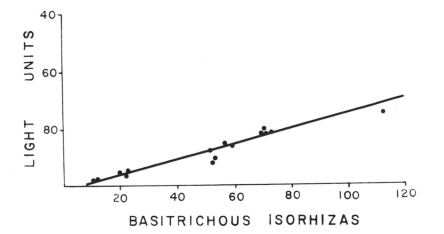

Fig. 1 Counts of <u>Anemonia</u> tentacle basitrichous isorhiza nematocysts discharged on a coated coverslip and plotted against microphotometer readings. Sixteen plotted values are shown as well as
a best-fit line. r= - 0.97070

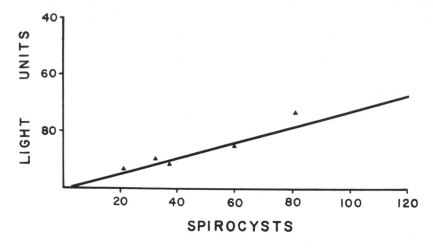

Fig. 2 Counts of <u>Anemonia</u> tentacle spirocysts discharged on a coated coverslip and plotted against microphotometer readings. Five
plotted values are shown as well as a best-fit line. r= - 0.97328

     Discharge was measured before stimulation and the experimental
anemones given either of two treatments: (1) an anemone was jabbed
sharply six times in succession in the mid-column (although not hard
enough to penetrate the tissue) using a blunt probe, and (2) a
well-adhered anemone was gently pried off the substrate with a fine
spatula. Coated coverslips were presented to the anemone's tentacles immediately following the probing, and fifteen seconds after
the release of the pedal disk from the substrate.

     In order to standarize for possible uneven contact of the tent-
acles with the coverslip, only the two densest traces were selected.
Each trace was then measured with the photometer at 400X magnification in areas of high, low and mean basitrich or spirocyst density.
These areas of variable density within the trace were chosen by a
combination of visual estimation and photometer readings. The
numerical average of the six light readings (three from each trace)
was then calculated and the data analyzed at the Florida State
University Computer Center.

                              RESULTS

     Stimulation of the column of <u>Anemonia</u> <u>sargassensis</u> with a
blunt probe caused a significant increase in the discharge of both

basitrichous isorhiza nematocysts and spirocysts (Table 1). Before
mechanical stimulation, the mean number of basitrichs discharged to
the coverslips was 85.4, but after stimulation this figure rose to
a mean of 117. Similarly, before stimulation of the column, the
mean number of spirocysts discharged to the coverslips was 53.4,
but after stimulation an average of 71.2 spirocysts were discharged.
These figures were equivalent to an average increase in the number
of basitrichs discharged of 37% and an average increase in the number of spirocysts discharged of 33%.

     When the pedal disk of Anemonia was removed from the substrate,
an even more dramatic increase in the numbers of cnidae discharged
was observed (Table 2). The numbers of basitrichs discharged increased from a mean of 76.6 to 106.6 for an average increase of 39%.
However, by far the greatest increase occurred for spirocysts

Table 1 Numbers of Anemonia tentacle basitrichous isorhizas and
spirocysts discharged to a coated coverslip before and after stimulation with a blunt probe. A single anemone was used in each trial.

Trial	# of Basitrichs			# of Spirocysts		
	Before stimu- lation	After stimu- lation	% Change	Before stimu- lation	After stimu- lation	% Change
1	87	122	+40	54	71	+31
2	72	104	+44	45	71	+58
3	72	104	+44	55	65	+18
4	98	126	+29	59	79	+34
5	98	129	+32	54	70	+30
Range	72-98	104-129		45-59	65-79	
Mean	85.4	117.0	+37	53.4	71.2	+33
Control mean (unstim- ulated)	89.3	85.3	-4	62.6	61.0	-2

Table 2 Numbers of <u>Anemonia</u> tentacle basitrichous isorhizas and spirocysts discharged to a coated coverslip before and after removal of the pedal disk from the substrate. A single anemone was used in each trial.

Trial	# of Basitrichs			# of Spirocysts		
	Before stimu-lation	After stimu-lation	% Change	Before stimu-lation	After stimu-lation	% Change
1	65	86	+32	63	107	+70
2	73	106	+45	71	125	+76
3	74	112	+51	58	131	+126
4	92	122	+33	92	141	+53
5	79	107	+35	43	98	+128
Range	65-92	86-122		43-92	98-141	
Mean	76.6	106.6	+39	65.4	120.4	+84
Control mean (unstim-ulated)	69.0	65.1	-5	61.0	60.0	-2

following this treatment. The number of spirocysts discharged increased from a mean of 65.4 to 120.4 for an average increase of 84%.

In all trials, the unstimulated control anemones showed no increase in the numbers of either spirocysts or basitrichs discharged between the start and finish of the experiments. In fact, there was a slight, but insignificant, decrease in the numbers of both cnidae discharged during this time period.

To determine if there was any difference between the numbers of basitrichs or spirocysts discharged in response to both types of mechanical stimulation, an analysis of variance was calculated (Table 3). The results show that the difference between the discharge of basitrichs and spirocysts measured before and after each mechanical stimulation was highly significant ($p \leq .001$). In

Table 3 Analysis of variance of numbers of <u>Anemonia</u> tentacle basitrichous isorhizas and spirocysts discharged to a coated coverslip
before and after mechanical stimulation. Anemones were either
probed in mid-column or had the pedal disk removed from the substrate (five trials each).

Source of variation	Degrees of freedom	F-test	Significance
Control (no stim-ulation) cnidae vs experimental cnidae	1	196.11	.001 (highly sig.)
Experimental cnidae before stimulation vs after stimulation	1	285.99	.001 (highly sig.)
Type of cnida (nema-tocyst or spirocyst) vs type of stimulation (probing column or removal of pedal disk)	1	28.45	.001 (highly sig.)
Type of cnida (nema-tocyst or spirocyst) vs stimulation (probing column or removal of pedal disk) vs time of measurement (before or after stimulation)	1	88.34	.001 (highly sig.)

addition, the number of basitrichs or spirocysts discharged was
dependent upon the type of mechanical stimulation (probed column or
removal from substrate) as well as the time of measurement (before
or after stimulation) ($p \leq .001$).

DISCUSSION

The present study is the first to demonstrate a significant increase in the number of nematocysts and spirocysts discharged by a
cnidarian in response to mechanical stimulation. Previous studies,
both in our laboratory and others, have shown a significant decrease
in the number of nematocysts and spirocysts discharged, or number of

prey captured, in cnidarians fed to repletion. It seems clear that
the ability of a cnidarian to discharge fewer cnidae when satiated
would be advantageous both to the individual and the species. since
this would permit conservation of a complex intracellular secretion
product which presumably represents a significant energy investment.

What adaptive advantage would accrue, however, to an organism
capable of increasing its tentacle nematocyst and spirocyst discharge in response to mechanical stimulation? Several functional·
possibilities come to mind: (1) a defensive function, (2) an adhesive function during prey capture, (3) an adhesive function during
symbiotic interactions, and (4) adhesive function during substrate
attachment.

Concerning a defensive function, it is thought that cnidarians
employ nematocysts during defense (e.g., see Mariscal, 1974). We
have frequently noted a marked increase in stickiness and/or the
stinging capabilities of the tentacles of anemones when disturbed.
It presumably would be of adaptive advantage to an anemone to be
able to actively discharge more nematocysts when disturbed to
discourage a potential predator.

In the past we have assumed that the ability to subdue large,
active prey was primarily a function of the number of tentacles
which could be brought to bear, thus increasing the number of cnidae
discharged. However, based on the present results, it seems possible
than an anemone could also directly increase the number or type of
nematocysts discharged by the same number of tentacles. In addition,
the number of spirocysts discharged during prey capture might also
be increased in order to aid prey capture.

Another interesting use of nematocysts and spirocysts is during
shell-changing behavior of symbiotic sea anemones. These anemones
normally live attached to gastropod shells inhabited by hermit crabs.
Mariscal (1972) and McFarlane and Shelton (1975) have shown that
both nematocysts and spirocysts are involved in shell adhesion by
the tentacles of Calliactis during the change from one shell to
another, but to varying degrees. The hermit crab commonly stimulates the anemone mechanically to cause the anemone to release the
pedal disk and to facilitate re-attachment to the new shell (Ross,
1974 and personal observations). Increased nematocyst and spirocyst
discharge could also occur in this situation after stimulation by
the crab, based on our results.

One of the most striking observations of the present study concerns the role of nematocysts and especially spirocysts in substrate
attachment and adhesion. It was noted that after experimental removal of the pedal disk of Anemonia from the substrate, there was
an average increase of 84% in the number of spirocysts discharged

over those discharged when attached. By comparison, there was
"only" a 39% average increase in the number of basitrichs discharged under the same conditions. Since Anemonia sargassensis
normally lives attached to drifting sargassum weed or on rocky
substrates in high energy wave areas, it would be of obvious adaptive advantage for a free anemone to be able to increase the discharge of cnidae and facilitate re-attachment. Our data suggest
that the spirocysts are the primary adhesive organelle in such
situations, at least for Anemonia.

Based on the observations of the present study and those of
recent studies of cnidarian feeding (see Mariscal, 1974 for a review), it seems likely that at least some nematocysts and spirocysts are not independent effectors in the classical sense. An
independent effector in regard to nematocysts was originally defined by G. H. Parker (1916, 1919) as one which responded directly
to various stimuli, but was "not under nervous influence."

Concerning the nature of this influence or control over nematocyst and spirocyst discharge, both nervous and non-nervous mechanisms have been suggested (Mariscal, 1974) but at present there is
little direct evidence which would allow a decision to be made.

What seems clear now is that animals such as cnidarians are
capable of much more complex behavior than has been previously
thought possible and in fact, are capable of responses more characteristic of many "higher" organisms. Prosser (1973) has pointed
out in this regard that "cnidarian systems may serve as models for
more complex systems but are in no sense simple" and we would agree.

LITERATURE CITED

Bouchet, C., 1961. Le contrôle de la décharge nématocystique chez
     l'Hydre. C. R. Acad. Sci., 252: 327-328.
Burnett, A. L., Lentz, T., and Warren, M., 1960. The nematocysts
     of Hydra (Part I). The question of control of the nematocyst
     discharge reaction by fully fed hydra. Ann. Soc. Roy. Zool.
     Belgium, 90: 247-267.
Carlgren, O., 1945. Further contributions to the knowledge of the
     cnidom in the Anthozoa especially in the Actiniaria. Lunds.
     Univ. Arsskr., Avd. 2 (N.S.) 41: 1-24.
Davenport, D., Ross, D. M., and Sutton, L., 1961. The remote con-
     trol of nematocyst-discharge in the attachment of Calliactis
     parasitica to shells of hermit crabs. Vie Milieu, 12: 197-209.
Ellis, V. L., Ross, D. M., and Sutton, L., 1969. The pedal disc of
     the swimming sea anemone Stomphia coccinea during detachment,
     swimming, and resettlement. Can. J. Zool., 47: 333-342.

Ewer, R. F., 1947. On the functions and mode of action of the nema-
    tocysts of hydra. Proc. Zool. Soc. London, 117: 365–376.
Jones, C. S., 1947. The control and discharge of nematocysts in
    hydra. J. Exp. Zool., 105: 25–60.
Lentz, T. L., 1966. The cell biology of hydra. Wiley, New York.
Mariscal, R. N., 1973. The control of nematocyst discharge during
    feeding by sea anemones. Pub. Seto Mar. Biol. Lab., Proc.
    Second Int. Symp. Cnidaria, 20: 695–702.
Mariscal, R. N., 1974. Nematocysts. Pages 129–178 in L. Muscatine
    and H. M. Lenhoff, Eds., Coelenterate biology: reviews and
    new perspectives. Academic Press, New York.
Pantin, C. F. A., 1942a. Excitation of nematocysts. Nature, 149:
    109.
Pantin, C. F. A., 1942b. The excitation of nematocysts. J. Exp.
    Biol., 19: 294–310.
Parker, G. H., 1916. The effector system of actinians. J. Exp.
    Zool., 21: 461–484.
Parker, G. H., 1919. The elementary nervous system. Lippincott,
    Philadelphia.
Parker, G. H., and Van Alstyne, M. A., 1932. The control and dis-
    charge of nematocysts, especially in Metridium and Physalia.
    J. Exp. Zool., 63: 329–344.
Picken, L. E. R., and Skaer, R. J., 1966. A review of researches
    on nematocysts. Symp. Zool. Soc. London, 16: 19–50.
Prosser, C. L., 1973. Central nervous systems. Pages 633–708 in
    C. L. Prosser, Ed., Comparative animal physiology. W. B.
    Saunders Co., Philadelphia.
Ross, D. M., 1974. Behavior patterns in associations and interac-
    tions with other animals. Pages 129–178 in L. Muscatine and
    H. M. Lenhoff, Eds., Coelenterate biology: reviews and new
    perspectives. Academic Press, New York.
Ross, D. M., and Sutton, L., 1964. Inhibition of the swimming res-
    ponse by food and of nematocyst discharge during swimming in
    the sea anemone Stomphia coccinea. J. Exp. Biol., 41: 751–757.
Sandberg, D. M., 1972. The influence of feeding on behavior and
    nematocyst discharge of the sea anemone Calliactis tricolor.
    Mar. Beh. Physiol., 1: 219–237.
Sandberg, D. M., Kanciruk, P., and Mariscal, R. N., 1971. Inhibi-
    tion of nematocyst discharge correlated with feeding in a sea
    anemone, Calliactis tricolor (Leseur). Nature, 232: 263–264.
Smith, S., Oshida, J., and Bode, H., 1974. Inhibition of nematocyst
    discharge in hydra fed to repletion. Biol. Bull., 147: 186–
    202.

A COMPARISON OF PUTATIVE SENSORY RECEPTORS ASSOCIATED WITH NEMATOCYSTS IN AN ANTHOZOAN AND A SCYPHOZOAN

Richard N. Mariscal and Charles H. Bigger

Dept. of Biological Science, Florida State University

Tallahassee, Florida 32306

Although the cnidocil apparatus has long been considered to be the receptor for nematocyst discharge in hydrozoans, the corresponding structure in scyphozoans and anthozoans has been little studied (see Mariscal, 1974c for recent review). Westfall (1966a, b) has called this a "cnidocil" in a scyphozoan and Pantin (1942) has used the term "ciliary cones" to describe small structures associated both with cnidocytes and sensory cells in sea anemones. Using scanning electron microscopy, Mariscal (1974a, b) has shown a correlation between nematocyst-bearing regions and ciliary cones on the tentacles of certain sea anemones and corals. More recently, Mariscal, Bigger and McLean (1976) have reported that spirocysts are not associated with ciliary structures on the tentacles of various zoantharians, but rather occur in close association with two different types of microvilli.

Based on descriptions in the literature and the results of the present study, we shall use the above terms in the following ways: (1) the term "cnidocil apparatus" or "cnidocil complex" will be used for the putative sensory receptors associated with nematocysts in the Class Hydrozoa only; (2) the term "ciliary cone complex" will be used to refer to the putative sensory structures associated with nematocysts in the Class Anthozoa only; (3) the term "flagellum-stereocilia complex" or "F-S complex" will be used to refer to the putative sensory structures associated with nematocysts in the Class Scyphozoa.

The present study reports on the association of putative sensory receptors with two, and perhaps three, different types of nematocysts in an octocoral and a scyphozoan.

MATERIALS AND METHODS

Two species were examined in the present study: the gorgonian Leptogorgia virgulata (Class Anthozoa, S. C. Octocorallia)
from the north Florida Gulf Coast and the scyphistoma and medusa
stages of Cassiopea xamachana (Class Scyphozoa, O. Rhizostomeae)
from the Florida Keys and Caribbean. The bulk of the study involved use of the scanning electron microscope (SEM) supplemented
by observations of specific tissues with the transmission electron
microscope (TEM).

For SEM observations, Leptogorgia was anesthetized in a 1:1
solution of isotonic $MgCl_2$ in sea water and fixed overnight in cold
2% glutaraldehyde in Millonig's phosphate buffer. Cassiopea
scyphistomae were raised in the laboratory on a diet of Artemia
nauplii and fixed without anesthetization in Parducz's (1967) solution or with anesthetization in cold 2% glutaraldehyde in sea water
for up to three hours. Following fixation, the material was transferred to a 16% glycerol solution for up to 24 hours and then prepared for Freon critical point drying following the procedures given
in Mariscal 1974a, b). The prepared tissues were then mounted,
coated with gold-palladium in a Denton vacuum evaporator and examined in a Cambridge S4-10 scanning electron microscope.

For TEM observations, both Leptogorgia and Cassiopea tissues
were anesthetized in $MgCl_2$ and fixed in cold 2% glutaraldehyde for
2 1/2 hours followed by postfixation in 2% OsO_4 for 30 minutes,
staining in uranyl acetate for 15 minutes and embedding in Epon.
Sections were examined in a Phillips 200 transmission electron
microscope.

RESULTS

The pharynx and mouth region of Leptogorgia is covered with
long flagella, similar to what we have observed for other anthozoans
(unpublished). The oral disc surface, however, is covered by a dense
layer of microvilli interspersed with isolated patches of flagella
(Fig. 1). Some of the flagella appear to be grouped in small clusters lacking stereocilia, while others occurring singly appear to have
a circlet of stereocilia at their base.

Although the tips of what appear to be in situ nematocysts can
be observed on the oral disk associated with a ciliary cone complex
(Figs. 1, 2), these latter structures are more common on the pinnules
of the tentacles (Figs. 3, 4). Each ciliary cone complex consists
of a longer flagellum (length about 5-8 μm), sometimes called a
kinocilium, surrounded at its base by a ring of shorter stereocilia
(length about 1-2 μm). The apical tips of the nematocysts associated

with these ciliary cones can be seen with SEM to be projecting
slightly above the tentacle surface (Figs. 3, 4). Some cones in
association with their adjacent nematocyst form small hillocks on
the tentacle surface when viewed from the side.

Examination of the ciliary cone-nematocyst complexes on the
pinnules with transmission electron microscopy has confirmed that
not only are nematocysts associated with the cones, but that the
cnidocyte forms the central cell in what appears to be a three to
five cell complex of surrounding peripheral cells. Each peripheral
cell contributes the sterocilia to the cone complex. It is not yet
clear if the flagellum is associated with the central cnidocyte or
one of the peripheral cells.

The general surface of the pinnule surrounding the ciliary
cone complexes is covered with a dense layer of short microvilli
(length about .6-.9 μm), similar to the situation previously described for the oral disk (Figs. 1-4). Fine anastomosing branches
join the microvilli to each other, as well as to the bases of the
stereocilia (Figs. 2-4). Some of these fine branches also run
from the microvilli directly to a ring around the apical tip of the
in situ nematocyst (Figs. 2, 4).

Unlike most large cnidarian taxa, only a single type of nematocyst, the atrichous isorhiza, has been reported for the Subclass
Octocorallia (Weill, 1934). At least five different categories of
atrichs, however, have been described in octocorals by Weill (1934)
based on the shape of the undischarged capsule. Due to their small
size and reluctance to discharge, it is unknown at present just what
relationship capsule size and shape has to thread morphology and the
possibility exists that more than one type of nematocyst might be
present in the group.

Based on our observations with phase contrast and TEM, it
appears that the so-called "atrichs" of Leptogorgia do in fact
possess spines which are large enough to be seen with the light
microscope. Thus, based on the studies in our laboratory and elsewhere, it seems that many of the classically considered atrichs are
in fact holotrichous isorhizas due to the presence of small spines.
Although an attempt was made earlier to retain the category of
"atrichous isorhiza" (Mariscal, 1974c), it now seems that we should
set up several subcategories of holotrichs to incorporate both the
former "good" holotrichs and what used to be considered "atrichs."
This is also true of the so-called "atrich" found in the acrorhagi
of certain sea anemones (Bigger, this volume) as well as for
"atrichs" in other coelenterate species (Westfall, 1965), including
some other octocorals (Schmidt, 1972, 1974).

The polyp stage of the rhizostome scyphozoan, Cassiopea
xamachana occurs as a large scyphistoma (up to one cm. or more in

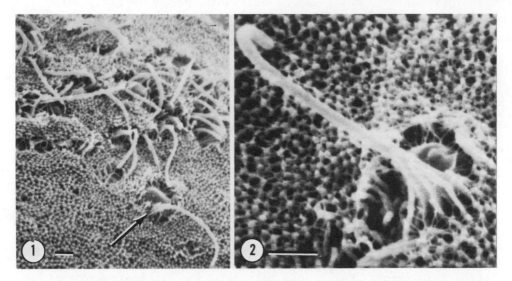

Fig. 1.   SEM of oral disk of <u>Leptogorgia virgulata</u> showing cluster
of flagella surrounded by microvilli and apparent nematocyst
associated with ciliary cone complex (arrow).   Scale bar: 1 µm.

Fig. 2.   SEM of single ciliary cone complex associated with apparent nematocyst on oral disk of <u>L</u>. <u>virgulata</u>.   Scale bar: 1 µm.

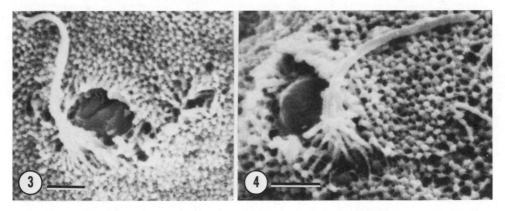

Fig. 3.   SEM of <u>L</u>. <u>virgulata</u> pinnule showing ciliary cone complex
with apparent nematocyst surrounded by microvilli.   Scale bar 1 µm.

Fig. 4.   SEM of another <u>L</u>. <u>virgulata</u> pinnule nematocyst associated
with ciliary cone complex.   Note interconnecting microvilli.
Scale bar: 1 µm.

length) which lives attached to various hard substrates. Two different types of easily identifiable nematocysts were observed, both on
the scyphistoma and the medusa during examination by SEM: the
heterotrichous microbasic eurytele and the holotrichous isorhiza.

    The discharged thread of the heterotrichous microbasic eurytele
nematocyst of Cassiopea was found emerging from within a flagellum-
stereocilia complex (F-S complex) similar to the ciliary cone complex of Leptogorgia (Fig. 5). However, the flagellum of the F-S
complex varied in length depending on the type of associated nematocyst. The length of the flagellum in the eurytele complex was
about 4 μm total length compared to an average length of about 6 μm
for the corresponding structure in Leptogorgia.

    The second type of nematocyst associated with the F-S complex,
both in the scyphistoma and medusa, was the holotrichous isorhiza
(Figs. 6, 7). The holotrich thread emerged from within the center
of a three to five cell rosette or cluster of peripheral cells
surrounding a central cnidocyte in both the scyphistoma and medusa
(Figs. 6, 7). Similar to the situation observed above for Leptogorgia, each peripheral cell in the rosette contributed a proportion
of the stereocilia to the F-S complex overlying the cnidocyte (Fig.
8). This was also confirmed by TEM. Although not entirely clear,
the flagellum in the F-S complex appeared to be associated with the
central cnidocyte. Interestingly enough, the flagellum associated
with the holotrich F-S complex was much longer than in the eurytele
complex, averaging about 15 μm in length. Thus it was possible to
identify tentatively the particular nematocyst type associated with
the F-S complex simply by examining the surface of the animal. This
is the first evidence that different ciliary cone or F-S complexes
may have a different structure, perhaps related to nematocyst type
or function.

    Because of the contribution of stereocilia by the peripheral
cells making up a rosette, the F-S complex was always found at the
junctions of adjacent cells. However, another flagellar structure
also occurred in the center of each peripheral cell. This consisted
of a single long flagellum surrounded at its base by a circlet of
short microvilli (Figs. 6, 7, 8), similar to those described for
other cnidarians (e.g., Westfall, 1965; Mariscal, 1974a, b; Chapman,
1974).

    Although there were several general similarities in surface
ultrastructure between Leptogorgia and Cassiopea, there were also
several striking differences. For example, the apical tips of the
in situ nematocysts were generally not visible on the surface of the
Cassiopea scyphistoma and medusa, but could be seen on the surface
of Leptogorgia. In addition, the surface of the tentacles of the
scyphistoma and the mouth fronds of the medusa generally lacked the
dense accumulations of microvilli and cilia which have been observed

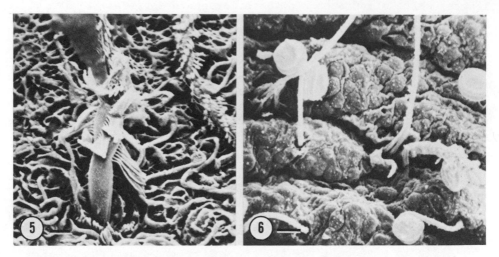

Fig. 5.   SEM of discharged heterotrichous microbasic eurytele
emerging from within flagellum-stereocilia complex near budding
zone of Cassiopea scyphistoma.   Scale bar: 1 μm.

Fig. 6.   SEM of holotrichous isorhiza emerging from F-S complex
on tentacle of Cassiopea scyphistoma.   Scale bar: 1 μm.

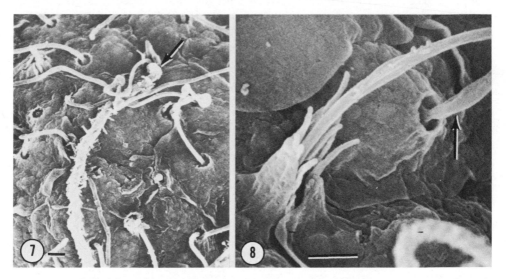

Fig. 7.   SEM of holotrich emerging from F-S complex on mouth fronds
of Cassiopea medusa (arrow).   Scale bar: 1 μm.

Fig. 8.   SEM of F-S complex (left) and single flagellum (arrow)
on mouth frond of Cassiopea medusa.   Scale bar: 1 μm.

on the surfaces of various anthozoans (e.g., Mariscal, 1974a, b, and this study).

DISCUSSION

The cnidocil apparatus has long been considered to be the receptor involved in nematocyst discharge in hydrozoans. Various workers have suggested that the cnidocil itself functioned as a mechanical "trigger" to initiate discharge following a prior chemical sensitization (e.g., Jones, 1947). Based on her examination of a single species (Aurelia aurita), Westfall (1966a, b) has stated that scyphozoans also possess a cnidocil apparatus. Blanquet and Wetzel (1975) have also used the term "cnidocil" to describe a flagellum-stereocilia complex associated with nematocysts in Chrysaora polyps.

Although there has been a good deal of TEM work on the ultra-structure of hydrozoan cnidocils (e.g., Chapman and Tilney, 1959a, b; Mattern, Park and Daniel, 1965; Slautterback, 1967; Bouillon and Levi, 1967 and Westfall, 1966a, 1970), there have not been enough comparable studies on either anthozoans or scyphozoans to allow us to understand the structure of the flagellum-stereocilia complex in either of these groups. Our SEM and TEM observations of the hydrozoan cnidocil apparatus (unpublished), the anthozoan ciliary cone complex and the flagellum-stereocilia complex of scyphozoans show a general resemblance between all three structures. In addition, we have found that nematocysts are closely associated with all three structures, both based on our work and that of others.

We have also shown, at least in Leptogorgia and Cassiopea, that the ciliary cone complex and the flagellum-stereocilia complex are multicellular structures, each apparently containing a central cnidocyte surrounded by a rosette of three to five peripheral cells which contribute the stereocilia to the complex. These findings represent the first evidence that the putative sensory complex associated with nematocysts is a multicellular one.

It is interesting to note here that the general arrangement of the above putative sensory complexes associated with nematocysts in both Leptogorgia and Cassiopea is somewhat similar to that of a ciliary cone sensory receptor complex described by Peteya (1975) for several anthozoans. However, in the cases described by Peteya, the central cell did not contain a nematocyst. It therefore seems likely that ciliary cones are associated both with nematocysts and sensory cells, as Pantin (1942) and others have previously pointed out. Earlier light microscopical studies of ciliary cones were unable to resolve the fact that these structures, rather than being associated with single cells, are in fact a multicellular complex, as both our work and Peteya's has shown. Similarly, Tardent and

Stoessel (1971) and Tardent and Schmid (1972) have described a new hydrozoan mechanoreceptor, which, had it been associated with a nematocyst, would have been called a cnidocil.

Based on a review of the literature, Mariscal (1974b) earlier suggested several possible arrangements of cells in a ciliary cone complex associated with nematocysts. These included a single-cell system similar to that suggested by some earlier light microscopical studies, a two-cell system similar to that described by Tardent and Schmid (1972) for the rigid hydrozoan mechanoreceptor and a three-or-more-cell system. The observations of the present study suggest that, in the case of Cassiopea and probably Leptogorgia, the three-or-more-cell system best fits the present ultrastructural evidence. Such a multicellular sensory system associated with nematocysts could have several possible arrangements including the nematocyst contained in the central cell (as with Cassiopea and apparently Leptogorgia), or nematocysts contained in the peripheral cells, or perhaps both.

This study was supported in part by NSF Grant GB-40547 to the senior author.

SUMMARY

Examination of the surface of the gorgonian Leptogorgia virulata and the scyphistoma and medusa of the scyphozoan Cassiopea xamachana with both SEM and TEM has shown the association of nematocysts with a ciliary cone complex or a flagellum-stereocilia complex. Each complex appears to be multicellular with the central cnidocyte surrounded by three to five peripheral cells, each of which contributes the stereocilia.

LITERATURE CITED

Blanquet, R. S. and B. Wetzel, 1975. Surface ultrastructure of the scyphopolyp, Chrysaora quinquecirrha. Biol. Bull., 148: 181-192.
Bouillon, J. and C. Levi, 1967. Ultrastructure du cnidocil de l'appareil cnidociliare de l'appareil perinématocystique et du cnidopode des nématocystes d'hydroides. Ann. Sci. Nat. Zool., 9: 425-456.
Chapman, D. M., 1974. Cnidarian histology. Pages 1-92 in L. Muscatine and H. M. Lenhoff, Eds., Coelenterate Biology: Reviews and New Perspectives. Academic Press, New York.
Chapman, G. B. and L. G. Tilney, 1959a. Cytological studies of the nematocysts of Hydra I. Desmonemes, isorhizas, cnidocils and supporting structures. J. Biophysic. Biochem. Cytol., 5: 68-78.

Chapman, G. B. and L. G. Tilney, 1959b. Cytological studies of the
    nematocysts of Hydra II. The stenoteles. J. Biophysic.
    Biochem. Cytol., 5: 79-84.
Jones, C. S., 1947. The control and discharge of nematocysts in
    hydra. J. Exp. Zool. 105: 25-60.
Mariscal, R. N., 1974a. Scanning electron microscopy of the
    sensory surface of the tentacles of sea anemones and corals.
    Z. Zellforsch., 147: 149-156.
Mariscal, R. N., 1974b. Scanning electron microscopy of the
    sensory epithelia and nematocysts of corals and a coralli-
    morpharian sea anemone. Proc. Second Internat. Coral Reef
    Symp., 1: 519-532.
Mariscal, R. N., 1974c. Nematocysts. Pages 129-178 in L.
    Muscatine and H. M. Lenhoff, Eds., Coelenterate Biology:
    Reviews and New Perspectives. Academic Press, New York.
Mariscal, R. N., C. H. Bigger and R. B. McLean, 1976. The form
    and function of cnidarian spirocysts 1. Ultrastructure of the
    capsule exterior and relationship to the tentacle sensory
    surface. Cell and Tissue Research (in press).
Mattern, C. F. T., H. D. Park and W. H. Daniel, 1965. Electron
    microscope observations on the structure and discharge of
    the stenotele of hydra. J. Cell. Biol., 27: 621-638.
Pantin, C. F., 1942. The excitation of nematocysts. J. Exp. Biol.,
    19: 294-310.
Parducz, B., 1967. Ciliary movement and coordination in ciliates.
    Pages 91-128 in G. H. Bourne and J. F. Danielli, Eds.,
    International Review of Cytology. Academic Press, New York.
Peteya, D. J., 1975. The ciliary-cone sensory cell of anemones and
    cerianthids. Tissue and Cell, 7: 243-252.
Schmidt, H., 1972. Die Nesselkapseln der Anthozoen und ihre
    Bedeutung für die phylogenetische Systematik. Helgoländer
    wiss Meeresunters., 23: 422-458.
Schmidt, H., 1974. On evolution in the Anthozoa. Proc. Second
    Internat. Coral Reef Symp., 1: 533-560.
Slautterback, D. B., 1967. The cnidoblast-musculoepithelial cell
    complex in the tentacles of Hydra. Z. Zellforsch., 79: 296-
    318.
Tardent, P. and V. Schmid, 1972. Ultrastructure of mechanore-
    ceptors of the polyp Cornye pintneri (Hydrozoa, Athecata).
    Exper. Cell. Res., 72: 265-275.
Tardent, P. and F. Stoessel, 1971. Die Mechanorezeptoren der
    Polypen von Coryne pintneri, Sarsia reesi und Cladonema
    radiatum (Athecata, Capitata). Rev. Suisse Zool., 78: 680-688.
Weill, R., 1934. Contribution a l'étude des cnidaires et de leurs
    nématocystes. Trav. Stn. Zool. Wimereux, 10-11: 1-701.
Westfall, J. A., 1965. Nematocysts of the sea anemone Metridium.
    Amer. Zool., 5: 377-393.

Westfall, J. A., 1966a.   Fine structure and evolution of nemato-
     cysts.   Proc. Sixth Internat. Congress Electron Microscopy,
     2: 235-236.
Westfall, J. A., 1966b.   The differentiation of nematocysts and
     associated structures in the Cnidaria.   Z. Zellforsch., 75:
     381-403.
Westfall, J. A., 1970.   The nematocyte complex in a hydromedusan,
     Gonionemus vertens.   Z. Zellforsch., 110: 457-470.

Behavioral Physiology

A VIEW OF THE EVOLUTION OF CHEMORECEPTORS BASED ON RESEARCH WITH

CNIDARIANS

Howard M. Lenhoff, Wyrta Heagy and Jean Danner

Departments of Developmental and Cell Biology, and of
Molecular Biology and Biochemistry
University of California, Irvine, 92717, U.S.A.

Since Loomis (1955) discovered that reduced glutathione (GSH)
was the specific activator of feeding in Hydra littoralis, research
on cnidarian chemoreceptors has taken two lines. One, involving
mostly work on hydra, is concerned with elucidating the mechanism
by which the receptor is activated (see Lenhoff, 1974). The other
is of a more comparative nature showing that numerous compounds
activate feeding in a wide range of cnidarians.

In this paper, we use data from both lines of research to develop a hypothesis concerning the evolution of cellular receptors
to amino acids and their derivatives. For example, those receptors
may be involved in chemoreception, neurotransmission, and hormone
activation. We propose that such receptors in higher forms evolved
from primitive receptors which responded to a broad spectrum of
amino acids and peptides, and which originally functioned in cell
drinking and feeding.

SURVEY OF CHEMICAL ACTIVATORS OF FEEDING AMONG CNIDARIANS

Inspection of Table 1 shows a number of patterns which support
our hypothesis. The data show that cnidarians tested from every
class and most families elicit a feeding response to either one or
a few small molecules. The molecules found most commonly to initiate a feeding response are the tripeptide GSH and the imino acid
proline. In the Hydrozoa the feeding response of each organism
investigated was induced by a single specific molecule. Proline
seems to be especially prevalent as an activator among the Families
Clavidae and Williidae of the Suborder Gymnoblastea. For example,

571

all members of these groups thus far tested, i.e., <u>Cordylophora</u>,
<u>Pennaria</u>, <u>Tubularia</u> and <u>Proboscidactyla</u>, responded only to proline.
All members of the Hydridae (sub-order Gymnoblastea) tested responded
only to GSH. The only other hydrozoans tested were one calyptoblast
and two siphonophores; all responded only to GSH.

Turning to the Anthozoa, we see three trends which correspond
somewhat to the three Orders of the subclass Zoantharia. In general, although more than one compound triggers feeding behaviors, the
animal's receptors still exhibit a fair degree of specificity. For
example, the most specificity is seen among members of the Order
Actinaria: <u>Boloceroides</u> responds primarily to valine, <u>Anthopleura</u>
to GSH, <u>Haliplanella</u> to leucine, <u>Actinia</u> to glutamic acid and <u>Cal-
liactis</u> to GSH. The specificity broadens when we consider the Order
Zoanthidia; whereas <u>Zoanthus</u> responds primarily to GSH, <u>Palythoa</u>
responds to relatively high concentrations of either GSH or proline,
or to low amounts of these two activators acting synergistically.
Lastly, corals (Order Madreporaria) seem to respond best to proline
alone or GSH alone, as well as to a varying number of other amino
acids at relatively higher concentrations.

<u>Chrysaora</u>, the only example of the Scyphozoa tested, seems to
respond to GSH and to a wide number of amino acids. More scyphozoans need to be tested before we can make any generalizations about
that group.

DO THE GLUTATHIONE AND PROLINE RECEPTORS HAVE A COMMON ORIGIN?

Fulton has suggested (1963) that the evolution of a receptor
site for glutathione into one for the α-imino acid proline may have
proceeded by means of slight structural changes in the receptor site.
His postulate was based on the knowledge that one of the possible
cyclized forms of glutathione in solution is close in structure to an
α-imino acid. Because proline is also present in the fluids released
from wounded prey organisms, the change in structure of the receptor
site was not disadvantageous to <u>Cordylophora</u> but, under some circumstances, advantageous, and so persisted.

In support of Fulton's suggestion are a number of cases reported in Table 1. For example, let us assume that the earliest cnidarians responded to a wider range of amino acids and to GSH, as has been
reported for <u>Chrysaora</u> (Loeb and Blanquet, 1973). Possibly there
evolved a cnidarian whose receptor was modified so that it would
recognize both GSH and proline. Such a receptor might exist in
<u>Palythoa</u> (Reimer, 1971b), although it is also possible that <u>Palythoa</u>
has two distinct receptors for each of those activators. Possibly
organisms with this proposed "GSH and proline" receptor might then
evolve into organisms having two distinct receptors, one to GSH and

another to proline. Such a situation appears to exist in <u>Cyphastrea</u>
and in other organisms listed in Table 1. And, eventually organisms
evolved with only GSH receptors, as seen in hydra and siphonophores,
or proline receptors, as found in the colonial gymnoblasts.

A similar argument could be constructed regarding the evolution of receptors to amino acids having apolar side chains. For
example, the valine receptor of <u>Boloceroides</u> and the leucine receptor of <u>Haliplanella</u> may have evolved from a receptor with a specificity for apolar amino acids in general. Possibly receptors for
other amino acids may have followed similar pathways from the original primitive cnidarian receptor which responded to many amino acids.

EVOLUTION OF CELLULAR RECEPTOR SITES IN GENERAL

Among the earliest receptor sites to evolve were probably
those associated with the induction of pinocytosis in single cells.
In recent years the chemical induction of pinocytosis has been
well studied in ameba (Chapman-Andresen, 1962) and in white blood
cells (Cohn, 1967). Both kinds of cells respond to a range of
small charged molecules; of the amino acids, aspartate and glutamate are particularly effective. In general, it might be said that
single cells depend on external chemical cues which stimulate the
uptake of nutrients from their environment; hence, these cells may
have evolved receptor sites with broad specificity such as might
prove useful to guarantee the cell sufficient food to survive.

It thus seems reasonable to suppose that cnidarian cells utilizing pinocytosis to take up nutrients from their gastrovascular
cavity also respond to a broad range of molecules. In accord with
this supposition is Slautterback's finding that certain amino acids
could stimulate the immediate formation of a large network of microvilli at the apical end of the endodermal digestive cells that line
the gut of hydra. Among the most active amino acids were the isomers of tyrosine, with <u>m</u>-tyrosine the most active. Phenylalanine
was ineffective. Other amino acids showing activity were cysteine
and glutamate (D. Slautterback, personal communication).

Slautterback's finding that tyrosine stimulates microvilli
formation in hydra endoderm cells takes on particular importance
in light of a discovery by Blanquet and Lenhoff (1968). They showed
that hydra have a receptor on the surface of cells lining the gut
that is activated by tyrosine, and leads to the hydra exhibiting
a "neck response," i.e., a constriction of the upper third of the
body tube. These neck constrictions apparently allow hydra to retain previously ingested food in the gut while swallowing newly
captured prey. No other natural amino acid, including phenylalanine, could substitute for tyrosine. It is interesting to note,

TABLE 1

CHEMICAL ACTIVATORS OF FEEDING IN THE CNIDARIA

	ACTIVATOR	REFERENCE
I. CLASS HYDROZOA		
A. Order Hydroida		
Hydra littoralis	GSH	Loomis (1955)
	Tyrosine[1]	Blanquet and Lenhoff (1968)
		Heagy et al (unpublished)
H. attenuata	GSH	Lenhoff (unpublished)
H. pirardi	GSH	Lenhoff (unpublished)
H. pseudoligactis	GSH	Mariscal (1971)
H. viridis	GSH	Fulton (1963)
Cordylophora lacustris	Proline	Pardy and Lenhoff (1968)
Pennaria tiarella	Proline	Spencer (1974)
Proboscidactyla flavicirrata	Proline	Rushforth (these Proceedings)
Tubularia sp.	Proline	Lenhoff and Schneiderman (1959)
Campanularia flexuosa	GSH	
B. Order Siphonophora		
Physalia physalis	GSH	Lenhoff and Schneiderman (1959)
Nanomia cara	GSH	Mackie and Boag (1963)
II. CLASS ANTHOZOA		
A. Order Actiniaria		
Anthopleura elegantissima	GSH[2]	Lindstedt (1971b)
	Asparagine[2]	
Boloceroides sp.	Valine	Lindstedt et al (1968)
	Isoleucine (?)[3]	
Actinia equina	Glutamate	Steiner (1957)
Haliplanella luciae	Leucine	Lindstedt (1971a)
Calliactis polypus	GSH, Proline	Reimer (1973)
	9 amino acids	

TABLE 1 (cont'd)

B. Order Madreporaria		
Fungia scutaria	Proline, GSH	Mariscal and Lenhoff (1968)
Cyphastrea ocellina	Proline, GSH	Mariscal and Lenhoff (1968)
Pocillopora damicornis	Proline, GSH[4]	Mariscal and Lenhoff (1968)
	Phenylalanine	Mariscal (1971)
Pocillopora meandria	Proline, GSH	Mariscal (1971)
Porites compressa	Proline, GSH	Mariscal (1971)
Leptastrea bottae	Proline, GSH	Mariscal (1971)
Montastrea cavernosa	Glutamate, Aspartate	Lehman and Porter (1973)
	Proline, Arginine	
Tubastrea manni	Proline, GSH	Mariscal (1971)
C. Order Zoanthidea		
Palythoa psammophilia	Proline, GSH[5]	Reimer (1971b)
	Proline & GSH[5]	
	Alanine, Serine, Lysine	
Zoanthus pacificus	GSH	Reimer (1971a)
	Glycine	
III. CLASS SCYPHOZOA		
Chrysaora quinquecirrha	20 amino acids	Loeb and Blanquet (1973)
	GSH	
	glycylglycine	

1. See text.
2. Asparagine stimulates tentacle bending and GSH stimulates mouth opening.
3. Possibly a weak activator, but also effective as a competitive inhibitor.
4. Caused a slight mouth opening, but no contraction of tentacles.
5. Act synergistically.

however, that m-tyrosine was more active than the other tyrosine isomers in activating both the neck response and in microvilli formation.

Recognizing that in response to tyrosine hydra display these two events, a cellular one and an organismic one, we can pose a number of intriguing questions: Is the same receptor site used to trigger both events? If not, did the receptor for neck formation evolve from the receptor for microvilli formation? It would appear simpler for hydra to use an existing receptor for two functions rather than evolve another.

This reasoning may be stretched even further to postulate that there exists a direct line of evolution of receptor sites from those found on single cells inducing pinocytosis or phagocytosis, to those coordinating feeding responses in such simple "tissue-level" organisms as cnidarians, and finally to the receptors for neurotransmitters and peptide hormones in higher organisms. For example, because dopamine and norepinephrine are formed directly from tyrosine, it seems simpler and more efficient from an evolutionary standpoint for organisms to retain and utilize modifications of a primitive tyrosine receptor to recognize structurally related compounds rather than to evolve new receptor sites for each of these "analogs."

Would not the same argument apply to the evolution of receptors for such neurotransmitters as glutamic acid or glycine? Are there similar evolutionary relationships between the glutathione receptors involved in activating the cnidarian feeding response, and the glutathione site associated with the γ-glutamyl transpeptidase mechanism of amino acid transport (Meister, 1973) (see below)? In any of these cases, it would seem simpler for organisms during evolution to modify existing receptors to control new tasks rather than to develop completely new receptor-effector systems.

SUMMARY, FUTURE DIRECTIONS AND CONCLUSIONS

Data summarized in Table 1 show that a number of groups among the Cnidaria have their own characteristic type of activating system for their respective feeding behaviors. For example, the hydras respond to GSH, the colonial gymnoblasts to proline, the sea anemones to one specific amino acid, etc. It is tempting to use these findings to formulate ideas on the evolution of the various groups of the Cnidaria, but we feel that there is not sufficient data at this time to warrant such speculations.

On the other hand, we do feel the current information justifies speculations on the evolution of receptor proteins as described on the preceding pages because such views are analagous to our ideas of the conservative view of the evolution of proteins in general.

The difficulty, of course, lies in proving these speculations. Actually, there are a number of ways that evidence can be obtained. These ways, however, depend upon the development of methods for investigating receptors. One method, for example, would involve the isolation of receptor proteins and the analysis of their amino acid sequence. Based upon the current status of receptor purification, however, such an approach will not be possible for years. A second approach would consist of a search for different kinds of receptors to amino acids and neurotransmitters among members from all the phyla. From such a comparative study it might be possible to determine at which point in evolution the different receptors arose.

We are using a third approach, that of comparing the relative activities of two proteins that might be related to each other, to chemical analogs. In our case, we are interested in proteins that recognize the tripeptide reduced glutathione. Previously we have reported (see Lenhoff, 1974) that the GSH receptor of hydra is specific for the γ-glutamyl moiety of GSH and can tolerate some changes in the thiol of the cysteinyl moiety. Hydra, for example, can respond to GSH which has its thiol methylated. At that time, no other protein combining with GSH was known to have such an unusual specificity for the molecule. Recent work with the enzyme γ-glutamyl transpeptidase ("GTP"), however, has shown a similar specificity of this enzyme for GSH.

We were intrigued by this similar specificity of GTP, especially since this enzyme is implied to act in the transport of amino acids into cells (Meister, 1973). Could it be that the GSH receptor of hydra evolved from the GTP thought to be used in amino acid transport? Could they be the same protein? Next we analyzed hydra for this enzyme. We found it present in significant concentrations, and its requirement for amino acid acceptors similar to that reported for the same enzyme activity as found elsewhere.

To answer the question of whether or not the GSH receptor and the GTP from hydra are the same, we decided to look deeper into their relative specificities for GSH. We were fortunate, in this case, to be working with proteins specific for GSH rather than for single amino acids because it is possible to make a large variety of structural analogs of this tripeptide. Using analogs synthesized by our colleagues M. H. Cobb and G. Marshall of Washington University, we were able to distinguish between the two activities. Whereas both proteins had about the same specificity for the γ-glutamyl-cysteinyl part of GSH, they differed with respect to their activities with analogs having substitutions at the glycine moiety; on one hand the GTP reacted poorly with analogs having amino acids with large side chains substituted for the glycine in GSH, while on the other hand, the GSH receptor was activated by those same analogs (Danner et al, 1976; and in preparation).

Such experiments show how those two proteins in hydra have similar and yet different specificities for GSH. Possibly the GSH receptor evolved from a mutation in a repeating unit of a gene controlling the synthesis of GTP. We eventually plan to purify both these proteins and to compare them more closely.

The above speculations, like most concerning evolution, are difficult to prove. But they may help to make us aware that unifying concepts, tacitly assumed in the case of enzymes and cell organelles, also may apply to the basic aspects of chemoreception. Specifically, such speculations emphasize that the behavioral responses of lower invertebrates to a peptide or an amino acid may have many fundamental features in common with some hormonal and neurotransmitter responses in higher organisms. By focusing on the primary events of the combination of the activator with the receptor to initiate a series of coordinated activities, we may find new approaches and new insights into universal, yet little understood, chemical control mechanisms.

ACKNOWLEDGEMENTS

We thank Dr. David Slautterback for his comments and for allowing us to quote from his unpublished material on microvillus formation. The research discussed was supported by grants from the National Institutes of Health and the National Science Foundation.

LITERATURE CITED

Blanquet, R., and H. M. Lenhoff. 1968. Tyrosine enteroreceptor of hydra: its function in eliciting a behavior modification. Science 159: 633-634.

Chapman-Andresen, C. 1962. Pinocytosis in amoebae. C. R. Trav. Lab. Carlsberg 33: 73-264.

Cohn, Z. A. 1967. The regulation of pinocytosis in mouse microphages. II. Factor inducing vesicle formation. J. Exp. Med. 125: 213-232.

Danner, J., M. Houston-Cobb, W. Heagy and G. R. Marshall, 1976. γ-Glutamyl transpeptidase in hydra. Fed. Proc. 35:7.

Fulton, C. 1963. Proline control of the feeding reaction of Cordylorphora. J. General Physiology 46: 823-837.

Lehman, J. T., and J. W. Porter. 1973. Chemical activation of feeding in the Caribbean reef building coral Montastrea cavernosa. Biol. Bull. 145: 140-149.

Lenhoff, H. M., and H. A. Schneiderman. 1959. The chemical control of feeding in the Portuguese man-of-war, Physalia physalis L., and its bearing on the evolution of the Cnidaria. Biol. Bull. 116: 452-460.

Lenhoff, H. M. 1974. On the mechanism of action and evolution of
     receptors associated with feeding and digestion. In Coelenter-
     ate Biology: Reviews and New Perspectives: 1974, L. Muscatine
     and H. Lenhoff, eds., pp. 359-384. New York: Academic Press.
Lindstedt, K. J., L. Muscatine, and H. M. Lenhoff. 1968. Valine
     activation of feeding in the sea anemone Boloceroides. Comp.
     Biochem. Physiol. 26: 567-572.
Lindstedt, K. J. 1971a. Chemical control of feeding behavior.
     Comp. Biochem. Physiol. 39: 553-581.
Lindstedt, K. J. 1971b. Biphasic feeding response in a sea anem-
     one: control by asparagine and glutathione. Science 173:
     333-334.
Loeb, M., and R. Blanquet. 1973. Feeding behavior in polyps of the
     Chesapeake Bay sea nettle, Chrysaora quinquecirrha (Desor,
     1848). Biol. Bull. 145: 150-158.
Loomis, W. F. 1955. Glutathione control of the specific feeding
     reactions of hydra. Ann. New York Acad. Sci. 62: 209-228.
Mackie, G. O., and D. A. Boag. 1963. Fishing, feeding and diges-
     tion in siphonophores. Pubbl. Staz. Zool. Napoli 33: 178-196.
Mariscal, R. N., and H. M. Lenhoff. 1968. The chemical control of
     feeding behavior in Cyphastrea ocellina and some other Hawaiian
     corals. J. Exp. Biol. 49: 689-699.
Mariscal, R. N. 1971. The chemical control of the feeding behavior
     in some Hawaiian corals. In Experimental Coelenterate Biology:
     1971, H. M. Lenhoff, L. Muscatine, and V. Davis, eds., pp. 100-
     118. Honolulu: University of Hawaii Press.
Meister, A. 1973. On the enzymology of amino acid transport.
     Science 180: 33-39.
Pardy, R. L., and H. M. Lenhoff. 1968. The feeding biology of the
     gymnoblastic hydroid, Pennaria tiarella. J. Exp. Zool. 168:
     197-202.
Reimer, A. A. 1971a. Feeding behavior in the Hawaiian zoanthids
     Palythoa and Zoanthus. Pacific Sci. 25: 512-520.
Reimer, A. A. 1971b. Chemical control of feeding behavior in
     Palythoa (Zoanthidea, Coelenterata). Comp. Gen. Pharm. 2:
     383-396.
Reimer, A. A. 1973. Feeding behavior in the sea anemone Calliactis
     polypus (Forskÿl, 1775). Comp. Biochem. Physiol. 44: 1289-
     1301.
Spencer, A. N. 1974. Behavior and electrical activity in the
     hydrozoan Proboscidactyla flavicirrata (Brandt). Biol. Bull.
     146: 100-115.
Steiner, G. 1957. Über die chemische Nahrungswahl von Actinia
     equina (L.). Naturwissenschaften 44: 70-71.

CHEMORECEPTION AND CONDUCTION SYSTEMS IN SEA ANEMONES

I.D. LAWN

Bamfield Marine Station
Bamfield, B.C., V0R 1B0
Canada

INTRODUCTION

In an earlier review of reception in Cnidaria, Ross (1966) stressed the supreme importance of chemoreception: "If we are to understand the behaviour of these animals we must recognize the primacy of chemoreception in determing what they do." Much effort has been directed towards the identification of the chemical activators involved in certain responses, especially feeding behaviour (see review by Lindstedt, 1971). These studies have relied heavily on assays in which behavioural observations have been utilized as the measured output of the sensory response.

In recent years, several advances in electrophysiological techniques (Josephson,1966; McFarlane,1969; Lawn,1975) have opened up a new line of approach to the study of chemoreception in cnidarians. The advantages of using sea anemones for this type of work include their (generally) large size, good survival rate, ability to undertake complex behaviour patterns, and our consider- able background knowledge of their morphology, physiology and behaviour. One major drawback is that the amplitude of pulses recorded from sea anemones tends to be very small - no more than 4 microvolts in some cases. This makes exacting demands on the recording circuitry employed. A second drawback is that excitable cells in sea anemones are too small for intracellular recordings to be made, certainly with present techniques. As far as sensory responses are concerned, this makes it impossible to determine whether the activity detected by the extracellular electrodes is a true representation of the primary-order response. Despite these reservations, much useful information on chemosensory activity in

sea anemones has recently emerged and this will form the subject of
this review.

CONDUCTION SYSTEMS

Suction electrodes attached to the tentacles of a sea anemone
can detect pulses associated with at least three diffuse conduction
systems. One pulse type is associated with the familiar through-
conducting nerve-net (Pantin,1935; McFarlane,1969) and travels at
a velocity of up to 120 cm/sec. The two other pulse types are associated with separate slow-conduction systems. One of these seems to
be situated exclusively in the ectoderm (McFarlane,1969) and
conducts with a velocity of 3-14 cm/sec. This system is called the
SS1 (slow system 1). The other, termed SS2, is confined to the
endoderm and conducts at a velocity of 3-5 cm/sec (McFarlane,1969).
Other pulse types may also be detected under certain conditions.
Many of these undoubtedly represent muscle action potentials associated with the contraction of muscles underlying the electrodes.
The spread of conduction is usually localized and pulse amplitude
varies considerably. To date, no recorded pulses have been assigned
unequivocally to a localized nerve-net, although it seems likely
from histological and behavioural evidence that this must exist
(Robson,1963; Lawn, 1976a).

The analysis of recorded sensory activity is often limited by
the observer's ability to distinguish between the different pulse
types. For this reason, simultaneous recordings from different
regions of the anemone are usually required. This compounds the
problem of keeping the electrodes attached during a recording
sequence, especially one in which the anemone undertakes complicated manoeuvres, such as swimming or shell-climbing. Activity in
any one system may be distinguished from that in the others by
comparing the conduction velocities, thresholds of excitation and
characteristic waveforms of the recorded pulses (Lawn,1976a).

CHEMORECEPTION INVOLVING THE SS1

Preparatory Feeding Behaviour in Tealia felina

It was noted almost a century ago that sea anemones in a
rock pool respond to nearby food by opening and expanding their
oral discs and tentacles (Pollock,1883). In Metridium senile
(Parker,1919; Pantin,1950) and Tealia felina (McFarlane,1970;
McFarlane and Lawn,1972) subsequent stages involve extension and
slow swaying movements of the column. This constitutes the pre-
feeding response and clearly increases the food-capture range of
the tentacles.

The chemosensory activity accompanying this response in
T. felina has been closely studied (McFarlane,1970; McFarlane and
Lawn,1972; Lawn,1975). Recording electrodes detect an increase in
SS1 activity when the anemone is exposed to dissolved food substances. This response typically consists of 20 or more SS1 pulses.
The mean inter-pulse interval may vary between 25sec and 1min 30sec
and the interval between pulses increases as the response progresses.
Lawn (1975) has shown that this is a genuine process of sensory
adaptation and does not result from fatigue in the conduction
system. In a typical response, the activity ceases within 30min,
but in rare cases may continue for up to 1hr. The most significant
feature of these responses is that the duration of sensory activity
and the accompanying process of sensory adaptation operate over
much longer time-scales than comparable phenomena encountered in
higher animals. This seems to be a characteristic feature of chemosensory responses in sea anemones and may prove to be equally
characteristic of other lower and predominantly sessile invertebrates.

Antifacilitation of pulse amplitude when the interval between
pulses decreases to less than about 20sec is also evident in these
responses (Lawn,1975). The cause of this has yet to be established,
but a similar effect can be reproduced by repetitive stimulation of
the SS1 (McFarlane and Lawn,1972). It is possible that this effect
is brought about by a decrease in the number of spiking cells under
the recording electrode as a consequence of repetitive firing in
the SS1. This has been proposed in Shelton's (1975) model for
SS1 conduction in Calliactis parasitica. The model assumes that the
SS1 is a neuroid system but the same effect could be obtained from
a suitably arranged network of nervous elements. The nature of the
SS1 (and SS2), therefore, still remains unknown.

To locate the chemoreceptors involved in the pre-feeding
response, it was necessary to devise a more precise method of applying the soluble food extract than simply adding it to the surrounding
sea water. A polyethylene suction electrode (chemical-stimulating
electrode) with wide tip-diameter (I.D.=1mm) was used for this
purpose (Lawn,1975). A total of 0.8ml of extract could be contained
within the electrode which was then attached to various regions of
the anemone. This arrangement may be used for both electrical and
chemical stimulation of the selected site. Multi-channel recordings
verified that the sensory response to the food extract was generated
at the site of the applied chemical stimulus (Lawn,1975).

The location of the receptor sites was determined by placing
the chemical-stimulating electrode on different regions of the
anemone and counting the number of SS1 pulses evoked for each
placement. Mean values were calculated for each region tested to
provide an index of responsiveness. The pre-feeding chemoreceptors
were found to be dispersed throughout the column ectoderm, being

particularly abundant at the column base. The pedal disc, oral disc, tentacles and pharynx are apparently devoid of these particular receptors, even though such regions are capable of conducting SS1 pulses (Lawn,1975).

Swimming in Stomphia coccinea

The SS1 also seems to be responsible for triggering the remarkable swimming behaviour of Stomphia coccinea (Lawn,1976a). Upon contact with the starfish Dermasterias imbricata, the anemone first withdraws then expands, rapidly detaches from the substrate and undertakes a series of swimming motions involving flexions of the column (Yentsch and Pierce,1955). The detachment process usually takes less than 10sec from the initial contact and the swimming flexions may continue for upto 5min.

Recording electrodes attached to the anemone detect a relatively rapid train of SS1 pulses following contact of the starfish with the tentacles. This activity usually continues for about 8sec but like all chemosensory responses encountered in sea anemones, there is considerable variation in this respect. The typical mean inter-pulse interval is about 600msec and intervals approaching the absolute refractory period of the SS1 (275msec in Stomphia at 9°C) may occasionally be encountered, especially between the first few pulses of the train. This is the most rapid response known for a sea anemone, although it is still slow relative to similar responses encountered in higher animals. Sensory adaptation and antifacilitation of pulse amplitude are again evident. No activity in the nerve-net or the SS2 appears at any stage of the response and SS1 activity usually ceases before or during detachment. Electrodes on the tentacles detect no activity related to the column flexions suggesting that these may be controlled by a local system, perhaps the network of large multi-polar nerve cells found in the endoderm of the column (Robson,1963).

CHEMORECEPTION INVOLVING THE SS2

Shell-settling in Stomphia coccinea

After its swimming response has ceased, Stomphia often displays a positive reaction to molluscan shells (Ross,1965; Ross and Sutton, 1967). The anemone emerges from its post-swimming torpor and if a tentacle is contacting a Modiolus shell, it may lightly adhere to it. There then follows a sequence in which the oral disc is brought close against the shell, the mid-column circular muscles expand then contract and the pedal disc begins to inflate. Eventually, the base of the pedal disc contacts the surface of the shell and sticks to it. The entire sequence may take from 3-8min to complete.

Recording electrodes detect a short complex burst of pulses when a responsive tentacle first adheres to the shell (Lawn and McFarlane,1976). These pulses remain confined to the region of contact and are not through-conducted. They may, therefore, be associated with localized movements of the tentacles or, perhaps, activity in a local conduction system. The SS2 then begins to fire repetitively: typically, a total of about 16 SS2 pulses at a mean inter-pulse interval of 5sec will be elicited. With the use of multiple electrodes (Lawn,1975) these were shown to originate at the site of shell-tentacle contact. This sensory response continues through the various stages of shell-settling until the pedal disc begins to inflate. A period of complex electrical activity then begins, accompanied by local bending movements of the column and tentacles. Usually, the SS2 pulses become lost in this activity and can no longer be monitored, although there is some evidence that the SS2 is turned off at this stage (Lawn and McFarlane,1976). This may be due to simple sensory adaptation, or to some type of inhibitory feedback mechanism operating from stretch receptors in the pedal disc. The complex activity declines as the pedal disc begins to deflate and the anemone relaxes. If the pedal disc does not contact the shell during these manoeuvres, the whole cycle may be repeated until successful contact is made.

The sensory response itself displays no obvious sensory adaptation; antifacilitation of pulse amplitude is only occasionally discernible. This illustrates how even the general properties of sensory responses may vary between species, although relatively long time-scales are again involved. Successful responses were obtained in cases where only one tentacle was in contact with the shell. If this contact were broken, the train of SS2 pulses would cease. An incomplete shell response would result if only a few pulses had been elicited.

Feeding in <u>Stomphia coccinea</u> and <u>Calliactis parasitica</u>

If food particles or dissolved food extracts are placed on to the margin of the mouth of a <u>Stomphia</u> already attached to a shell or some other substrate, the mouth opens widely. This is accompanied by an increase in SS2 activity from a resting level of 1 pulse every 60-70sec up to a level of 1 pulse every 4-10sec (Lawn and McFarlane, 1976). This increased activity originates from the site of the applied food and therefore constitutes a genuine chemosensory response. In this case, sensory adaptation occurs, suggesting a difference in the activation kinetics of the SS2-shell receptors and the SS2-food receptors. If food particles are placed on the mouth of a

detached <u>Stomphia</u>, similar sensory activity may be recorded. A
curious outcome of this, however, is that instead of mouth opening,
the entire shell-settling behaviour is released.

In <u>Calliactis parasitica</u> a similar feeding response is evoked
(McFarlane,1975), although the SS2 seems to fire at lower frequencies than in <u>Stomphia</u>. When food is placed on the tentacles or
around the mouth, the SS2 frequency will typically increase from a
resting level of 1 pulse every 70sec to 1 every 30sec. Activity in
the SS1 and through-conducting nerve-net may also be encountered.
The nerve-net pulses are not always seen and probably arise from
stimulation of ectodermal mechanoreceptors. SS1 pulses are often
seen early in the response, but many of these originate in the
column, presumably due to substances diffusing from the food, and
therefore represent pre-feeding activity. It is possible that there
are also SS1 chemoreceptors in the tentacles and pharynx but activity
from these regions may also be generated by SS1 mechanoreceptors
responding to solid food. An interesting feature of the feeding
response is the very long duration of sensory activity. In <u>Calliactis</u>,
the pulse frequency of the SS2 may not return to its resting level
for more than 4hr after ingestion (McFarlane, 1975).

CHEMORECEPTION INVOLVING BOTH SS1 AND SS2

The shell-climbing behaviour of <u>Calliactis parasitica</u> has been
well documented (see descriptions by Ross,1960; Ross and Sutton,
1961). When the tentacles contact certain molluscan shells, they
will adhere, provided that the pedal disc is not already attached
to an active shell. A behavioural sequence ensues whereby the pedal
disc detaches from the substrate, its margin expands and the column
thins then bends to bring the foot over towards the shell. If the
pedal disc contacts the shell successfully, it adheres, the tentacles
release their hold and the anemone gradually assumes its normal
posture. The entire sequence may take from 10-30min to complete.

The chemosensory response accompanying this behaviour has been
recently analyzed (McFarlane,1976) and seems to involve activity
in both the SS1 and SS2. This demonstrates how two essentially
similar behaviour patterns (shell-settling in <u>S. coccinea</u> and
shell-climbing in <u>C. parasitica</u>) may be coordinated by different
means, even in animals as simply organized as these.

Activity in both systems commences when a responsive tentacle
first adheres to the shell. As with <u>Stomphia</u>, (Lawn and McFarlane,
1976) the sensory response is abolished as soon as shell-tentacle
contact is broken. The mean inter-pulse intervals are 7sec (SS1)
and 6sec (SS2) in a typical response, though once again there is
much variation. Sluggish responses showed a lower pulse frequency

than active ones. The sensory train itself usually exhibits irregular firing, suggesting that mean frequency is the most important parameter for the animal. There is some evidence of sensory adaptation in both systems, although the complete response has never been successfully monitored beyond the stage where the pedal disc starts moving towards the shell. Most of the pulses originate from the site of shell-tentacle contact, indicative of a true sensory response. The SS1 and SS2 pulses are not phase-locked, suggesting that the shell chemoreceptors in the ectoderm may feed into each system separately.

THE THROUGH-CONDUCTING NERVE-NET

In the foregoing responses, no evidence of chemoreception involving the through-conducting nerve-net has been obtained. This in itself, however, does not imply that this system is exclusively concerned with mechanoreception or pacemaker activity, functions already attributed to it by various authors (Passano and Pantin, 1955; McFarlane, 1974; Lawn, 1976b,c). Indeed, some responses in which chemical activators cause symmetrical contractions of the musculature may well be coordinated by the through-conducting nerve-net. This is probably the case with Anthopleura elegantissima in its response to the alarm pheromone anthopleurine (Howe and Sheikh, 1975). Electrophysiological recordings should confirm the involvement of the through-conducting nerve-net in this response and resolve the question of whether or not chemical irritation of the underlying musculature is occurring.

ECTODERMAL-ENDODERMAL CONNECTIONS

The SS2 is an endodermal conduction system (McFarlane, 1969). In those chemosensory responses where the SS2 is involved (shell-climbing and true feeding behaviour in both Stomphia and Calliactis) it is apparent that ectodermal chemoreceptors located on the tentacles and oral disc must connect with the endoderm through the mesogloea in those regions. In Mimetridium cryptum, individual neurites may pass from the endoderm of the mesenteries to the ectoderm of the oral disc and tentacles (Batham, 1965). These form part of the through-conducting nerve-net, however, and probably activate those muscles involved in the protective closure response. It is not known how the ectodermal receptors connect with the SS2, for this system may be neuroid (McFarlane, 1969). Recent ultrastructural studies on Stomphia coccinea (Peteya, personal communication) may help to resolve this.

A similar problem exists in explaining how the ectodermal and perhaps neuroid SS1 (McFarlane, 1969) feeds into the endodermal elements. This occurs in later stages of pre-feeding responses (swaying of the elongated column), in shell-climbing of <u>Calliactis parasitica</u> (thinning and bending of the column) and in swimming of <u>Stomphia coccinea</u> (column elongation and swimming flexions). All of these stages involve contractions of endodermal musculature.

The swimming response of <u>S. coccinea</u> provides an excellent means of studying the inputs from ectoderm to endoderm, as the behaviour is stereotyped and repeatable. In addition, it has been shown that the SS1 feeds into and triggers a 'swimming programme' that is maintained and coordinated by a localized motor system, probably endodermal (Lawn, 1976a). The SS1 may be selectively activated by electrical stimulation of ectodermal flaps cut in the column (McFarlane, 1969; McFarlane and Lawn, 1972). Preliminary experiments (Lawn, unpublished observations) in which the ectoderm above and below a mid-column flap was carefully removed, show that this will block the passage of the SS1 from the flap to the oral and pedal discs. Stimulation of the SS1 at the flap elicits certain phases of the swimming response, such as flexions of the endodermal parieto-basilar muscles. This indicates that transmesogloeal connections must exist in the column of <u>Stomphia</u>, a condition that could never be demonstrated by light-microscope studies on other anemones (Batham, Pantin and Robson, 1961). It remains to be seen whether these connections in <u>Stomphia</u> will be visualized under the electron microscope.

CHEMICAL ACTIVATORS

The chemical activators for the pre-feeding response in <u>Tealia</u>, shell-climbing in <u>Calliactis</u> and <u>Stomphia</u>, and swimming in <u>Stomphia</u> remain unidentified. In the case of true feeding behaviour of <u>Calliactis parasitica</u>, 10^{-5}M reduced glutathione and 10^{-3}M proline, arginine, serine and valine were shown to elicit sensory activity in the SS2 (McFarlane, 1975). Obviously, the identification of the chemical activators involved in the other responses is a major task for future research.

SUMMARY

Recent studies reveal several features related to the phenomenon of chemoreception in sea anemones. It appears that the duration of chemosensory activity and the accompanying process of sensory adaptation occur over much-extended time-scales compared with the same phenomena encountered in higher animals. The SS1 alone, the SS2 alone, and both slow systems together may mediate chemosensory responses in these animals. Although the through-conducting nerve-net has not yet been implicated in chemoreception, it seems likely that this system could be involved with the 'alarm' behaviour of <u>Anthopleura elegantissima</u> when responding to the pheromone anthopleurine. Ectodermal receptors may link with endodermal elements not only through the mesogloea of the oral disc and tentacles but also of the column. The chemical activators that trigger many complex behaviour patterns in sea anemones have yet to be identified. Electrophysiological studies should provide a more precise method of studying the mechanisms of chemoreception in these animals than that provided by behavioural observations alone.

REFERENCES

Batham, E.J. 1965. The neural architecture of the sea anemone Mimetridium cryptum. Am. Zool., 5: 395-402.

Batham, E.J., C.F.A. Pantin and E.A. Robson 1961. The nerve-net of Metridium senile: artifacts and the nerve-net. Q.J. microsc. Sci., 102: 143-156.

Howe, N.R. and Y.M. Sheikh 1975. Anthopleurine: a sea anemone alarm pheromone. Science, N.Y., 189: 386-388.

Josephson, R.K. 1966. Neuromuscular transmission in a sea anemone. J. exp. Biol., 45: 305-319.

Lawn, I.D. 1975. An electrophysiological analysis of chemoreception in the sea anemone Tealia felina. J. exp. Biol. 63: 525-536.

Lawn, I.D. 1976a. A slow conduction system triggers swimming in the sea anemone Stomphia coccinea. Nature (Lond)., in press.

Lawn, I.D. 1976b. The marginal sphincter of the sea anemone Calliactis parasitica. I. Responses of intact animals and preparations. J. comp. Physiol., 105: 287-300.

Lawn, I.D. 1976c. The marginal sphincter of the sea anemone Calliactis parasitica. II. Properties of the inhibitory response. J. comp. Physiol., 105: 301-311.

Lawn, I.D. and I.D. McFarlane 1976. Control of shell-settling in the swimming sea anemone Stomphia coccinea. J. exp. Biol., in press.

Lindstedt, K.J. 1971. Chemical control of feeding behaviour. Comp. Biochem. Physiol. 39A: 553-581.

McFarlane, I.D. 1969. Two slow conduction systems in the sea anemone Calliactis parasitica. J. exp. Biol. 51: 377-385.

McFarlane, I.D. 1970. Control of preparatory feeding behaviour in the sea anemone Tealia felina. J. exp. Biol. 53: 211-220.

McFarlane, I.D. 1973. Spontaneous contractions and nerve net activity in the sea anemone Calliactis parasitica. Mar. Behav. Physiol. 2: 97-113.

McFarlane, I.D. 1975. Control of mouth opening and pharynx protrusion during feeding in the sea anemone Calliactis parasitica. J. exp. Biol. 63: 615-626.

McFarlane, I.D. 1976. Two slow conduction systems coordinate shell-climbing behaviour in the sea anemone Calliactis parasitica. J. exp. Biol., in press.

McFarlane, I.D. and I.D. Lawn 1972. Expansion and contraction of the oral disc in the sea anemone Tealia felina. J. exp. Biol. 57: 633-649.

Pantin, C.F.A. 1935. The nerve net of the Actinozoa. I. Facilitation. J. exp. Biol. 12: 119–138.

Pantin, C.F.A. 1950. Behaviour patterns in lower invertebrates. Symp. Soc. exp. Biol. 4: 175–195.

Parker, G.H. 1919. The elementary nervous system. Philadelphia: Lippincott.

Passano, L.M. and C.F.A. Pantin 1955. Mechanical stimulation in the sea anemone Calliactis parasitica. Proc. Roy. Soc. Lond. B. 143: 226–238.

Pollock, W.H. 1883. On indications of the sense of smell in Actiniae. With an addendum by G.J. Romanes. J. Linn. Soc. (Zool) 16: 474–476.

Robson, E.A. 1963. The nerve-net of a swimming anemone, Stomphia coccinea. Q.J. microsc. Sci. 104: 535–549.

Ross, D.M. 1960. The association between the hermit crab Eupagurus bernhardus (L) and the sea anemone Calliactis parasitica (Couch). Proc. zool. Soc. Lond. 134: 43–57.

Ross, D.M. 1965. Preferential settling of the sea anemone Stomphia coccinea on the mussel Modiolus modiolus. Science, N.Y. 148: 527–528.

Ross, D.M. 1966. The receptors of the Cnidaria and their excitation. Symp. zool. Soc. Lond. 16: 413–418.

Ross, D.M. and L. Sutton 1961. The response of the sea anemone Calliactis parasitica to shells of the hermit crab Pagurus bernhardus. Proc. Roy. Soc. Lond. B, 155: 266–281.

Ross, D.M. and L. Sutton 1967. The reponse to molluscan shells of the swimming sea anemones Stomphia coccinea and Actinostola new species. Can. J. Zool. 45: 895–906.

Shelton, G.A.B. 1975. The transmission of impulses in the ectodermal slow conduction system of the sea anemone Calliactis parasitica (Couch). J. exp. Biol. 62: 421–432.

Yentsch, C.S. and D.C. Pierce 1955. A 'swimming' anemone from Puget Sound. Science, N.Y., 122: 1231–1233.

INTERACTIONS BETWEEN CONDUCTING SYSTEMS IN THE SEA ANEMONE CALLIACTIS PARASITICA

I.D.McFARLANE & A.J.JACKSON

GATTY MARINE LABORATORY, UNIVERSITY OF
ST ANDREWS, ST ANDREWS, FIFE, SCOTLAND

Three conducting systems have been described in sea anemones (Pantin, 1935; McFarlane, 1969a,b, 1974a; McFarlane & Lawn, 1972). The through-conducting nerve net (TCNN) excites fast and slow muscle contractions. Slow systems 1 & 2 (SS1 & SS2), both possibly non-nervous, inhibit inherent contractions of ectodermal and endodermal muscles respectively. The SS1 coordinates pedal disk detachment in Calliactis parasitica.

Recent studies have begun to show how the conducting systems control behaviour. Suction electrode recordings from tentacles show, for example, SS2 pulses during shell settling behaviour of Stomphia coccinea (Lawn & McFarlane, 1976) and both SS1 and SS2 pulses during shell climbing in Calliactis (McFarlane, 1976). Electrical stimulation of the appropriate conducting system(s) elicits at least parts of these behaviour patterns, in the absence of the normal stimulus. We do not, however, understand how the conducting systems coordinate effector action. For example, the SS1 and SS2 inhibit muscles, yet simultaneous SS1 and SS2 stimulation of Calliactis evokes complex movements of many muscle groups. Also, we do not know how local movements, such as the column bending seen during shell climbing, are controlled. None of the three known systems shows a restricted spread of pulses.

There is now evidence for a fourth conducting system in Calliactis (Jackson & McFarlane, in preparation). We consider here possible interactions between the four conducting systems and we suggest that the fourth system connects with other conducting systems and also provides local control of muscles.

591

METHODS

Calliactis were supplied by the Plymouth Marine Laboratory. Technical procedures have been previously described (McFarlane, 1969a, 1974a).

RESULTS

The evidence for a fourth conducting system is not based on direct recordings of electrical activity in the system.

A single shock to the column always evokes single TCNN, SS1 and SS2 pulses (early pulses), but often also elicits a burst of delayed SS1 pulses. The early pulses come less than 1 sec after the shock, whereas the burst is delayed by 5 - 28 sec. A single shock may also evoke a delayed burst of TCNN pulses (McFarlane, 1974b). SS1 and TCNN bursts can occur together, apparently without interaction.

Delayed repetitive SS1 pulses may come from pacemaker cells in a fourth conducting system linked to the SS1. The evidence for this is firstly that the stimulus threshold of delayed SS1 pulses exceeds that of the three known conducting systems. Secondly, delayed pulses arise distant to the stimulating electrode. Thirdly, the pathway between the stimulated region and the point of origin of delayed SS1 pulses is endodermal, whereas the SS1 is ectodermal. The delay is largely due to delayed pulse initiation, either in the fourth system itself or at its links with the SS1. Hence we term the system the Delayed Initiation System (DIS). Electrical and mechanical stimulation of other regions can also elicit delayed SS1 pulses. Stimulation of the pharynx, however, evokes just a single delayed SS1 pulse, never a burst. The pulse arises close to the point of stimulation (McFarlane, 1975) and has a shorter delay (3 - 8 sec) than that seen for column stimulation.

The DIS may be a local conducting system. Stimulation of the pharynx evokes a single delayed SS1 pulse. If DIS activity could spread freely we should often see multiple SS1 pulses for the DIS pulse(s) could pass from the pharynx to the pacemakers affected by column stimulation. In the column the DIS is in the body wall endoderm, and thus is near the circular and parietal muscles. The DIS may control the local parietal muscle contractions responsible for column bending.

An endodermal conducting system, the DIS, connects with the SS1. The connexion may also work in the opposite direction as SS1 stimulation causes column extension (McFarlane, 1974a). Extension is due to endodermal circular muscle contraction, so

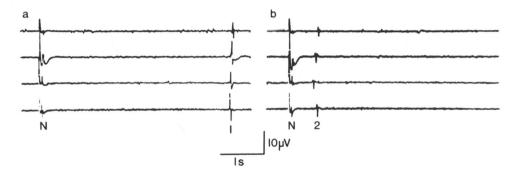

Fig. 1. A single shock to the pharynx does not always evoke a
delayed SS1 pulse. The early SS2 pulse may inhibit the DIS
pacemakers. Four recording electrodes, attached to tentacles.
N: TCNN pulse; 1: delayed SS1 pulse; 2: early SS2 pulse.

the ectodermal SS1 may connect with an endodermal system,
possibly the DIS, which directly excites this contraction.
Lawn (personal communication) has demonstrated SS1 to endoderm
connexions in Stomphia, mainly in the upper column. Some DIS
to SS1 links occur in the same region in Calliactis. The fact
that SS1 stimulation causes symmetrical column extension does
not conflict with the suggestion that the DIS is a local system
because multiple inputs from the SS1 would convert the DIS into
an apparent through-conducting system.

     Similarities between TCNN and SS1 bursts suggest that the
DIS pacemakers may give rise to both events. Both bursts show
a similar pulse initiation delay after a single shock, shift
of pulse origin during the burst, and rapid fatigue. TCNN
bursts differ from SS1 bursts in that they occur spontaneously,
are evoked at TCNN threshold, and they often contain more
pulses. The following experiment suggests a further similarity
between the pacemakers. Some DIS to SS1 connexions occur in
the pharynx. Pharynx stimulation sometimes gave a delayed SS1
pulse that arose close to the stimulating electrode but not
every shock evoked a delayed pulse. The responses to
stimulation close to SS2 threshold fell into two sets. In one,
there was a delayed SS1 pulse but not an early SS2 pulse (Fig.
1a). In the other, there was an early SS2 pulse but not a
delayed SS1 pulse (Fig. 1b). The early SS2 pulse may prevent
the initiation of the delayed SS1 pulse. SS2 activity also
inhibits the TCNN pacemakers, causing increased inter-pulse
intervals (McFarlane, 1974b). The TCNN and DIS pacemakers
appear to respond similarly to SS2 activity. Note that an
early SS1 pulse was seen only with stimulation near the mouth.

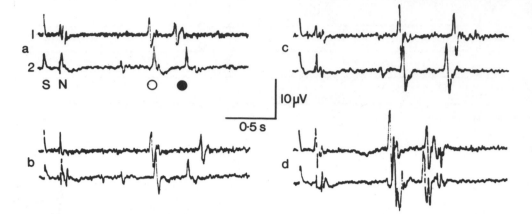

Fig. 2. Two early SS1 pulses to single shocks. Two recording
electrodes, on tentacles 2 cm apart. Stimulating electrode on
the column. (a,b): The second SS1 pulse may not always come
from the stimulus site. (c,d): SS1 pulses without and with
following complex activity, apparently associated with outward
sweep of the tentacles. S: stimulus artefact; N: TCNN pulse;
open and closed circles: first and second SS1 pulses.

        Further examples of conducting system interactions come
from observations on a single animal. Although this anemone
showed consistently atypical responses to stimulation, similar
events have been seen, albeit infrequently, in other specimens.
This anemone gave two early SS1 pulses to a single shock. The
double pulses were not produced by multiple firing in the SS1
for the second pulse did not always come from the point of
stimulation. In Fig. 2a the second SS1 pulse, like the first,
arrived at electrode 1 before electrode 2. In Fig. 2b, however,
the second pulse reached electrode 2 first. The second pulse
was identified as an SS1 pulse because it had a similar
conduction velocity to the SS1 and because its size was reduced
when it closely followed the first SS1 pulse - the passage of
a single SS1 pulse results in fatigue so that following SS1
pulses are smaller (McFarlane, 1969a). The second pulse may
come from DIS pacemakers that are activated with a delay far
shorter than normal. The responses were, however, probably not
due to direct stimulation of the DIS because spontaneous double
pulses were also seen. It was also noted that the TCNN pacemakers in this anemone showed a greater than normal tendency
to fire after single shocks. The anemone was very excitable
and most TCNN pulses were accompanied by an inward sweep of all
the tentacles. The double SS1 pulses were also often accompanied
by tentacle contraction, but this was an outward sweep, away

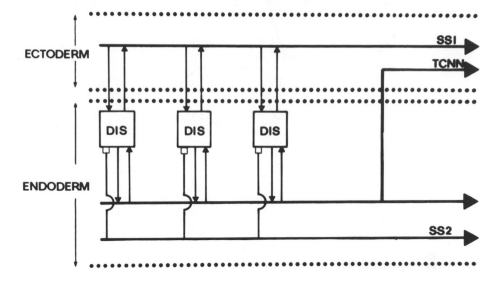

Fig. 3.   Possible interconnexions between the conducting systems.

from the mouth. When this contraction did not occur there was
little or no activity following the SS1 pulses (Fig. 2c). When
there was a contraction, complex potentials were seen after both
SS1 pulses (Fig. 2d), or sometimes after the second pulse only.
The contraction sometimes occured in a restricted region and
the complex activity was then seen only at electrodes within
the area of contraction. These results show that the SS1 can
have a loose excitatory coupling with certain muscles, probably
the ectodermal longitudinals in the tentacles. The SS1 may
directly excite the muscles or may act via an intermediate
motor system. It appears that the muscles on one side of the
tentacles receive a different 'innervation' from those on the
other side. A similar situation exists in the dactylozooid
of Hydractinia (Stokes, 1974).

DISCUSSION

The scheme for conducting system interconnexions (Fig. 3)
shows the DIS as a population of pacemaker cells connected to
both the SS1 and the TCNN. The similarity between SS1 and TCNN
bursts may of course simply reflect a basic similarity of the
pacemakers, not that the systems share the same pacemakers.
Individual cells in the DIS might form links with only one
system. The SS2 is shown to make inhibitory connexions with
DIS pacemakers.

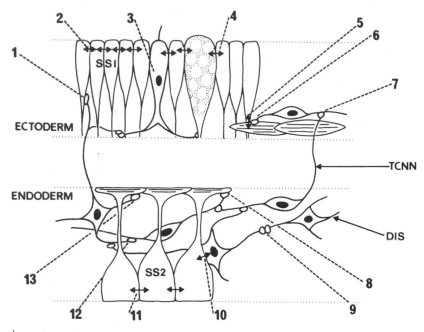
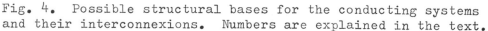

Fig. 4. Possible structural bases for the conducting systems
and their interconnexions. Numbers are explained in the text.

        Fig. 4 is a speculative translation of this scheme into
structure. We suggest that the DIS consists of multipolar
nerve cells. Such cells in Stomphia may be the pacemakers of
the swimming contractions (Robson, 1963). Similar cells in
Calliactis have been proposed as the pacemakers of the TCNN
(McFarlane, 1974b). Thus the column may contain two functionally separate but interacting nerve nets - the TCNN and the DIS
(multipolar nerve net). Pantin (1935) presented evidence for
two nets in the column. One of these ('mesenteric' net) is the
TCNN and the other ('primary' net) may be the DIS. Pantin
believed that the primary net shows interneural facilitation
and can thus coordinate local contractions. Two interacting
nerve nets have been proposed in the pennatulid Renilla
(Anderson & Case, 1975). Fig. 4 shows connexions (12) between
the DIS and the TCNN. The DIS may make transmesogleal links (1)
with the SS1 and also connexions (9) with other DIS cells. The
SS2 may inhibit DIS pacemakers, possibly by direct connexions
(10) although this effect might be indirect, via a non-contact
effect like the Extrinsic Hyperpolarizing Potential seen in the
Mauthner cells of fish (Furukawa & Furshpan, 1963). Conduction
in the SS2 may be by low resistance pathways between endodermal
musculo-epithelial cells (11). Similarly the SS1 may involve
cell-to-cell conduction in the supporting layer of the ectoderm

(2). Connexions (4) between the SS1 and a gland cell are shown because the pedal disk detachment and column mucus shedding seen after SS1 stimulation are probably secretory events (McFarlane, 1969b). Although a nerve net has not been convincingly described in the column ectoderm there do appear to be sense cells present in this region (Hertwigs, 1879-80). The sense cells (3) may connect with the SS1 or possibly with the DIS. SS1 activity inhibits ectodermal muscles (McFarlane & Lawn, 1972). A link (5) is shown between the epithelium and the muscle cells (which are present only in the oral disk and tentacles), although again such inhibition may be indirect. The diagram omits endodermal sense cells, which are apparently connected to the TCNN (Passano & Pantin, 1955). Muscle excitation derives from TCNN innervation of ectodermal (6) and endodermal (8) muscles. The ectodermal and endodermal TCNNs may be joined by links (7) across the mesogloea (Robson, 1965). The DIS may also innervate muscles (13) to control local contractions. Robson (1971) has proposed a model for control of local contractions in Gonactinia that involves the nerve net acting via multipolar nerve cells which function as intermediate motor units.

SUMMARY

There may be four conducting systems in Calliactis: the through-conducting nerve net (TCNN), the delayed initiation system (DIS), and two slow systems (SS1 & SS2). We suggest that the DIS is a multipolar nerve net which connects with the SS1. DIS pacemakers may also connect with the TCNN and may themselves be inhibited by the SS2. The DIS may coordinate local muscle contractions. The SS1 has a loose excitatory coupling with tentacle muscles. A scheme is proposed for the structural bases of the conducting systems and their interactions. This considers the TCNN and DIS as nerve nets and the SS1 and SS2 as non-nervous systems.

REFERENCES

ANDERSON, P.A.V., AND J.F. CASE, 1975. Electrical activity associated with luminescence and other colonial behavior in the pennatulid Renilla köllikeri. Biol. Bull., 149: 80-95.

FURUKAWA, T., AND E.J. FURSHPAN, 1963. Two inhibitory mechanisms in the Mauthner neurons of goldfish. J. Neurophysiol., 26: 140-176.

HERTWIG, O. AND R. HERTWIG, 1879-80. Die Actinien anatomisch und histologisch mit besonderer Berücksichtigung des Nervenmuskelsystems untersucht. Jena Z. Naturw., 13: 457-640; 14: 39-89.

LAWN, I.D. AND I.D. McFARLANE, 1976. Control of shell settling
    in the swimming sea anemone Stomphia coccinea. J. Exp.
    Biol., 419-429.
McFARLANE, I.D., 1969a. Two slow conduction systems in the sea
    anemone Calliactis parasitica. J. Exp. Biol., 51: 377-385.
McFARLANE, I.D., 1969b. Coordination of pedal-disk detachment
    in the sea anemone Calliactis parasitica. J. Exp. Biol.,
    51: 387-396.
McFARLANE, I.D., 1974a. Excitatory and inhibitory control of
    inherent contractions in the sea anemone Calliactis
    parasitica. J. Exp. Biol., 60: 397-422.
McFARLANE, I.D., 1974b. Control of the pacemaker system of the
    nerve net in the sea anemone Calliactis parasitica. J.
    Exp. Biol., 61: 129-143.
McFARLANE, I.D., 1975. Control of mouth opening and pharynx
    protrusion during feeding in the sea anemone Calliactis
    parasitica. J. Exp. Biol., 63: 615-626.
McFARLANE, I.D., 1976. Two slow conducting systems coordinate
    shell-climbing behaviour in the sea anemone Calliactis
    parasitica. J. Exp. Biol., 64: 431-445.
McFARLANE, I.D. AND I.D. LAWN, 1972. Expansion and contraction
    of the oral disc in the sea anemone Tealia felina. J.
    Exp. Biol., 57: 633-649.
PANTIN, C.F.A., 1935. The nerve net of the Actinozoa. II.
    Plan of the nerve net. J. Exp. Biol., 12: 139-155.
PASSANO, L.M. AND C.F.A. PANTIN, 1955. Mechanical stimulation
    in the sea-anemone Calliactis parasitica. Proc. Roy. Soc.
    Lond. B, 143: 226-238.
ROBSON, E.A., 1963. The nerve-net of a swimming anemone,
    Stomphia coccinea. Q. J. Microsc. Sci., 104: 535-549.
ROBSON, E.A., 1965. Some aspects of the structure of the
    nervous system in the anemone Calliactis. Am. Zool., 5:
    403-410.
ROBSON, E.A., 1971. The behaviour and neuromuscular system of
    Gonactinia prolifera, a swimming sea-anemone. J. Exp.
    Biol., 55: 611-640.
STOKES, D.R., 1974. Physiological studies of conducting systems
    in the colonial hydroid Hydractinia echinata I Polyp
    specialization. J. Exp. Zool., 190: 1-18.

SLOW CONDUCTION IN SOLITARY AND COLONIAL ANTHOZOA

G.A.B. Shelton and I.D. McFarlane

Department of Zoology, South Parks Road, OXFORD, England
and
Gatty Marine Laboratory, ST ANDREWS, Scotland

INTRODUCTION

The term 'slow conduction system' has appeared many times in the recent literature on the electrophysiology of the Anthozoa (e.g. McFarlane, 1969a, 1969b, 1970, 1974a, 1974b, 1975; McFarlane and Lawn, 1972; Lawn, 1975; McFarlane and Shelton, 1975; Shelton, 1975a, 1975b, 1976; Shelton and McFarlane, 1976). In the sea anemone Calliactis parasitica, two slow conduction systems have been described - the SS1 (slow system 1) and the SS2 (slow system 2) (McFarlane, 1969a). A third system, the 'nerve net' as described by e.g. Pantin (1935), Josephson (1966) and Robson and Josephson (1969) is also present. Slow systems have an important role in many aspects of the behaviour of sea anemones (see below) and recent work (Shelton, 1975c; Anderson and Case, 1975; Shelton and McFarlane, 1976; Shelton, 1976) indicates that they may be of general occurrence in the phylum. It is important therefore, to define what is meant by 'slow conduction', to describe the physiological properties of slow conduction systems and, as far as possible, to try to account for these properties in terms of the actions of particular cells in the animal. This paper describes some of the properties of the best known slow conduction system, the SS1 of Calliactis, and compares them with other known slow conduction systems, nerve nets and neuroid systems.

SLOW CONDUCTION IN CALLIACTIS

Extracellular polythene suction electrodes, differential pre-amplifiers and a storage oscilloscope were used to monitor electrical activity (McFarlane, 1969a; Shelton, 1975c). Fig.1 (A) shows pulses

599

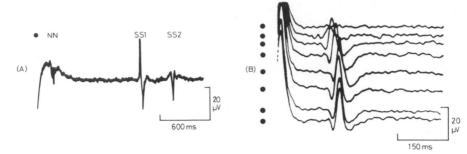

FIGURE 1. Electrophysiological recordings from <u>Calliactis</u> <u>parasitica</u>.
(A) Nerve net (NN) and two slow conduction system pulses (SS1) and
(SS2) recorded from a tentacle following a single electrical stimulus
to the column ●. (B) Effect of repetitive stimulation of the column
nerve net. Note the increases in delay of successive pulses.
Stimulus frequency 1 per 800ms.

in each of the three known conduction systems in <u>Calliactis</u>. These
were evoked by applying a single 15V 1ms electrical stimulus to the
column of an intact animal at $12^{o}C$. The nerve net pulse is readily
distinguishable as a rapidly-conducted pulse, often compound; two or
more within say one second evoke a large muscle action potential.
The SS1 pulse is more slowly-conducted (4-10 cm s^{-1}), more liable to
fatigue and of longer duration. The SS1 is ectodermal. The third
pulse represents activity in the SS2 (endodermal). It is small and
even more slowly-conducted. All three systems conduct non-
decrementally but may be distinguished on the basis of conduction
velocity, pulse size and shape, excitability and the ease with which
they may be stimulated and recorded from different parts of the
animal.

Repetitive stimulation of the SS1 in <u>Calliactis</u> (McFarlane,
1973; Shelton, 1975b) at e.g. 1 per 3 seconds has two effects. There
is a marked increase in conduction delay (100% or more) and a
decrease in pulse amplitude. As the system fatigues, the pulse
amplitude and duration become very variable (Shelton, 1975b). It is
difficult to measure the conduction velocity of the SS1 down or
across the column owing to the problems associated with recording
from regions other than the top of the column, the oral disc and
tentacles. The column is generally tough and inflexible and this
does not permit satisfactory electrode contact though it is sometimes possible to record from the base of the column. Recordings
made from the base do not show a significant difference in conduction
velocity from oral disc to pedal disc compared with the velocity from
pedal disc to oral disc. In an attempt to obtain some data on circumferential conduction of SS1 pulses, Laukner and Green (unpublished)
performed a number of cutting experiments on the columns of intact

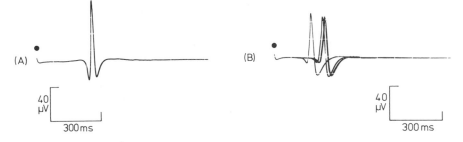

FIGURE 2. Electrophysiological recordings from the pneumatosaccus of Physalia. (A) Response to a single electrical stimulus ●. (B) Response to repetitive stimulation at 1 per 6 seconds. Traces superimposed.

and half-animals. The simplest preparation involved making shallow scrapes with a razor blade to isolate an 'L'-shaped area of column ectoderm with the vertical part of the 'L' parallel to the longitudinal axis of the animal and the horizontal part parallel with the margin of the pedal disc. Narrow horizontal strips were found to conduct very poorly, if at all. This contrasts with data on narrow vertical strips (Shelton, 1975b) and suggests that there may be preferential longitudinal conduction in the column SS1. Preferential longitudinal conduction has also been described in Hydra (Kass-Simon, 1972) and in Gonactinia (Robson, 1971). If this is so for the SS1, it is necessary to modify the model proposed by Shelton (1975b). This suggested that the conducting units of the SS1 were connected together in a hexagonal array. It may be that, in fact, there are many longitudinal connections but only a small number of horizontal ones. Much more data is needed on this. If there is a smaller number of horizontal connections it is important to know if they are evenly distributed or mainly confined to particular areas. It is conceivable that an SS1 pulse could pass up one side of the column but cause much less excitation to the SS1 on the other side of the column. Clearly this could have some bearing on the control of behaviour. In the oral disc and tentacles, SS1 pulses conduct at a greater velocity than in the column (McFarlane, 1969a) but it is not known if there are any preferred pathways of conduction.

Though it may be that the SS1 model needs modification, the hypothesis is still valid. This states that the decreasing amplitude, increasing duration and decreasing conduction velocity of SS1 pulses following repetitive stimulation, can be explained in terms of a decreasing number of active, low-threshold conducting units. Spencer (1974b) has suggested a similar mechanism for presumed neuroid systems in some Hydrozoa. Whether the conducting units of the SS1 are nervous, non-nervous or a combination of the two is still not clear. Fatigue in the SS1 shows many similarities to fatigue in

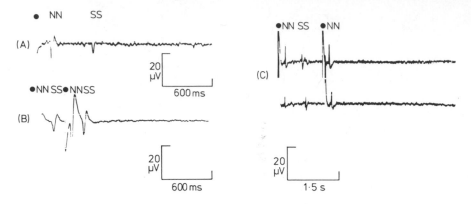

FIGURE 3. Eusmilia fastigiata. (A) Nerve net (NN) and slow system
(SS) responses. (B) Effect of two electrical stimuli (recording made
from the stimulated polyp). A muscle response and a larger and more
delayed slow pulse follow the second stimulus. (C) Two stimuli as
(B) but two-channel recordings made from a second polyp. No slow
system pulse follows the second stimulus.

known neuroid systems (see below). Studies with drugs e.g. menthol,
chloretone and magnesium chloride have shown that all three systems
are anaesthetised in the order SS1, SS2, nerve net (McFarlane, 1973b).
Intracellular recording has so far failed due to the small size of
the column ectoderm cells in Calliactis and the large amount of mucus
which is secreted. A thorough electron microscopical study of the
column ectoderm is required. The small amount which has been done
has not revealed clearly identifiable nerve cells. There is, however,
some evidence for low-resistance electrical junctions between ectodermal cells (Shelton, 1975b). The situation is quite different in
cerianthids, of course, which have a well-developed ectodermal nerve
net and musculature (Peteya, 1973).

    Nerve nets, for example the endodermal net in the column of
Calliactis, may also show changes in conduction velocity with repetitive stimulation (Fig. 1 (B)) (Shelton, 1975a, 1975d). In
Calamactis praelongus, successive pulses in the nerve net may speed
up or slow down depending on pulse number and frequency (Pickens,
1974). It has been suggested that facilitation of conduction velocity may produce greater synchronisation of muscle contraction
(Pickens, 1974). Increasing response delay following repetitive
stimulation has been reported for a number of conduction systems
e.g. the colonial pulse system of Proboscidactyla (Spencer, 1974a),
the skin impulse system in Oikopleura (Bone and Mackie, 1975) and
the epithelial system in Stomotoca (Mackie, 1975). Fig. 2 shows
extracellular recordings made from the pneumatosaccus of Physalia.
The float of Physalia consists of an outer case (pneumatocodon) and
an invaginated inner tube (pneumatosaccus). There is a layer of

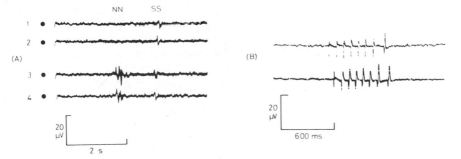

FIGURE 4.   (A) <u>Pennatula</u> <u>phosphorea</u>. Four successive electrical
stimuli. The slow system (SS) has the lower threshold. On raising
the stimulus intensity (3,4), nerve net pulses (NN) are evoked.
(B) <u>Virgularia</u> <u>mirabilis</u>. Two-channel recording of a stimulated
burst of nerve net activity.

ectoderm, a layer of mesogloea and a layer of endoderm in both
pneumatosaccus and pneumatocodon. Though muscle fibres are present
in all four layers of epithelium, nerves have only been found in the
pneumatocodon ectoderm (Mackie, 1965). This suggests, as Mackie
(1965) predicted, that the pulses recorded from the pneumatosaccus
could be epithelial (neuroid) pulses. Note the decrease in conduction velocity and pulse amplitude with repetitive stimulation. Many
neuroid or suspected neuroid systems show this property in common
with the SS1.

SLOW CONDUCTION IN OTHER ANTHOZOA

     Recent electrophysiological recordings made from the madrepor-
arian coral <u>Eusmilia</u> <u>fastigiata</u> (Shelton and McFarlane, 1976) have
shown the presence of a colonial slow conduction system in this
species. Fig. 3 shows some typical recordings. Like the SS1, pulses
are most easily recorded from the tentacles and oral disc, though
supra-threshold stimulation of any part of the outside of the animal
evokes a slow response. Two or three slow pulses may be produced in
response to a single stimulus. At low frequencies of stimulation
(1 per 10 seconds) the slow system conducts without decrement to all
other polyps in the colony. The refractory period of the system is
very long, however, (3 seconds or more) in interpolyp regions. This
precludes colonial conduction of high-frequency slow pulses. In
tentacles and oral disc, the refractory period is much shorter (less
than a second). The nerve net has a uniformly short refractory
period (less than 100 ms) and conducts without decrement to all
polyps, though this may not be the case in all species. Repetitive
stimulation of the slow system leads to increased response delay and
<u>increased</u> pulse amplitude. Preceding slow pulses facilitate

the amplitude of succeeding slow pulses. The mechanism of this
facilitation is not known.

     There is also evidence for slow conduction systems in the other
sub-class of Anthozoa, the Octocorallia (Shelton, 1975c; Anderson
and Case, 1975). In the sea pen Pennatula phosphorea, the system
through-conducts to all parts of the colony (Fig. 4). It has a
lower threshold to stimulation than the nerve net. Following repetitive stimulation, there is an initial decrease in response delay
followed by a steady increase and the final failure of the system
to conduct. The SS1 model proposed by Shelton (1975b) would explain
this in terms of an initial fall followed by a steady rise in the
number of high-threshold cells. The nerve net of the sea pen
Virgularia mirabilis (Fig. 4 (B)) also shows a facilitation of conduction velocity following repetitive stimulation. The greatest
increase in conduction velocity is produced over long distances at
high frequencies of stimulation (Shelton, 1975b). When the nerve
net is 'fatigued', response delay increases with successive stimuli.

     Josephson (1965) has described a 'slow system' in the hydroid
Tubularia. The system is through-conducting and often fires several
times following a single stimulus (Cf the slow system in Eusmilia).
During repetitive slow system activity, the second slow pulse has a
smaller amplitude and a lower conduction velocity than the first
(Cf the SS1). Conduction velocity is comparable to that of the SS1
(mean values from 4.5 to 6.9 cm s^{-1}) (Josephson, 1965). No behavioural output for the slow system in Tubularia has been described.

                SLOW CONDUCTION AND THE CONTROL OF BEHAVIOUR

     Some progress has been made concerning the relationship between
slow system activity and the behaviour of anthozoans. The SS1 responds to dissolved food juice (e.g. crushed Mytilus) (Lawn, 1975)
and in detached specimens responds to tentacle contact with perio-
stracum (McFarlane, 1969b). SS1 activity causes oral disc expansion
in the sea anemone Tealia felina (McFarlane and Lawn, 1972). This
seems to be due to relaxation of the oral disc radial muscles. In
Calliactis parasitica, high-frequency SS1 activity leads to pedal
disc detachment (McFarlane, 1969b). In addition, the old mucous
coat is shed from the column. Both effects may be due to the release
of ectodermal secretions. The SS2 causes inhibition of inherent
contractions of the endodermal column circular and parietal muscles
(McFarlane, 1974a). Nerve net pacemaker activity is also inhibited
(McFarlane 1974b). Continued SS2 stimulation evokes mouth opening
and pharynx protrusion (McFarlane, 1975). McFarlane and Shelton
(1975) have suggested that nematocyst and spirocyst discharge
thresholds may also be influenced by the SS1. The roles of slow
conduction systems in other Anthozoa are not so well understood.

In <u>Eusmilia</u>, the slow system responds to the presence in the water of dissolved food juice and causes oral disc and tentacle expansion.

Slow conduction appears to be of common occurrence in many Anthozoa, both Octocorallia and Hexacorallia. The discovery of these systems has necessitated a complete re-appraisal of the methods of behavioural control in the Anthozoa. Slow conduction seems to have an important role in the co-ordination of symmetrical and asymmetrical responses (McFarlane, in press). The slow conduction systems so far described in <u>Calliactis</u>, have been associated with the epithelial layers. Conduction velocity is low (of the order of 10 cm s^{-1}). Pulse size is small (less than a millivolt) but this may be due to complications caused by poor electrode contact and high level of mucus production. Pulse duration is long (of the order of 100 ms) and pulse shape is usually smooth. Refractory period may be up to several seconds. Slow systems have shown both facilitation and defacilitation of pulse size and latency. The term 'slow conduction' has been adopted mainly because it has not yet proved possible to classify the phenomenon with certainty under either of the more precise headings 'neuroid system' or 'nerve net'. The evidence presented here shows that slow conduction has some properties which could be assigned to either category but similarities in lay-out and function may lead to similarities in the physiological properties of nerve nets and slow systems. The task should be made easier however, as more physiological data becomes available.

ACKNOWLEDGEMENTS

The award of a Royal Society John Murray Travelling Studentship, a Royal Society Parliamentary Grant-in-aid of research and travel funds from the University of Oxford is very gratefully acknowledged.

SUMMARY

Suction electrode recordings of electrical activity are described for a number of species of Anthozoa.
The characteristics of the ectodermal slow conduction system of the sea anemone <u>Calliactis</u> <u>parasitica</u> are compared with those of nerve nets and neuroid systems of other Cnidaria.
The physiology of slow conduction in colonial Anthozoa (Octocorallia and Hexacorallia) is discussed.
The role of slow conduction is considered with reference to the control of behaviour in Anthozoa.

REFERENCES

ANDERSON, P.A.V. AND J.F. CASE, 1975.   Electrical activity
   associated with luminescence and other colonial behaviour in the
   pennatulid Renilla kollikeri. Biol.Bull., 149 : 80-95.
BONE, Q. AND G.O. MACKIE, 1975.   Skin impulses and locomotion in
   Oikopleura (Tunicata: Larvacea). Biol.Bull., 149 : 267-286.
JOSEPHSON, R.K., 1965.   Three parallel conducting systems in the
   stalk of a hydroid. J. exp. Biol., 42 ; 139-152.
KASS-SIMON, G., 1972.   Longitudinal conduction of contraction
   burst pulses from hypostomal excitation loci in Hydra attenuata.
   J. comp. Physiol., 80 : 29-49.
LAWN, I.D., 1975.   An electrophysiological analysis of chemo-
   reception in the sea anemone, Tealia felina. J. exp. Biol.,
   63 : 525-536.
McFARLANE, I.D., 1969a.   Two slow conduction systems in the sea
   anemone Calliactis parasitica. J. exp. Biol., 51 : 377-385.
McFARLANE, I.D., 1969b.   Co-ordination of pedal disc detachment
   in the sea anemone Calliactis parasitica. J. exp. Biol., 51 :
   387-396.
McFARLANE, I.D., 1970.   Control of preparatory feeding behaviour
   in the sea anemone Tealia felina. J. exp. Biol., 53 : 211-220.
McFARLANE, I.D., 1973a.   Spontaneous electrical activity in the
   sea anemone Calliactis parasitica. J. exp. Biol., 58 : 77-99.
McFARLANE, I.D., 1973b.   Spontaneous contractions and nerve net
   activity in the sea anemone Calliactis parasitica. Mar. Behav.
   Physiol., 2 : 97-113.
McFARLANE, I.D., 1974a.   Excitatory and inhibitory control of
   inherent contractions in the sea anemone Calliactis parasitica.
   J. exp. Biol., 60 : 397-422.
McFARLANE, I.D., 1974b.   Control of the pacemaker system of the
   nerve net in the sea anemone Calliactis parasitica. J. exp. Biol.,
   61 : 129-143.
McFARLANE, I.D., 1975.   Control of mouth opening and pharynx
   protrusion during feeding in the sea anemone Calliactis parasitica.
   J. exp. Biol., 63 : 615-626.
McFARLANE, I.D. AND I.D. LAWN, 1972.   Expansion and contraction of
   the oral disc in the sea anemone Tealia felina. J. exp. Biol.,
   57 : 633-649.
McFARLANE, I.D. AND G.A.B. SHELTON, 1975.   The nature of the
   adhesion of tentacles to shells during shell-climbing in the sea
   anemone Calliactis parasitica. J. exp. mar. Biol. Ecol., 19 :
   177-186.
MACKIE,G.O., 1965.   Conduction in the nerve-free epithelia of
   siphonophores. Amer. Zool., 5 : 439-453.
MACKIE, G.O., 1975. Neurobiology of Stomotoca. 11. Pacemakers and
   conduction pathways. J. Neurobiol., 6 : 357-378.
PANTIN, C.F.A., 1935.   The nerve net of the Actinozoa. 1.
   Facilitation. J. exp. Biol., 12 : 119-138.

PETEYA, D.J., 1973.   A light and electron microscope study of the
  nervous system of Ceriantheopsis americanus (Cnidaria, Ceriantharia).
  Z. Zellforsch., 141 : 301-317.
PICKENS, P.E., 1974.   Changes in conduction velocity within a nerve
  net. J. Neurobiol., 5 : 413-420.
ROBSON, E.A., 1971.   The behaviour and neuromuscular system of
  Gonactinia prolifera, a swimming sea-anemone.   J. exp. Biol.,
  55 : 611-640.
ROBSON, E.A. AND R.K. JOSEPHSON, 1969.   Neuromuscular properties of
  mesenteries from the sea anemone Metridium. J. exp. Biol., 50 :
  151-168.
SHELTON, G.A.B., 1975a.   Electrical activity and colonial behaviour
  in anthozoan hard corals. Nature, Lond., 253 : 558-560.
SHELTON, G.A.B., 1975b.   The transmission of impulses in the ecto-
  dermal slow conduction system of the sea anemone Calliactis
  parasitica (Couch). J. exp. Biol., 62 : 421-432.
SHELTON, G.A.B., 1975c. Colonial conduction systems in the Anthozoa:
  Octocorallia. J. exp. Biol., 62 : 571-578.
SHELTON, G.A.B., 1975d.   Colonial behaviour and electrical activity
  in the Hexacorallia. Proc. R. Soc. Lond.B, 190 : 239-256.
SHELTON, G.A.B., 1976.   Co-ordination of behaviour in cnidarian
  colonies. Spec. Vol. Syst. Ass. (G.P. Larwood, ed.), in press.
SHELTON, G.A.B. AND I.D. McFARLANE, 1976. Electrophysiology of two
  parallel conducting systems in the colonial Hexacorallia.
  Proc. R. Soc. Lond. B, 193 : 77-87.
SPENCER, A.N., 1974a.   Behaviour and electrical activity in the
  hydrozoan Proboscidactyla flavicirrata (Brandt). 1. The hydroid
  colony. Biol. Bull., 145 : 100-115.
SPENCER, A.N., 1974b.   Non-nervous conduction in invertebrates and
  embryos. Amer. Zool., 14 : 917-929.

AN ELECTROPHYSIOLOGICAL ANALYSIS OF BEHAVIOURAL INTEGRATION IN COLONIAL ANTHOZOANS

Peter A. V. Anderson

University of California, Santa Barbara

Santa Barbara, California 93106 U.S.A.

Coloniality has apparently arisen independently on several occasions in the Cnidaria (Hyman, 1940). In anthozoans alone there is evidence of four distinct colonial groups. The polyps that comprise anthozoan colonies, while having morphological variations consistent with the groupings of Octocorallia and Hexacorallia, all closely follow the anthozoan body plan. This structural uniformity is reflected in the constancy of certain behaviours. Thus, retraction, expansion and feeding are carried out in basically the same way by octocorallian, hexacorallian and solitary polyps (Horridge, 1957). Indeed, Buisson (1973) suggests that pennatulid polyps are even capable of the slow rhythmical contractions of column musculature typical of actinians (Batham and Pantin, 1950).

When studied from the viewpoint of the individual polyp, it is easy to imagine evolution of a colonial anthozoan from a budding, solitary ancestor consequent upon failure of separation or upon subsequent fusion of the newly formed buds. However, the behaviour of many existing colonies is inconsistent with such origins. If the nerve net of the original polyp were maintained intact during budding so as to extend throughout the colony, one would expect the whole colony to behave as a single unit. Alternatively, if all interpolyp connections within the nerve net were lost, one would expect all the polyps to behave independently. Modifications of this basic concept must be made to include other, possibly epithelial (McFarlane, 1969), conduction systems which might be continuous through a colony composed of polyps with segregated nerve nets. However, if behaviour known to be controlled solely by the nerve net, i.e., polyp retraction (Robson and Josephson, 1969) is considered, it is apparent that a great many colonies display both independent and coordinated behaviour.

Normally any one polyp is behaviourally independent of all other polyps in the colony (Horridge, 1957), except during a colonial response when their activities are coordinated. The commonest colonial response, polyp retraction, in fact involves behaviour normally carried out autonomously by individual polyps. It should be pointed out that there are colonies with no colonial behaviour, e.g. *Lobophyllia* (Horridge, 1957), but whether these represent a primitive condition or a later development is unknown. The presence, in the majority of colonies, of two levels of behaviour, colonial and polyp, requires additional control mechanisms not available from the act of budding. Recent studies (Anderson and Case, 1975; Anderson, 1976a) provide information on how this autonomy could be attained without loss of a colonial response capability.

The organ-pipe coral, *Tubipora musica* (subclass Alcyonaria, order Stolonifera), is composed of polyps capable of both autonomous and coordinated behaviour. Individual polyps display a variety of tentacle and column movements which culminate in the withdrawal response. These autonomous activities can all be evoked with gentle mechanical stimulation. More intense stimuli evoke colonial polyp retraction which spreads across the colony as a series of radiating waves. A single electrical stimulus produces only scattered tentacle waving from various polyps but each additional shock makes each polyp twitch and partially retract in sequence. Electrophysiological recordings (for methods see Anderson and Case, 1975) reveal that colonial polyp retraction is controlled by a colonial through-conduction system. Each suprathreshold electrical shock evokes an impulse detectable from all polyps of the colony (Fig. 1a,b). This impulse has two components, a small unit evoked by every stimulus and a second, facilitating unit evoked by the second and all subsequent stimuli (Fig. 1a). The second unit increases in size and duration during a series. Since these impulses are coincident with polyp retraction, it is probable that the first component represents activity in a conduction system with the second unit being a muscle action potential from retractor muscles, facilitated and activated by that system. Through-conducted impulses are recorded spontaneously at a frequency of 1-2 per minute.

A second class of spontaneous impulses (1-2/min.) similar in shape to the through-conducted ones, is recorded from individual polyps. These impulses are not through-conducted, even between adjacent polyps. Low intensity focal stimulation of a polyp that evokes autonomous behaviour evokes these non-through-conducted or "isolated" impulses. Isolated impulses contain both components described earlier. That the through-conducted and isolated impulses are basically the same is revealed by experiments in which the through-conduction system was stimulated immediately after an isolated impulse had been recorded. A single stimulus to the

through-conduction system evokes an impulse composed of the single
unit described previously (Fig. la,b). However, if the evoked
impulse closely follows an isolated impulse (Fig. lb) it contains
both the initial component and the slow facilitating muscle action

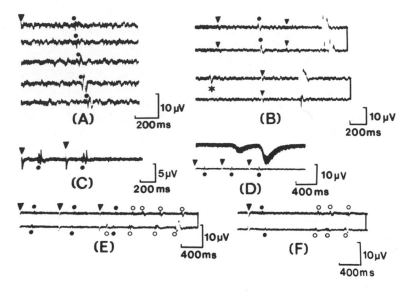

Fig. 1 (A) Electrical activity from *Tubipora*, evoked by 5 stimuli
(▼) at 2 sec. intervals. ● marks the initial invarient component.
(B) Upper traces, simultaneous recordings from two *Tubipora* polyps
of activity evoked by two stimuli (▼). Lower traces, effect of a
stimulus applied after an isolated impulse (✻). (C) Electrical
activity from a *Renilla* polyp. Each of two stimuli (▼) evokes
two impulses. The NN impulses are marked ● . (D) Simultaneous
recordings of luminescence and electrical activity from *Renilla*.
Each stimulus (▼) evokes a NN impulse (●) which facilitates the
luminescence (upper trace). (E) Recordings of bursts of ZNN
impulses (o) from two *Renilla* polyps. The NN impulses which trigger
the burst are marked ● . (F) A burst of ZNN impulses recorded
simultaneously from the top (lower record) and bottom (upper record)
of a *Renilla* polyp. The stimulus (▼) is the last of a series
which all evoked NN impulses (●). (A and B from Anderson, 1976a;
C, D, E, F, from Anderson and Case, 1975.)

potential. This ability for both the through-conducted and isolated impulses to facilitate and activate the same muscles argues
that they have a common input to those muscles, an arrangement
supported by the fact that a single isolated impulse and the first
component of the through-conducted impulses are similarly shaped
(Fig. 1b).

    The evidence suggests that isolated impulses are responsible,
at least in part, for the behaviour of individual polyps. Colonial
responses are coordinated by through-conducted impulses which
invade all polyps. Even though the isolated impulses and through-
conducted impulses share a common pathway, within the polyp, the
isolated impulses cannot invade the colonial conduction system
unless a polyp has been severely stimulated. This asymmetry between
the colonial conduction system and the polyp system can be best
explained by some sort of rectification. Speculation as to how
such rectification could occur is pointless, largely because the
cellular bases of the various conduction systems remain obscure.
If *Tubipora* is placed in a 50% solution of isotonic $MgCl_2$ in sea
water, normal colonial activity and normal facilitating impulses
can be recorded for up to 50 minutes. This insensitivity to Mg^{++}
ions is unusual in the Anthozoa, where exposure to high Mg^{++} concentrations usually results in rapid abolition of all activity
(McFarlane, 1969; Anderson and Case, 1975) and is suggestive of
epithelial conduction. Any interpretations of data from *Tubipora*
should therefore be made with this possibility in mind. This
rectifying process gives *Tubipora* polyps a measure of independence,
yet allows colonial retraction to occur when necessary. A somewhat
similar, but more complex arrangement is found in another alcyonar-
ian *Renilla köllikeri* (order Pennatulacea). *Renilla* colonies are
composed of two classes of polyp, autozooids (or feeding polyps)
and siphonozooids (responsible for water intake and circulation),
that are capable of a variety of autonomous behaviours. Because
of their larger size, the autozooids are more suitable for behavioural studies and unless stated otherwise, the term *Renilla* polyp
will hereafter refer to the autozooid.

    There are four separate colonial responses in *Renilla* (Parker,
1920). Parker's suggestion that three of these, luminescence,
polyp retraction and rachidial contraction were under the control
of a common nerve net has been confirmed (Anderson and Case, 1975).
There are two colonial conduction systems in *Renilla* (Anderson and
Case, 1975). Both can be evoked by electrical (Fig. 1c) or
mechanical stimuli but to date, only the nerve net (NN) has been
linked with any behaviour and identified conduction pathway
(Anderson and Case, 1975; Satterlie, Anderson and Case, in preparation). Simultaneous recordings of electrical activity and
luminescence reveal (Fig. 1d) that a NN impulse always precedes the
onset of luminescence. The luminescent tissue in *Renilla* is
confined to the tentacles and calices of the autozooids and to the

siphonozooids. Despite this location of the effectors in the
polyps, luminescence is a purely colonial response, never having
been observed as part of the behaviour of a single polyp. This
colonial function for luminescence is reflected in the direct
manner by which it is controlled by the NN. A similar argument
could be forwarded for rachidial contraction, a contraction of the
rachis which converts it from its normally flat shape to a domed
one. Considered to be part of an escape response (Kastendiek,
1975), this behaviour is also under direct NN control in a frequency
dependent manner. In contrast, polyp retraction occurs either
colonially or individually. Autozooids retract upon gentle mechanical stimulation without affecting other polyps in the colony.
More severe stimulation of a polyp or electrical or mechanical
stimulation of the rachis evokes colonial polyp retraction. In
contrast to colonial polyp retraction in *Tubipora* and the luminescent response of *Renilla*, which spread as regular concentric
waves, colonial polyp retraction in *Renilla* spreads in a very
random manner. No patterns of spread can be identified. Instead,
widely separated polyps may begin to retract almost synchronously
while the onset of retraction in adjacent polyps may be separated
by several seconds. This irregularity reflects the means by which
this response is controlled.

Electrophysiological recordings reveal the involvement of two
separate conduction systems in the colonial polyp retraction
response. The electrical stimuli evoke NN impulses which invade
all the polyps. After a series of NN impulses have entered a
polyp, a burst of impulses from a second conduction system, the
Zooid Nerve Net (ZNN), is triggered. The ZNN impulses triggered
in one polyp do not invade other polyps and consequently resemble
the isolated impulses of *Tubipora*. They are similarily composed
of two components, a small, invariant initial component and a
second facilitating unit. Since the appearance of ZNN impulses
coincides with polyp retraction, it is probable that the two components have the same significance as suggested for *Tubipora*,
namely, an impulse from a conduction system and a second muscle
action potential from retractor muscles. In this case, the presence
of a well-developed nerve net on the surface of the mesenteric
retractor muscles (Fig. 2) suggests that retraction may be effected
through a nerve net in the same manner as in sea anemones. Hence
the term Zooid Nerve Net for the polyp conduction system of
Renilla is used.

The interaction between the ZNN and the NN is not fully understood, but appears to be the mechanism for maintaining autonomy.
Simultaneous recordings from two different levels of the column of
a polyp reveal that the NN impulses travel up the column and the
ZNN impulses travel down (Fig. 1f), suggesting that the interaction
between the two systems is in the region of the crown or upper
column. Little is known about the nature of the interaction or

the requirements for the ZNN burst. However, recent morphological
descriptions of the through-conducting nerve nets of several
pennatulids (Satterlie, Anderson and Case, in preparation) may
allow morphological identification of the point, and perhaps, the
nature of the interaction. The behaviour of an individual polyp
can be controlled, at least partially, by the actions of the ZNN,
without colonial responses being evoked. When a colonial response
is required, simultaneous activation of the ZNN in all the polyps
could be effected by NN impulses. The interaction is not, however,
the simple 1:1 arrangement found in *Tubipora*. Instead, a variable
number of NN impulses are necessary to trigger the ZNN burst in
different polyps. The function of the additional independence so
obtained by the polyp is unclear but it is conceivable that the
presence of two additional colonial responses with protective functions [luminescence (Bertsch, 1968) and rachidial contraction (Kastendiek, 1975)] may reduce the necessity for total polyp retraction.

Renilla and *Tubipora* are colonies with through-conducted
control of colonial behaviour. In contrast, many colonies have
colonial responses which spread incrementally. That is, a single
stimulus evokes polyp retraction only in a limited area surrounding
the stimulating site, but each additional stimulus increases the
area of response (Horridge, 1957). Three types of incremental
spread have been identified (Horridge, 1957): those in which the
increments increase in size (e.g. *Sarcophyton*), those in which the
increments are of equal sizes (e.g. *Palythoa*) and those in which
the increments decrease in size (e.g. *Porites* and *Goniopora*). In
the latter case, the continual diminution in increment size leads

Fig. 2 Normarski optics light micrograph of a *Renilla* autozooid
mesentery, showing a network of bipolar (arrows) and occasional
tripolar neurons on the surface of a muscle sheet (M).

to a finite maximum area of response. If one assumes that the
conduction system responsible for colonial coordination is a nerve
net and that the spread of polyp retraction mirrors the spread of
activity through that nerve net, the question of how a conduction
system composed of all-or-nothing conducting units could propagate
activity incrementally, must be answered. Various models (Horridge,
1957; Josephson *et al.*, 1961) have been proposed but until recently
little evidence has been available to validate the above assumptions
and test the models.

     Through-conducted electrical activity can be recorded from
Porites (Shelton, 1975a,b) a madreporarian coral with a decreasing
increment colonial response. This finding invalidated the second
of the above assumptions and, at the time, suggested that the
various models were inadequate. As an alternative, Shelton (1975
b) suggested that conduction delay increases for consecutive
impulses, as they were propagated across the colony, were responsible for the observed behaviour by way of the effect the increased
interpulse intervals would have on neuromuscular facilitation in
the polyps.

     Contrasting results have been recently obtained from another
madreporarian coral, *Goniopora lobata*, which has a colonial
response similar to that of *Porites* (Anderson, 1976b). Activity in
the colonial conduction system responsible for colonial coordination spreads incrementally (Fig. 3a) in the same manner as the
behavioural response. That is, if two recording electrodes are
placed in a straight line with the stimulating electrode, and a
train of stimuli applied, initially no activity is recorded. With
repeated stimulation, impulses are first recorded from the nearer
electrode and then the farther. The activity spreads farther and
faster with higher stimulus frequencies, but has never been
recorded from points outside the maximum area of response of the
species. This mirroring of the behavioural response with activity
confirms one of the assumptions made in the early models. Of these
various models, only the computer simulation of Josephson *et al.*
(1961) has successfully produced results comparable to some of those
obtained behaviourally by Horridge (1957). Of the numerous variables tested in this model, the most important proved to be
interneural facilitation, which in that study had been given
variable decay time constants. The results obtained electro-
physiologically (Anderson, 1976b), while not confirming the
presence of interneural facilitation, did not refute it. Since it
is the only known mechanism by which an incremental spread of
activity could be attained, interneural facilitation deserves
inclusion in this and any other models with the provision that it
still awaits direct physiological confirmation.

     The role of interneural facilitation in producing incremental
spread of activity is augmented by another factor, termination of

a pathway by the passage of an impulse. Normally, a train of
stimuli evokes impulses which "spread" to the recording electrodes
(Fig. 3a). However, in 5 percent of all experiments involving
stimulus trains, a single "early" impulse was recorded after the
first stimulus. The early impulse does not apparently affect the
spread of subsequent impulses which are not recorded again until
the correct number of stimuli for the conduction distance and
stimulus frequency have been applied (Fig. 3b). The appearance,
in a conduction system which normally operates incrementally, of
an impulse after the first stimulus is surprising, but it can be
explained on the basis of pathways facilitated by spontaneous
activity (see Anderson, 1976b). To explain the absence of further
impulses after the early one, it is necessary to conclude that the
pathway it traveled must be unavailable to a following impulse.
The duration of the terminating factor must be long enough to

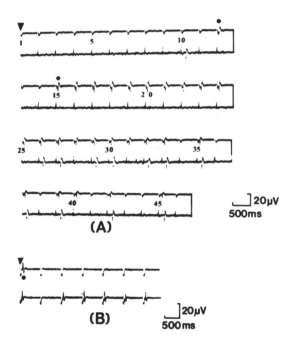

Fig. 3 (A) Electrical activity from *Goniopora*. The stimulating
site and the two recording sites formed a straight line. Initially,
the stimuli (▼) do not evoke recordable impulses. After additional
stimulation, impulses (●) appear first at the nearer recording site
(upper record) and then at the farther site. (B) An "early"
impulse evoked by the first of a series of stimuli. The two
records are continuous. (A and B from Anderson, 1976b.)

exclude a pathway to consecutive impulses evoked at stimulus frequencies of 1-2/sec. While refractoriness of the conducting units could be involved it is unlikely that it could attain values of the order of 1000 or 500 msec. Alternatively, fatigue of the junctions within that pathway could be longer lasting and therefore more effective.

A train of impulses applied to a single point on a *G. lobata* colony eventually produces retraction of all the polyps in a circle 3 cm. in diameter. With *G. columna*, the maximum area of response is 6 cm. in diameter, while other unidentified species have similar variations in their maximum areas of response. Since the spread of polyp retraction is effected by the action and interaction of interneural facilitation, pathway termination and, perhaps, other variables, any species specific variations in the values of these parameters would produce species specific variations in the spread of the colonial response. In their computer simulation, Josephson *et al.* (1960) found similar results when the values of their several variables were changed.

The function of incremental spread is apparently to restrict the spread of polyp retraction to small areas of the colony on a frequency dependent basis, rather than having the whole colony responding to focal stimulation. The extreme size of some madreporarian colonies emphasizes this. In a large colony, such limitations on the extent to which a colonial response spreads could be sufficient to negate the necessity for mechanisms to maintain the autonomy of individual polyps. However, studies of the behaviour of individual *Goniopora* polyps reveal that autonomous behaviour is also characteristic of these colonies, although the mechanism by which it is attained was not studied in this species.

These various results indicate that, at least in corals which display behavioural coloniality, there are special colonial conduction systems linking the individuals in the colonies. In cases where the conduction system operates incrementally, it appears that inherent properties of the conduction system function to produce an incremental spread of activity. In others, where activity is through-conducted, effectors involved in colonial responses are connected directly to the colonial conduction system while effectors used in both autonomous and colonial behaviour are linked indirectly, with the nature of the indirect linkage serving to dictate the patterns of colonial behaviour. The requirement for colonial coordination in anthozoans appears to be the development of at least one purely colonial conduction system. The theory that colonial behaviour could be coordinated by a continuous extension of the nerve net of the original budding ancestor does not appear to be valid.

This work was supported by O.N.R. Contract N00014-75-C-0242 and N.S.F. Grant BMS 72-01971 to Dr. James F. Case, and N.S.F. Grants OFS 74-01830 and OFS 74-23242 to the Alpha Helix Southeast Asian Bioluminescence Expedition (Scripps Institute of Oceanography).

REFERENCES

Anderson, P.A.V., 1976a. The Electrophysiology of the Organ-Pipe Coral, *Tubipora muscia*. In press, *Biol. Bull*. 150.

Anderson, P.A.V., 1976b. An Electrophysiological Study of Mechanisms Controlling Polyp Retracting in Colonies of the Scleractinian Coral *Goniopora lobata*. In press, *J. Exp. Biol*.

Anderson, P.A.V. and J.F. Case, 1975. Electrical Activity Associated with Luminescence and Other Colonial Behavior in the Pennatulid *Renilla köllikeri*. *Biol. Bull*., 149:80-95.

Batham, E. J. and C.F.A. Pantin, 1950. Inherent Activity in the Sea Anemone *Metridium senile* (L). *J. Exp. Biol*., 27:290-301.

Bertsch, H., 1968. Effects of Feeding by *Armina californica* on Bioluminescence of *Renilla kollikeri*. *Veliger*, 10:440-441.

Buisson, B., 1973. Données sur le Comportement Rythmique du Polype Isolé de la Colonie de *Veretillum cynomorium* Pall. (Cnidaria Pennatularia). *C. R. Acad. Sci. Paris*, 277:1541-1544.

Horridge, G.A., 1957. The Co-ordination of the Protective Retraction of Coral Polyps. *Phil. Trans. Roy. Soc. Lond. B.*, 675: 495-529.

Hyman, L.H., 1940. The Invertebrates. Protozoa Through Ctenophora. McGraw-Hill Book Company, New York.

Josephson, R.K., R.F. Reiss and R.M. Worthy, 1961. A Simulation Study of a Diffuse Conducting System Based on Coelenterate Nerve Nets. *J. Theoret. Biol*., 1:460-486.

Kastendiek, J., 1975. The Behavior, Distribution and Predator-Prey Interactions of *Renilla köllikeri*. Ph.D. Thesis, University of California, Los Angeles. 235 pp.

McFarlane, I.D., 1969. Two Slow Conduction Systems in the Sea Anemone *Calliactis parasitica*. *J. Exp. Biol*., 51:377-385.

McFarlane, I.D., 1973. Spontaneous Contractions and Nerve Net Activity in the Sea Anemone *Calliactis parasitica*. *Mar. Behav. Physiol*., 2:97-113.

Parker, G.H., 1920. Activities of Colonial Animals. II. Neuro-muscular Movements and Phosphoresence of *Renilla*. *J. Exp. Zool*., 31:475-515.

Robson, E.A. and R.K. Josephson, 1969. Neuromuscular Properties of Mesenteries from the Sea Anemone *Metridium*. *J. Exp. Biol*., 50:151-168.

Shelton, G.A.B., 1975a. Electrical Activity and Colonial Behaviour in Anthozoan Hard Corals. *Nature, Lond*., 253:558-560.

Shelton, G.A.B., 1975b. Colonial Behaviour and Electrical Activity in the Hexacorallia. *Proc. Roy. Soc. Lond. B.*, 190:239-256.

MORPHOLOGY AND ELECTROPHYSIOLOGY OF THE THROUGH-CONDUCTING SYSTEMS IN PENNATULID COELENTERATES

Richard A. Satterlie, Peter A. V. Anderson and James Case

Dept. of Biological Sciences, University of California

Santa Barbara, California 93106

Recent electrophysiological investigations of colonial anthozoan nervous systems (Anderson and Case, 1975; Shelton, 1975) have supplemented behavioral studies which found that colonial control was either through-conducted (Parker, 1920; Nicol, 1955; Horridge, 1957; Buck, 1973), or restricted to a finite portion of the colony with subsequent incremental spread (Horridge, 1957). In the Pennatulacea, colonial nervous systems have been found to be through-conducting (Anderson and Case, 1975; Shelton, 1975). A possible morphological basis for such coordination, namely ecto-dermal, mesogleal, and endodermal nerve nets, has been described within the rachis of some pennatulids (Titschack, 1968, 1970; Buisson, 1970). However, there is no conclusive evidence to link these nerve nets with through-conducted colonial coordination, and the exact morphological circuitry has yet to be elucidated. Here we correlate the morphology of possible conducting elements with the colonial behavior and the electrophysiology of the through-conducting systems of five species of pennatulids as a contribution towards determination of the anatomical basis for through-conduction.

MATERIALS AND METHODS

Specimens of *Acanthoptilum, Ptilosarcus, Renilla, Stylatula,* and *Virgularia* were collected locally by divers. Techniques for animal maintenance and electrophysiological recording are reported in Anderson and Case (1975). Specimens for histological examination were either fixed in Bouin's fluid and stained by the One-Step Trichrome method of Gabe (1968), or fixed and stained according to the Rapid Golgi technique (Polyak, 1942). Tissue for electron microscopy was fixed in 2 percent glutaraldehyde and 1 percent OsO_4

in cacodylate buffer. Small tissue pieces were decalcified overnight by the method of Dietrich and Fontaine (1975), dehydrated in
a graded ethanol series to propylene oxide, and embedded in
araldite. Sections were cut with glass knives on a Porter Blum
MT-2 ultramicrotome and stained with uranyl acetate and lead
citrate prior to examination (Siemens Elmiskop 1A).

RESULTS

The extent and speed of rachidial contraction varies among the
pennatulids and provides a basis for separation of colonial withdrawal responses into two general categories: (1) rapid, complete
withdrawal of the entire rachis into the substrate; and (2) slow,
continuous rachidial contractions which, once sufficiently initiated,
proceed without further stimuli to culminate in deflation of the
colony. *Stylatula* and *Virgularia* represent the former category and
Acanthoptilum, Renilla, and *Ptilosarcus*, the latter.

Both kinds of colonial contraction are apparently mediated by
well developed longitudinal musculature associated with flattened
endodermal tubes, embedded in mesoglea and surrounding the central
water canals of the rachis (Figure 1a). Exceptions are *Acanthoptilum*, in which poorly developed longitudinal muscle bundles limit
rachidial contractility, and *Renilla*, in which longitudinal muscle
bundles radiate in a fan-like manner from the dorsal track of the
rachis and surround the endodermal solenia whose side branches
interconnect the polyp chambers. By comparison with the longitudinal
muscle bundles, the circular musculature of the pennatulid rachis
is poorly developed. Circular muscle cells comprise the inner
portion of the longitudinal muscle tubes, and sometimes the outermost layer of the water canal endoderm (Figure 1a). The circulars

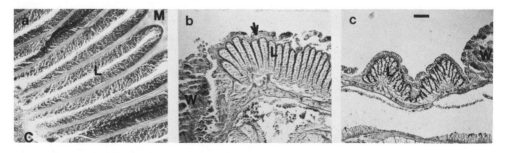

Figure 1. Regional differences in *Virgularia* rachis. (a) Zone 4,
(b) Zone 2, (c) Zone 1. C-circular muscle bundles, L-longitudinal
muscle bundles, M-mesoglea. Arrow shows position of sub-ectodermal
mesoglea. Cross sections. Scale = 0.1 mm.

appear to act against the hydrostatic pressure of the central
canals during rachidial peristalsis and are jointly active with the
longitudinals during protective contractions, aiding in rapid
rachidial water evacuation.

Regional differences in external and internal anatomy, and in
physiological responses were immediately evident in the rachis of
Stylatula and *Virgularia*. Consequently we defined rachidial zonation on the basis of external characters as follows;Zone 1, distal
rachis bearing mature pinnae; Zone 2, distal rachis bearing
developing and immature pinnae; Zone 3, proximal rachis devoid of
pinnae but centrally impaled by the axial spine; Zone 4, peduncle.
These differences are subtle in *Acanthoptilum* and *Ptilosarcus*.
This zonation is most pronounced in the development of longitudinal
musculature of the rachis and peduncle. The muscle tubes of Zone 4
are extensive, occupying approximately one half of the rachidial
radius in cross section, and completely encircling the central
canals in a spoke-like array (Figure 1a). The muscle tubes of Zone
3 also surround the water canals but are reduced, occupying only
about one fourth of the rachidial radius. The musculature of the
distal rachis, Zones 2 and 1, is restricted to dorsal and ventral
clusters which become smaller in the more distal regions (Figures
1b, 1c). Zone 3 and especially the peduncle are responsible for
rapid retraction of *Stylatula* and *Virgularia* during protective
behavior.

The rachidial musculature of pennatulids is activated by
through-conducted impulses which, whether elicited by electrical or
mechanical stimuli, are conducted either up or down the rachis of
sea pens, and in all directions in the rachis of *Renilla*. Nerve
net impulses were recorded from the rachis (*Stylatula* and *Virgularia*),
the siphonozooid strip (*Ptilosarcus*), and the pinnae (*Acanthoptilum*).
Impulse characteristics are shown in Figure 2 and Table 1 (*Renilla*
data from Anderson and Case, 1975). Conduction velocity measurements are presented as latency recorded between two electrodes with

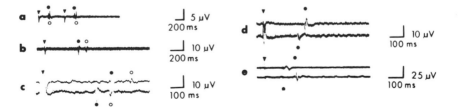

Figure 2. Suction electrode recordings of through-conducted nerve
net impulses. (a) *Renilla*, (b) *Acanthoptilum*, (c) *Ptilosarcus*,
(d) *Stylatula*, (e) *Virgularia*. (▼)-stimulus artifact, (•)-nerve net
impulse, (o)-impulse from second conduction system.

the interelectrode distance adjusted to that of a relaxed animal
to compensate for rachidial contraction. Measurements from
Stylatula and *Virgularia* are listed according to zonation. No
recordings were made from the peduncle due to its highly contractile
nature.

All five species possess nerve elements in the ectoderm,
mesoglea, and endoderm. In the ectoderm, nerve cells run in two
distinct orientations. The majority are directed inward toward
the center of the rachis, penetrating the mesoglea in tissue
bridges (Figure 3). In addition, a few nerve cells run along the
ectoderm-mesoglea interface. These latter nerves, which occasionally enter the mesoglea, appear to link the inward directed
connectives. *Acanthoptilum* has, in addition, a loose accumulation
of nerve cells well within the ectodermal layer. Synapses are
present, although few in number (Figure 8c).

In the mesoglea, nerve elements are well developed with nerve
cells usually in bundles wrapped by amoebocytes, and having two
main orientations, similar to the ectodermal nerves. Some bundles
run inward toward the center of the rachis (Figures 4, 5a) in
channels between the longitudinal muscle tubes. From these bundles
smaller nerve cells diverge to enter mesogleal slips between the
longitudinal muscle cells (Figure 6). Some of these nerves penetrate the muscle layer while others appear to terminate as nerve
endings in the mesoglea adjacent to the muscle cell processes
(Figure 8d). The main nerve bundles continue inward to the circular
muscle layer and form connections similar to those in the longitudinal muscle bundles (Figure 5a).

Table 1. Colonial nerve net impulse characteristics. The nerve
net impulse was used in those animals with two conducting systems.

		Maximum Impulse Height (µV)	Conduction Velocity (cm/sec)	
Renilla		4	7–8	(16°C)
Acanthoptilum		11	10–15	(19°C)
Ptilosarcus		15	19–23	(18°C)
Stylatula	Zone 1	40	75–94	(16–18°C)
	Zone 2	40	95–118	(16–18°C)
	Zone 3	30	99–142	(16–18°C)
Virgularia	Zone 1	120	74–88	(15–18°C)
	Zone 2	80	81–94	(15–18°C)
	Zone 3	50	33–43	(15–18°C)

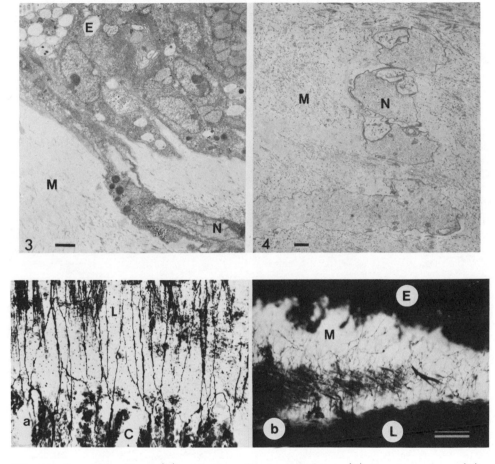

Figure 3. Neurites (N) passing from ectoderm (E) to mesoglea (M) in *Virgularia* cross section. Scale = 1μ.

Figure 4. Mesogleal nerves (N) directed toward centrally located longitudinal muscle bundles of rachis in *Virgularia*. Longitudinal section just under ectoderm. Scale = 1μ.

Figure 5. Mesogleal nerve cells. Golgi impregnation. (a) Longitudinal section of rachis-Zone 3 (*Virgularia*), (b) longitudinal section of rachis-Zone 3 (*Stylatula*). C-circular muscles, E-ectoderm, L-longitudinal muscles, M-mesoglea. Scale = 0.1mm.

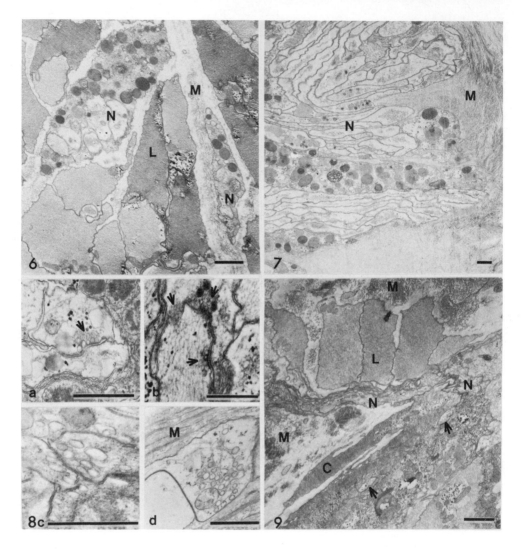

Figure 6. Mesogleal channels (M) between longitudinal muscle cells (L) in *Renilla*. Cross section. N-neurite bundles.

Figure 7. Nerve bundles (N) in the mesoglea of *Ptilosarcus*. Cross section.

Figure 8. Synaptic types within the mesoglea (a,b,d) and ectoderm (c) of *Virgularia* (a,b), *Acanthoptilum* (c), and *Stylatula* (d). Arrows indicate presumed direction of transmission.

Figure 9. Mesogleal nerve cells (N) penetrating endodermal tissue in *Renilla*. Cross section. C-circular muscle cells, L-longitudinal muscle cells, arrows indicate neurites of endodermal nerve net. Scales = 1μ.

Between the ectoderm and the longitudinal muscle layer, the
nerve net ramifies throughout the mesoglea in all parts of the
rachis, pinnae and polyps (Figure 5b). This is best seen in
Acanthoptilum, Virgularia (except zone 4), and *Stylatula*. Zone 3
of *Virgularia* has well developed inward directed nerve bundles, but
lacks well developed longitudinally oriented nerves (Figure 5a).
In *Renilla*, the nerve net is distributed throughout the mesoglea of
the coenenchyme with only a slight increase in density near the
dorsal and ventral ectodermal layers. *Ptilosarcus* has very few
nerve cells immediately under the ectoderm although large accumulations of nerves are seen deep in the fleshy rachis (Figure 7).

Interneural synapses are present throughout the mesoglea, the
most common type being polarized, "en passant" contacts. Polarized,
reciprocal synapses are occasionally seen; symmetrical synapses are
rare. Frequently, accumulations of vesicles, seemingly identical
to those found at synapses are found, but without evidence of intercellular contacts. Representative synaptic structures are shown in
Figure 8.

While nerve elements are present in the endoderm beneath the
circular muscle layer, they are rare. No neuromuscular synapses
have been found. Mesogleal nerves penetrate the endoderm (Figure 9)
although their relationship to the endodermal nerve cells is unclear.

                              DISCUSSION

The basic theme of neural architecture in pennatulid coelenterates is a combined ectodermal-mesogleal synaptic nerve net which
innervates the entire rachidial polyp population and the colonial
musculature, as well as a poorly developed endodermal nerve net
receiving connections from the mesogleal nerves. This is in general
agreement with results of other authors (Titschack, 1968, 1970;
Buisson, 1970). The mesoglea and ectoderm of the rachis, with
associated neural elements, are continuous with the same elements
of the pinnae and polyps. In addition, distinct polyp nervous
systems are present (Anderson and Case, 1975).

Electrical recordings from the rachis and polyps of *Acanthoptilum, Ptilosarcus, Stylatula,* and *Virgularia* confirm previous
reports that the colonial nervous systems of pennatulids are
through-conducting (Anderson and Case, 1975; Shelton, 1975). A
single stimulus applied anywhere on the rachis produces an impulse
which invades the entire colony. The through-conducted impulses of
Acanthoptilum and *Ptilosarcus* are followed by a second through-
conducted impulse separable from the first by threshold and conduction
velocity. Similar results were obtained from *Renilla* (Anderson and
Case, 1975) and *Pennatula phosphorea* (Shelton, 1975). Evidence for
two colonial conduction systems was not obtained for *Stylatula* or

Virgularia, or in a different species of *Virgularia* (Shelton, 1975).
Of the pennatulids tested to date, those with two conduction
systems exclusively fall into the category with slow colonial
withdrawal responses, while those with one recordable conduction
system are classified as fast.

Several factors can determine the conduction velocity in a
nerve net. Most apparent are nerve cell density, orientation,
length, shape (bi-, tri-, or multipolar), neurite diameter, and
synaptic properties (number, type, physical and chemical characteristics). The maximum neurite diameter of mesogleal nerve elements
in the pennatulids is closely correlated with conduction velocity
and peak-to-peak impulse height (Table 2). Although impulse size
and shape are variable with present recording techniques, a consistent pattern is nevertheless present. The correlation extends, as
well, to those pennatulids studied by other investigators.

The effects on conduction velocity of nerve cell density and
orientation are evident from examination of different regions of
Stylatula and *Virgularia.* In Zones 1 and 2, the density of nerve
cells in the subectodermal mesoglea is reduced, as nerves diverge
and connect to nerves of the pinnae and polyps. This possibly
contributes to the reduced conduction velocity in the more distal
regions of the rachis. Zone 3 of *Virgularia* has a conduction
velocity much lower than might be expected were it not for the
predominant orientation of nerves inward toward the longitudinal
muscles and the resultant decreased density in the longitudinal axis
of the rachis.

Distributional and structural differences in the ectodermal-
mesogleal nerve nets of various pennatulids, as well as in different

Table 2. Correlation between electrophysiological and morphological
characteristics of through-conducting systems of pennatulids.
(a) Anderson and Case, 1975; (b) Shelton, 1975; (c) Buisson, 1970
(neurite diameter calculated from figures).

	Impulse Height (µV)	Conduction Velocity (cm/sec)	Neurite Diameter (µ)
Renilla köllikeri	3-4[a]	6-8[a]	1.75
Pennatula phosphorea	7-8[b]	7.5[b]	----
Acanthoptilum gracile	8-11	10-15	1.66
Veretillum cynomorium	----	14-15[c]	±1.50[c]
Ptilosarcus guerneyi	10-15	19-23	2.20
Virgularia mirabilis	10-15[b]	40-60[b]	----
Stylatula sp	30-40	75-142	3.70
Virgularia sp	50-120	33-94	5.10

zones of the same organism, are consistent with the demonstrated electrophysiological characteristics of their through-conduction systems. This correlation suggests that the observed mesogleal-ectodermal nerve net is in fact the cellular basis for through-conduction in the various species studied and may prove to be a useful base from which to study the organization of conducting systems which are not through-conducting.

This research was supported by NSF grant BMS 72-01971.

REFERENCES

Anderson, P.A.V., and J.F. Case, 1975. Electrical activity associated with luminescence and other colonial behavior in the pennatulid *Renilla köllikeri*. *Biol. Bull.*, 149:80-95.

Buck, J., 1973. Bioluminescent behavior in *Renilla*. I. Colonial responses. *Biol. Bull.*, 144:19-42.

Buisson, B., 1970. Les supports morphologiques de l'integration dans la colonie de *Veretillum cynomorium* Pall. (Cnidaria, Pennatularia). *Z. Morph. Tiere*, 68:1-36.

Dietrich, H.F., and A.R. Fontaine, 1975. A decalcification method for ultrastructure of echinoderm tissues. *Stain Tech.*, 50: 351-354.

Gabe, M., 1968. *Techniques Histologiques*. Masson and Cie, Paris.

Horridge, G.A., 1957. The co-ordination of the protective retraction of coral polyps. *Phil. Trans. Roy. Soc. Lond. B.*, 240:495-529.

Nicol, J.A.C., 1955. Nervous regulation of luminescence in the sea pansy *Renilla köllikeri*. *J. Exp. Biol.*, 32:619-635.

Parker, G.H., 1920. Activities of colonial animals. II. Neuro-muscular movements and phosphorescence in *Renilla*. *J. Exp. Zool.*, 31:475-515.

Polyak, S., 1942. *The Retina*. University of Chicago Press, Chicago.

Shelton, G.A.B., 1975. Colonial conduction systems in the anthozoa: Octocorallia. *J. Exp. Biol.*, 62:571-578.

Titschack, H., 1968. Über das Nervensystem der Seefeder *Veretillum cynomorium* (Pallas). *Z. Zellforsch.*, 90:347-371.

Titschack, H., 1970. Histologische Untersuchung des mesogloealen Nervenplexus der Seefedern *Pennatula rubra* (Ellis) und *Pteroides griseum* (Bohadsch). *Vie Milieu*, 21:102.

PROBABLE FUNCTIONS OF BIOLUMINESCENCE IN THE PENNATULACEA

(CNIDARIA, ANTHOZOA)

James G. Morin

Department of Biology, University of California

Los Angeles, California, U.S.A. 90024

Within the coelenterates the ability to emit light is widely distributed; most of the major coelenterate taxa contain both luminescent and non-luminescent species (Harvey, 1952; Morin, 1974). Production of light within the luminescent species is biochemically, structurally, and physiologically complex (Harvey, 1952; Nicol, 1960; Herring, 1972; Buck, 1973; Morin, 1974; Morin and Reynolds, 1974; Cormier et al., 1975; Anderson and Case, 1975). The complicated mechanisms responsible for the well-controlled light emission suggest that the light serves some functional role in the life of these organisms. In this paper I discuss the characteristics of the luminescent signal in four pennatulaceans (sea pens) and then consider the ways that the light is potentially employed by these pennatulaceans, and possibly by all luminescent benthic cnidarians.

CHARACTERISTICS OF THE BIOLUMINESCENT SIGNAL

The following section describes the ways that bioluminescence is expressed in space and time for four representative pennatulaceans from southern California: Ptilosarcus gurneyi, Acanthoptilum gracile, Stylatula elongata, Renilla kollikeri.

Spatial Orientation of the Luminescent Tissues

Light emission results from a complex intracellular reaction within specific luminescent cells (photocytes) (Morin and Hastings, 1971a,b; Morin, 1974; Cormier et al., 1975). These cells can be visualized in living or appropriately preserved pennatulaceans through the use of fluorescence microscopy (Titschack, 1966; Morin,

1974; Morin and Reynolds, 1974; Morin et al. a, in preparation).
The photocytes are endodermal cells, 10-20 μm in diameter, and they
usually possess one or more neurite-like cytoplasmic projections.
The cells are often aggregated in particular sites where they
anastomose and interdigitate profusely. These distinct clusters
lie in specific parts of the autozooids and siphonozooids but nowhere else in the colony. In each species the rachidial morphology
and the photocyte distribution profoundly affect the way in which
the light is outwardly projected. Sea pens are usually found
anchored in relatively homogeneous soft sediment substrates in areas
where there are significant bidirectional (short-term, surge generated) or unidirectional (long-term, flowing in one direction for
minutes, hours or days) water currents. These currents also influence the direction of light emission (Fig. 1).

     Ptilosarcus gurneyi (Fig. 1a). This sea pen is fleshy and
robust (to 40 cm in height). Many semitransparent autozooids are
borne on one edge of each of about 30 paired lateral leaves of the
upright rachis. Two rows of numerous fixed siphonozooids are
vertically located on the rachidial stalk on the side opposite the
autozooids. In southern California Ptilosarcus is found in deep
water (>30m) where the dominant currents are unidirectional. The
colony orients with the autozooid surface of the leaves directed
away from the prevailing current and the siphonozooids directed
into the current. Both of the polyp types are brilliantly luminescent. Because of the orientation on the vertical colony, the
siphonozooids, which have numerous photocytes in each chamber,
project their light into the current. The photocytes within the
autozooids are situated in the lateral sides of each of the tentacles. This morphology and the orientation of the colony cause
the light from the autozooids to be directed primarily downcurrent
and laterally with some light shining upcurrent between the leaves.
Since the whole colony often bends over in strong currents, the
light from the autozooids is also directed somewhat downward toward
the substrate.

     Acanthoptilum gracile (Fig. 1b). This species is very long
(to 1m), slender and flexible. As in Ptilosarcus the autozooids
in Acanthoptilum are located on numerous paired leaves. The siphonozooids are rather sparcely distributed on the opposite side and
are more abundant near the base of the rachis. Acanthoptilum
occurs in deep water (>25m) or in shallow embayments where water
currents are primarily unidirectional. As with Ptilosarcus,
Acanthoptilum orients with the autozooids facing away from the
current and the siphonozooids directed into the current. The photocytes in the siphonozooids are abundant in the laterally orienting
chambers but absent in the dorsal and ventral chambers. Therefore,
the light from the siphonozooids is directed into the current and
somewhat laterally. Luminescence in the autozooids is associated
with the nonretractile, semitransparent calyx which surrounds each

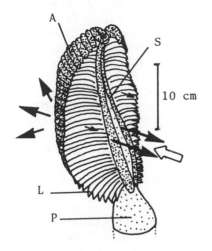

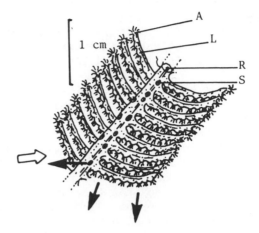

a. Ptilosarcus gurneyi   b. Acanthoptilum gracile

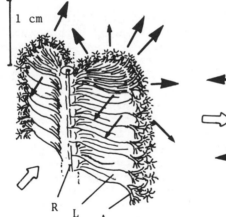

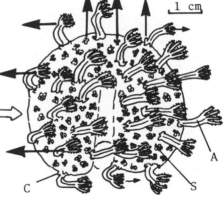

c. Stylatula elongata   d. Renilla kollikeri

Figure 1. Orientation of luminescence in pennatulaceans in relation to current flow. a. Whole Ptilosarcus gurneyi with peduncle in
substrate. b. Portion of an Acanthoptilum gracile rachis showing
nine pairs of leaves. c. Portion of a Stylatula elongata rachis showing eight pairs of leaves. d. Renilla kollikeri colony from above
and showing only half of the autozooids and siphonozooids.
Solid arrows indicate direction of luminescence, open arrows indicate
direction of water current. A, autozooids; C, calyx teeth of autozooid; L, leaf; P, peduncle; R, axial rod; S, siphonozooids.

polyp like a long, rigid collar. The photocytes are restricted to
the two basally directed chambers of each calyx. The effect is
that the light is primarily directed downward toward the substrate.
The marked flexibility of the stalk causes the colony to bend over
even in relatively weak currents so that the light from the autozooids is also directed into the current.

   Stylatula elongata (Fig. 1c). This sea pen is very rigid and
erect (up to 60 cm in height) so that it remains straight even in
strong currents. The multiple leaves form an approximate hemisphere
around the stalk with the numerous autozooids splaying out along
the outward edge of each leaf. There are no siphonozooids. Stylatula occurs in relatively shallow water (5 to 40m) where either uni-
or bidirectional currents prevail. In bidirectional currents the
colony pivots so that the autozooids consistently face away from
the oncoming current. The photocytes in the semitransparent polyps
lie in a cluster along the upper column of each mesentery. This
orientation results in a light emission in all directions from the
colony, especially downcurrent, laterally, and upcurrent between
the leaves. There is also a substantial upward and downward
illumination.

   Renilla kollikeri (Fig. 1d). The rachis of Renilla (sea pansy)
is disk-shaped and the autozooids and clusters of siphonozooids are
distributed over the upper rachidial surface. Renilla is found in
shallow water (intertidal to 20m) where it is exposed to moderate
to strong bidirectional wave surge (up to several m/sec). The
Renilla rachis avoids the surge by being flattened next to the sand
surface. The autozooids, however, project several centimeters
above the rachis into the current. With each passing surge the
semitransparent autozooids bend over and the tentacles form a cone
above the mouth away from the prevailing current and often touch
the rachis. The photocytes are primarily situated in most of the
chambers of the siphonozooid clusters, in the calyces at the base
of the autozooids, and in the aboral axial regions of the tentacles
of the elongate and flexible autozooids. When luminescence is
elicited from Renilla this distribution of photocytes provides
light output upward from the siphonozooid clusters and calyces of
the autozooids, and into the prevailing water currents and slightly
laterally from the backward directed photocytes at the base of the
tentacles of the autozooids. Owing to the autozooid transparency,
some light is also directed orally, that is downcurrent.

   The observed spatial effect of the luminescent emission in all
four species is for a brilliant and conspicuous pattern to be projected away from the colony in particular orientations depending on
the animal. However, owing to the transparency of particular tissues,
some light is sent in most directions.

Kinetics of the Light Emission

Light emission from all four pennatulacean species shows some temporal similarities. 1) Luminescence is intermittent and is emitted almost exclusively in response to some external stimulation. 2) Local stimulation of a single polyp produces luminescence from only the stimulated polyp. However, with increased intensity of stimulation or stimulation of the rachis, all of the photocytes of the colony will luminesce as a wave of through conducted excitation, carried by the nerve net system (Anderson and Case, 1975), passes along the colony in all directions from the point of stimulation. Thus, depending on the point and degree of stimulation, the colony can respond with restricted local luminescence or a colony-wide conducted wave of luminescence. As the initial wave passes over the colony the autozooids retract into the rachis. 3) With electrical stimulation there is usually a one to one relationship between the stimulus and the response; rarely is there repetitive activity following each stimulus. 4) Each stimulus, subsequent to the one which first elicits light, usually produces a wave of light which is brighter than the previous one. This effect is caused by a facilitation of light output from each photocyte and not recruitment. The effect of these features in the four pennatulaceans is the production of travelling waves of light which are proportional to the input stimulus; strong stimuli initiate highly facilitated waves of light while weak stimuli initiate weak waves of light or even only local luminescence from individual polyps.

The major differences between the four species of sea pens are in 1) the duration of the luminescent flash, 2) its conduction velocity, and 3) the resultant width of the luminescent wave (Table 1, Fig. 2). At one extreme is Acanthoptilum where the duration of the flash is short while the conduction velocity is slow. The effect of these two characteristics yields a narrow band of light slowly travelling the length of the colony away from the point of stimulation (Fig. 2b). Compared to Acanthoptilum, Renilla shows a slightly wider band of luminescence but a similar slow conduction velocity (Fig. 2a). The wider band is a result of a considerably

Species	Flash Duration (msec)	Conduction Velocity (cm/sec)	Width of Luminescence (cm)
Acanthoptilum gracile	100	4-6	0.6
Renilla kollikeri	275	5-7	2.0
Ptilosarcus gurneyi	140	25	3.5
Stylatula elongata	250-300	125-140	40

Table 1. Kinetics of luminescence in some pennatulaceans.

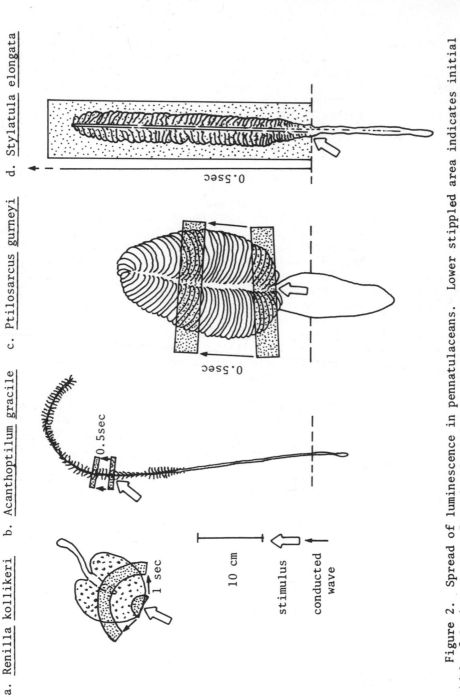

Figure 2. Spread of luminescence in pennatulaceans. Lower stippled area indicates initial width of luminescence. Second stippled area above indicates the location of the luminescent wave after the time indicated. In Stylatula (d) the wave would have travelled off scale.

longer flash duration. On the other hand, Ptilosarcus shows a width
of luminescence similar to that of Renilla but achieves it by the
converse combination of a fairly fast conduction velocity and a
short flash (Fig. 2c). Thus, both Renilla and Ptilosarcus show a
fairly wide band of luminescence but the wave travels faster in
Ptilosarcus. At the opposite extreme is Stylatula which has a
slow flash and a very rapid conduction velocity which produces a
very wide band of luminescence (Fig. 2d). The result is that the
entire colony is luminescent at the same time. This complete
luminescence in Stylatula is accompanied by a rapid withdrawal of
the whole colony into the sand. None of the other three pennatulaceans has the ability to withdraw rapidly into the substrate.

FUNCTIONS OF THE BIOLUMINESCENCE

The overall result of the spatial and temporal luminescent
displays from these pennatulaceans is a nocturnal signal which
1) is intermittent, 2) is simple, 3) is bright and conspicuous,
4) shows a maximum contrast with the dark surround, 5) is predictable, 6) demonstrates a distinct orientation, 7) shows movement
which leads away from the stimulus, 8) indicates the size of the
colony, and 9) is proportional to the stimulus. This well-controlled
and distinct luminescence from the emitter clearly suggests that the
light is a signal directed outwardly toward other organisms. The
light signal could be directed toward any of three primary receivers:
1) other members of the same species, 2) potential prey organisms,
and/or 3) potential predators.

Any receiver of the signal must have sufficient photoreceptive
equipment to collect and process the information, and, since the
signal is elicited upon contact, the receiver must also possess a
degree of motility. Therefore, other members of the same species
would not seem to be the target organisms since pennatulaceans are
sedentary and do not frequently physically contact one another, do
not have well-developed photoreceptors, and do not appear to respond
to induced luminescence from other individuals or artifical flashes
of light. On the other hand, visually orientating, positively
phototatic, motile prey organisms might be attracted to the light.
However, since luminescence is elicited only intermittently upon
direct contact and not continuously, the signal would not appear
to be used for attracting prey. Further more, there is evidence
that alcyonarians have difficulty feeding on motile prey organisms
but instead selectively prey on nonmotile, current-carried phytoplankton and detritus (Roushdy and Hansen, 1961; Kastendiek, 1975).

There are a number of predators which are known to prey on
cnidarians (Salvini-Plawen, 1972). Often these associations are
highly specific and frequently the cnidarians have evolved special

mechanisms for avoiding or reducing the effects of the predation
(Ross, 1974). Populations of Ptilosarcus gurneyi and Renilla
kollikeri have been studied for their interactions with their
asteroid and gastropod predators (Birkeland, 1974; Kastendiek,
1975). These predators are slow moving, feed both day and night,
and apparently do not locate their prey by visual means. Both
Ptilosarcus and Renilla can successfully avoid these specialist
predators by means of tactile defenses (presumably nematocysts),
specific behavioral activities, or spatial refuges (Birkeland, 1974;
Kastendiek, 1975). The nudibranch Armina californica, a specific
pennatulacean predator, does elicit luminescence from Renilla upon
attacking it but is not deterred by the luminescence (Bertsch, 1968).
Thus, bioluminescence is probably of little value in protecting sea
pens from specialized, continuously feeding, nonvisually orienting
predators such as asteroids, gastropods and other similar predators.

On the other hand, many predatory demersal fishes and decapod
crustaceans, which often orient visually and are also highly motile,
become significantly more active on soft sediment bottoms at night
than during the day (Morin et al. b, in preparation). These organisms must frequently confront sea pens at night. However, most of
the nocturnal, sand bottom, demersal fish that have been studied
feed principally on small crustaceans, annelids and molluscs (Ford,
1965; Greenfield, 1968; Frey, 1971), but they potentially could
produce substantial damage if they were to intentionally or
accidentally prey on a pennatulacean colony. In a few cases sea
pen remains have been found in the stomachs of nocturnally active
demersal fish (Ford, 1965; N. Davis, personal communication), and
pennatulaceans with minor damage to the colony attributable to these
fish or possibly to crabs have occasionally been observed in the
field (personal observation; Kastendiek, 1975). Nocturnal fish
are very light sensitive (Woodhead, 1966; Blaxter, 1975), and any
significant impact on a pennatulacean by one of these organisms
would initiate a detectable bioluminescent response commensurate
with the intensity of the encounter. This light signal would have
the required orientation, contrast, conspicuousness and movement
to be received by the intruder as a direct signal.

The induced bioluminescent signal could deter subsequent
predation by nocturnal fish in any or all of three possible ways.
First, the potential predator could be startled or frightened by
the bright and moving signal. Secondly, if the flashes were sufficiently bright the predator could be momentarily blinded and dis-
oriented. Since the pennatulaceans are usually found in significant currents such a disorientation, along with the cessation of
the luminescence, would separate the two organisms and make subsequent relocating more difficult. Thirdly, the light could be
intended as a warning illumination to the predator. Since pennatulaceans possess nematocysts which are probably effective against

asteroids (Kastendiek, 1975) it is reasonable to assume that they are also effective against fish. Luminescence could be used as a first line of defense, warning the intruder of subsequent danger by neamtocysts. An added complexity could be that the light might also be used as a "burglar alarm" (Burkenroad, 1943), where second order predators are alerted by the light to the presence of their prey which have attacked the pennatulacean. While there are very little data on the feeding habits and visual sensitivity of nocturnal predatory crustaceans, they would probably elicit and respond to pennatulacean bioluminescence in similar ways. The anatomical and physiological arrangement of the luminescent system of sea pens is uniquely designed to be responsive to stimuli from both large predators such as crabs and fish, which could induce colony-wide luminescence, or small organisms such as predatory planktonic crustaceans, which would induce local polyp luminescence.

I wish to thank Anne Harrington and Dr. John Kastendiek for many enlightening discussions. Anne Harrington assisted in the preparation of the manuscript. This work was supported in part by USPHS Grant NS9546.

SUMMARY

The complexity of the bioluminescent system in pennatulaceans suggests that the light must serve as a signal to other organisms. Some important characteristics of the luminescent signal are that it is intermittent, simple, conspicuous, predictable, proportional to the stimulus; shows contrast and movement away from the stimulus; and demonstrates a distinct orientation with respect to the colony and its position in the water currents. On the basis of these characteristics, the probable requirements of the receiver, and available ecological information, it is concluded that the light signal is directed toward potential predators which are motile, nocturnally active, and orient visually, such as fish and crustaceans. They probably respond to the signal by being startled, temporarily blinded, and/or warned of other possible pennatulacean defenses.

LITERATURE CITED

Anderson, P.A.V. and J.F. Case. 1975. Electrical activity associated with luminescence and other colonial behavior in the pennatulid, Renilla köllikeri. Biol. Bull. 149:80-95.
Bertsch, H., 1968. Effect of feeding by Armina californica on the bioluminescence of Renilla köllikeri. Veliger, 10: 440-441.
Birkeland, C., 1974. Interaction between a sea pen and seven of its predators. Ecol. Monog., 44: 211-232.
Blaxter, J.H.S., 1970. Light-Fishes. Pages 213-320 in O. Kinne, Ed., Marine Ecology, Vol. I, pt. 1. Wiley-Interscience, London.

Buck, J., 1973. Bioluminescent behavior in Renilla. I. Colonial
    responses. Biol. Bull., 144: 19-42.
Burkenroad, M.D., 1943. A possible function of bioluminescence.
    J. Mar. Sci., 5: 161-164.
Cormier, M.J., J. Lee and J.E. Wampler, 1975. Bioluminescence:
    Recent advances. Ann Rev. Biochem. 44: 255-272.
Ford, R.F., 1965. Distribution, population dynamics and behavior
    of a bothid flatfish, Citharichthys stigmaeus. Ph.D. thesis,
    Univ. of Calif., San Diego. 243p.
Frey, H.W. (ed.), 1971. California's living marine resources and
    their utilization. State of Calif., Dept. of Fish and Game.
    148p.
Greenfield, D.W., 1968. Observations on the behavior of the
    basketweave cusk-eel, Otophidium scrippsi. Calif. Fish and Game.
    54: 108-114.
Harvey, E.N., 1952. Bioluminescence. Academic Press, New York.
    647p.
Herring, P.J., 1972. Some developments in the study of luminescent
    marine animals. Proc. Roy. Soc. Edinburgh B. 73: 229-238.
Kastendiek, J.E., 1975. The role of behavior and interspecific
    interactions in determining the distribution and abundance of
    Renilla kollikeri, a member of a subtidal sand bottom
    community. Ph.D. thesis, Univ. of Calif., Los Angeles. 194p.
Morin, J.G., 1974. Coelenterate bioluminescence. Pages 397-438
    in L. Muscatine and H. Lenhoff, Eds., Coelenterate Biology:
    Reviews and New Perspectives. Academic Press, New York.
Morin, J.G., and J.W. Hastings, 1971a. Biochemistry of the bio-
    luminescence of colonial hydroids and other coelenterates.
    J. Cell. Physiol. 77: 305-312.
Morin, J.G., and J.W. Hastings, 1971b. Energy transfer in a bio-
    luminescent system. J. Cell Physiol. 77: 313-318.
Morin, J.G., and G.T. Reynolds, 1974. The cellular origin of bio-
    luminescence in the colonial hydroid Obelia. Biol. Bull. 147:
    397-410.
Nicol, J.A.C., 1960. The regulation of light emission in animals.
    Biol. Rev. 35: 1-42.
Ross, D.M., 1974. Behavior patterns in associations and inter-
    actions with other animals. Pages 218-312 in L. Muscatine and
    H. Lenhoff, Eds., Coelenterate Biology: Reviews and New
    Perspectives. Academic Press, New York.
Roushdy, H.M., and V.K. Hansen, 1961. Filtration of phytoplankton
    by the octocoral Alcyonium digitatum. Nature 190: 649-650.
Salvini-Plawen, L.V., 1972. Cnidaria as food sources for marine
    invertebrates. Cahiers Biol. Mar. 13: 385-400.
Titschack, H., 1966. Uber die Lumineszenz und ihre Lokalisation
    bei Seefedern. Zool. Anz. Suppl. 29: 120-131.
Woodhead, P.M.J., 1966. The behavior of fish in relation to light
    in the sea. Oceanogr. Mar. Biol. Ann. Rev., 4: 337-403.

STRATEGIES FOR THE STUDY OF THE COELENTERATE BRAIN

L.M. Passano

University of Wisconsin - Madison

Madison, Wisconsin, 53706, U.S.A.

Some years ago it was fashionable to hope that the nervous system of some coelenterate would be simple enough so that it could be 'described completely', either functionally, morphologically or, ideally, both. But now, for at least three reasons, this approach has palled. In the first place, the nervous system seems more complex, with the passing of the simplistic view that the functional units are always neurons, linked together by standardized junctions. At least in other, 'higher' forms, there are many different kinds of synapses, even within single neurons; single neurons can subserve more than one function in an organism, even at the same time. Secondly, non-nervous components in coelenterate coordinating systems are now known to be basic elements in behavioral control (Mackie, 1970). Finally, behavioral studies always reveal greater complexity than was previously imagined.

We must continue to develop our understanding of all of the functional properties of the individual neurons that make up the coelenterate nervous system, but a simple summation of these properties is insufficient to understand the function of the coelenterate brain. The many possible types of interconnections of individual cells predict that even 'simple' neural systems will have emergent properties that cannot be predicted intuitively from knowledge of individual elements. However, I believe that the properties of the jellyfish brain cannot be determined by a 'wiring diagram' alone. They can be fully determined only if we have a complete ethology of the animal along with the electrophysiology. The interactions between brain components are too varied and too rich to permit us to predict output from input with just a wiring diagram.

639

I am using the word 'brain' in a deliberately provocative manner, but there is precedent. A "survey of the current status of all the life sciences" sponsored by the National Academy (Kuffler, et al., 1970) showed a figure of the "phylogeny of the brain" with Hydra representing the least complex central nervous system. To most biologists, neuroscientists and students of behavior, the coelenterate nervous system is the representative (and misrepresented) 'simplest' nervous sytem. We must try to make their 'model' accurate and valuable for them. In turn, we can learn from the neurobiologists, especially those working on other invertebrates.

Let us define a brain as that part of the animal that acts like the vertebrate brain, that controls the animal's behavior. The coelenterate brain then, includes the nervous system with its one to several functional units (called 'systems', but probably sub-systems would be better), conducting epithelia, and specialized properties of both sensory receptors and effectors. Compared to 'other animals', coelenterates have poorly developed ganglionic assemblages, although these do occur in rudimentary form. The nerve cells are unsheathed by glia, although perhaps sheathed by epithelia instead.

Like other animals, coelenterates show patterned spontaneous activity not directly related to motor output. But compared to other animals, endogenous activity is simple in form. Coelenterates have a much slower time scale than other animals, and they lack the barrage of sensory information seen in most forms.

What can we, as coelenterate specialists, contribute to the science of neurobiology? What can we learn in return from this newest specialization within our science? Due to George Mackie's pioneering work, the possibilities of non-nervous conduction are now widely recognized. What other contributions are possible? Coelenterates have few functional elements that generate simple but highly adaptive behavior with little or no proprioceptive feedback. I believe that the central role of nervous tissue in coelenterates is to generate spontaneous behavior, which sensory input shapes and modifies but does not initiate. If this role can be firmly established, current ideas on the phylogeny of CNS function would have to be revised. In return, we would benefit from the infusion of new ideas. For instance, we might look for central control of sensory input. The nematocysts might act as receptors as well as effectors, and their sensitivity could be set centrally. We might look at the 'pacemaker systems' as possible non-specific activating systems controlling both perceptual sensitivities and motivational states of the animals. We might look for evidence that immediate past events influence the 'behavior' of neuronal systems, just as Wiersma (1974) has suggested for the functioning of individual neurons.

	Epithelial Pulse (EP)	Pre-Swim Pulse (PSP)	Marginal Pulse (MP)	Tentacle Pulse (TP)	Radial Pulse	Ring Pulse (RP)	Swallowing Pulse (SP)
Anthomedusae:							
Sarsia tubulosa[1]	X	X	X	X	X	(x)	X
Spirocodon saltatrix[2]	X	X	X	X	X		
Stomotoca atra[3]	X	X	X	(x)	X	X	X
Leptomedusa:							
Phialidium hemisphaericum[1]	X	X	X				
Limnomedusa:							
Proboscidactyla flavicirrata[4]	X	X	X	X			
Trachymedusa:							
Liriope tetraphylla[1]		X	X				

1) Passano, 1965; 1973; unpublished.   3) Mackie and Singla, 1975; Mackie 1975.
2) Ohtsu and Yoshida, 1973.   4) Spencer, 1975.

Table 1. Electrical pulse types in various hydromedusae.

The coordinating mechanisms of a number of hydrozoan jelly-fish have now been examined. In spite of the regrettable tendency for each investigator to develop a new nomenclature for the various coordinating-conducting systems, it does seem possible to homolo-gize the elements of hydromedusan brains, as indicated in Table 1. Epithelial conduction pathways, and two pacemaker systems, the marginal pulse (MP) and the pre-swim pulse (PSP) systems, are of general occurrence. Both pacemaker systems (each made up of an undetermined number of individual interacting pacemakers) seem to be nervous, but both interact with various epithelial pathways. In addition there are local tentacle systems, and velar, manubrial and radial muscle systems. The cryptic "ring pulses" (Mackie, 1975) of Stomotoca, of unknown function, seem to occur in Sarsia as well (Passano, unpublished), and are probably also of general occurrence. I expect that "swallowing pulses" will also be found in other hydro-medusae.

We know something about the differences in behavior between these species, but not yet enough to correlate the continuous swimming of Stomotoca, for instance, to specific characteristics of its PSP system. Likewise, how does Sarsia generate its 'fishing' behavior, where a vigorous swim bout alternates with still, tentacle-extended sinking? There is both a general similarity and very specific species differences that need to be explained.

One early general conclusion, that the different pacemaker systems are arranged in hierarchies (Passano, 1965), seems still to hold. In Sarsia, for instance, the MP pacemaker output strongly

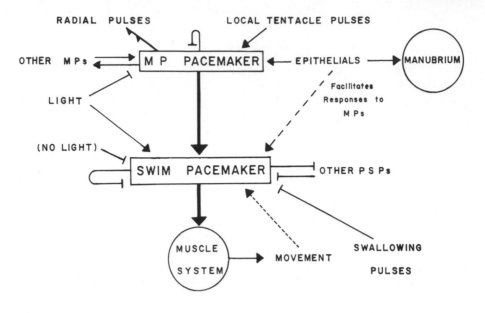

Figure 1. Schematic model showing relationships between the two
pacemaker systems in Sarsia (from Passano, 1973; modified).

interacts with the PSP pacemaker system, whereas there is no evidence that the PSPs influence MP pacemakers. Figure 1 illustrates
the kinds of interactions and directionality of influence that we
see within the coelenterate brain. Each pacemaker is self-limiting,
each has specific sensory input pathways, as well as separate output conduction pathways. The coupling between these two can be
very loose or very tight, depending on unknown circumstances. The
degree of coupling has a profound affect on Sarsia's behavior.
Probably it is more nearly correct to say that the specific behavior of the animal is determined by the ways in which this pacemaker coupling is modified.

    Figure 2 shows a situation in which there is very tight coupl-
ing. The pattern is highly regular in this pinned-down half Sarsia
from which both ocelli have been removed (18° C; in dark). A similar but less extreme coupling is shown by the 'blinded' half in
Figure 3, but here the repeated pattern is much less regular. Such
tight coupling can occur in normal 'control' preparations as well,
and is often associated with a 'facilitation' effect where the PSP
initially requires two MPs in the majority of cases (see Figs. 1
and 2, Passano, 1973).

    Many hydromedusae have discrete photoreceptors, and still
more are light sensitive. In Sarsia removing the ocelli inhibits

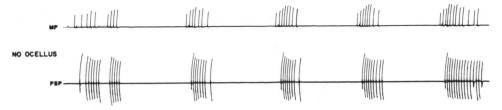

Figure 2. Tightly coupled pacemakers, Sarsia (time scale as Fig. 3).

some responses but may enhance others (Fig. 3). The animal also
shows distinctive 'day' (i.e., illuminated) and 'night' (in the
dark) activity patterns in the laboratory which seem to have counter-
parts in Nature. Since light has different effects on the MP and
PSP pacemakers, and since the former can 'drive' the latter, the
two interacting systems would seem to have a greater behavioral
potential. If the ocelli are directionally sensitive (which seems

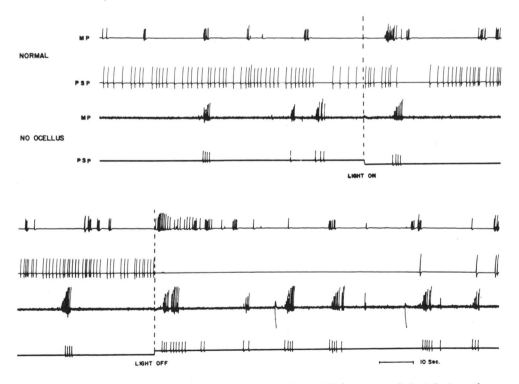

Figure 3. Light response to normal and ocelli-removed half-Sarsia
preparations. PSPs are blocked for a minute or so after the light
is turned off, if ocelli are present, while MPs are enhanced.

likely on anatomical grounds), we have a mechanism that gives a
different behavioral output dependent upon the ratio of light
striking the ocellus to ambient illumination. Still to be explained, however, is the fact that shading, which blocks PSPs in
Sarsia, augments PSPs and MPs in Spirocodon and Stomotoca. Microelectrode probing of the Sarsia nerve tract leading from the ocellus
have shown small, discrete spikes upon illumination; yet experimental evidence such as that illustrated in Fig. 3 clearly shows
that some 'off' signal is being received by the PSP pacemakers in
the marginal nerve ring area. Therefore light is just one of several environmental 'modalities' known to influence behavior. For
instance, there is a change in the pattern of endogenous output
when temperature changes by as little as one degree. One expects
that such responses to temperature, light, and other environmental
variables also interact with each other in complex, behaviorally
meaningful ways.

    Fashions change in science just as they do in clothing, pop-
ular music and morals. Our models change; the desk calculator is
no longer a mechanical 'adding machine', but is a miniaturized,
complex electronic instrument with capabilities (seemingly) limited only by how much you are willing to pay. Now, with these new
'models' in front of us, we can imagine greater complexity within
small packages. Our expectations of what our 'simple' animals can
do are influenced by our tools and toys. Just as we must continue
with basic electrophysiological probing by newer, more powerful
techniques (see articles by Kass-Simon and Spencer in this volume),
we must also gather quantitative behavioral information to demonstrate the capabilities of the coelenterate brain. Comparative
studies, and developmental studies from polyp to newly-formed medusa
to adult jellyfish are needed. We should study the interactions
of the component sub-systems, such as the MP and PSP pacemakers,
while experimentally manipulating 'environmentally relevant' factors such as water movements, light, temperature, density or salinity and chemical stimulation in trace amounts (as might occur as
the result of the presence of predators, food organisms or con-
specifics). We may be able to experimentally manipulate epithelial 'electronic' activity while monitoring pacemaker output, to
test George Mackie's hypothesis (see article, this volume) of their
interaction.

    Success will be measured by the degree of predictability that
we achieve. Understanding of the coelenterate brain can only be
achieved when we characterize the animal's behavior in terms of the
interactive functioning of its component parts. The very nature
of this brain is that it generates variability. Recognition of
this variability as a basic biological feature of the brain is one
important conclusion. We must characterize this variability, quantify it and compare it between individuals, between species and

and between different morphotypes of these polymorphic organisms
Most importantly, we must probe further into the brain to determine wherein this variability is generated.

Nerve cells and the nervous system are just one component out
of many in the 'brain' of coelenterates. They are not required for
conduction, although they may allow signals to pass more rapidly
from one part of the organism (or superorganism) to another. They
are, however, essential for behavior, as the work with nerve-free
Hydra (Schwab, et al., this volume) so elegantly demonstrates.
Therefore, the things that characterize nerve cells to us, their
shape, constituents, location and functions, are collectively responsible for their biological role or roles. The nerve cells, in
turn, are a fundamental component of the coelenterate brain; there
are no simpler animals for which such a statement can be made with
any certainty. We need to determine the role of nerve cells in
coelenterates both to guide neuro-biologists and to learn more about
our own animals.

References

Kuffler, S., E.V. Evarts, E.R. Kandel, I.J. Kopin, V.B. Mountcastle,
    W.J.H. Nauta, S.L. Palay, and W.A. Spencer, 1970. The nervous
    system. Pages 324-378 in P. Handler, Ed., Biology and the future
    of man. Oxford Univ. Press, New York.
Mackie, G.O., 1970. Neuroid conduction and the evolution of conduc-
    ting tissues. Quart. Rev. Biol., 45: 319-332.
Mackie, G.O., 1975. Neurobiology of Stomotoca. II. Pacemakers
    and conduction pathways. J. Neurobiol., 6: 357-378.
Mackie, G.O., and C.L. Singla, 1975. Neurobiology of Stomotoca.
    I. Action systems. J. Neurobiol., 6: 339-356.
Ohtsu, K., and M. Yoshida, 1973. Electrical activities of the
    anthomedusan, Spirocodon saltatrix (Tilesius). Biol. Bull., 145:
    532-547.
Passano, L.M., 1965. Pacemakers and activity patterns in medusae:
    homage to Romanes. Am. Zoologist, 5: 465-481.
Passano, L.M., 1973. Behavioral control systems in medusae: a com-
    parison between hydro- and scyphomedusae. Publ. Seto Marine Biol.
    Lab., 20: 615-645.
Spencer, A.N., 1975. Behavior and electrical activity in the hydro-
    zoan Proboscidactyla flavicirrata (Brandt). II. The medusa. Biol.
    Bull., 149: 236-250.
Wiersma, C.A.G., 1974. Behavior of neurons. Pages 419-431 in F.O.
    Schmitt, and F.G. Worden, Eds., The neurosciences third study pro-
    gram. MIT Press, Cambridge.

THE CONTROL OF FAST AND SLOW MUSCLE CONTRACTIONS

IN THE SIPHONOPHORE STEM

G. O. Mackie

Biology Department
University of Victoria
Victoria, British Columbia, Canada

The stem of the siphonophore Nanomia cara is the only coelenterate preparation so far found which allows intracellular recordings to be made both from muscles (Spencer, 1971) and nerves (Mackie, 1973); as such it offers unique possibilities for the analysis of neuro-muscular interactions at the cellular level. The stem muscle is also interesting as an example of an effector capable of responding by rapid, twitch-like contractions (both local and general) and also by slower, sustained contractions (Mackie, 1964). Such muscles are found in all three classes of the Cnidaria (reviewed by Josephson, 1974) but the ways in which the two sorts of response ('fast' and 'slow') are differentiated remains problematical. Only one histological type of muscle fibre seems to be involved; it is assumed that there are two excitation pathways. In sea anemones, the electrical correlates of fast contractions are known, but slow contractions can occur in the absence of recordable signals (McFarlane, 1973). In Nanomia, however, both fast and slow contractions have distinctive electrical correlates, the analysis of which is our present concern.

This work was done at the same time as a study on the giant axons mediating escape behaviour in Nanomia (Mackie, 1973) and is essentially Part II of the same account.

METHODS

In addition to the techniques noted in Mackie (1973) a method for stabilizing the stem for intracellular recordings was used here. This involved passing a glass rod up the lumen of the stem and stretching the stem along it by suction tubes attached at the two ends, these tubes doubling as stimulating or recording electrodes.

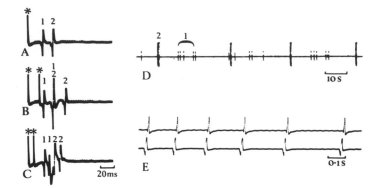

Fig. 1   Activity in the two nerve nets (1,2) recorded extracellu-
         larly in 1:10 Mg/SW.   In A, a single shock (*) fires the
         two nets.   N1 conducts faster than N2 so a second impulse
         in N1 can be made to catch up with (B) or overtake (C) a
         previously launched N2 impulse, demonstrating independence
         of the two pathways.

         Both systems are spontaneously active, firing in bursts (D).
         The N2 burst shown expanded in E is from the same recording.

Pilloried stems show similar activity to normal stems.   Magnesium
ions, which elevate the muscle response threshold to nerve input
were used in some experiments to subdue hyperactive stems.   Even
1:15 Mg/SW (1 part of isotonic magnesium chloride plus 15 parts of
sea water) has a perceptible damping effect on twitch activity.
In the extracellular recordings, negative is up, positive down,
except for Fig. 7 where the polarities were inadvertently reversed.

RESULTS

The Two Nervous Conducting Systems, N1 and N2

        There are two independent, unpolarized through-conducting
nerve nets in the stem ectoderm, each associated with a rapidly
conducting giant fibre.   Conduction velocity in each system is
related to giant fibre diameter.   Microelectrode recordings were
obtained from these fibres and it was shown that impulses could
pass around regions where the giant fibres were cut (presumably
travelling in the two nerve nets) and reinvade the giant axons on
the other side of the cut (Mackie, 1973).   Fig. 1 documents the
independence of the two nervous systems, the difference in their
conduction velocities and their spontaneous activity.   Both systems
are photoexcitable.   There are no nerves in the endoderm.

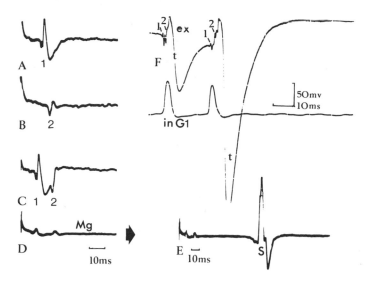

Fig. 2 Nerve spikes and associated twitch potentials recorded
extracellularly. In sea water, nerve spikes are over-
shadowed by twitch potentials (A,B,C) but the latter are
blocked in 1:5 Mg/SW (D). The slow system (S) continues
unaffected (D,E). In (F), facilitation of the twitch (t)
response to the second of the two N1-N2 pairs is recorded by
a suction electrode on the stem surface (upper trace) while
a microelectrode in the giant axon (G1) associated with the
N1 records pure N1 action potentials simultaneously (lower
trace, and amplitude calibration).

Twitch Response

As noted in the earlier report, events in N1 and N2 act
synergistically upon the ectodermal muscle, producing graded, fast
contractions (twitches). In extracellular recordings, the nerve
spikes are augmented by after-events of variable magnitude
representing locally induced myoepithelial potentials (Fig. 2A-C).
In 1:5 Mg/SW the after-events are greatly reduced, although conduction of the nerve spikes is unaffected (Fig. 2D,E). The after
potentials show facilitation during rapid firing of one or both nerve
nets together (Fig. 2F,4) and the amplitude of the response judged
visually varies accordingly. In nature, twitches would occur
whenever the combined activity of the two nerve nets produced a
barrage of several spikes in a tight frequency relationship and
this might occur either spontaneously, in response to direct
stimulation of the stem, or indirectly, to stimulation of one of
the appendages. The effect would be a local or general withdrawal
response. Strong responses are also accompanied by locomotion

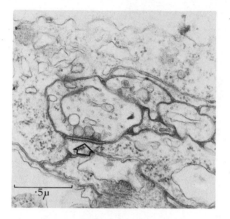

Fig. 3   Neuromuscular junction in the stem ectoderm.   Such junctions
         occur in the superficial part of the myoepithelium.

(Mackie, 1964). For the most part, twitch amplitude is determined
by the numbers and frequencies of nerve spikes, but twitches may
be augmented or prolonged by input from another source, the slow
system, as we note below.

     Neuromuscular junctions (Fig. 3) are very numerous in the
stem ectoderm, and doubtless represented chemical synapses mediating
twitch responses. Twitches cannot propagate in the muscle itself,
so unless there is limited local spread between the myoepithelial
cells, we should assume that each cell is innervated, receiving
input from either N1 or N2 or both. It has not been possible to
differentiate the two nets or their junctions histologically.

     Intracellular recordings from the muscle cells support this
interpretation. When the extracellular electrode picks up a nerve
spike, the intracellular electrode in the muscle records a small
depolarization of about 4-6mv (Figs. 5,6) which is evidently an
excitatory junctional potential (nEJP). Inputs from the two nets
look similar, sum together, and facilitate one another. The amplitude of the muscle twitch potential evoked by a given input varies
according to the physiological state and immediate past history of
the preparation. This was shown in Fig. 2F and is also apparent
in the intracellular records, e.g. Fig. 5B-D. None of the intracellular records used in this paper show a major twitch response,
because such responses dislodge the electrode. Judging from the
baseline deflection of the extracellular recordings, twitches
commonly grade up to 50mv or more in preparations where both nerve
nets have been made to fire at high frequencies along with the S
system.

     Following stimulation at any point, nerve impulses are

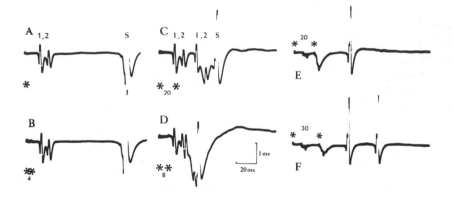

Fig. 4   Extracellular records of muscle twitches and slow (S) system
         responses. The response to two shocks 4 ms apart (B) is
         the same as to a single shock (A). A second shock >7 ms
         after the first is effective, producing a second N1-N2
         pair, and a baseline deflection representing the evoked
         twitch depolarization of the myoepithelium (C,D). Twitch
         amplitude is a function of both number and frequency of N
         inputs. Reduction of S conduction time (see in text, 'piggy-
         back effect') is also seen.

         E and F are from a similar preparation in which N events are
         recorded at lower amplitude. Each shock fires an N1-N2
         pair as before. The twitch response to the second pair is
         markedly facilitated in E (shocks 20 ms apart), less so in
         F (30 ms). Note also the long refractory period of the S
         system.

initiated which propagate all the way along the stem in both directions, generating twitches as they go. If a section of the stem is
placed in a magnesium bath, the twitch is abolished in this region,
but appears again on the other side (Fig. 7). Thus, twitches are
not propagated in the muscle itself, but are generated sequentially
along the stem by the passage of nerve events, spreading at a
velocity related to that of nervous transmission (<3 m/s).

Slow Conduction System (S) and Slow Muscle Response

        Spencer (1971) described propagated events in the stem
whose spontaneous production and conduction at 0.2 m/s was not
blocked by magnesium. Extracellular recordings showed these events
as large potentials (>1mv) while intracellular recordings (the
first for any coelenterate tissue) from the ectoderm layer showed
them as unexpectedly small depolarizations arising from a 50-80mv
resting potential. In Spencer's Figs. 2A,D these events ('spon-

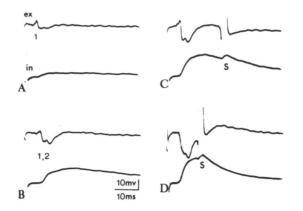

Fig. 5   Twitch depolarizations recorded intracellularly with a
         microelectrode in a muscle cell (lower trace and amplitude
         calibration) with simultaneous extracellular monitor (upper).
         In A, N1 fired alone, giving a single excitatory junction
         potential (nEJP) in the muscle.   B, C and D are from a
         sequence in which similar N1-N2 pairs were evoked each time
         but where the twitch depolarization level was found to vary.

         The S system fired each time but is off the record in A and
         B.   With larger twitches, (C,D) reduction of S conduction
         time is apparent.   The intramuscular correlate (sEJP) is
         seen mounted on the twitch depolarization, but distinct from
         it, indicating that the piggyback relationship had been lost
         earlier in the pathway.

taneous suprathreshold pulses') measure 10-12mv, but values up to
20mv are noted in the text. (The larger events shown in his Figs.
2B,C as responses to strong stimulation may include a twitch component.   Spencer's 'subthreshold pulses' (Fig. 2B) would appear
to correspond to my nEJPs).   The propagated events seen in extracellular recordings are here called S (slow, or Spencerian) pulses.
They are conducted at about 0.3 m/s in my recordings.   They seem to
occur spontaneously but not in consistent patterns.   Their insensitivity to levels of magnesium which block twitch responses is shown
in Figs. 2E,7.   The correlated events seen in intracellular recordings
from ectodermal muscle cells have amplitudes in the range 10-20mv,
and arise from a 75-85mv resting potential.   The small size of
these events (ca. 25% of the resting potential) suggests that they
are not being generated or propagated in the muscle itself.   Moreover, much larger (twitch) potentials in the same cells clearly do
not propagate.   What we are recording is evidently a second type of
junctional potential representing input from the slow conduction
system, which may be referred to as an sEJP.   The large size of the
extracellular S pulse, its persistence at full amplitude in magnesium and its slow conduction suggest that it is neither a nerve

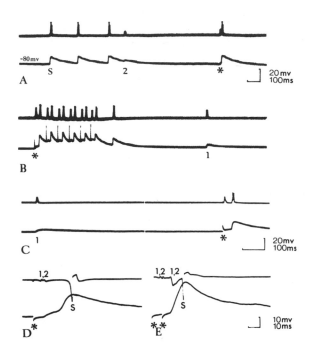

Fig. 6 Slow and twitch potentials recorded intra- (lower) and
        extra- (upper) cellularly, arrangement as in Fig. 5.
        In A,B and C the microelectrode registers nEJPs corres-
        ponding to spontaneous N1 and N2 events, and larger sEJPs
        corresponding to spontaneous S potentials. Shocks (*)
        evoke nEJPs followed by sEJPs. In B stimulation frequency
        is too low to permit twitch development and the mounting
        level of depolarization is due to summing of sEJPs, re-
        presenting a fairly strong contraction of the 'slow' type.

        D shows summation of nEJPs from N1 and N2 to which is added
        an sEJP. In E, four nEJPs and an sEJP have summed to
        produce a smooth 30mv twitch depolarization. (N.B.: the
        extracellular monitor which was slightly nearer the stimu-
        lating site, here records only the initial (positive) com-
        ponent of the S event due to amplifier overload. The event
        is really primarily negative-going, in contrast to twitches
        seen extracellularly, which allows it to be distinguished
        even where, as here, the two events coincide. In the intra-
        cellular record we see not the primary S event, but an sEJP
        which has the same polarity as the twitch and merges with it.)

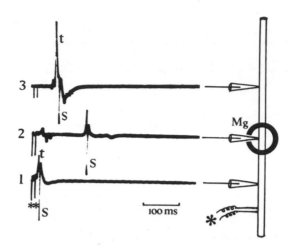

Fig. 7   Indirect production of twitches and slow potentials
         (polarities reversed). A shock pair produces two N1-N2
         pairs which propagate up the stem through a region containing
         1:10 Mg/SW. The twitch response (t) is blocked in the
         Mg bath (trace 2), but a twitch is again generated on the
         far side (trace 3) showing that twitches can be evoked by
         N combinations far from the site of stimulation.

         S potentials appear on both channels 1 and 3 in the piggy-
         back time relationship. It is apparent that the S event
         in channel 3 must have been generated on the far side of
         the magnesium block along with the regenerated twitch. The
         S event which eventually arrives on channel 2 was carried
         piggyback as far as the Mg bath after which, with the collapse
         of its carrier, it continued up the stem at its slow velo-
         city. It never appeared on channel 3 having been blocked
         by the regenerated S event coming <u>down</u> the stem.

spike nor a combination of a nerve spike with locally induced myoepithelial potential, like the twitch recorded extracellularly.
The S pulse is almost certainly an epithelial potential conducted
in the endoderm, although intracellular recordings from the layer
are not available to prove this. The cells are connected by numerous gap junctions (C.L. Singla, personal communication). Endodermal epithelia are known to conduct in hydromedusae (Mackie and
Passano, 1968) and impulses can propagate from ectoderm to endoderm, probably via the transmesogloeal tissue bridges which have
been located histologically in appropriate regions. In <u>Nanomia</u>,

transmesogloeal bridges also occur between the two layers (Fig. 8).
Although the histological evidence is still incomplete, it seems
probable that the two layers are electrically coupled via these
bridges, allowing slow pulses in the endoderm to be picked up as
electrotonic junctional potentials (sEJPs) in the ectodermal
muscle. In support of this interpretation, we may note that
muscle twitches recorded extracellularly show the opposite polarity
to S pulses, suggesting that the main components of the two events
are not both being generated in the same cell layer. The sEJP
however has the same polarity as the nEJP and is clearly recorded
in the same cells.

Activity in the S system can be evoked without accompanying
twitch activity (Fig. 6B). S bursts having interpulse intervals
of 100-400 ms are accompanied by slowly developing contractions
of the stem and of the bases of some of the zooids. Because of
its long refractory period (25-30 ms) the S system cannot fire
at more than about 30-40 Hz. When stimulated at these frequencies
summation of sEJPs to a maximum of ca. 35mv has been observed,
representing a strong, sustained contraction. N1 and N2 are fired
concurrently but fail to generate twitches at these relatively
low frequencies, so we are seeing an essentially pure slow muscle
response. With more rapid stimulation a twitch component starts
to develop and is added to the slow response. The ability of the
S system to fire spontaneously in the absence of twitches probably
accounts for the ability of the stem to maintain differing pos-
tural extensions in nature. Other possible roles for the S system
remain unexplored. There is no true endodermal muscle, merely a
few wisps of thin filaments in some regions.

When recorded concurrently with twitches, sEJPs may be seen
mounted on twitch depolarizations in the 20-30mv range. In this
position they show amplitudes similar to those of sEJPs recorded
at resting potential from the same preparation, which again
suggests electrical rather than chemical mediation.

Retroactivation of S System: Piggyback Effect

A striking feature of the stem preparation is the variable
conduction time of S pulses. In the absence of nerve spikes, or
where these are too few or far apart to generate twitches the S
pulse propagate at about 0.3 m/sec, but where twitches are evoked
concurrently with S pulses, the latter arrive earlier, this
'acceleration' being more marked the bigger the twitch, up to the
point where the two events appear simultaneously all along the
conduction route (Figs. 4,5). To explain this phenomenon, let us
suppose that the S events are being carried 'piggyback' on the
crest of a twitch depolarization for some or all of the distance
over which they are recorded. When they fall off the carrier

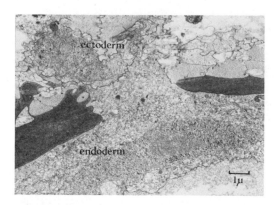

Fig. 8   Transmesogloeal tissue bridge. The endodermal process
         splits into branches in the ectoderm, contacting several
         ectodermal cells.

event (or where the latter is wiped out by magnesium, Fig. 7),
they continue to spread, but now their rate of spread is the (much
slower) propagation velocity of the S system itself. In support
of this interpretation we may invoke the same electrical coupling
presumed to mediate sEJPs. Electrical junctions are typically
non-rectifying, so retroactivation of the (endodermal) slow
conduction system would be a reasonable expectation when the
ectodermal muscle to which it is coupled fires in the twitch
response. (Thus, in Fig. 9 two-way communication is indicated
at the ecto-endodermal junctions.) We further assume that there
is a critical level of twitch depolarization (piggyback threshold)
above which coupling potentials invading the endoderm evoke S
spikes. Fig. 10 shows this level as a dotted line on the muscle
trace. S spikes would not be evoked with twitches below this
level (probably about 25mv) although it is probable that the
conduction velocity of an established S event would be increased
by the depolarizing effect of the coupling potentials on the cells
ahead of it in the conduction path. Thus a complete spectrum of
conduction 'velocities' between 0.3 m/s and nearly 3.0 m/s would
be expected, and is observed. It is not the velocity of S propagation which is variable, but the proportion of the conduction time
spent on and off the piggyback.

     Why should twitch amplitude decline with distance, allowing S
events to fall off their carrier potentials? A simple explanation
is to be found in the fact that N1 and N2 conduct at different
conduction velocities, so N-inputs to the muscle become spread out
with distance, and twitch amplitude, which is frequency-related,
will accordingly decline. In the deliberately oversimplified
example represented in Fig. 10, N1 and N2 are initially close

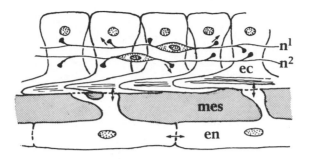

Fig. 9   Summary diagram of the histological relationships. The
        two nerve nets (N1, N2) synergistically fire the muscle
        through one-way synapses. The endoderm (en) is coupled
        to the ectoderm (ec) by transmesogloeal (mes) bridges,
        allowing two-way interactions to occur.

together, having been launched simultaneously. The further they
travel, the more they spread out and finally N1 gets so far ahead
of N2 that it barely facilitates the response to N2. Twitch
amplitude now fails to attain the critical threshold needed for
synchronous firing of the S system. S events however continue on
their own.

      The siphonophore has found an ingenious way of grading response
intensity with distance from the site of stimulation. We can see
how local twitch responses are feasible, even though the conduction
systems are through-conducting.

      It will be apparent that while S potentials can occur alone
producing 'pure' slow muscle responses, any substantial twitch
activity, whether local or general, will be accompanied by the
production of propagated S events which will have the effect of
reinforcing or prolonging the twitch contraction. Fast and slow
contractions are therefore not sharply distinct in this animal -
an untidy arrangement (nature is often untidy!) but one which
probably makes the system more versatile and hence more efficient
than would be the case if there were no way of mixing the two
sorts of input.

                            DISCUSSION

      Figures 9 and 10 summarize the relationships proposed in this
paper to explain fast and slow muscle contractions. Other con-
structions may be possible but I can think of none which come at

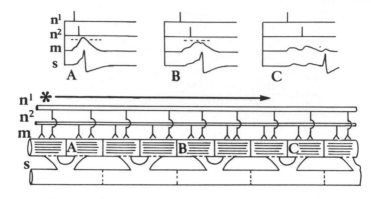

Fig. 10   Scenario to explain decline of muscle response intensity
with distance (A - C) from stimulus (*). Nl, N2 and S
events are through conducted. Nl conducts about twice
as fast as N2. The muscle (M) does not conduct. At
point A, Nl and N2 are still close together; Nl facilitates N2 producing a large twitch in the muscle, which
results in synchronous firing of the coupled S system.
At B and C Nl goes progressively further ahead of N2,
twitch level drops below the critical piggyback threshold
(dotted line) leaving S to auto-propagate. M and S records
are shown as intracellular recordings, but such records
have not been made in the case of S, and are inferred.

all near to fitting the evidence. The critical experiments needed
to substantiate the model are obvious: direct recordings from the
endoderm to show that S pulses are conducted there; simultaneous
recordings from the two cell layers to elucidate two-way coupling
interactions, transducer recordings showing tension fall-off along
the stem as a function of changing N-input relationships, etc.
Nanomia is not an easy animal to work with and is not always easy
to obtain, but it remains our best bet for this type of study.

My colleague, L.M. Passano, speaks of "the coelenterate way of
doing things" and this work illustrate the point that while we may
search for, and even find, neuromuscular mechanisms reminiscent of
other phyla, giant axons, dual excitatory nerve inputs, etc. no
facile comparisons can be made with animals whose muscles are arranged
in discrete units. A single long, cylindrical muscle poses quite
different problems from a series of short separate ones.

Siphonophores are colonies, the integration of whose component
members has been explored at the behavioral level in other papers
(e.g. Mackie, 1964). The way in which stem activity is triggered by
and in turn generates activity in the appendages needs to be re-

examined in the light of this analysis of the stem action system. Judging from my own (unpublished) observations, these input-output relationships are quite subtle but should yield many lessons of interest when critically examined. The morphological complexity of siphonophores may seem confusing, but the division of labour which accompanies polymorphism has resulted in certain striking simplifications of the functions of individual members which makes them useful for the investigation of the action systems one at a time, in isolation.

ACKNOWLEDGEMENTS

I thank the Director of the Friday Harbor Laboratories for providing space and facilities for the execution of this project, and Dr. L.M. Passano for searching criticisms and penetrating observations. The work was supported by the National Research Council of Canada.

REFERENCES

Josephson, R.K. 1974. Cnidarian neurobiology. Ch. VI. in "Coelenterate Biology" (ed. L. Muscatine and H.M. Lenhoff) Academic Press.

Mackie, G.O. 1964. Analysis of locomotion in a siphonophore colony. Proc. Roy. Soc. B, 159, 366-391.

Mackie, G.O. 1973. Report on giant nerve fibres in Nanomia. Publ. Seto Marine Lab. 20, 745-756.

Mackie, G.O. and Passano, L.M. 1968. Epithelial conduction in Hydromedusae. J. Gen. Physiol. 52, 600-621.

McFarlane, I.D. 1973. Spontaneous contractions and nerve net activity in the sea anemone Calliactis parasitica. Mar. Behav. Physiol. 2, 97-113.

Spencer, A.N. 1971. Myoid conduction in the siphonophore Nanomia bijuga. Nature 233, 490-491.

THE EFFECT OF ELECTRICAL STIMULATION AND SOME DRUGS ON CONDUCTING

SYSTEMS OF THE HYDROCORAL MILLEPORA COMPLANATA

H.A.M. de Kruijf

Caribbean Marine Biological Institute

P.O. Box 2090, Curacao, Netherlands Antilles

INTRODUCTION

Spontaneous electrical activity has been recorded with suction electrodes from polyps of the fire coral Millepora complanata (de Kruijf, 1976). Two types of pulses could be distinguished, contraction pulses (CPs) associated with contraction of the polyp and small positive pulses (PPs) not associated with any overt behaviour. CPs are intimately associated with the PPs but the latter can occur without being followed by a contraction pulse. There is evidence for a gastrozooid conducting system and a dactylozooid conducting system (de Kruijf, 1975, 1976). Both systems are of decremental nature and are clearly linked. In this report the results are described concerning experiments with Millepora complanata colonies using electrical stimulation and some drugs. These findings are related to earlier observations.

MATERIAL AND METHODS

For this study I used fresh Millepora complanata colonies collected from depths of 2 - 5 m. Details of recording techniques and other particulars were described earlier (de Kruijf, 1976).
The colonies were stimulated using a suction electrode attached to the base of a dactylozooid or a gastrozooid. The electrical stimuli were square pulses of 2-6 ms duration produced by a Grass SD9 stimulator.

RESULTS

Responses To Electrical Stimulation

Electrical stimulation of a zooid initiates a burst of potentials which can be recorded from any nearby polyp. Typical bursts
recorded in gastrozooids or dactylozooids have usually the following characteristics in common (Fig. 1). The first potentials as
reaction to an electrical stimulus are some very small PPs, often
somewhat irregular, every other one having a larger amplitude than
the preceding PP and of moderate duration, 40 - 80 ms. It is very
difficult to establish the actual number of PPs due to the small
size of the very first ones. These first pulses are usually interrupted by one, two, and sometimes three CPs which are always negative (Fig. 1 and subsequent figures). The next large potentials
(Burst Potential, BPs) are mostly biphasic starting positively and
of very regular shape. The negative phase can disappear within three
or four potentials. These pulses are by far the largest recorded in
a burst except for the CPs, the duration varies between 40 and 100
msec and the amplitude found, was maximum 4mV. The amplitude of the
next potentials decreases and maintains a size of 5 to 20% of the
largest potentials. These rather small pulses tended to be monopha-
sic and positive but in many trials the pulses were biphasic or even
negative and often of compound nature (Fig. 1). Similar bursts have
been occasionally produced spontaneously by polyps and can also be
initiated by mechanical stimulation. Small positive potentials are
usually found preceding the CPs as was found earlier in spontaneous records (de Kruijf, 1976). Similar combinations of CPs and PPs
are found in bursts (Fig. 2). These findings suggest that burst
potentials (BPs) are a separate set of pulses and neither positive

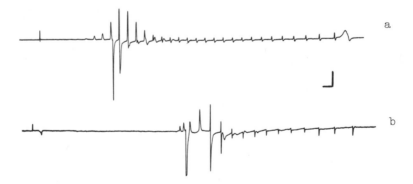

Fig. 1. Bursts initiated in polyps of <u>Millepora complanata</u> by single
electrical stimuli. (a) Gastrozooid; PPs, preceded by a minimal, irregular phenomenon, are followed by a CP. A series of burst potentials (BPs) complete the burst. Some of the BPs are of a compound
nature. (b) Dactylozooid; a similar type of burst. Hor. scale: 0.2
s; vert. scale: 0.5 mV.

pulses nor reversed contraction pulses.

The intervals within a burst show a fixed pattern. The last interval has the longest duration (340 ms) whereas the interval becomes shorter towards the beginning (Fig. 3). This pattern of relatively long intervals at the end and progressively shorter ones towards the

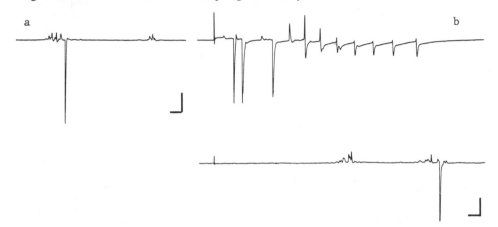

Fig. 2. (a) Spontaneous activity from a gastrozooid. A PPburst is followed by a CP and a PPburst without a CP. (b) Simultaneous record from two gastrozooids. The complete burst fails to appear at the far recording site (lower trace) but two PPbursts are initiated, one with a CP. In the nearest gastrozooid three CPs are evoked, in the lower trace but one. Hor. scales: 0.2 s; vert. scale: 0.5 mV.

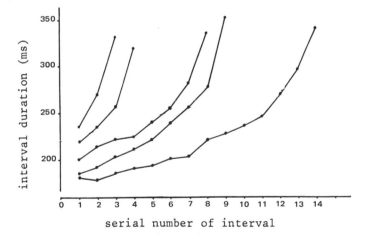

Fig.3. Distribution of interval duration in electrically evoked bursts. Bursts consisted of four Burst Potentials (n=20); five BPs (n=16); nine BPs (n=14); ten BPs (n=15); and fifteen BPs (n=12). The bursts were all recorded from gastrozooids.

beginning of the burst was found for all bursts regardless of their
actual number of potentials. The refractory period was not determined in these experiments.

Bursts were simultaneously recorded from two sites, the first
near the stimulating electrode, the second on varying distances from
the first recording sites. The three electrodes were always on a
straight line and attached either to gastrozooids or to dactylozooids.
When the two recording sites are very near, the bursts are similar:
the number of pulses is nearly equal and only a very short delay is
noticeable in gastrozooids as well as in dactylozooids. As the recordings are made progressively more distant the difference in number
of pulses becomes larger and larger, the far electrode having the
least number of pulses. If the distance is great enough, entire bursts
fail to appear at the far recording site (Fig. 2b). This was found
both in gastrozooids and dactylozooids (Fig. 4a and b). When bursts
are recorded at both sites, the intervals between the potentials of
the burst match when progressively earlier intervals are compared
from the last towards the beginning of the bursts. This indicates
that the initial potentials fail to invade the more distant parts,
but once pulse invasion is achieved, the burst pattern is maintained
and the system is through-conducting to that point. These observations
suggest that there are a series of facilitation sites distributed
within the conducting system of gastrozooids and dactylozooids. As
each burst travelling through a conducting system encounters a facilitation site the first pulse of a burst will fail to be propagated
beyond that site, but will enable succeeding pulses to do so. The net
effect is a loss with distance of the first pulse in a burst but a

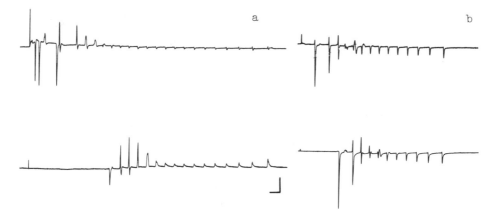

Fig.4. (a) Simultaneous record of bursts recorded in two gastrozooids
(distance: 6.8 mm) and evoked in a gastrozooid. In the gastrozooid
nearest to the stimulation site (upper trace) 6 BPs more are recorded
than at the far recording site. (b) Bursts recorded from dactylozooids
(distance 2.25 mm). There is a loss of two BPs in the lower record i.e.
from the far recording site. Hor.scale: 0.2 s; vert.scale: 0.5 mV.

through-conducting of the later potentials. The distance to be reached by a burst is therefore limited by the number of pulses in the
initiated burst and the distance between facilitation sites. By know-
ing the distance between the two recording sites and by counting the
numbers of pulses lost between the two sites it should be possible
to determine the approximate distance between the facilitation sites
(the facilitation unit distance). To do so three assumptions
must be made. 1) Each facilitation site blocks one potential only
and this allows other potentials to pass, 2) the duration of the specific facilitation at a site is sufficiently long so that successive
pulses within any given burst are unaffected, and 3) the facilitation
sites are evenly distributed throughout the colony. The loss of one
single pulse has been observed between two closely placed electrodes,
hence the first assumption is probably permissible. The second assumption probably is an oversimplification but even more so the third assumption. Even if the facilitation sites are regularly distributed in the
the canals then still large variations may be found due to irregular
distribution of the canals connecting the polyps and the varying length
of the canals. These factors may be, in part, responsible for the variation described below.

A plot of the pulses lost between recording electrodes (with stimulation near one of them) against the distance is shown in figs. 5a and
5b for gastrozooids and for dactylozooids. When the distance of two recording electrodes is within 10 mm the correlation is high for both
dactylozooids and gastrozooids. The facilitation unit distance calculated
from the same set of experiments, was 1.06 ± 0.24 mm (S.D.) for gastrozooids but 1.47 ± 0.28 mm for the dactylozooids. This difference is ob-

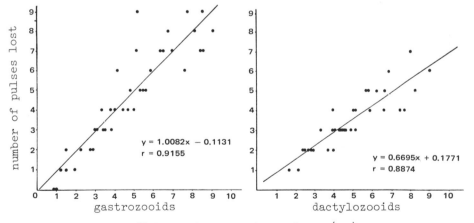

Fig.5. Relation of number of pulses lost between two polyps and the
distance between them. (a) Plot for loss of BPs between two gastrozooids
the mean distance per lost pulse is 1.06 mm. (b) Similar plot for
dactylozooids, the mean distance is 1.47 mm.

viously significant to a very high degree. There was no difference
between the conducting velocity of through-conducted pulses which was
about 20 cm per sec (9-41 cm/sec, 139 measurements) for both types or
polyps. However, this velocity does not determine the actual response
velocity of the behaviour of the colony i.e. the velocity with which
a colony or a number of polyps actually react upon a stimulus. That
time is determined by the duration of the lost pulses (40 - 80 ms) and
the intervals between the potentials (a mean of 230 - 270 ms). The
behavioral response velocity is about 2-8 cm per sec. The same experiments indicated that the number of potentials per burst may depend on
which type of polyp was stimulated. Additional experiments were carried
out with recording from two gastrozooids and stimulation was first applied through a dactylozooid and next through a gastrozooid or the reverse. The distance between stimulation site and first recording site
was always about equal. The same was done for dactylozooid recordings
which were also stimulated through each type of polyp. The number of
pulses in the first recording polyps were counted for each trial and
compared with its counterpart (table 1). The results presented in
table 1 show that gastrozooids produce longer bursts than dactylozooids. (Fig. 6).

The relation between the gastrozooid and the dactylozooid conducting systems seems to be beyond doubt, so an additional series of
experiments were done to obtain more information on the relation of
the two systems with respect to burst coordination. Recording electrodes were placed on a gastrozooid and the stimulation electrode was

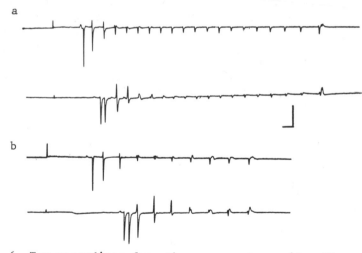

Fig. 6. Two recordings from the same gastrozooids. Electrical stimulation was given through a gastrozooid (a) and a dactylozooid (b). The
distance between the stimulation polyp and the nearest gastrozooid
was 1.53 mm (upper trace) and 1.16 mm lower records). In both records
the first gastrozooid reacts with one CP, the second gastrozooid in
both cases with two CPs. Hor. scale: 0.2 s; vert. scale: 1 mV.

Table 1. Number of pulses in burst recorded in gastrozooids and evoked with electrical stimuli in each of the two types of polyps. The statistical tests were done with Student's t-test for means. The last column shows the distance from the first recording site to the point where theoretically the burst would not be recorded any more. This highly theoretical value is found by multiplying the mean number of pulses recorded in a polyp and the mean facilitation distance for the stimulated polyp.

Stim. polyp	rec. polyp	exp.	mean of pulses $\pm$ s.d.	d_s-d	g_s-g	d_s-g	mean cond. distance (mm)
g_s	d	33	14.0 ± 7.2	$p \leq 0.01$	$p \geq 0.05$	$p \leq 0.05$	15.8
d_s	d	33	9.0 ± 5.0	---	$p \leq 0.01$	$p \geq 0.05$	13.2
g_s	g	22	17.7 ± 6.2	---	---	$p \leq 0.01$	18.8
d_s	g	22	10.1 ± 3.9	---	---	---	14.8

placed on a gastrozooid very near to one of the recording sites,and, in the second trial, near the other site.

The results are somewhat ambivalent. A burst first recorded in a gastrozooid and then in a dactylozooid lost one pulse per 0.93 mm on the average (0.93 ± 0.23 mm, n=38). A burst first recorded in a dactylozooid then in a gastrozooid lost one pulse per 0.98 mm (0.98 ± 0.31 mm, n= 38). The average facilitation unit distance appears to be much shorter when two different types of polyps are involved than in the case where only one type of polyp is involved. This would suggest that the transfer of a burst from one system to the other system costs about one extra pulse or maybe even more. However in that case one would expect a greater unit distance in the dactylozooid-gastrozooid recording compared to the all gastrozooid recording. This is because the burst in the last polyp did not change from one system to the other but the burst in the dactylozooid had lost at least one extra pulse to pay for the transfer from one system to the other. According to this idea the mean distance should have been larger than the mean distance in the gastrozooid system. I do not have an explanation for this phenomenon.

Responses To Drugs.

Four drugs have been tested for their possible effect on Millepora. Recording was done on two gastrozooids, about 1 cm apart. An initial 30 minutes in sea water preceded the experiments in seawater with the test solution. After 30 minutes the test solution was replaced by fresh seawater. Porpranolol, an adrenergic blocking substance, and lidocaine (or xylocaine which blocks nerve conduction in mammals) had no appreciable effect. 5- Hydroxytryptamine (5-HT, 0.001M) produced two effects, in the test solution PPs and CPs could be electrically evoked but not so with BPs, and during the recovery the rate of spontaneous firing was about 2 to 3 times higher than in the control per-

iod (Fig.7a). The increase of spontaneous activity faded away in 20-30
minutes. The quaternary ammonium compound, cetrimide (tetradecyl trimethyl
ammonium bromide) had a remarkable effect on the electrical activity
of Millepora (Fig. 7b). In solution with high concentration (0.005 M
and higher) extremely long bursts were recorded with very short intervals between the bursts. The pulses became rapidly smaller and after
10 to 20 minutes could not any more be distinguished from the noise.
There was no recovery. Lower concentrations produced fewer bursts but
still very long lasting and the colony could keep that going on for
the full test period. Recovery was also slow and poor.

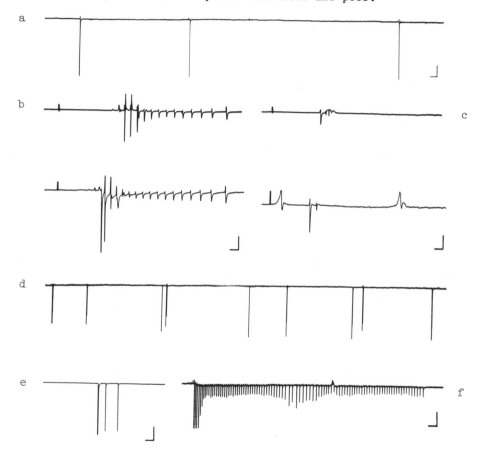

Fig. 7. Effect of the cetrimide and 5-HT on Millepora complanata. Spontaneous activity (a) and evoked burst (b) in a gastrozooid prior to
test solution; evoked activity in 0.001 M5-HT (c); spontaneous activity after return to fresh seawater (d); spontaneous activity before and
in the presence of 0.0005 M cetrimide (e) and (f). Hor.scale:5s(a) and
(d), 1 s (e) and (f), 0.2 s (b) and (c); vert. scale: 0.5 mV.

DISCUSSION

Electrical stimulation of Millepora complanata polyps can evoke bursts
consisting of three types of potentials, small positive potentials
(PPs), burst potentials (BPs), and contraction potentials (CPs). The
absence of BPs and the presence of PPs in records of spontaneous activity and the occasional occurrence of PPs without BPs after electrical
stimulation suggest that these potentials are of different origin.
But careful examination of many records of electrically evoked bursts
did not exclude the possibility of PPs being precursors of BPs originating in the same conducting system. These potentials will be considered
separately below. Nevertheless they are even more intimately related
to each other than PPs and CPs (de Kruijf, 1976) spontaneous electrical
activity.
      Bursts have been recorded in other hydroids like Cordylophora
(Josephson, 1961), Obelia (Morin and Cooke, 1971 a and b), Hydra (Rushforth and Burke, 1971) and in a solitary Tubularia species (de Kruijf,
in preparation). The bursts in Cordylophora and Obelia were initiated
by electrical stimuli, the two other animals produced bursts spontaneously. Obelia's LP-system resembles the Millepora burst system as
to the presence of facilitation sites. In considering what the nature
of a conducting system may be one can arbitrarily apply the properties
of the colonial pulse system in Proboscidactyla (Spencer, 1974) as
criteria. That system is most probably of epithelial nature. Two of
the properties or criteria are a) the conducting system shows no
consistent facilitating properties and b) multiple firing to a single
stimulating shock is not seen. The Millepora conducting system clearly
does not fit these two descriptions nor does the Obelia LP-system
(Morin and Cooke, 1971a) although that system is considered to be epithelial in nature. A third property or criterion is the low conduction velocity in the epithelial conducting system of Proboscidactyla
whereas in Millepora and Obelia the velocity is around 20 cm/sec. The
conduction system in the stolon of Cordylophora is considered to be
epithelial (Josephson 1974) too like Probosidactyla's colonial pulse
system. Based on the various differences the Proboscidactyla properties (consistent facilitation, multiple firing, and conduction velocity) the LP-system in Obelia and the burst system in Millepora
is a neuronal network having interneural junctions which require to
be facilitated and are non-polarized.
      Bursts are initiated by the quaternary ammonium compound, cetri-
mide. The known action of such compounds is on cholinergic mechanisms
but cetrimide also can act as a typical cationic surfactant on membranes lowering the surface tension (Blacow, 1972). Cholinergic mechanisms
have not yet been shown to exist in any coelenterate and extrapolating
this statement it seems unlikely that Millepora should have such a mechanism. Consequently, cetrimide's action must be on membranes of the
conducting system. The actual burst conducting system is, as we have
seen, most probably neuronal. The neuronal nature indicated by the experiments as well as the membrane effect by cetrimide are not necessarily in conflict, but it remains uncertain yet how the results should
be interpreted.

I thank dr. J.P. Groenhof for supplying the drugs and Ingvar Kristensen and Cocky de Kruijf for criticism. Attendance at the Conference has been made possible by a grant of the Netherlands Foundation for the Advancement of Tropical Research (WOTRO).

SUMMARY

Upon single electrical stimuli Millepora complanata polyps produce bursts of pulses consisting of positive potentials, burst potentials, and contraction potentials. Facilitation sites in gastrozooids are 1.06 ± 0.24 mm apart, in dactylozooids 1.47 ± 0.28 mm. Conduction of through-conducted pulses is 20 cm/sec in both types of polyps. The data suggest that the burst system is of neuronal nature. 5-HT inhibits BPs and increases the rate of spontaneous firing during recovery. Cetrimide, a quaternary ammonium compound, induces continuous firing of bursts over long periods.

REFERENCES

Blacow, N. editor (1972). Extra Pharmacopoeia. 26th edition, Pharmaceutical Press, London, pp. 171-173.
Josephson, R.K., 1961. Repetitive potentials following brief electric stimuli in a hydroid. J. Exp. Biol. 38, 579-593.
Josephson, R.K., 1974. Cnidarian Neurobiology. In Coelenterate Biology, Reviews and new Perspectives ed. by L. Muscatine and H.M. Lenhoff Academic Press, New York, 245-280.
de Kruijf, H.A.M. (1975). General morphology and behaviour of gastrozooids and dactylozooids in two species of Millepora (Milleporina, Coelenterata). Mar. Behav. Phys. 3, 181-192.
de Kruijf, H.A.M. (1976) Spontaneous electrical activity and colonial organization in the hydrocoral Millepora complanata (Milleporina, Coelenterata). Mar. Behav. Physiol. 4, in press.
Morin, J.G. and I.M. Cooke (1971a). Behavioral physiology of the colonial hydroid Obelia. II Stimulus-initiated electrical activity and bioluminescence. J. Exp. Biol. 54, 707-721.
Morin, J.G. and I.M. Cooke (1971b). Behavioral physiology of the colonial hydroid Obelia. III. Characteristics of the bioluminescent system. J. Exp. Biol. 54, 723-735.
Parmentier, J. and J. Case (1973). Pharmacological studies of coupling between electrical activity and behaviour in the hydroid Tubularia crocea (Agassiz). Comp. gen. Pharm. 4, 11-15.
Rushforth, N. and D.S. Burke (1971). Behavioral and electrophysiological studies of Hydra. II Pacemaker activity of isolated tentacles Biol. Bull. 140, 502-519.
Spencer, A.N. (1974). Non-nervous conduction in invertebrates and embryos. Amer. Zool. 14, 917-929.

THE REACTIONS OF HYDRA ATTENUATA PALL. TO VARIOUS PHOTIC STIMULI

Pierre Tardent, Ernst Frei, and Markus Borner

Zoological Institute, University of Zurich

Künstlergasse 16, 8006 Zurich / Switzerland

INTRODUCTION

According to various investigations and occasional observations (Trembley, 1744; Wilson, 1891; Haase-Eichler, 1931; Haug, 1933; Singer et al., 1963; Passano and McCullough, 1965; Tardent and Frei, 1969) the freshwater Hydra perceives photic stimuli and reacts to them. The reaction-patterns to sudden light-intensity changes consist of extensions and/or contractions of the body and the tentacles or of phototactically motivated locomotory activities oriented to or from a permanent light source.

In the present work, which aims at a better understanding of these behaviors we have investigated the reactions of Hydra attenuata Pall. to sudden and repeated intensity changes, to electronic flashes and to monochromatic light, and have studied the effect of dark- and light adaptation on photosensitivity.

MATERIAL AND METHODS

Polyps of Hydra attenuata Pall. were either kept permanently in complete darkness (dark-adapted) or exposed continuously (light-adapted) to an illumination of 3000 Lux (1.10^4 erg/cm²/sec.). The handling of the dark-adapted animals was done under a weak red light. The animals were kept in Loomis' solution (Loomis and Lenhoff, 1956) at

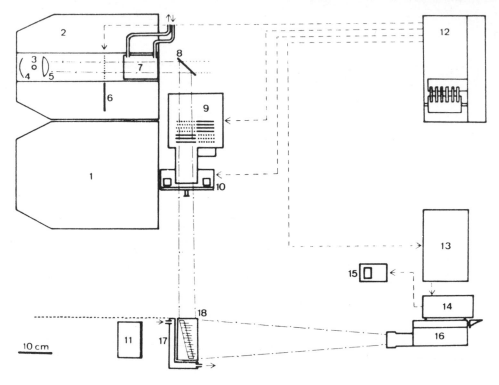

Fig. 1: Experimental set-up seen from above and drawn in
        scale. 1 = rectifier, 2-5 = xenon high-pressure
        lamp, 6 = electrically controlled diaphragm, 7 =
        cooling device, 8 = infra-red absorbing mirror
        (Balzers, Lichtenstein), 9,10 = device for the
        automatic control of neutral-density- and mono-
        chromatic filters, 11 = electronic flash, 12 =
        programmable control-unit, 13 = power supply,
        14 - 16 = recording movie-camera (Bolex) with
        pulse-counter.

17-18°C and fed fresh-water copepods twice a week. Specimens selected for an experiment were given food every
2nd day during the week preceding the start of an experiment. Each animal was used for only one experimental run.

     The experimental set-up (Fig. 1) in which 14 polyps
were exposed to various light programs of desired duration,
and by which their reactions were recorded quantitatively,
consisted of the following elements: 1) The light-source

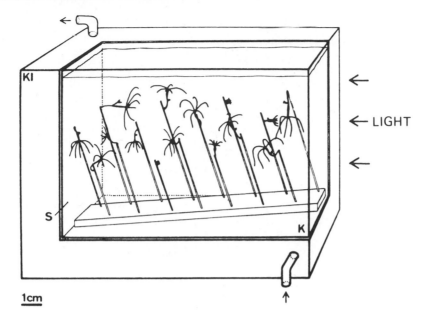

1cm

Fig. 2: Plexiglas-aquarium (K; 12 x 8 x 6 cm) with coo-
        ling mantle (Kl) and mirror (S).

was a xenon high-pressure lamp of known emission spectrum.
The intensity and quality of the light-beam were regulated
by means of neutral-density (Kodak) and monochromatic
(Balzers, Lichtenstein) filters, which could be introduced into the light-path under electrical control. The
light-intensities were measured with a Lux-meter (Metro-
watt, Nürnberg), and the energies transmitted by the
various filters were actinometrically determined according
to Wegner and Adamson (1966). 2) For each experimental run
14 polyps were placed on the tips of 14 upright glass-
needles, which were attached to the bottom of a small
plexiglass aquarium (Fig. 2) in such a way that each animal was in the light-beam. The water temperature of this
container, which was protected from vibration was kept
constant by a mantle through which water of the desired
temperature was circulating (Fig. 2). 3) The behavior of
the animals was recorded with a Bolex movie-camera. For
taking pictures during dark periods the shutter of the
camera was synchronized with an electronic flash. The
light-source, the manipulation of the filters, the electronic flash and the camera-shutter were all supervised

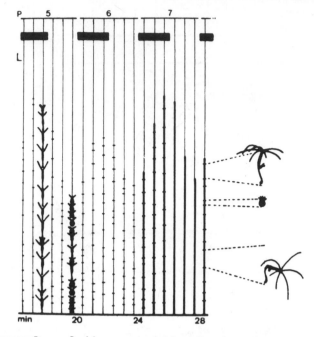

Fig. 3: Example of the quantitative evaluation of the
        state of contraction of 14 polyps as recorded
        by single movie camera pictures in a alternating
        dark-light program (see Fig. 6).

by a control unit, which could be programmed such as to
make the presence of an observer unnecessary. The state
of contraction or extension of the 14 polyps was recorded
by projecting each picture on to a screen and by adding
up the body lengths of all 14 animals to a column (Fig. 3).
Most of the illumination programs consisted of sequences
in which 2 different light-intensities were alternating
in periods of 2 min. each (Fig. 6-8).

RESULTS

The contraction of a <u>Hydra</u> elicited by a light-stimu-
lus (Fig. 4b) is, unlike the contractions following
electrical or mechanical stimulations (Fig. 4 c-e) dis-
continuous, and resembles in this respect the spontaneous
contractions (Fig. 4a), which are endogenously controlled
by 2 pace-maker centers (Passano, 1962; Passano & McCullough
1962, 1963, 1964, 1965). This suggests that the reactions
to sudden light-intensity-changes are merely modifications

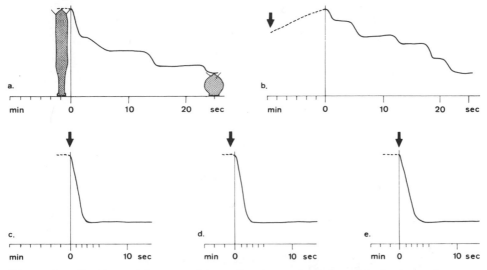

Fig. 4: Contraction patterns recorded by high-speed
movie-camera (3,5 pictures per sec.). a = spontaneous contraction; responses to: b = light-
stimulus, c = electrical pulse (alternating
current), d = electrical pulse (continuous current)
e = mechanical stimulus. Arrows indicate the onset of the stimulus (from Borner and Tardent,
1971).

of these activities. The average frequency of the spontaneous contractions increases with the light-intensity
(Borner and Tardent, 1971), whenever the animals are exposed to relatively long periods (2-24 hours) of unchanged intensities (Fig. 5). In these particular cases
(Borner and Tardent, 1971) every sudden change to another
intensity-level is answered by a typical behavioral
pattern: during the first 40-50" all behavioral responses
seem to be inhibited (Passano and McCullough, 1963). This
is followed by a period of elongation lasting about 6
min. after which the animal resumes its contraction-ex-
tension-activity, which shows a 3 fold acceleration as
compared to the rhythm proper to the previous illumination
period. This state of hyperactivity lasts for 30' to 60'
after which the frequency falls back to a level which is
characteristic for the new intensity-level (Fig. 5).

The animals can, however, be maintained in a permanent state of hyperactivity, if they are exposed to in-

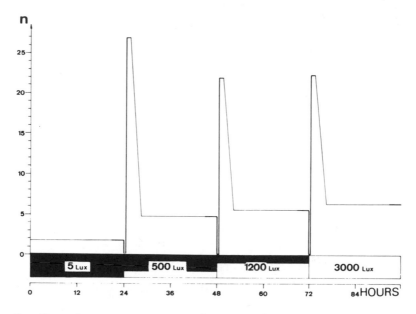

Fig. 5: Spontaneous, pacemaker-controlled contraction
activities at 4 different light intensities. Each
change of intensity is answered by a short period
of hyperactivity. n = number of contractions per
hour (from Borner and Tardent, 1971).

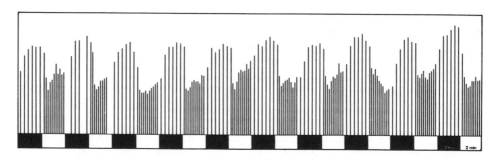

Fig. 6: Contraction-extension behavior (see also Fig. 3)
of 14 dark-adapted animals synchronized with an
alternating and repeated dark-light-program. Each
period lasted 2'. The light intensity in the light
period was 90 000 Lux.

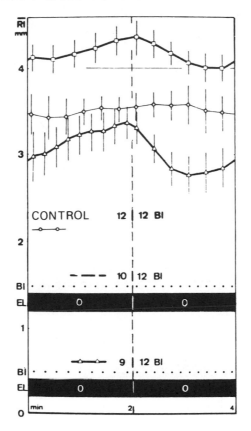

Fig. 7: Responses of 14 animals (average curves) to 3
different programs of electronic flashes (BL) given
in total darkness at varying intervals. In the controls the sequence of flashes (12 per 2 min) was
not altered throughout the experiment, while in
the other 2 series the sequence was altered according to the figures given in the graph.

tensity changes repeated at intervals of 2' each (Figs.
6-8). In this case the extension-contraction activity
becomes perfectly synchronized with the program. The polyps elongate during the period of lower intensity and
start contracting at the onset of the period of higher
intensity (Fig. 6). By combining different pairs of in-

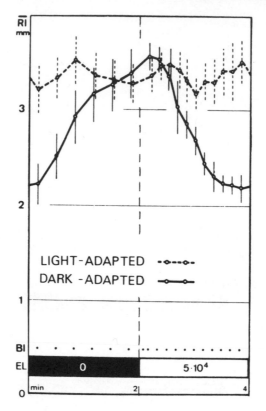

Fig. 8: Comparison between the responses of dark- and
         light adapted animals to an alternating dark-
         light-program. The curves represent the average
         of 10 subsequent dark-light periods of 2' each
         (EL = erg/cm^2/sec; BL = time sequence of electro-
         nic flashes, Rl = lengths of the body columns
         of 14 animals).

tensities in such a standardized program we found that
<u>Hydra</u> <u>attenuata</u> Pall. is capable of perceiving and re-
acting to sudden intensity changes of the magnitude of as
little as 8%. These synchronized activities last for as
long as the program is running; there is no detectable
habituation to photic stimuli even if the latter are regularly repeated for several hours.

Although <u>Hydra</u> does not react to a single electronic
flash (1/2000 sec; 9.10^3 erg/cm^2) by which it is hit in

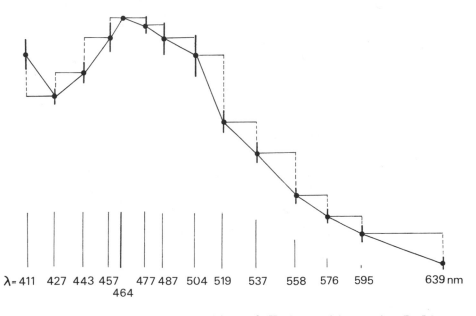

λ= 411   427 443 457   477 487 504 519   537   558   576   595   639 nm
464

Fig. 9: Spectral sensitivity of Hydra attenuata Pall.

complete darkness it will respond to a series of repeated
flashes, when these are given at a constant frequency for
2' and then at a different frequency for the next 2' (Fig.
7). Surprisingly enough the polyps are capable of distinguishing between flash frequencies of 9 and 12 flashes regularly given in periods of 2' each.

Pretreatment of the experimental animals was found to
be of the utmost importance: previously dark-adapted polyps
responded differently, i.e. much more sensitively, to
light-stimuli than light-adapted specimens (Fig. 8). Animals kept for weeks in complete darkness remain in a semi-
contracted state and will stretch when suddenly illuminated
(Passano and McCullough, 1963; Frei, 1973). This stretching,
the speed of which is proportional to the light-intensity
given after darkness, is followed by a fast contraction

(Fig. 4b). The state of adaptation is reversible: dark-
adapted specimens will, when constantly illuminated for
at least 3 days, behave like light-adapted polyps and vice-
versa (Frei, 1973).

For testing the spectral sensitivity the same experimental set-up was used. In each series 2 different wavelengths were checked against each other in a 2' alternating program. When testing 2 different wave lengths against
each other, care was taken always to apply the same transmission energy of 203 erg/cm^2/sec, by calibrating the
monochromatic filters combining them with neutral-density-
filters (Wegner and Adamson, 1966). The wave length at
which <u>Hydra</u> <u>attenuata</u> Pall. exhibited the highest degree of
spectral sensitivity is in the blue range (470 nm), thus
confirming the previous findings of other authors (Haase-
Eichler, 1931; Singer et al., 1963; Passano and McCullough,
1964). Synthetic carotenoids of different absorption properties, such as β-carotene and canthaxanthine, when injected into the gastric cavity and taken up by the entodermal cells (Frigg, 1970) did not alter the pattern of
spectral sensitivity (cfr. Feldman and Lenhoff, 1960).

                            DISCUSSION

The method for recording the reactions of <u>Hydra</u> to
a number of different photic stimuli has proved to be satisfactory, because, besides being fully automatic it
allows an objective, quantitative evaluation of the con-
traction-extension-behavior of as many as 14 animals at
a time.

In accordance with other observations (see Rushforth,
1973) the responses to light stimuli are nothing more than
modifications of the endogenous, pace-maker controlled activities (Passano and McCullough, 1962, 1964, 1965). The
simplest influence exerted by light consists of a dependence of the frequency of the basic contraction pattern
on the light intensity to which the animals are exposed
over periods longer than one hour (see Fig. 5, Passano and
McCullough, 1964). A second kind of modification of the
endogenous rhythm concerns the behavior following a sudden
change of the light-intensity: when dark-adapted, mostly
half-contracted animals are illuminated there is, as already shown by Passano and McCullough (1963, 1964) a period
of inactivity (<u>H</u>. <u>littoralis</u> = 5-10", <u>H</u>. <u>attenuata</u> =

40-60") preceding the expansion of the polyps, which is
then followed by an elongation and a fast contraction
(Frei, 1973) leading to a state of hyperactivity (Fig. 5).
The latter can be seen after any intensity change (Fig.
5), unless this change is not less than 8%; i.e. Hydra
attenuata is capable of perceiving intensity changes as
small as 8%. When, after one intensity change the intensity-level is kept constant the induced hyperactivity
will stop and the animal will continue at a new frequency
the spontaneous contractions, suggesting that in this particular case a sort of habituation has taken place (Borner and Tardent, 1971). No signs of habituation can be observed if the polyps are kept in a constant state of hyperactivity by repeated short-term intensity changes given at
regular intervals of 2' (Figs. 6, 8). In this case the
contraction-extension pattern synchronizes itself with the
alternating light-program (Fig. 6) such that the lower intensity elicits an extension, the higher a contraction of
the animal. It seems therefore that repeated photic stimuli
keep the pace-maker centers permanently at a higher level
of activity.

H. attenuata, which does not respond to a single
electronic flash, is capable of integrating series of
flashes (Fig. 7) and of reacting, when the intervals of
these series are slightly altered, as if they were subjected to a regular dark-light program (Fig. 8).

Nothing is known so far at which level adaptation to
permanent light-conditions takes place, nor can we say
which types of cells are capable of perceiving photic stimuli (Weber and Tardent, 1976) nor how light acts upon the
pace-maker-centers and influences their activities (Passano and McCullough, 1962, Rushforth, 1973). Nothing is
known about the photochemical base of the spectral sensitivity. That the mass of carotenoids that is always present in the entodermal cells does not act as a screening
filter, since the pattern of spectral sensitivity could
not be altered by injecting carotenoids of different absorption properties.

This work was supported by the Swiss National Science
Foundation (Grant No. 3.711.72). We thank Prof. I. Deak
for reading the manuscript.

REFERENCES

Borner, M. and P. Tardent, 1971. Der Einfluss von Licht
    auf die Spontanaktivität von Hydra attenuata
    Pall. Rev. suisse Zool., 78: 697-704.
Feldman, M. and H.M. Lenhoff, 1960. Phototaxis in Hydra
    littoralis: Rate studies and localization of
    the "photoreceptor". Anat. Rec., 137: 354-355.
Frigg, M., 1970. Vorkommen und Bedeutung der Carotinoide
    bei Hydra. Z. vergl. Physiol., 69: 186-224.
Haase-Eichler, R.H., 1931. Beiträge zur Reizphysiologie
    von Hydra. Zool. Jb., Abt. Zool., 50: 265-312.
Haug, G., 1933. Die Lichtreaktionen der Hydren (Chloro-
    hydra viridissima und Pelmatohydra oligactis).
    Z. vergl. Physiol., 19: 246-303.
Loomis, W.F. and H.M. Lenhoff, 1956. Growth and sexual
    differentiation of Hydra in mass culture. J. Exp.
    Zool., 132: 555-573.
Passano, L.M., 1962. Neurophysiological study of the co-
    ordinating systems and pacemakers of Hydra.
    Amer. Zool., 2: 435-436.
Passano, L.M. and C.B. McCullough, 1962. The light respon-
    se and the rhythmic potentials of Hydra. Proc.
    Nat. Acad. Sci. (Wash.), 48, 1376-1382.
Passano, L.M. and C.B. McCullough, 1963. Pacemaker hier-
    archies controlling the behavior of Hydra.
    Nature (London), 199: 1174-1175.
Passano, L.M. and C.B. McCullough, 1964. Coordinating
    systems and behavior in Hydra. I. Pacemaker
    system of the periodic contractions. J. Exp.
    Biol., 41: 643-644.
Passano, L.M. and C.B. McCullough, 1965. Coordinating
    systems and behavior in Hydra. II. The rhythmic
    potential system. J. Exp. Biol., 42: 205-231.
Singer, R.H., N.B. Rushforth and A.L. Burnett, 1963. The
    photodynamic action of light on Hydra. J. Exp.
    Zool., 154: 169-174.
Tardent, P., 1966. Zur Sexualbiologie von Hydra attenuata
    Pall. Rev. suisse Zool., 73: 357-381.
Tardent, P. and E. Frei, 1969. Reaction patterns of dark-
    and light-adapted Hydra to light stimuli. Expe-
    rientia, 25: 265-267.
Trembley, A., 1744. Mémoires pour servir à l'histoire d'un
    genre de polypes d'eau douce à bras en forme de
    cornes. Leyden, J. and H. Verbeck.
Weber, Ch. and P. Tardent, 1976. A qualitative and quantita-
    tive inventory of nervous cells in Hydra attenua-
    ta Pall., see this book.

Wegner, E.E. and A.W. Adamson, 1966. Photochemistry of
          complexions. III. Absolute quantum yields for
          the photolysis of some aqueous chromium (III)
          complexes. Chemical actinometry in the long
          wavelength visible region. J. Amer. Chem. Soc.,
          88: 394-404.
Wilson, E.B., 1891. The heliotropism of Hydra. Amer. Nat.,
          25: 413-433.

ANALYSIS OF HYDRA CONTRACTION BEHAVIOUR

Cloe Taddei-Ferretti, L. Cordella and S. Chillemi

Laboratorio di Cibernetica del C.N.R.
Via Toiano 2, 80072 Arco Felice, Napoli, Italia

INTRODUCTION

It has been pointed out (Rushforth, 1971, 1973) that the effect of light on contractions of Hydra can be: 1) excitatory, when a contraction occurs within some minutes after a light pulse and is followed by a positive phototropic movement (Haug, 1933); 2) inhibitory, when a light pulse interrupts a contraction in progress, anticipating the next one (Passano and McCullough, 1964) or when the contraction frequency momentarily decreases after a transition from darkness to light (Passano and McCullough, 1964: Borner and Tardent, 1971). The contraction-relaxation behaviour is due to interactions among different pacemakers: excitatory interactions among tentacle pulses and their excitation of contraction pulses (CP's) (Rushforth and Burke, 1971; Rushforth, 1973) and mutual inhibition between CP's and rhythmic potentials (RP's) (Passano and McCullough, 1963). The polarity of Hydra is reflected also in the location of the pacemakers of RP's and CP's in the lower column and in the sub-hypostome, respectively (Passano and McCullough, 1963). At this point, we felt that it would be interesting to record the reaction time after single or repetitive light pulses of different polarity, intensity and duration and to investigate the details of the inhibitory effect of light. In addition one could examine the shape of the bioelectric events at the two ends of the animal in undisturbed conditions and under electrical stimulation of either polarity.

MATERIAL AND METHODS

Three experimental set-ups were used: 1) where not otherwise

stated, the base of the animal was lightly sucked into a polyethylene suction Ag/AgCl electrode and another Ag/AgCl electrode
was used as the indifferent one; 2) the animal was allowed to adhere naturally to the uninsulated tip of an Ag/AgCl electrode (elsewhere insulated) with an indifferent electrode as above; 3) the
animal was drawn into the inside of a polyethylene tube, 0.5 mm
I.D., with a particular part of its column in contact with a hole
made in the wall of the tube with a needle point; the tube was filled
with culture solution and immersed in it, except at the two ends; 3
Ag/AgCl electrodes were then placed inside the two ends of the tube
and just over the hole, so that any one of them could be made the
indifferent electrode and the two others the active ones feeding
into two recording channels.

In all cases the solution in the electrodes was the same as the
Ham, Fitzgerald and Eakin (1956) culture solution in which the experiments were conducted. Bioelectric activity was fed directly from the
electrodes to a Tektronix 5031 oscilloscope (input impedance 1MΩ)
and in series to a YEW 3047 pen recorder. Recordings were made either
ac or dc. Electric pulses were provided by a Grass S48 stimulator
linked to a Grass CCU1 constant current generator, the two endings of
which were attached directly to the recording and indifferent electrodes. The polarity of the pulses could be varied: where the positive output of the generator was attached to the active electrode
we refer to Case A, where the negative, to Case B. Further details
of the experimental conditions are reported elsewhere (Taddei and
Cordella, 1975). The frequency, fH, of the contraction pulse trains
(CPT's) during a time period is here defined as the reciprocal of the
average value, $PH = \Sigma_{i=1}^{n-1} PHi/(n-1)$, of the time intervals PHi between
the beginning of each two adjacent CPT's in that period; fS and PS are
the frequency and the inter-stimulus interval for repetitive stimulation.

RESULTS

After a positive light step fH momentarily decreases, after a
negative one it increases, and the fH variation increases with the
relative, not absolute, light level. A CPT in progress can be
immediately interrupted by a darkness pulse and by a positive or
negative light step (and by a positive pulse, as already known).
A light step can also affect the intensity and number of CPs inside
a CPT.

If an animal is exhibiting a regular CPT pattern, whether
spontaneously or brought on by a single positive photic stimulus
(step or pulse), one can observe the dependence of the response time
to a stimulus (interval, θ, between the time of application of the
stimulus and the time of occurrence of the first CP of the next
CPT) on the phase of stimulus application in the inter-CPT's interval

(time, T<PH, of application of the stimulus after the first CP of
the past CPT): both for a light pulse delivered to an animal in
complete darkness and a dark pulse for one in light (Figs. 1a, b
respectively), θ is not a constant but is a function of T, which
depends on the pulse polarity; or, in other words, the phase
response curve (dependence of the phase shift, ΔΦ, of the next
CPT after the pulse on T, being +ΔΦ advances and -ΔΦ delays) is not
expressed by ΔΦ = -T+A (being A = const $\overline{<}$ PH) (Fig. 1,c). For low
values of T, the positive or negative pulse (which is shifting the
next CPT) arrives during a CPT in progress and has the described
inhibiting effect on it.

     CPT's are rapidly entrainable with repetitive light pulses
(in a 1:1 ratio between CPT's and pulses if fS does not differ too
much from fH, otherwise more than 1 CPT alternating with 1 stimulus
or 1 CPT alternating with more than 1 stimulus), and the phase
relation established between CPT's and stimuli depends on (fH-fS)
(Fig. 1,d). After a period of such stimulation, the CPT's activity
remains locked to fS for a time of the order of 5-40 PS, during
which time PHi variations are much smaller than before the
stimulation. If the pulse duration is very long (of the order of
PH/2, being fS of the order of 1/PH), CPT's begin in darkness at
a fixed time after the end of each stimulus. With repetitive stimuli
of both short and long duration this time interval increases
with increase of the intensity of the stimuli; in the latter
case (long duration) the interval can exceed the off stimulation
period when the off level is higher than complete darkness (cfr.
Tardent et al., 1976). With repetitive (short) darkness pulses, as
a rule entrainment is achieved for a period of time after which the
CPT series slowly shifts in relation to the series of stimuli.

     The resting potential on which CPT's grow (recorded dc) is not
a steady one but consists of a slow oscillation or 'big slow wave'
(BSW) of some millivolts amplitude and period PH, with a shape
changing slightly according to the individual tested and at the
falling phase of which CP's occur, preceded by RP's (Fig. 1,e). The
same observation has also been made with electrode set-up 2. With
this set-up correlations both between CPT's and contraction behaviour
and between the inter-CPT portion of the BSW and relaxation behaviour
have been observed. Occasionally, for some animals, the portion
of the BSW coincident with the CPT changes so rapidly that it can be
observed in ac recordings (Fig. 1,f). A light pulse affects the BSW
in a way dependent on the phase at which it occurs: given when a
CPT is in progress it replaces the downward with an upward slope;
given after a CPT the upward slope is momentarily replaced by a
horizontal or downward trend. Occasionally the BSW amplitude was
higher after a negative light step and lower during stimulation
with repetitive light pulses. In addition to the BSW, a 'little
slow wave' (LSW) with period equal to the inter-CP period has been

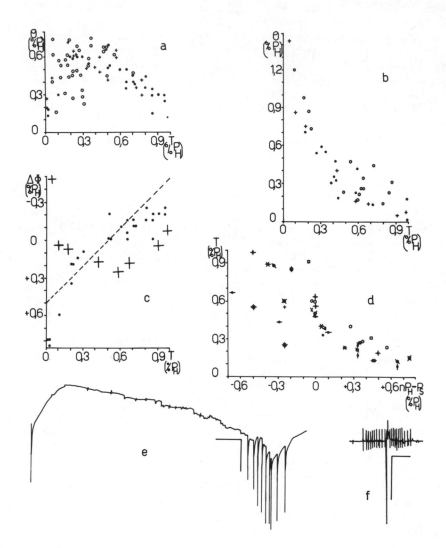

Fig. 1 a) Θ versus T for 10 s 3000 lux pulses (on animal in darkness); b) Θ versus T for 10 s darkness pulses (in 3000 lux); c) ΔΦ
versus T; --- hypothetical fixed time of response, i.e., ΔΦ =
-T+PH/2; ... 10 s 3000 lux pulses (in darkness); +++ 10 s darkness
pulses (in 3000 lux); d) T versus nPH-PS; n = number of CPT's
occurring during 1 PS. In a-d all times are expressed as fractions
of PH; PH of all animals are normalized to 1; each animal one sym-
bol; e) dc recording of a BSW with RP's and a CPT; f) ac recording
of BSW derivative during a CPT. Scale: 30 s, 1 mV in e; 30 s,
250 µV in f.

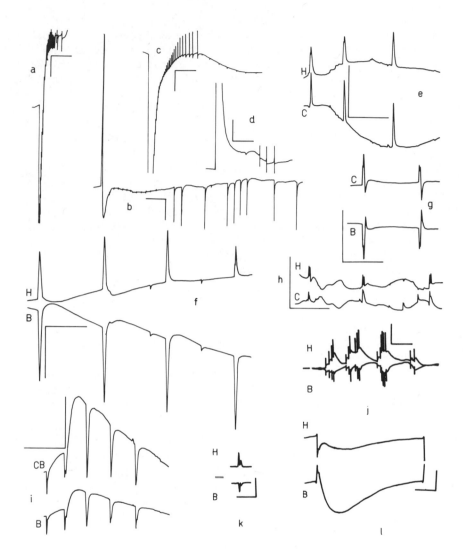

Fig. 2 a-d) dc recording of bioelectric activity after 0.5 s (in
a, b) or 5 s (in c, d) 150 V pulse with stimulator setting A (in
b, d) or B (in a, c); base line at the same level in all cases;
the actual voltage across the animal is much smaller due to the
high resistence in series with the animal; no CCUl is used in this
case. e-l) BSW, LSW, CP, CPT recorded dc at the base, column beside
the base, medium part of the column, hypostome (B, CB, C, H). Scale:
30 s, 2 mV in a-d; 6 s, 500 μV in e-i; 2 min, 500 μV in j; 1.5 s,
500 μV in k, l. Reference level: -2mV in j,k.

observed (Fig. 2 e-1).

An electric pulse elicits two different types of both behavioural and bioelectric responses, depending on the direction of the current flow along the animal's long axis. With electrode set-up 2, relaxation or contraction of the animal occurs with stimulator setting A or B respectively. With electrode set-up 1, RP's (and subsequently CP's after some minutes) are elicited with stimulator setting A, while an immediate and long series of CP's is elicited with stimulator setting B (Fig. 2 b, a). On increasing either the intensity or the duration of the stimulus, in Case A the CP's occurrence is anticipated, in Case B more CP's appear and at their end RP's occur (Fig. 2 d, c). With electrode set-up 3 the same type of bioelectric response is obtainable with stimulator setting A or B using the base or hypostome electrode respectively as the stimulating-recording one (alternatively, the other type of response is obtainable with Case B at the base electrode or with Case A at the hypostome electrode). With prolonged dc stimulation, the on part of a positive stimulus or the off part of a negative one at the base, or the off part of a positive stimulus or the on part of a negative one at the hypostome, elicited the same type of response.

The polarity of the BSW, LSW's and CP's recorded (with electrode set-up 3) at the upper, medium and medium-low part of the column is the same as the polarity recorded at the hypostome, and is the opposite of that recorded at the lower part of the column and at the base (Fig. 2 e-1). The polarities of the three types of bioelectric events do not all reverse at the same point (Fig. 2 e, h).

<h2 style="text-align:center">DISCUSSION</h2>

CPT behaviour under photic stimulation indicates that the CPT's triggering system has the properties of an oscillator. Postulating that a single stimulus causes an immediate phase shift of the biological clock regulating the rhythm (Pittendrigh and Minis, 1964), the phase relations between repetitive stimuli and rhythmic biological events during entrainment can be predicted by the phase response curve for a single stimulus (Pittendrigh, 1965) and vice-versa (Pavlidis, 1973). The good correspondence of the trend of $\Delta\Phi$ on T curve (Fig. 1, c, curve ...) and the trend of T on nPH-PS curve (Fig. 1, d) rotated 90° clockwise for positive light pulses, as well as the short period during which entrainment can be maintained as a consequence of the long tract with negative slope of the phase response curve for negative light pulses (Fig. 1, c, curve +++), indicate that the hypothesis that a stimulus causes an immediate phase shift along the time axis holds also for the mechanism generating CPT's. Such a hypothesis can also explain the two contrary effects of light (inhibition of a CPT in progress and excitation of a contraction within few minutes) if considered

together with the hypothesis that it is not a single triggering
pulse which causes a train of CP's but that a train of triggering
pulses causes a CPT: as soon as a positive stimulus is given to
the CPT's triggering system, either during or after the
occurrence of the train of the triggering pulses, the state of the
system immediately changes as shown by Fig. 1, c (curve ...); the
above train of triggering pulses can be identified with the Rushforth (1971, 1973) Multiple Event Generator. The CPT's triggering
system is presumably nervous (Schwab et al., 1976) and its
rhythmicity could originate either endogenously in the nervous
system or by a secondary nervous process involving transduction of
primary biochemical oscillations acting as depolarizing input to
the nervous system itself (Taddei, Cordella and Chillemi, 1976).
Further experiments are needed. The BSW could reflect a rhythmic
metabolic activity, possibly related to an ion transport mechanism
in which, e.g., Na^+ and Ca^{2+} fluxes are mutually dependent: an
influence is known (Macklin and Josephson, 1971) both of external
Na^+ concentration on the amplitude of CP's (caused by Ca^{2+} influx
into epitheliomuscular cells) and of external Ca^{2+} concentration
and of CPT occurrence on transepithelial potential (caused by
active Na^+ transport by epitheliomuscular cells); moreover periodic
contractions (to which CPT's are related), causing liquid material
to be periodically expelled by the mouth, are a part of the osmo-
regulatory behaviour involving such active Na^+ transport (Benos and
Prusch, 1973). Na^+ - Ca^{2+} interaction is known in other systems
(Lüttgau and Niedergerke, 1958; Blaustein and Wiesmann, 1970;
Rubin, 1970; Baker, 1972; Yoshikami and Hagins, 1973) as well as an
action of light on Na^+ flux similar to the action of Ca^{2+} (Yoshikami and Hagins, 1973). Further experiments are needed. The
differing effects of electrical current flow in the two directions
along the body axis (which is in accordance with Jennings (1906))
could possibly be explained by the different conductance values of
the two cations of the solution, Na^+ and Ca^{2+} (that, at infinite
dilution, are respectively 50.1 and 59.5 (Kortüm and Bockris, 1951);
this means that Ca^{2+} migrates from cathode to anode at a rate about
20% higher than that of Na^+. As the pacemakers of CP's and RP's,
which inhibit each other in the generation of the periodic contraction-relaxation behaviour, are located at different ends of the
animal (and if Na^+ and Ca^{2+} play a role in such behaviour), the
different relative mobility of these ions could explain the effect
of the direction of the stimulating current (as in the case of
Paramecium ciliary reversal (Jahn, 1962)). Examples of reversal of
the polarity of bioelectric events depending on the site of the
recording electrode are known in Coelenterates (Josephson, 1967;
Ohtsu and Yoshida, 1973; Spencer, 1974; Mackie, 1975), fish
electrocytes (e.g., Keynes and Martins-Ferreira, 1953) and Vertebrate
retina (Tomita, 1950; Brown and Wiesel, 1961). Also for Hydra one has
to postulate an electrogenic zone, possibly related to the animal's
polarity, so that recording at the proximal and distal ends of Hydra

is equivalent to recording at the opposite ends of an electric
dipole generating an electromotive force, whose intensity varies
during PH (Fig. 2, j) perhaps reflecting metabolic events. Further
experiments are needed to provide insight into the significance
of these findings.

ACKNOWLEDGEMENTS

We are indebted to Dr. L.M. Passano for much advice, fruitful
suggestions and criticism, to Drs. R.K. Josephson, G. Kass-Simon
and A.N. Spencer for stimulating discussions, to Dr. D.M. Ross for
the revision of the English manuscript, to Mr. A. Cotugno for
continuous valuable assistance, to Mr. F. Forte for help in equipment preparation and to Mr. S. Piantedosi for photographic help.

SUMMARY

The amount and polarity of phase shift in sequences of
contraction pulse trains (CPTs) is a function of the phase of
application and of the polarity of a light pulse: predictions can
be made for stimulus - CPT phase relations during entrainment with
repetitive stimulation. CPT's arise at a definite phase of a
slow bioelectric wave. Depending upon the direction of a current
flow along the animal's long axis, its behavioural and bioelectric
responses vary. The polarity of bioelectric signals is linked
to the body polarity.

LITERATURE CITED

Baker, P.F., 1972. Transport and metabolism of calcium ions in
    nerve. Progr. Biophys. molec. Biol. 24: 177-223.

Benos, D.J., and R.D. Prusch, 1973. Osmoregulation in Hydra:
    column contraction as a function of external osmolality.
    Comp. Biochem. Physiol. 44A: 1397-1400.

Blaustein, M.P., and W.P. Wiesmann, 1970. Effect of sodium ions
    on calcium movements in isolated synaptic terminals. Proc.
    Nat. Acad. Sci. 66: 664-671.

Borner, M., and P. Tardent, 1971. Der Einfluss von Licht auf die
    Spontanaktivität von Hydra attenuata Pall. Rev. Suisse Zool.
    78: 697-704.

Brown, K.T., and L.T.N. Wiesel, 1961. Analysis of the intra-
    retinal electroretinogram in the intact cat eye. J. Physiol.
    158: 229-256.

Ham, R.G., D.C. Fitzgerald Jr. and R.E. Eakin, 1956. Effect of
    lithium ion on regeneration of Hydra in a chamically defined
    environment. J. Exp. Zool. 133: 559-572.

Haug, G., 1933. Die Lichtreaktionen der Hydren (Chlorohydra viridissima und Pelmatohydra oligactis). Z. vergl. Physiol. 19: 246-303.

Jahn, T.L., 1962. The mechanism of ciliary movement. II. Ion antagonism and ciliary reversal. J. Cell Comp. Physiol. 60: 217-228.

Jennings, H.S., 1906. Behavior of the lower organisms. pp 208-210 of 1962 edition, Indiana University Press, Bloomington.

Josephson, R.K., 1967. Conduction and contraction in the column of Hydra. J. Exp. Biol. 47: 179-190

Keynes, R.D., and H. Martins-Ferreira, 1953. Membrane potentials in the electroplates of the electric eel. J. Physiol. 119: 315-351.

Kortüm, G., and J.O'M. Bockris, 1951. Textbook of Electrochemistry. Elsevier Press, Amsterdam.

Lüttgau, H.C., and R. Niedergerke, 1958. The antagonism between Ca and Na ions on the frog's heart. J. Physiol. 143: 486-505.

Mackie, G.O., 1975. Neurobiology of Stomotoca. II. Pacemakers and conduction pathways. J. Neurobiol. 6: 357-378.

Macklin, M., and R.K. Josephson, 1971. The ionic requirements of transepithelial potentials in Hydra. Biol. Bull. 141: 299-318.

Ohtsu, K., and M. Yoshida, 1973. Electrical activity of the antho-medusan, Spirocodon saltatrix (Tilesius). Biol. Bull. 145: 532-547.

Passano, L.M., and C.B. McCullough, 1963. Pacemaker hierarchies controlling the behaviour of hydras. Nature 199: 1174-1175.

Passano, L.M., and C.B. McCullough, 1964. Co-ordinating systems and behaviour in Hydra. I. Pacemaker systems of the periodic contractions. J. Exp. Biol. 41: 643-664.

Pavlidis, T., 1973. P. 46 of Biological oscillators: their mathematical analysis. Academic Press, New York.

Pittendrigh, C.S., 1965. On the mechanism of entrainment of circadian rhythm by light cycles. Pages 277-297 in J. Aschoff, ed., Circadian clocks. North-Holland Publishing Co., Amsterdam.

Pittendrigh, C.S., and D.H. Minis, 1964. The entrainment of circadian oscillations by light and their role as photoperiodic clocks. Am. Naturalist 98: 261-294.

Rubin, R.P., 1970. The role of calcium in the release of neurotransmitter substances and hormones. Pharmacol. Rev. 22: 389-428.

Rushforth, N.B., 1971. Behavioral and electrophysiological studies of Hydra. I. Analysis of contraction pulse patterns. Biol. Bull. 140: 255-273.

Rushforth, N.B., 1973. Behavioral modifications in Coelenterates. pp 123-169 in W.C. Corning and J.A. Dyal, eds., Invertebrate learning, Vol. 1 Plenum Press, New York-London.

Rushforth, N.B., and D.S. Burke, 1971. Behavioral and electro-physiological studies of Hydra. II. Pacemaker activity of isolated tentacles. Biol. Bull. 140: 502-519.

Schwab, W.E., R.K. Josephson, N.B. Rushforth, B.A. Marcum and R.D. Campbell, 1976. Excitability of nerve-free Hydra. Abstracts of this Symposium.

Spencer, A.N., 1974. Behavior and electrical activity in the hydro-zoan Proboscidactyla flavicirrata (Brandt) I. The hydroid colony. Biol. Bull. 145: 100-115.

Taddei-Ferretti, C., and L. Cordella, 1975. Modulation of Hydra attenuata rhythmic activity. Photic stimulation. Arch. it. Biol. 113: 107-121.

Taddei-Ferretti, C., L. Cordella and S. Chillemi, 1976. Hydra simple nervous system and behaviour. In press, in Proc. 3rd Eur. Meet. Cyb. Syst. Res., Vienna.

Tardent, P., E. Frei, M. Borner and F. Zürcher, 1976. The reactions of Hydra attenuata Pall. to various photic stimuli. This volume.

Tomita, T., 1950. Studies on the intraretinal action potential I. Relation between the localization of micro-pipette in the retina and the shape of the intraretinal action potential. Jap. J. Physiol. 1: 110-117.

Yoshikami, S., and W.A. Hagins, 1973. Control of the dark current in Vertebrate rods and cones. pp 245-255 in H. Langer, ed., Biochemistry and physiology of visual pigments. Springer-Verlag, Berlin.

THE ULTRASTRUCTURAL BASIS FOR THE ELECTRICAL COORDINATION BETWEEN EPITHELIA OF HYDRA

Linda Hufnagel and G. Kass-Simon

Depts. of Microbiology and Zoology, University of

Rhode Island, Kingston, Rhode Island   02881

Previously published electrophysiological and ultrastructural studies suggest that the coordinated behavior of Hydra depends on both nervous and non-nervous conduction (c.f. Josephson and Macklin, 1967; Kass-Simon, 1972; Wood, 1961; Hand and Gobel, 1972). As discussed in another paper in this volume (Kass-Simon, 1976), hydra's behavioral repertoire seems to require the controlled passage of information between the two muscular cell layers, the ectoderm and the endoderm, which are separated from each other by a thick acellular layer, the mesoglea. To help reveal a morphological basis for this interaction, we undertook a light and electron microscopic study of Hydra, directed exclusively toward defining how the two cell layers interact morphologically. Three separate but related topics were addressed: 1) direct interaction of muscular cells of the two epithelia via gap junctions; 2) bridging of the mesoglea by nerve cell processes, and 3) a third type of cellular process which has been reported to occur within the mesoglea of Hydra, but which is actually a bacterial symbiont.

METHODS

Starved Swiss Hydra attenuata were relaxed with menthol, fixed with 4% glutaraldehyde in 0.084M cacodylate buffer, pH 7.4, at room temperature for two hrs, washed in buffer, post-osmicated, stained en bloc for 2 hrs with 1% uranyl acetate at RT, dehydrated through alcohols and flat-embedded in Spurr's standard

medium. Observations were made on silver to gray ultrathin
sections, unstained or stained with uranyl acetate and lead citrate,
and on thick (1/2-1µ) sections stained with Toluidine Blue. Extensive observations were made on both transverse and longitudinal
sections from the hypostomal and basal thirds of five individuals.
In addition, a few observations have been made on the mid-body
region, on transverse sections, and on the region at the base of
the tentacles, on longitudinal sections. Electron micrographs
were taken with RCA 3G, Philips 201 and Hitachi HS-9 electron
microscopes.

OBSERVATIONS

Hydra relaxed with menthol failed to respond when touched
with the tip of a pipette. However, response to a tactile stimulus
returned within 5 min when the animals were transferred to fresh,
non-mentholated medium. There were no apparent effects of the
menthol treatment on the ultrastructure of Hydra.

In all regions of Hydra examined, direct contact between
epitheliomuscular cells of the two layers via processes containing
gap junctions was seen. Frequently, the endodermal cell process
was in the form of a calyx, surrounding a finger-like extension
from an ectodermal cell (see fig 1). On several occasions, the
reverse configuration was seen, with the ectodermal cell forming
a calyx around the endodermal cell (see fig 2). Sometimes, a
process from one cell layer spanned the entire mesoglea and im-
pinged directly on a cell body of the apposing layer, or the apposing processes lay side-by-side or took on a more complex relationship. In transverse sections, each endodermal muscular cell
possessed several processes, each of which made contact with a
different ectodermal muscular cell (see fig 3). In longitudinal
sections, ectodermal cells were seen to contact more than one
endodermal muscular cell. A diagram summarizing these observations is given in fig 5.

In all instances, gap junctions were seen between the epitheliomuscular cell processes. No other type of junction was recognized. The gap junctions were 7-layered, typical of gap junctions
stained en bloc with uranyl acetate (see fig 4). The distance
between outer dense layers averaged 184 Å. Usually the gap junctions occupied a large portion of the area of apposing membranes.
However, some regions were devoid of junctioning.

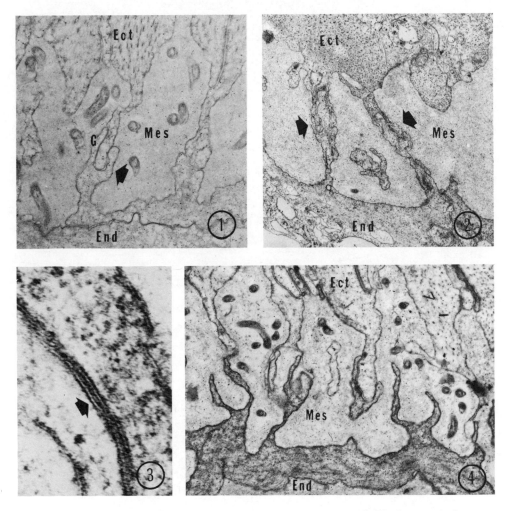

Fig. 1. An endodermal muscular cell process forming a calyx
   around an ectodermal muscular cell extension (Arrow) (End-
   endoderm: Ect-ectoderm; Mes-mesoglea; G-gap junction).
   X 20,354.

Fig. 2. Calyx formation by ectodermal cell processes (arrows).
   X 8,910.

Fig. 3. High resolution micrograph of a gap junction between
   epitheliomuscular cell processes, stained en bloc with uranyl
   acetate. Note 7-layered structure. X 200,000.

Fig. 4. A single endodermal cell can have several extensions,
   each of which contacts a separate ectodermal cell. X 15,261.

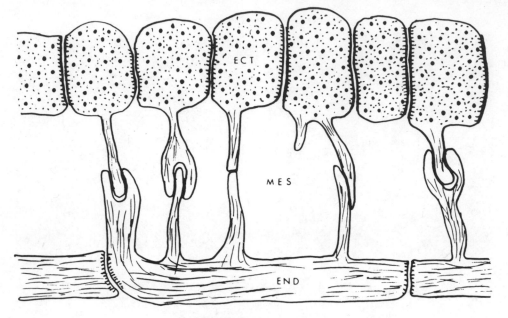

TRANSVERSE SECTION

Fig. 5. A diagram summarizing the kinds of relationships which
can exist between ecto- and endodermal muscular cells.

    The muscular cell processes usually contained fibrillar
material oriented parallel to the process itself; occasionally, a
few microtubules were found, in the ectodermal cell processes.
Generally, the ectodermal processes were more electron transparent than the endodermal processes. Often, the tips of the
processes seemed swollen and electron transparent.

    When we compared different regions of the animal, we ob-
served a marked asymmetry in the frequency of bridging of the
mesoglea by epithelial cell processes. Near the anterior end of
the hypostome, epithelial bridges were quite numerous, while
toward the posterior end of the hypostome the processes were
much less frequent (compare figs 6 and 7 with 8 and 9). In the
mid-body region and the peduncle, muscular cell processes were
quite scarce (fig 10). However, at the basal disc, there again
seemed to be an increased number of processes (fig 11). The
relative frequency of gap junction-containing bridges between
epithelia in different regions of the animal is diagrammatically
portrayed in figure 12.

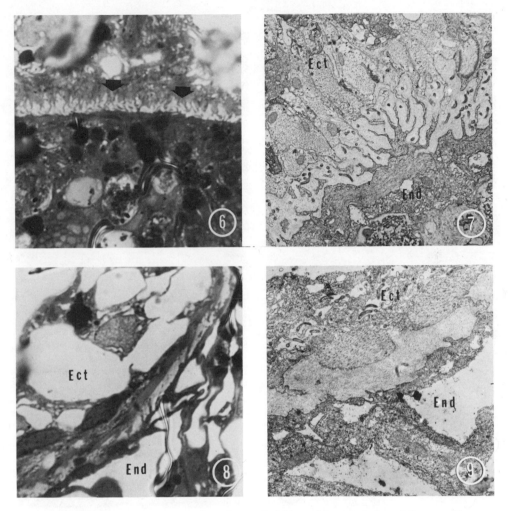

Fig. 6.   Light micrograph of a toluidine blue-stained 1μ section, from the tip of the hypostome.   Note frequent cell processes crossing the mesoglea (arrows).   Compare with Fig. 8. X 1,625.

Fig. 7.   Electron micrograph of an ultrathin section from the tip of the hypostome, showing the numerous muscular cell processes within the mesoglea.   Compare with Fig. 9. X 5,060.

Fig. 8.   Light micrograph of a thick section from the base of the hypostome.   Note the relatively infrequent occurrence of cellular processes across the mesoglea.   Compare with Fig. 6. X 1,625.

Fig. 9.   Electron micrograph from the base of the hypostome. Note that transmesogleal processes are scarce.   Compare with Fig. 7.   X 5,062.

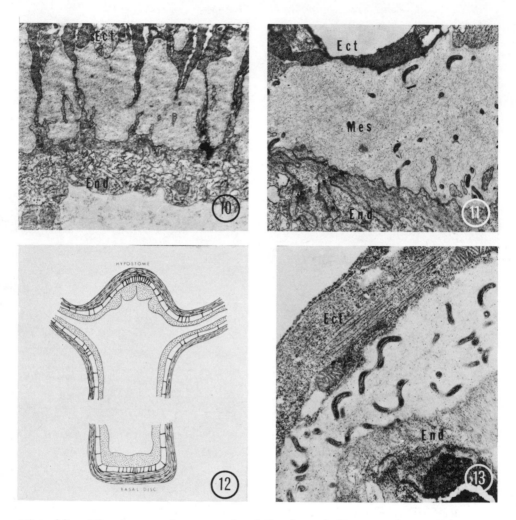

Fig. 10. Electron micrograph of the basal disc region. Note abundance of transmesogleal processes. Compare with Fig. 11. X 7,560.

Fig. 11. Electron micrograph from the peduncle region. Note the scarcity of transmesogleal processes. Compare with Fig. 10. X 7,560.

Fig. 12. A diagram summarizing the relative abundance of inter-epithelial processes in various regions of hydra's body.

Fig. 13. Electron micrograph of spiral bodies found in the meso-glea and in intercellular spaces and identified as spirochete bacteria. X 12,250.

We also directed our attention toward whether the epithelia communicate via nerve cell processes. We found no convincing instances where neurites crossed the mesoglea, although such processes were easily identified within the epithelia. We did encounter a number of ambiguous cases where it was impossible to identify the cell type from which a process arose, or with which it made contact. We have also seen neurites in contact with the mesoglea, where they could be available for contact with processes from the other side. Even if the two cell layers do communicate via nerve cell processes, it is clear from our observations that this occurs relatively infrequently, compared with direct epitheliomuscular contacts.

Within the mesoglea, we also observed elongated structures, 1000-1500 Å in diameter and many microns in length (see fig 13). Similar structures also were found in the spaces between epitheliomuscular cells. These bodies were spiral in form, and consisted of a central cylinder, and outer sheath and an axial filament oriented helically around the central cylinder, within the sheath, (Hufnagel, manuscript in preparation). We identified them as spirochete bacteria (see discussion).

DISCUSSION

Hadzi (1909), in a light microscopical study, first recognized that the mesoglea of <u>Hydra</u> is occasionally traversed by cytoplasmic processes. Wood (1961), using electron microscopy, found that extensions of the muscular cells cross the mesoglea and contact muscular cells of the other layer by means of specialized junctions. Bridging of the mesoglea by epitheliomuscular cell processes has been reported on a number of other occasions since (Haynes, Burnett and Davis, 1968; Davis and Haynes, 1968; Hand and Gobel, 1972; Westfall, 1973; Bonnefoy and Kolenkine, 1975; and Hufnagel, Borsay and Kass-Simon, 1975).

The presence of gap junctions between muscular cell processes was recognized by Hand and Gobel (1972), who characterized them quite extensively, following staining with lanthanum hydroxide and ruthenium red, and who also suggested that they are important for coordination of epidermal and gastrodermal contraction (Hand, 1971). They were also identified by Westfall (1973), Bonnefoy and Kolenkine (1975), and by us (Hufnagel et al., 1975).

Recently, the occurrence of the gap junctions within cup-like regions of the cell extensions was reported by Bonnefoy and Kolenkine (1975) and by us (Hufnagel et al., 1975). We reported that the endodermal process forms a calyx-like expansion around the tip of the ectodermal process, while Bonnefoy and Kolenkine reported the opposite configuration, with the calyx at the end of the ectodermal process. In our studies reported here, we now find that both arrangements are possible; however, the endodermal calyx appears to occur more frequently.

We feel that the most significant result of our current study has been the discovery that mesogleal muscular cell processes are asymmetrically distributed within the body of Hydra, occurring most requently near the tip of the hypostome and at the basal disc. Since it is a generally accepted fact that epithelial cells are migratory, we feel it likely that these processes and their gap junctions are being constantly formed and taken apart. However, what we are seeing at one moment in time is probably typical of their relative frequency at any other time, and so the asymmetrical distribution is significant.

We were surprised at first to find no convincing evidence that nerve cell processes cross the mesoglea. However, there appears to be some controversy in the literature over this important question. Westfall (1973) reported that neurites occasionally traverse the mesoglea, particularly in the hypostomal region, but her published picture was of a differentiating cell. Lentz, (1966), stated unequivocally that "neurites of nerve cells do not cross or extend into the mesoglea", but also said that Burnett, in a personal communication, reported seeing small differentiating interstitial cells crossing the mesoglea. Thus, it seems possible that differentiating nerve cells migrate across the mesoglea for developmental purposes, but whether they provide a path for conduction between epithelia, as Westfall (1973) suggests, is still not clear.

In a study of the mesoglea of Hydra pirardi, Davis and Haynes described long, dense objects, about 750-800 Å in diameter, which they identified as cellular extensions. We have not observed structures of this type. Larger structures, about 1,500-2000 Å in diameter, also identified as tubular cellular processes, were seen by Bonnefoy and Kolenkine (1975), in Hydra vulgaris and Pelmatohydra oligactis. We observed similar structures in Hydra attenuata, but conclude that they are not cellular processes, but symbiotic spirochete bacteria, since they exhibit a number of morphological features common to members of the Order Spirochaetales. These include a spiral form, the presence of a central cylinder and an outer sheath and the presence within the outer sheath of an axial filament. Davis and Haynes (1968) also observed bacteria in the mesoglea of all species of Hydra. In Hydra attenuata, at least, there does not appear to exist a system of tubular cell processes which could provide another means for communication between epithelia.

In conclusion, we believe that the excitatory communication between the epithelia must occur primarily, if not exclusively, via gap junctions between muscular cells. Furthermore, our latest observations suggest that the strength of this communication is a function of the frequency of gap junctions in different regions of the animal. Since processes containing gap junctions occur most frequently at the tip of the hypostome and the basal disc region, these may be places where the two epithelia are most able to talk to each other electrically. The significance of these observations, in relation to the complex behavior of Hydra, will be discussed in another paper (Kass-Simon, this volume). As far as we know, this is the first time that such an asymmetry in distribution of potential electrical synapses has been demonstrated in a coelenterate.

ACKNOWLEDGEMENTS

This study was performed under a URI Grant-in-Aid to L. Hufnagel and an NSF Grant #BMS75-02293 to G. Kass-Simon. The authors wish to express their gratitude to Doranne Borsay, who helped to initiate this study, and to Lynn Myhal and Ronnie Diesl, who assisted with technical aspects of the research.

REFERENCES

BONNEFOY, A-M. and X. KOLENKINE, 1975. Ultrastructure et signification des "corps tubulaires" et "particules" de la mesoglée chez les Hydres. C. R. Acad. Sc. Paris, Ser D. 280: 2673-2676.

DAVIS, L. E. and J. F. HAYNES, 1968. An ultrastructural examination of the mesoglea of Hydra. Zeitschr. f. Zellf. 92: 149-158.

HADZI, J., 1909. Ueber das Nervensystem von Hydra. Arb. zool. Inst. Wien. 17: 225-268.

HAND, A. R. and S. GOBEL, 1972. The structural organization of the septate and gap junctions of Hydra. J. Cell Biol. 52: 397-408.

HAND, A. R., 1971. Observations on the substructure of septate and gap junctions in Hydra. Anat. Rec. 169: 333-334 (abstr).

HAYNES, J. F., A. L. BURNETT and L. E. DAVIS, 1968. Histological and ultrastructural study of the muscular and nervous systems in Hydra. I. The muscular system and the mesoglea. J. Exp. Zool. 167: 283-294.

HUFNAGEL, L., D. BORSAY and G. KASS-SIMON, 1975. Ultrastructural basis for direct electrical coordination of ectodermal and endodermal epitheliomuscular cells in Hydra. J. Cell Biol. 67: 185a.

JOSEPHSON, R. K., and M. MACKLIN, 1967. Transepithelial potentials in Hydra. Science 156: 1629-1631.

KASS-SIMON, G., 1972. Longitudinal conduction of contraction burst pulses from hypostomal excitation loci in Hydra attenuata. J. comp. Physiol. 80: 29-49.

LENTZ, T. L., 1966. The Cell Biology of Hydra, North-Holland, Amsterdam

WESTFALL, J. A., 1973. Ultrastructural evidence for a granule-containing sensory-motor-interneuron in Hydra littoralis. J. Ultr. Res. 42: 268-282.

WOOD, R. L., 1961. The fine structure of intercellular and mesogleal attachments of epithelial cells in Hydra. LENHOFF, H. M. and LOOMIS, W. F., Eds., pages 51-67 in The Biology of Hydra. Univ. Miami Press, Coral Gables, Florida.

COORDINATION OF JUXTAPOSED MUSCLE LAYERS AS SEEN IN HYDRA

G. Kass-Simon

Department of Zoology

University of Rhode Island, Kingston, R.I.   02881

Fish guts, rabbit oviducts, worms, blood vessels and Hydra
share with each other the problem of coordinating juxtaposed layers
of muscle. In Hydra, as in the viscera of vertebrates, there is now
compelling evidence that this coordination is essentially mediated
by nerves acting on sheets of electrically conducting muscle cells.
In fact, the best way to look at Hydra, is probably to see it as a
miniature vertebrate gut with nematocysts and tentacles. Looking
at it in this way immediately causes much of its behavioral system
to fall into place.

Hydra's two muscle layers are arranged so that the cell processes of each layer lie at right angles to each other. Slow,
gradual contractions of the internal, circular muscle layer are
responsible for the elongation of the animal and are accompanied
by regularly occurring rhythmic potentials (Shibley, 1969). These
pulses, which were first described by Passano and McCullough (1963)
and which were originally thought to be produced and conducted in
the endodermal nerve net, are reminiscent in their regularity of
the rhythmic pulses often described for mammalian small intestine.

Opposed to this slow, internal contraction is the faster,
periodically occurring contraction of the external longitudinal
musculature which compresses the animal into a compact ball. These
periodic contractions were found by Passano and McCullough (1964)
to be triggered by large, somewhat faster potentials, which usually
occurred in bursts of 8 to 10 pulses and which they postulated were
produced and conducted in the ectodermal nerve net.

It has been demonstrated that contraction burst pulses are indeed ectodermally conducted while rhythmic potentials are conducted
over the endoderm (Kass-Simon and Passano, 1969, and in preparation).
As illustrated in Figure One, when an animal is turned inside out

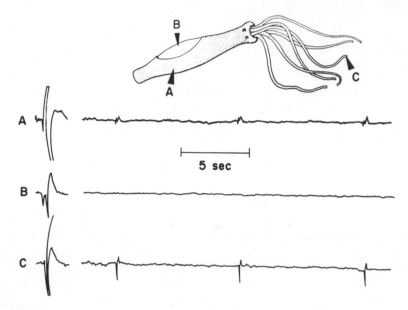

Figure One.
Recordings from tentacle tip, endoderm and ectoderm of a <u>Hydra</u> that
has been turned inside out. The large contraction burst pulse at the
left is recorded on all three channels. Rhythmic potentials do not
appear at the electrode which is on the ectoderm alone. (Arrow heads
indicate electrode placement. Shaded area represents endoderm).

and a portion of the endoderm scraped away, recordings (made with
polyethylene suction electrodes) from the endoderm-free area fail to
disclose rhythmic potentials, although these are readily picked up by
the electrodes on the endodermal surface and on the tentacle tip.
The large contraction burst pulse, seen at the left, is recorded by
all three electrodes.
        For a long time, no one really knew what <u>Hydra</u> was doing when
it alternately stretched and contracted in this way. But now, Macklin has presented persuasive evidence that the animal was expelling
the water which was always seeping into its hyperosmotic gut (Macklin,
1973).
        According to Macklin, the contractions are apparently triggered
by an increased hydraulic pressure within the enteron. And the
point to be noted here is that either there is direct mechanical stimulation of the ectodermal muscle layer, because of the circular expansion of the enteron, or the endodermal layer alone is being
directly stimulated and is in turn causing the ectodermal layer to
contract.

A similar, apparently excitatory influence of endoderm on ectoderm is seen in Hydra's famous somersaulting response to a light stimulus. Passano and McCullough (1963) showed that when Hydra's base is stimulated with light, the animal first elongates. This elongation is accompanied by an increase in the frequency of the internally conducted rhythmic potentials, followed by a shift in the site of origin of these pulses from base to hypostome, and culminates finally in an intense body contraction caused by the externally produced contraction burst pulses. Upon conclusion of the contraction burst, the initiation of rhythmic potentials returns to the basal region, and their frequency reverts to a previous, slower rate.

Set against these instances of mutual intermural excitation, are examples of mutual inhibition. Rushforth, et al. (1963) have shown that when a Hydra is exposed to reduced glutathione, the animal stretches out with a concomitant suppression of both spontaneous and light-induced contraction behavior. Similarly, if you insert an electrode into the basal pore of a Hydra, there is an apparent reduction of contraction burst activity, while rhythmic potentials appear with great regularity and perhaps even with an increase in frequency (Kass-Simon, unpublished observations).

Thus, if we were now to summarize Hydra's behavior, we would be forced to construct sentences like these: a) When Hydra is stretching, it is not also contracting; but b) either stretching or contracting ultimately results in the opposite event: either contracting or stretching. This interaction becomes somewhat difficult to understand physiologically, since the two layers must be interacting with one another in both an excitatory and inhibitory fashion.

Fortunately, the number of possible ways in which the layers can do this is limited by morphology. One really has only two alternatives: 1) either the layers are interacting directly by means of the epitheliomuscle cells themselves, and this includes mechanical or electrical interaction either between individual cells or between the two entire sheets of tissue; or, 2) the layers are interacting by means of nerve cells, or 3) both of the above.

There is rather strong morphological and physiological evidence that within each layer, at least some parts of the conducting path are in fact non-nervous. In the endoderm, nerves are so sparsely scattered that they do not form a complete net (McConnell, 1931, Lentz, 1966), while in the ectoderm, as has been pointed out elsewhere (Kass-Simon, 1970, 1972, 1973), the straight-line longitudinal bias of the conducting path does not appear to coincide with the histological picture of a diffusely distributed nerve net. Further, the extremely fast conduction velocity of many of the contraction burst pulses (Kass-Simon, 1973) strongly speaks in favor of an electrically conducting epitheliomuscular system. This conclusion has often been reached on other grounds by other people (Josephson

and Macklin, 1967, Horridge, 1968, among others). The situation in
both layers, therefore, seems to be analogous to what is going on
in functional muscle syncitia as seen in the guts and hearts of
vertebrates.

   But in contrast to the evidence for epitheliomuscular conducting paths, there are convincing arguments that impulse initiation
is due to activity produced by the nervous system as has been
classically proposed. Indeed, it would seem that Hydra represents
at least partially a neurogenic muscle system which is quite specifically controlled by cholinergic neurons (Kass-Simon and Passano,
in preparation).

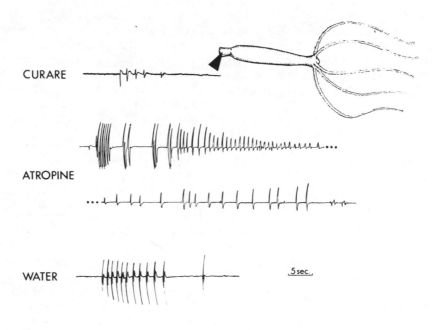

Figure Two.

Effect of curare and atropine on the contraction burst system in
Hydra. The records shown illustrate typical changes in the contraction burst pattern in animals treated with either curare or
atropine.

   If Hydra are placed in a solution of curare, there is a distinct
inhibition of body contraction behavior. The number of periodic
contractions that an animal undergoes falls from a mean of 11 to 12
bursts per hour to one of about 5 per hour and the sizes of the

contraction burst pulses are often reduced as can be seen in Figure Two. Further, there is a significant reduction in the number of pulses per burst, which fall from a mean of 11 pulses per burst in fresh water to one of 4 in curare. Curare, it will be recalled, is the specific cholinergic inhibitor for the nicotinic receptors of vertebrate striated muscle.

On the other hand, placing Hydra in a solution of atropine raises havoc with the contraction burst system. Atropine is the specific cholinergic inhibitor for muscarinic junctions in vertebrate smooth muscle. Instead of the ordinary number of pulses per burst, there are now some 20 to 30 pulses, and instead of the burst lasting for its usual 20 odd seconds, it now lasts for as long as a minute or more; the pulses finally die out in apparent exhaustion.

Now, either atropine has in some way excited the contraction burst system, or it has blocked an inhibition which would normally stop the contraction burst. That is, atropine would be interfering with those cholinergic nerves having muscarinic junctions whose effect is to inhibit the contraction burst activity of the external muscle layer.

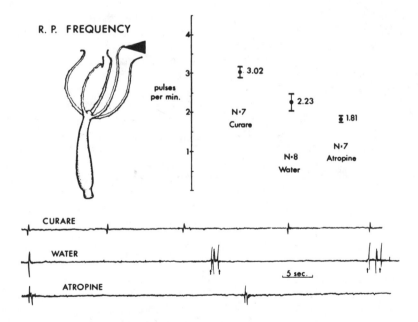

Figure Three.
Effect of curare and atropine on the frequency of rhythmic potentials. There is an apparent increase in frequency for those animals treated with curare, while the frequency for those in atropine declines. Values represent means and standard errors of the means.

A clue to what may be happening can be found in the effects of curare and atropine on endodermal activity. As expected, when animals are placed in atropine and their rhythmic potentials monitored, we find that there is a general tendency to reduce the frequency of rhythmic potential occurrence. (See Figure Three.) In contrast, the rhythmic potential frequency in curare appears to be increased. (The difference between atropine and curare is significant at p = 0.0016, while those among atropine, curare and fresh water are not; undoubtedly, this is because of the large variance in the fresh water control.) In other words, the effect of atropine and curare on the endodermal electrogenic system seems precisely the reverse of that on the ectodermal system.

So, if atropine inhibits the rhythmic potential system at the same time as it enhances the contraction burst system, and if curare inhibits the contraction burst system at the same time as it increases rhythmic potential output, it is very likely that the correct interpretation of these data is this: blocking the excitatory nicotinic junctions in the ectoderm releases the endoderm from an ectodermal inhibitory influence, while blocking the excitatory muscarinic junctions in the endoderm releases the ectoderm from endodermal inhibition.

Precedent for this sort of dual cholinergic control exists in the gut of the fresh water fish, the tench. Frey (1928) demonstrated that stimulating the vagus causes the tench gut to respond with a fast twitch of its external striated layer and a slow contraction of its internal smooth muscle layer. Upon application of curare, the fast component of the response is abolished, while the smooth muscle response is abolished by atropine. More importantly, Romanes (1885) had already shown that upon exposure to atropine, Sarsia responds with convulsive swimming and that application of nicotine causes a violent, continuous spasm in the animal. A similar dual system is probably operating in Hydra.

Further evidence for nervous participation in producing the electrical activity of Hydra's muscle layers is obtained from intracellular recordings made from dissociated endodermal epitheliomuscular cells (Kass-Simon and Diesl, 1976, in preparation). As can be seen in Figure Four, potentials which appear similar to junctional potentials recorded from other muscle systems, can be recorded from identified epitheliomuscular cells. In the top record spontaneous positive and negative potentials of about 1 mV are recorded in a cell which has a 2.5 mV positive resting potential. In the lower record, positive potentials are elicited by stimuli administered through the recording electrode. Although it is possible that these are direct responses of the membrane to the stimuli, it is more likely that these potentials are junctional potentials produced by nerves which have been antidromically stimulated. This is probable since the potentials are of varying polarity and size and appear either to sum or facilitate.

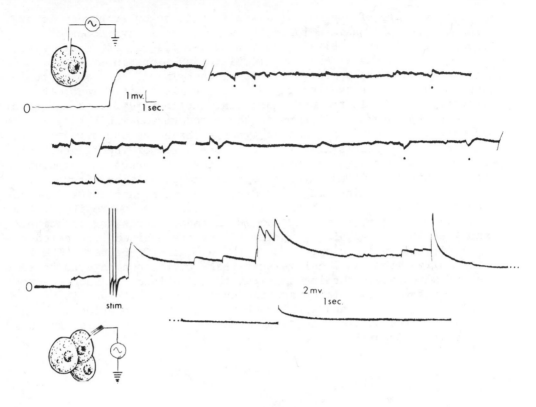

Figure Four
Intracellular recordings of presumed junctional potentials in dissociated endodermal epithelial cells.

        But to return to the original question. How do the two muscle
layers interact.
        There is now strong morphological evidence (Hufnagel, Borsay &
Kass-Simon, 1976, Hufnagel and Kass-Simon, present symposium) that
there is direct epithelial interaction between the two layers. Processes bearing calices which contain gap-junctions are found with high
frequency at both the hypostome and basal region of the Hydra. Nerve
connections through the mesoglea are rare, if present at all (Hufnagel
and Kass-Simon, present symposium). And it is precisely at the hypostome and base that the mutual excitatory interaction of the two cell
layers is seen during the light response. Therefore, it would seem
that these intermural bridges are likely candidates for the substrate
of such interactions. A similar excitatory mechanism via electric-
ally-coupled cells has recently been postulated for rabbit oviduct.
(Marshall, 1976, verbal communication).

The problem of the mutual inhibition is somewhat harder to explain. From the pharmacological studies, one of two modes of inhibition seems likely. Either, as in the top diagram in Figure Five,
there are both muscarinic and nicotinic junctions in each cell layer
and these have the opposite effect in each layer. Or, as in the
lower diagram, the two layers of muscle cells inhibit each other
directly. Thus, in the upper drawing, cholinergic fibres making nicotinic synapses are excitatory in the ectoderm and inhibitory in the
endoderm; while the muscarinic synapses have the opposite effect.
(A variant of this is illustrated in the inset of the figure, which
is meant to represent the possibility that there are two populations
of cholinergic neurons in each layer; one population enters into
nicotinic and one into muscarinic synapses with each layer.) When
the nicotinic synapsing fibres are active, the ectoderm is excited,
while the endoderm is inhibited. When these fibres are inactive,
and when the muscarinic synapsing fibres are excited, the ectoderm
is inhibited, while excitation in the endoderm increases. The
difficulty with this model is that there would then have to be some
way of alternately turning on first the nicotinic synapsing neurons
and then the muscarinic synapsing neurons. In addition, it would
imply that endoderm and ectoderm have membrane properties of both
smooth and striated muscle, and although this is possible, it adds
an additional complication.

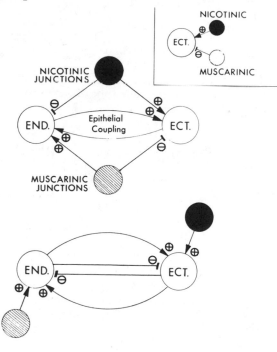

Figure Five.
Two ways in which ectoderm and endoderm might interact.

A simpler model and one that also fits the behavioral data is given in the lower drawing of Figure Five. Muscarinic and nicotinic junctions are both excitatory; the first acting on the endoderm in analogy to a vertebrate smooth muscle system, the other acting on the ectoderm in analogy to a striated muscle system. When either layer is active, it directly inhibits the other layer, either electrically or mechanically. As the level of excitation in the active layer increases, this inhibition is overridden directly by the excitation which is transmitted across the gap-junction-bearing bridges of the base and hypostome. The inactive layer now becomes active, and in turn temporarily inhibits the originally active layer.

In summary, then, excitation of each layer is likely to occur in two ways: by neuronal initiation through cholinergic nerves of epitheliomuscular potentials, and by direct electrical stimulation by the aposing layer across the mesoglea at the base and hypostome.

Inhibition, on the other hand, is likely to occur as a less specifically localized event; resulting from either direct electrical or mechanical interaction of the entire muscle sheets.

ACKNOWLEDGEMENTS

The work summarized here was done a) in the laboratory of L.M. Passano under NSF Grant No. BO12421 b) in the laboratory of L. Hufnagel, under University of Rhode Island Research Grant-in-Aid to L. Hufnagel, and c) in the author's laboratory under NSF Grant No. BMS75-02293. I would like to thank R.P. McCormick, Jr. for executing the diagrams.

BIBLIOGRAPHY

Frey, E. 1928. Giftwirkungen an dem quergestreiften Schleiendarm. Arch. f. exper. Pal. u. Pharmakol. 138: 228-239.

Horridge, G.A., 1968. Interneurons. W.H. Freeman & Co., London & San Francisco.

Hufnagel, L., Borsay, D., and Kass-Simon, G. 1975. Ultrastructural basis for direct electrical coordination of ectodermal and endodermal epitheliomuscular cells in Hydra. J. Cell. Biol. 67: 185a.

Josephson, R.K. and Macklin, M. 1967. Transepithelial potentials in Hydra. Science 156: 1629-1631.

Kass-Simon, G. 1970. Multiple excitation sites and straight-line conduction in the contraction burst system of Hydra. Amer. Zool. 10: 505.

Kass-Simon, G. 1972. Longitudinal conduction of contraction burst
    pulses from hypostomal excitation loci in Hydra attenuata.
    J. Comp. Physiol. (Z. Vergl. Physiol.) 80: 29-49.

Kass-Simon, G. 1973. Transmitting Systems in Hydra. Proc. 2nd
    Int. Symp. Publ. Seto Mar. Biol. Lab. Vol. 20.

Kass-Simon, G. and Passano, L.M. 1969. Conduction pathways in Hydra.
    Amer. Zool. 9: 113.

Lentz, T.L. 1966. The Cell Biology of Hydra. John Wiley and Sons,
    Inc.,New York.

Macklin, M. 1973. Water Excretion by Hydra. Science 179: 194-195.

McConnell, C.H. 1931. A detailed study of the endoderm of Hydra. J.
    Morphol. 53: 249-263.

Passano, L.M. and McCullough, C.B. 1963. Pacemaker hierarchies con-
    trolling the behavior of Hydra. Nature (Lond.): 1174-1175.

Passano, L.M. and McCullough, C.B. 1964. Co-ordinating systems and
    behavior in Hydra. I. Pacemaker system of the periodic con-
    tractions. J. Exp. Biol. 41: 643-644.

Passano, L.M. and McCullough, C.B. 1965. Co-ordinating systems and
    behavior in Hydra. II. The rhythmic potential system. J. Exp.
    Biol. 42: 205-231.

Romanes, G.J. 1885. Jelly-fish, star-fish, and Sea Urchins being a
    research on Primitive Nervous Systems. D. Appleton & Co.
    New York.

Rushforth, N., Krohn, I., Brown, L. 1964. Behavior in Hydra: Inhi-
    bition of the Contraction Responses in Hydra pirardi. Science
    145: 602-603.

Shibley, G. 1969. Gastrodermal contractions correlated with rhyth-
    mic potentials and prelocomotor bursts in Hydra. Amer. Zool.
    9: 586.

IONIC REQUIREMENTS OF TRANSEPITHELIAL POTENTIALS IN ISOLATED CELL

LAYERS OF HYDRA

Martin Macklin and Gary Westbrook

Department of Biomedical Engineering
Case Western Reserve University
Cleveland, Ohio 44106

Because hydra is one of the few fresh water coelenterates, it requires a unique system to maintain ionic and osmotic equilibrium. Several studies in the past have begun to elucidate the mechanism of ionic and osmotic regulation in hydra (Josephson and Macklin, 1969; Macklin, 1967; Macklin and Josephson, 1971; Marshall, 1969). These studies have demonstrated that there is a maintained electrical potential across the hydra epithelium with the gut positive relative to the external medium and it was shown by Macklin and Josephson (1971) that this sustained positive potential was related to an active sodium transport mechanism. Further, superimposed on this sustained positive potential, there are negative going spikes termed "contraction pulses" (CPs) which had been shown to relate to contraction of the body column (Josephson, 1967). Various studies of the osmotic properties of hydra cells and tissue (Benos and Prusch, 1972; Koblick and Yu-Tu, 1967; Lilley, 1955; Marshall, 1969; Steinbach, 1963), have demonstrated that hydra tissue is isosmotic to a solution with an osmotic strength of 40 to 60 milliosmol, and that the gut of the animal is approximately isosmotic to the hydra tissue. These experimental results have led to the conclusion that ionic and osmotic regulation of the intact hydra is maintained by an active transport of sodium with passive movement of water and an anion and with isosmotic flow from the tissue into the gut.

It was also demonstrated that hydra eliminate excess fluid from the gut cavity by contracting the body column and expelling excess water through the mouth of the intact animal (Macklin, Roma and Drake, 1973).

Because it was still uncertain which body wall layer of the animal was involved in osmotic regulation and the production of contraction pulses, and because it has been suggested previously that the cell layers might be interchangeable in terms of their osmotic regulatory properties, experiments were conducted on isolated cell layers to ascertain what properties of the individual cell layers might be important to the intact animal.

MATERIALS AND METHODS

All experiments were conducted at 20° to 23° with Hydra oligactus starved for 24 hours prior to use. Hydra were reared and separated into individual cell layers as described in Macklin (1976). Electrical measurements were also performed as described in Macklin (1976). A number of solutions varying in ionic concentration were used in this study (Table I). They are similar to those previously used (Macklin and Josephson, 1971). As previously described, the method for establishing representative values of transepithelial potential was one in which the animal was held in a test solution for five minutes after solution change; the potential was measured at 3, 4, and 5 minutes in the test solution; and the average of these three values was used as a measure of the resting potential for that regenerate at that experimental condition. The reported results are from the analysis of data from multiple animals for each experiment.

Solution changes were such that we were sure of a greater than 99.9% exchange of solution. The experimental chamber used is the same one previously used in experiments reported in Macklin and Josephson (1971).

RESULTS

The potentials reported here are generally lower than those previously reported for hydra by Macklin and Josephson (1971) and Josephson and Macklin (1969). This difference can be attributed to at least three factors. First, because we are working with regenerates, the nutritional state of the animal is probably poorer than that of intact animals used for experiments. Secondly, the small size of the regenerates frequently resulted in damage to the regenerates during the experiment, and the animals were so fragile that frequently they would be torn apart during solution changes. Thirdly, the animals were generally not studied for a total of 48 hours from their last feeding; that is, tissue layer separation was done on 24 hour starved animals and then the regenerate was allowed to rest for 24 hours prior to experimental use. It has been shown in unpublished work that the transepithelial resting

TABLE I

Experimental solutions used. All concentrations are in mM/1.

Solution name	Na^+	K^+	$Tris^+$	Ca^{++}	Mg^{++}	Cl^-	HCO_3^-	SO_4^-	Sucrose
Normal	1.5			1.5		3.0	1.5		
Na-free-K		1.5		1.5		3.0	1.5		
Na-free-Tris			1.5	1.5		3.0	1.5		
Ca-free-Sucrose	1.5						1.5		4.5
Haynes	12.0	0.6		12.6	1.8	37.2	0.6	1.8	

potential generally declines after the first 24 hours in starved animals. There was also variation from one experimental series to another and this may have been accounted for by skill of tissue layer separation or general nutritional state of the animals.

Effect of Ions on Transepithelial Resting Potential

The ions that were previously studied for their effect on intact animals, that is, sodium, calcium, bicarbonate, and chloride were studied with the regenerates, both epidermal and gastrodermal cell layer isolates. However, because of the preparation, it was not possible to perfuse the gut cavity. All solutions were tested for their effect on the transepithelial resting potential for both epidermal and gastrodermal isolates. Neither ouabain nor acetazolamide in concentrations up to 10^{-4} M had any effect on the transepithelial resting potential in ectodermal or gastrodermal regenerates. Also, anions were without effect on resting potential in the experiments conducted with isolated cell layer regenerates. Therefore, only results achieved with cations are reported here although anions were tested in the same manner as described in Macklin and Josephson (1971).

For ectodermal cell layer isolates (Figure 1), it can be seen that when sodium was replaced with potassium or tris (trishydroxymethylamino methane), there was a reduction in transepithelial potential. These results are similar to those previously reported in which the transepithelial potential was on the average 20 mv lower in the sodium free tris solution compared to the sodium free K solution (Macklin and Josephson, 1971). However, for intact animals the decrease in potential was much greater when sodium was replaced by potassium or tris. For the ectodermal pieces, the effect of sodium removal was not significantly manifest unless sodium was replaced by tris. In a second series of experiments (Figure 1), it was demonstrated that calcium removal from the external

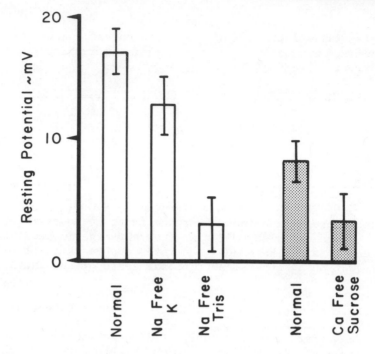

Figure 1. Relation between external medium and resting potential for ectodermal isolates. For the first series 12 animals
were used. All values were compared using Student's t-test. The
first two potentials did not differ significantly (P>0.1). Both of
the first two values differ from the Na free tris value (P<0.001).
For the second series of experiments 11 animals were used and the
two values differ (P<0.001). Sample periods were five minutes.

medium also caused a significant fall in transepithelial potential
in ectodermal pieces.

In a series of experiments to determine the effect of sodium
concentration on transepithelial resting potential in ectodermal
pieces (Figure 2), sodium concentrations were varied in decades in
the same manner as in Macklin and Josephson (1971) for intact animals. Sodium was replaced by tris for lower concentrations and
Na_2SO_4 was added to obtain 15 mM sodium. The same significant
change in resting potential is observed, however, the low concentration point is not as low as one would predict, and at high concentrations there seems to be a falling off of potential rather

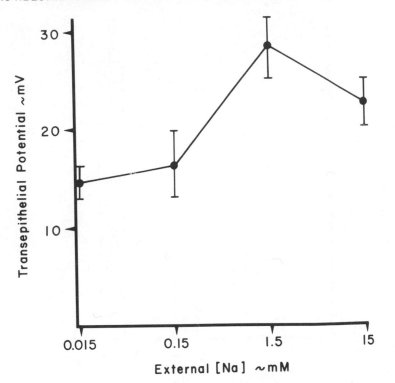

Figure 2.   Relation between transepithelial potential and external sodium concentration.   All solutions had equal osmotic pressure up to 1.5 mM Na.   The higher concentration medium used sodium sulfate to provide the additional sodium.

than a saturation as would be predicted.   This result is probably due to one or more of the following factors: that is, there is mechanical damage to the explant which would tend to decrease the overall magnitude of the resting potential in all solutions.   The regenerate probably has a low metabolic rate which would decrease the slope of the plotted values.   In the experiments the explant was placed on a small agar platform to protect the glass microelectrode used for impalement, and it is likely that the agar platform absorbed and retained sodium from the medium and thereby affected the transepithelial resting potential at the lower sodium concentrations.

When gastrodermal explants were studied, similar results to those obtained with ectodermal explants were seen (Figure 3).   However, because of the fragile nature of the gastrodermal explants, it was not possible to do as many experimental sequences with the

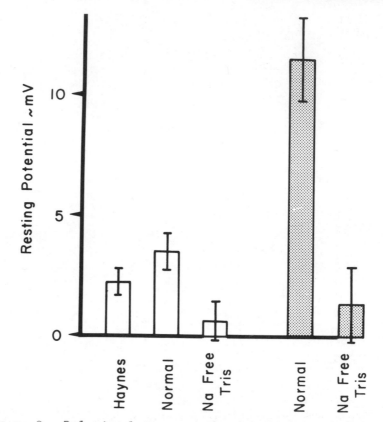

Figure 3. Relation between transepithelial resting potential and solution for gastrodermal isolates. For the first series, 22 isolates were used with test periods of five minutes. Each of the three values differ from each other (P<0.001). For the second series, six animals were used. These values differ significantly (P<0.001).

gastrodermal explants and, consequently, we were restricted to studies in which only two or three different solutions were used. In the first series of experiments (Figure 3), the normal gastrodermal regeneration culture solution (Haynes solution, Haynes and Burnett, 1963) was replaced with the normal solution and then sodium free tris solution. These results are different from previous results in which it was demonstrated that the resting potential for an intact animal was independent of the osmotic strength of the external medium up to an osmolarity approximately isosmotic to the animal tissue. However, in this case, decreasing the osmotic strength of the external medium caused an increase in the transepithelial rest-

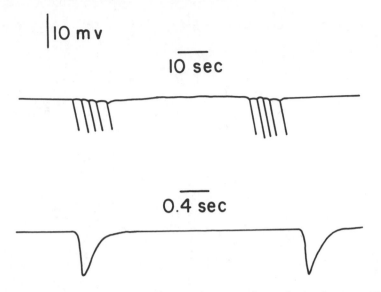

Figure 4. Electrical activity in ectodermal isolate. The first tracing shows two bursts and the second tracing shows the shape of individual contraction pulses. This record and those in the other figures we traced from original recordings for increased clarity.

ing potential of the isolate, suggesting that the gastrodermal isolate regenerating in a higher osmotic strength medium is capable of rapidly increasing its transport of ions when the osmotic strength of the external medium is lowered.

In a second series of experiments with gastrodermal pieces (Figure 3), the pieces were allowed to equilibrate with normal medium for one hour prior to experimental use. In this case, the maximal potential in the normal medium is much greater than in the solutions in which only five minutes were allowed for equilibration. The difference may also be due to the nutritional state of the animals used.

Effect of External Ions on the Contraction Pulses

Both regenerate types had spontaneous electrical activity which resembled contraction pulses as seen in intact animals. Although it is not possible to say that these are the same contraction pulses recorded from intact animals, their shape and pattern of occurrence is the same as those seen with intact animals. For ectodermal pieces (Figure 4), the contraction pulses had a shape and bursting pattern

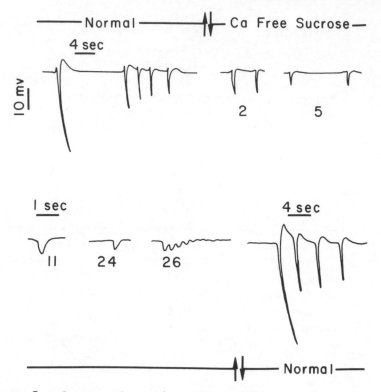

Figure 5.   Contraction pulses in a gastrodermal isolate.   Note how the contraction pulses diminish in the calcium free sucrose solution and return to the pretest value two minutes after the reintroduction of normal medium.   The numbers are minutes after the introduction of the calcium free sucrose medium.

which is very reminiscent of that seen in intact animals.   Contraction pulses of 20 mV were frequently recorded from ectodermal pieces. Watching ectodermal pieces during recording sessions confirmed the fact that these contraction pulses occurred simultaneously with contractions of the regenerate.   To ascertain which ions in the external medium were related to contraction pulses, they were each replaced. No ions studied had an effect on contraction pulses in the external medium for ectodermal pieces.

It has been generally assumed that contraction pulses are related to contraction of the ectodermal longitudinal musculature. Therefore, it would be expected that they would not be recorded from gastrodermal tissue.   However, gastrodermal isolates also demonstrated contraction pulses (Figures 5 and 6).   Whereas ectodermal

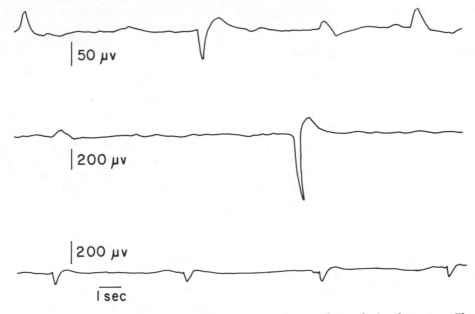

Figure 6. Electrical activity in gastrodermal isolates. The top two tracings were selected to show simultaneously both small magnitude CP activity as well as electrical pulses resembling RPs. The third and lower tracing is recorded from an isolate two hours after isolation and shows early electrical activity. (RPs are positive going and CPs are negative going pulses in intact animals.)

pieces had contraction pulses whose magnitudes were independent of external ion concentration, this was not the case for gastrodermal isolates. Gastrodermal contraction pulses were independent of sodium or anion concentration, however, they were sensitive to calcium concentration (Figure 5). Recordings from gastrodermal isolates produced magnitudes of the same order of those recorded from epidermal isolates; that is, contraction pulses of 20 mV were common for gastrodermal isolates. There was no observable behavioral correlate for contraction pulse in the gastrodermal isolate. That is, observing the gastrodermal isolate in the microscope during recording sessions, distinct contractions could not be observed during contraction pulse electrical activity as was the case with ectodermal pieces. Indeed, the only observed activity of the gastrodermal isolates was a slow pulsating movement.

Because it has been previously suggested by Passano and McCullough (1964) that RPs (rhythmic potentials) originate in the gastrodermal layer and it was also further suggested by Kass-Simon at

this symposium that RPs originate in the gastrodermal layer, high
sensitivity recordings of the gastrodermal layer were made to look
for RPs (Figure 6). The top two recordings show what appears to be
RPs in a gastrodermal isolate 24 hours after separation. It should
be noted that the two negative going potentials in this recording
which resemble CPs are low in magnitude. These are not representative CPs recorded from gastrodermal pieces but a segment of the record selected to demonstrate simultaneously low magnitude CPs and
RPs. The lower record is of greater interest because this was done
two hours post isolation, that is, prior to the time that any regeneration of epidermal tissue could have taken place. It is seen here
that there is spontaneous and rhythmic electrical activity originating in the gastrodermal layer prior to any significant regeneration.

DISCUSSION

Of the ions examined, sodium is most directly involved in the
maintenance of the transepithelial potential for both the ectodermal
and the gastrodermal isolates. This is the same result previously
reported for intact hydra. The increase in resting potential with
sodium concentration for the ectodermal explants (Figure 2) resembles that for intact animals (Macklin and Josephson, 1971). There
is a correspondence between the results showing a dependence of the
transepithelial resting potential on sodium for both the ectodermal
and the gastrodermal explants and for intact animals. Therefore,
it is likely that the transepithelial resting potential in explants
as well as in intact animals is a consequence of the active movement
of sodium from the external medium into the gut cavity. One would
not expect each cell layer to be independently involved in ionic regulation of the intact animal. That is, if the ectodermal layer as
previously suggested (Macklin and Josephson, 1971; Macklin, 1967) is
responsible for the active movement of sodium and thereby regulates
the ionic concentration of the tissue, then the gastrodermal layer
need not actively move sodium. Concentration of sodium in the hydra
tissue by the ectoderm and the passive movement of water could account for the osmotic concentration of tissue as well as the isosmotic flow of solution into the gut cavity. What then can be made
of the fact that when isolated, the gastrodermis also has a transepithelial resting potential and is sensitive to external sodium
concentration? The most likely explanation is that both cell layers
respond to the external medium and each can transport sodium if exposed to a low osmotic strength medium. Each cell layer apparently
can concentrate sodium; however, the cell layer found to lie on the
outside of the animal is the one which normally is responsible for
sodium accumulation.

In a previous report (Macklin, 1968), it was shown that when a
piece of tissue is pinned out flat, what was previously the inner

cell layer accumulates mucous droplets. This would suggest that the exposure of the surface which normally is adjacent to the gut cavity will result in this surface behaving as if it were the outside surface of the animal. In the same fashion, separating the two cell layers causes the gastrodermal cell layer to behave as if it were the external cell layer and thereby concentrate sodium.

The experiments with anions produced negative results with the isolated pieces in contrast to experiments conducted with intact animals. This absence of a response in the isolates might merely be the result of the poor condition of the isolated cell layers in contrast with the animal and the general low level of transepithelial potentials recorded in these experiments.

The effect of calcium removal on the resting potential for the external cell layer is analogous to that seen with intact animals. However, the effect of removing calcium from the external medium for the inner cell layer was without result; that is, calcium does not affect the transepithelial resting potential of the gastrodermal cell layer, whereas it does affect the transepithelial resting potential of the ectodermal layer. Why this differential result to calcium and sodium should exist is not known.

Previously we felt it particularly meaningful that removing calcium from the enteron had a significant effect on CPs, whereas removing calcium from the external medium was without effect on contraction pulses. This gave credence to the hypothesis that the CPs were calcium spikes and that an inward facing surface was involved in the production of CPs. In the present experiments, it was found that removing calcium from the external medium from the ectodermal pieces was without effect on CPs. This was consistent with previous results with intact animals. However, with gastrodermal explants, an unexpected result occurred. Gastrodermal explants had CP-like electrical activity, and this electrical activity was sensitive to external calcium concentration. This result suggests that the membrane surfaces adjacent to the mesoglea, that is the inner surface of the ectoderm and the outer surface of the gastroderm, are both capable of producing CP-like electrical activity and this activity is a calcium spike phenomena. The adaptive significance of this may be to protect an electrically active surface from the environment.

In sum, this series of experiments supports our previous conclusions and adds data about the regulatory behavior of hydra. Our results to date suggest and support the following model of osmotic regulation in hydra. Sodium is actively concentrated from the medium by the cell layer exposed to the external medium. Anions follow passively and water is osmotically drawn into the hydra tissue. The hydra tissue is isosmotic to a solution which is approximately 40 milliosmolar. The flow of water and solute into the gut is approximately isosmotic to the hydra tissue, and the animal periodically contracts to eliminate excess solution from its gut cavity. If the animal is intact, it opens its mouth to eject excess fluid; however, regenerates must burst to eliminate excess fluid.

This work was supported by National Institutes of Health Grant GM 20541.

SUMMARY

The resting potential across the ectodermal and gastrodermal isolated hydra cell layers depends on external sodium concentration. Calcium is only involved in the ectodermal transepithelial potential. Contraction pulses are spontaneously seen in ectodermal and endodermal hydra explants. These CPs are not dependent upon external calcium or sodium concentration for the ectodermal explants. However, the gastrodermal explant CPs are sensitive to external calcium concentration. CPs in ectodermal explants have visible behavioral correlates, whereas gastrodermal pieces do not have visible correlates for CPs. RPs are observed in gastrodermal explants and electrical activity is observed as early as two hours after separation of cell layers.

LITERATURE CITED

Benos, D. J., and R. D. Prusch, 1972. Osmoregulation in fresh-water hydra. Comp. Biochem. Physiol., 43A: 165-171.

Haynes, J. F., and A. L. Burnett, 1963. Dedifferentiation and redifferentiation of cells in Hydra viridis. Science, 142: 1481-1483.

Josephson, R. K., 1967. Conduction and contraction in the column of hydra. J. Exp. Biol., 47: 179-190.

Josephson, R. K., and M. Macklin, 1969. Electrical properties of the body wall of hydra. J. Gen. Physiol., 53: 638-665.

Koblick, D. C., and L. Yu-Tu, 1967. The osmotic behavior of digestive cells of Chlorohydra viridissima. J. Exp. Zool., 166: 325-330.

Lilly, S. J., 1955. Osmoregulation and ionic regulation in hydra. J. Exp. Biol., 32: 423-439.

Macklin, M., 1967. Osmotic regulation in hydra: Sodium and calcium localization and the source of the electrical potential. J. Cell. Physiol., 70: 191-196.

Macklin, M., 1968. Reversal of cell layers in hydra: a critical re-appraisal. Biol. Bull., 134: 465-472.

Macklin, M., 1976. The effect of urethan on hydra. Biol. Bull.,

150: in press.

Macklin, M., and R. K. Josephson, 1971. The ionic requirements of transepithelial potentials in hydra. Biol. Bull., 141: 299-318.

Macklin, M., T. Roma, and K. Drake, 1973. Water excretion by hydra. Science, 179: 194-195.

Marshall, P. T., 1969. Towards an understanding of the osmo-regulation mechanism of hydra. School Sci. Rev., 51: 857-861.

Passano, L. M., and C. B. McCullough, 1964. Co-ordinating systems and behavior in hydra. I. Pacemaker system of the periodic contractions. J. Exp. Biol., 41: 643-664.

Steinbach, H. B., 1963. Sodium, potassium and chloride in selected hydroids. Biol. Bull., 124: 322-336.

ELECTROPHYSIOLOGICAL CORRELATES OF FEEDING BEHAVIOR IN TUBULARIA

Norman B. Rushforth

Department of Biology, Case Western Reserve University

Cleveland, Ohio 44106

INTRODUCTION

Several studies have described spontaneous electrical activity in hydroids. Some of the observed pulses have common features in that they are relatively large in amplitude, on the order of several millivolts, have long duration, up to several hundred milliseconds, and may be externally recorded from large blocks of tissue. Such potentials, which are associated with simple recurrent behavioral activity, were first observed in Tubularia by Josephson (1962), in Hydra by Passano and McCullough (1962), in Obelia by Morin and Cooke (1971), in Corymorpha by Ball (1973).

In two hydroids, Hydra and Tubularia, some of the electrical pulses and associated behavioral events are inhibited during feeding. Rushforth and Hofman (1972) noted that feeding stimuli inhibit Contraction Pulses (CP's) and associated column contractions in Hydra. Such stimuli also inhibit bursts of Tentacles Contraction Pulses (TCP's) and contractions in isolated hydra tentacles (Rushforth and Burke, 1971). In Tubularia, concerts which are correlated with Hydranth Pulses (HP's), are suppressed during feeding (Josephson and Mackie, 1965) or exposure to food extracts (Rushforth, 1969). The present study was undertaken to investigate concert inhibition by feeding stimuli in Tubularia, and to learn something of its mechanism.

A Tubularia polyp has two sets of tentacles, small distal tentacles which surround the mouth, and larger proximal tentacles located around the base of the hydranth. Below the base of the proximal tentacles, the hydranth is attached by a contractile neck region to a more rigid proximal stalk. Distally the hydranth consists of a mobile proboscis to which are attached the mouth, the distal tenta-

729

cles, and hollow racemes which bear the gonophores.

The polyp shows considerable spontaneous activity, including individual or coordinated tentacle movements, bending or contraction of the hydranth and contraction of the neck region. Periodically, twice a minute or so, all the proximal tentacles give a concerted oral flexion, (a concert) followed by a peristaltic wave of contraction which sweeps down the proboscis. The concerts of a polyp are graded: in some the tentacle movements are brief and stop when the tentacles are only partially elevated (weak concerts); in others the tentacle movements are more vigorous, enveloping the proboscis, and are associated with contraction of all the gonophores and the neck (full concerts). The concert movements of the polyp seem to be primarily digestive activities, resulting in the mixing of the contents of the proboscis cavity and partial transport of digestion products to the stalk and gonophores.

The concert behavior of a <u>Tubularia</u> polyp is correlated with activity in several pacemaker systems. Full concerts result from the interactions among two principal pacemaker systems (Josephson and Mackie, (1965): (1) the NP System, localized in the neck region of the stalk, which produces NP's; and (2) the HP System in the hydranth, which produces HP's. Firing of the HP System usually evokes concurrent firing of pacemakers in the proximal and distal tentacles, and pacemakers in the gonophores. Respective pulses produced by these systems are termed PTP's, DTP's and GP's.

External electrical recordings have revealed the presence of three non-decremental, non-polarized conducting systems in the <u>Tubularia</u> stalk (Josephson, 1965). One of them is termed the Slow System (SS) because of its slow conduction velocity, approximately 5cm/sec. This system is quite labile and has as yet no known function. A second faster conducting system (approx. 17cm/sec) is called the Triggering System (TS), since its activation can trigger the NP System to fire. A third system, with a similar conduction velocity (approx. 15cm/sec), is designated the Distal Opening System (DOS), since it activates outward flaring of the distal tentacles. Electrically stimulating the DOS inhibits spontaneous firing of the NP and HP Systems (Josephson and Rushforth, 1973). Some of the complex behavioral sequences of the <u>Tubularia</u> polyp result from interactions among the various pacemakers and activity in the TS and DOS. In this report some electrophysiological correlates of feeding behavior will be presented and discussed in terms of known properties of pacemaker and conducting systems.

METHODS

The animals used were mature <u>Tubularia</u> collected at Woods Hole. They were maintained at 16-18°C within the laboratory and studied at this temperature within two days of their being collected. Some polyps

were cultured on glass slides in circulating sea water in an aquarium held at 15-17°C. They were fed every two days, and specimens
used for experiments were not fed the previous day. Visual observations of the feeding behavior were made using a dissecting microscope, and behavioral events were manually recorded using telegraph
keys which activated channels of an event recorder. The methods
used for electrically stimulating and recording electrical activity
in Tubularia polyps have been described previously (Josephson and
Mackie, 1965; Josephson and Rushforth, 1973).

Single Artemia salina nauplii were sucked into a glass holder
made from a pasteur pipette. One end of the pipette was drawn out
into a fine bore, and the other end was attached to flexible plastic
tubing. By releasing the suction gradually, the nauplius was gently
directed onto the selected region of the hydranth, creating minimal
mechanical disturbance to the preparation. An extract of Artemia was
prepared from a dense 1ml suspension of nauplii in 5ml of sea water.
The suspension was homogenized in a small tube using a ground glass
plunger, and centrifuged at 3000 rpm for 15 minutes. Aliquots of 1ml
of the filtered supernatant were diluted ten fold in sea water and
samples of this solution were administered to the hydranth using a
pasteur pipette.

RESULTS

A. EFFECTS OF FEEDING STIMULI

Experiments were conducted to determine the effects of feeding
stimuli on the concert activity of intact polyps. Interconcert intervals were measured for a group of ten polyps during a ten minute
period, and then each polyp was exposed to a single Artemia nauplius.
The length of the interconcert interval from the concert immediately
before prey capture to the concert following engulfment was determined. In addition, interconcert interval lengths were measured for
concerts in a ten minute period subsequent to the engulfment. For
another group of ten polyps, the same regime was used except that
an extract of Artemia was directed onto the tip of the proboscis
rather than a live Artemia nauplius. The results of these exposures
are given in Table 1 (series 1). The interconcert interval was significantly longer during exposure to live Artemia or an extract,
compared with that prior to exposure (P<.01). After exposure, the
interconcert interval was restored to prefeeding values. Exposure
of the polyps to small inert objects or a slight stream of directed
sea water did not affect the interconcert interval.

In a second series of experiments a year later, the number of
concerts were recorded for ten minute periods prior to, during, and
after exposure of intact polyps to live Artemia nauplii fed one at a
time. The number of concerts was dramatically reduced by this treatment, but during a ten minute period following cessation of feeding,

Table 1. Effects of Feeding Stimuli on Concert Activity
in Tubularia

Series 1   Interconcert Interval Length (Secs.)
Mean ± S.E. (n = 10)

Exposure to	Before	During	After
Single Artemia Nauplius	31.2 ± 2.4	61.6 ± 9.5*	31.5 ± 2.7
Exposure to Artemia Extract	30.3 ± 2.5	74.6 ± 6.3*	29.8 ± 3.2

Series 2   Number of Concerts /10 mins
Mean ± S.E. (n = 6)

Exposure to Multiple	Before	During	After
Artemia Nauplii	12.3 ± 1.3	0.5 ± 0.2*	15.8 ± 1.0*

* P < .01 compared with Before Exposure

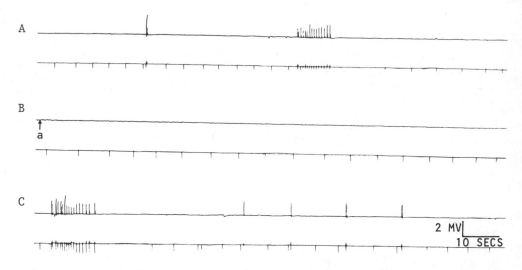

Figure 1. Dual electrical recording from an intact Tubularia polyp.
A-prior to feeding; B- during capture and engulfment of single Artemia nauplius (attachment at ↑a to a proximal tentacle); C- following engulfment. The upper channel of each record is a recording of
HP's with an electrode placed on the mid-proboscis region. The
lower channel is a recording of NP's with a second electrode on the
neck region.

the number of concerts was significantly greater than during the control period before feeding (P<.01, Table 1, series 2). Thus, concerts are inhibited during exposure to feeding stimuli, and in the case of multiple Artemia fed during a ten minute period, there is an increase in the frequency of concerts after feeding.

Individual polyps were fed single Artemia, while monitoring the electrical activity of the HP and NP Systems. It was observed that during capture and engulfment of the prey, HP's were inhibited (figure 1B) but the NP System continued to fire, giving primarily single pulses. After engulfment of the Artemia a full concert occurred, associated with concurrent firing of the HP and NP Systems in a burst. This was followed by restoration of the normal patterns of spontaneous activity (figure 1C). Similar results were obtained using Artemia extracts.

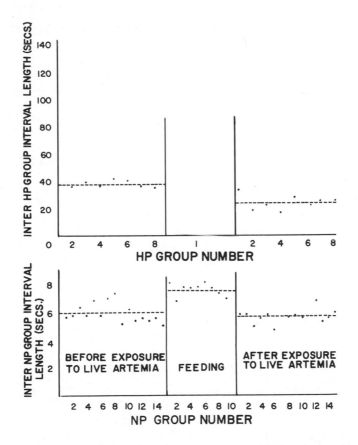

Figure 2. Inter-group interval lengths for HP's (upper) and NP's (lower) before, during, and after exposure to five Artemia nauplii, fed one at a time to intact Tubularia polyps (n=10).

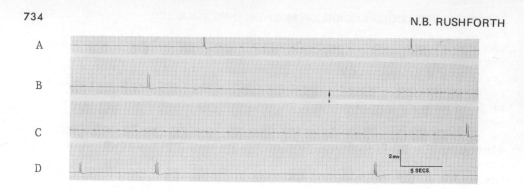

Figure 3. Electrical recording of HP's from an isolated hydranth preparation of <u>Tubularia</u>, before, during, and after capture of a single <u>Artemia</u> nauplius (attachment to a proximal tentacle at ↑a in B).

The interval lengths between groups of pulses (single or multiple events) were determined for HP's and NP's in intact polyps from records similar to those seen in figure 1. The mean interval lengths for a group of ten polyps prior to, during capture, and after engulfment of a five nauplii, fed one at a time, are plotted in figure 2. The marked suppression of the HP System is evident from the large inter- HP group interval during feeding (upper graph), while some inhibition of NP's is seen from an increased mean inter-NP group interval during this period (lower graph, P<.05). Post inhibitory excitability of the HP System is demonstrated by a significantly shorter inter-HP group interval after exposure to the live <u>Artemia</u> (upper graph, P<.05).

Inhibition of the HP System during feeding on <u>Artemia</u> was also observed in an isolated hydranth preparation (figure 3), produced by cutting off the stalk where it joins the hydranth. The NP System is localized in the upper stalk and removing the stalk allows examination of activity patterns in HP System without input from the NP System (Hofman and Rushforth, 1967). It was found that there is inhibitory rebound of the HP System following its suppression by feeding stimuli, using either live <u>Artemia</u> or extracts. During HP suppression, the mouth is induced to open by these stimuli.

If an <u>Artemia</u> nauplius is placed on a proximal tentacle, after a short latent period the tentacle flexes in the oral direction, and the proboscis bends towards the activated tentacle, usually bringing the prey in contact with the mouth region. The oral flexions of the proximal tentacle are associated with PTP's. This is most clearly observed in electrical recordings from a single isolated proximal tentacle cut from the parent hydranth. In this preparation spontaneous PTP's are recorded (figure 4, A), and are observed to increase in frequency on attachment of the tentacle to a live <u>Artemia</u> naup--

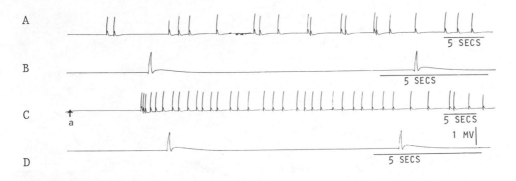

Figure 4. Electrical recording of PTP's from an isolated proximal tentacle preparation of Tubularia, before and after exposure to a single Artemia nauplius (attachment to the tentacle at ↑a in C). B and D are at a slower time scale to more clearly depict the shape of the pulses.

lius. These stimulated PTP's, correlated with tentacle flexions, similar to those in the intact preparation, gradually decline in frequency to prestimulation levels. The pulse shape of stimulated PTP's (figure 4D) is similar to that of spontaneous PTP's before Artemia attachment (figure 4B), indicating that the feeding stimulus affects only the pulse frequency. Similar responses are induced by Artemia extracts.

B.   Activation of feeding in Tubularia by Proline.

Early studies established that during capture of an Artemia nauplius, the DOSP System in Tubularia is activated (Rushforth, 1959). During the evoked response, one may record DOSP's and observe the correlated behavioral event of outward flaring of the distal tentacles. Since extracts of Artemia are found to produce these effects, the active stimulus is chemical in nature. The accumulating list of simple compounds, small peptides or amino acids, which have been shown to be stimulators of feeding behavior in coelenterates (see Lenhoff this volume), suggested a search for a feeding chemo-stimulator in Tubularia.

A general survey of the effects of a large number of amino acids revealed that only proline had the ability to induce a complete feeding reaction in Tubularia. Proline at concentrations of 2.5 x 10^{-6}M was found to initiate distal tentacle flaring and associated DOSP's, mouth opening, suppression of concerts, inhibition of HP's and reduced frequencies of NP's, and post inhibitory excitability of HP's and NP's. Inhibition was longer and post inhibitory rebound of HP's and NP's was greater at higher concentrations (e.g. 6 x 10^{-5}M, Rushforth and Fox, in preparation).

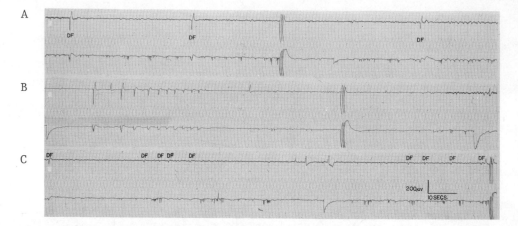

Figure 5. Activation of DOSP's and distal tentacle flaring by proline (for details see text).

    An example of DOSP's and associated distal tentacle flaring
induced by directing proline solution onto the distal portion of the
proboscis of an isolated hydranth of Tubularia, is seen in figure
5A. The upper channel of the dual recording is from an electrode
placed on the upper portion of the proboscis at the base of the distal tentacles. The lower channel is a recording from an electrode
at the base of the proboscis. On exposure of the proboscis to the
proline solution, a large DOSP was elicited together with a concerted aboral flare of all the distal tentacles (DF). The results of
three such exposures are seen in figure 5A, together with a short
burst of spontaneous HP's. It is seen that the evoked DOSP's on
successive stimulation by proline are reduced in amplitude. The
hydranth then was stimulated electrically with ten shocks using a
stimulating electrode placed on a gonophore. The evoked DOSP's
(designated with closed circles, figure 5B) are seen in addition to
a short burst of spontaneous HP's. A marked defacilitation of the
electrically stimulated DOSP's characteristic of the Distal Opening
System (Josephson, 1965), is evident in this two channel recording.
Following electrical stimulation of DOSP's the hydranth was exposed
to two live Artemia, which were captured and engulfed. Evoked DOSP's
and aboral flaring of the distal tentacles during the two feeding
sequences are recorded in figure 5C. The size of the DOSP's, particularly those after the first pulse, are attenuated considerably,
and a burst of HP's occurs after engulfment of the second Artemia.

    The continuity of the DOSP Systems between two different Tubu-
laria polyps present in a Y-colony (Josephson, 1965b), enabled experiments which confirmed the similar effects of proline and live
Artemia. In these studies, DOSP's were recorded from one polyp as

feeding stimuli were presented to the other polyp of the Y-colony pair (Rushforth and Fox, in preparation). A number of other compounds were utilized as possible stimulators of feeding in Tubularia. At concentrations in the range of 10^{-6} to 10^{-5}M, the following compounds did not initiate feeding responses: hydroxy-L-proline, reduced glutathione, glutamic acid, phenylalanine, aspartic acid, histidine, asparagine, cysteine, glycine, threonine, serine, cystine, alanine, valine, glutamine and leucine.

SUMMARY AND DISCUSSION

These studies demonstrate that live Artemia, Artemia extracts, and the amino acid proline initiate feeding responses in Tubularia. Such responses consist of: (1) activation of DOSP's and outward flaring of the distal tentacles; (2) mouth opening; (3) inhibition of HP's and NP's, and suppression of concerts; and (4) post-inhibitory excitability of HP's, NP's, and concerts. At higher concentrations of Artemia extract or proline the coordinated movements of the distal tentacles are supplanted by uncoordinated writhing movements, reminiscent of those seen in Hydra with Artemia extracts or the tripeptide reduced glutathione (Rushforth and Hofman, 1972).

Elements of the feedings response in Tubularia have been found to be evoked by electrical stimulation. Stimuli consisting of 1msec current pulses at approximately twice DOS threshold administered by a suction electrode attached to a gonophore, initiated DOSP's and aboral flaring of the distal tentacles. Repetitive stimulation at frequencies over the range of 1 per 2.5 secs. to 1 per 20 secs. inhibited spontaneous firing of the HP and NP Systems (Josephson and Rushforth, 1973). The degree of pacemaker inhibition was found to be greater the greater the stimulus frequency, as was the degree of rebound from inhibition at the end of the stimulation period. The latent period for NP inhibition was 0.4 to 0.6 secs. and the duration of inhibition was 4-5 secs. HP inhibition had a similarly long latent period and even longer duration, lasting 15-20 secs., successive stimuli producing a cumulative inhibitory effect.

Inhibition of HP's appears to be causally related to the reduced spontaneous concert movements during food capture and engulfment. The concerted tentacle movements and subsequent peristaltic contraction of the proboscis, which presumably would impede food ingestion, are suppressed, and the polyp undergoes the localized activities of feeding. Greatest inhibition occurs with HP's, which are directly associated with concerts, rather than NP's which are more loosely coupled with overt behavior.

The proline stimulated feeding response in Tubularia has some similarities to that elicited by reduced glutathione in Hydra (Rushforth and Hofman, 1972). Throughout the initial phases of feeding in Hydra the spontaneous contractions of the tentacle and

and body column are suppressed. These recurring simple behavioral
events which result from pacemaker activity, are supplanted by localized responses. In Hydra, reduced glutathione also inhibits the
production of electrical potentials associated with column contractions (CP's) and contractions of the tentacles (TCP's). These pulses
return at enhanced frequencies after inhibition, showing a similar
post-inhibitory rebound to that seen in Tubularia. Further studies
with hydroids and other coelenterates may demonstrate that such inhibitory phenomena have wide applicability as components of the feeding response.

REFERENCES

Ball, E.E., 1973. Electrical activity and behavior in the solitary
    hydroid Corymorpha palma. I. Spontaneous activity in whole
    animals and in isolated parts. Biol. Bull. 145:223-242.

Hofman, F. and N.B. Rushforth, 1967. Electrical activity in isolated
    parts of Tubularia. Biol. Bull. 133: 469.

Josephson, R.K., 1962. Spontaneous electrical activity in a hydroid
    polyp. Comp. Biochem. Physiol. 5: 45-58

Josephson, R.K., 1965. a. Three parallel conducting systems in the
    stalk of a hydroid. J. Exp. Biol. 42: 139-152

Josephson, R.K., 1965. b. The coordination of potential pacemakers
    in the hydroid Tubularia. Am. Zool. 5: 483-490

Josephson, R.K. and G.O. Mackie, 1965. Multiple pacemakers and the
    behavior of the hydroid Tubularia. J. Exp. Biol. 43: 293-332.

Josephson, R.K. and N.B. Rushforth, 1973. The time course of pace-
    maker inhibition in the hydroid Tubularia. J. Exp. Biol. 59:
    305-314.

Morin, J.G. and I.M. Cooke, 1971. Behavioral physiology of the colo-
    nial hydroid Obelia. I. Spontaneous movements and correlated
    electrical activity. J. Exp. Biol. 54: 689-706.

Passano, L.M. and C.B. McCullough, 1962. The light response and the
    rhythmic potentials of hydra. Proc. Nat. Acad. Sci. Wash. 48:
    1376-1382.

Rushforth, N.B., 1969. Electrophysiological correlates of feeding
    behavior in Tubularia. Amer. Zool. 9: 1114

Rushforth, N.B., and F. Hofman, 1972. Behavioral and electrophysio-
    logical studies of Hydra. III. Components of feeding behavior.
    Biol. Bull. 142: 110-131

INDEX

Note: A page number followed by a dash (e.g. 211-) means refer to that page and later in the same article.